Experimental Methods in the Physical Sciences

VOLUME 29A

ATOMIC, MOLECULAR, AND OPTICAL PHYSICS:
CHARGED PARTICLES

EXPERIMENTAL METHODS IN THE PHYSICAL SCIENCES

Robert Celotta and Thomas Lucatorto, *Editors-In-Chief*

Founding Editors

L. MARTON
C. MARTON

Volume 29A

Atomic, Molecular, and Optical Physics: Charged Particles

Edited by

F. B. Dunning
*Department of Physics,
Rice University, Houston, Texas*

and

Randall G. Hulet
*Department of Physics,
Rice University, Houston, Texas*

ACADEMIC PRESS

San Diego   New York   Boston   London   Sydney   Tokyo   Toronto

This book is printed on acid-free paper.

Copyright © 1995 by ACADEMIC PRESS, INC.

All Rights Reserved.
No part of this publication may be reproduced or transmitted in any form or by any means, electronic or mechanical, including photocopy, recording, or any information storage and retrieval system, without permission in writing from the publisher.

Academic Press, Inc.
A Division of Harcourt Brace & Company
525 B Street, Suite 1900, San Diego, California 92101-4495

United Kingdom Edition published by
Academic Press Limited
24-28 Oval Road, London NW1 7DX

International Standard Serial Number: 1079-4042

International Standard Book Number: 0-12-475974-2

PRINTED IN THE UNITED STATES OF AMERICA
95 96 97 98 99 00 BC 9 8 7 6 5 4 3 2 1

CONTENTS

CONTRIBUTORS	xiii
PREFACE	xv
LIST OF VOLUMES IN TREATISE	xvii

1. Spin-Polarized Electron Sources
by DANIEL T. PIERCE

1.1. Introduction	1
1.2. Background	3
1.3. The Photocathode	11
1.4. Incident Radiation	22
1.5. Operation of the Photocathode in an Electron Gun	24
1.6. Summary	30
Appendix A: Cleaning GaAs	33
Appendix B: Anodization of GaAs	34
References	35

2. Positron and Positronium Sources
by A. P. MILLS, JR.

2.1. Introduction	39
2.2. Positron Sources	39
2.3. Moderation	46
2.4. Transport, Lenses, and Mirrors	50
2.5. Remoderation	51
2.6. Trapping and Bunching	53

2.7. Positronium Formation	55
2.8. Summary and Conclusion	62
References	62

3. Sources of Low-Charge-State Positive-Ion Beams
G. D. ALTON

3.1. Introduction	69
3.2. Arc Plasma Discharge Ion Sources	74
3.3. Radiofrequency (RF) Discharge Ion Sources	116
3.4. Electron-Cyclotron Resonance and Microwave Ion Sources	123
3.5. Vacuum-Arc Ion Sources	129
3.6. Sources Based on Field Ionization and Field Evaporation	134
3.7. Surface Ionization and Thermal Emitter Ion Sources	147
3.8. Thermal Ionization Sources	159
References	162

4. Advanced Sources of Highly Charged Ions
RONALD A. PHANEUF

4.1. Introduction	169
4.2. The Electron-Beam Ion Source	172
4.3. The Electron–Cyclotron Resonance Ion Source	177
4.4. Other Techniques	185
References	187

5. Electron and Ion Optics
GEORGE C. KING

5.1. Introduction	189
5.2. Collimation and Definition of a Charged Particle Beam	191
5.3. Electrostatic Lenses	193
5.4. Designing Electrostatic Lens Systems	201
5.5. Computer Simulation Programs	204
References	207

6. Electron Energy Analyzers
J. L. ERSKINE

6.1. Introduction	209
6.2. Electron Energy Analyzing Systems	209
6.3. Trajectories and Focusing in Dispersive Deflection Analyzers	215
6.4. Fringing Fields, Terminations, and Guard Rings	220
6.5. Input Lenses and Operating Modes	221
6.6. Multichannel Energy Detection and Imaging	226
6.7. Construction Hints and Practical Details	227
References	228

7. Electron Polarimetry
TIMOTHY J. GAY

7.1. Introduction	231
7.2. Electron Polarimeters	234
7.3. Calibration Methods	243

7.4. Systematic Errors	245
7.5. Comparison of Polarimeters	247
References	250

8. Position Sensitive Particle Detection with Microchannel-Plate Electron Multipliers
KEN SMITH

8.1. Microchannel Plates	253
8.2. Position-Sensitive Readout Systems	259
8.3. Conclusion	270
References	270

9. Swarm Techniques
DAVID SMITH and PATRIK ŠPANĚL

9.1. Introduction	273
9.2. Afterglow Plasma Techniques	276
9.3. The Selected Ion Flow Tube Technique	285
9.4. Drift Tube Techniques	290
9.5. Swarm Experiments at Very Low Temperatures: The CRESU Technique	293
9.6. Concluding Remarks	295
References	295

10. Accelerator-Based Atomic Physics
C. R. VANE and S. DATZ

10.1. Introduction	299

10.2. Advantages of Accelerator-Based Atomic Physics (ABAP)	300
10.3. Types of Accelerators Used in Accelerator-Based Atomic Physics	306
10.4. Electrostatic Accelerators	307
10.5. Electrodynamic Accelerators	312
References	319

11. Ion Mass Analyzers
PETER W. HARLAND

11.1. Introduction	321
11.2. Motion of Charged Particles in Electric and Magnetic Fields	322
11.3. Uniform Electric Fields	323
11.4. Uniform Magnetic Fields	335
11.5. Superimposed Electric and Magnetic Fields	340
References	346

12. Ion Traps
HUGH A. KLEIN

12.1. Introduction	349
12.2. Types of Ion Trap	350
12.3. Trap Construction	352
12.4. Trapped Ion Detection, Diagnostics, and Cooling	356
References	359

13. Ultra-High-Resolution Mass Spectroscopy in Penning Traps
ROBERT S. VAN DYCK, JR.

13.1. Introduction	363
13.2. Trap Construction and Confinement	365
13.3. Basic Axial Resonance	368
13.4. Preparing the Ion Sample	371
13.5. Sideband Cooling Resonances	372
13.6. Magnetic Field Stability	374
13.7. Frequency-Shift Detector	376
13.8. Anharmonic Detection	378
13.9. The Ejection-Detection Method	380
13.10. The Pulse-and-Phase Method	382
13.11. Perturbations to the Cyclotron Frequency	384
References	387

14. Electron Beam Ion Traps
ROSCOE E. MARRS

14.1. Introduction	391
14.2. Principles of EBIT Operation	393
14.3. The 200-keV EBIT	400
14.4. X-ray Spectroscopy	402
14.5. Electron–Ion Collisions	410
14.6. Extraction of Highly Charged Ions from an EBIT	417
References	418

15. DC Current Measurements
THOMAS J. MEGO

15.1. Introduction	421

15.2. Description and Comparison of Current Amplifiers 422
15.3. Signal Quantification and Conversion 431
15.4. Practical Experimental Considerations 432
References 434

16. Signal Enhancement
JOHN R. WILLISON

16.1. Introduction 437
16.2. Noise Sources 438
16.3. Amplifiers 443
16.4. Signal Analysis 445
References 450

INDEX 453

CONTRIBUTORS

Numbers in parentheses indicate the pages on which the authors' contributions begin.

G. D. ALTON (69), *Oak Ridge National Laboratory, Oak Ridge, Tennessee, 37831-6368*
S. DATZ (299), *Physics Division, Oak Ridge National Laboratory, Oak Ridge, Tennessee, 37831-6377*
J. L. ERSKINE (209), *Department of Physics, University of Texas at Austin, Austin, Texas 78712*
TIMOTHY J. GAY (231), *Department of Physics and Astronomy, Behlen Laboratory, University of Nebraska, Lincoln, Nebraska 68588*
PETER W. HARLAND (321), *Chemistry Department, University of Canterbury, Christchurch, New Zealand*
GEORGE C. KING (189), *Department of Physics and Astronomy, University of Manchester, Manchester M13 9PL, United Kingdom*
HUGH A. KLEIN (349), *National Physical Laboratory, Teddington, Middlesex, TW11 OLW, United Kingdom*
ROSCOE E. MARRS (391), *Lawrence Livermore National Laboratory, Livermore, California 94550*
THOMAS J. MEGO (421), *Keithley Instruments, Inc., Solon, Ohio 44139*
A. P. MILLS, JR. (39), *AT&T Bell Laboratories, Murray Hill, New Jersey 07974*
PATRIK ŠPANĚL (273), *Department of Biomedical Engineering and Medical Physics, University of Keele, Stoke-on-Trent ST4 7QB, United Kingdom*
RONALD A. PHANEUF (169), *Department of Physics, University of Nevada, Reno, Reno, Nevada 89557-0058*
DANIEL T. PIERCE (1), *Electron Physics Group, National Institute of Standards and Technology, Gaithersburg, Maryland 20899*
KEN SMITH (253), *Departments of Physics and Space Physics and Astronomy, Rice University, Houston, Texas 77005*
DAVID SMITH (273), *Department of Biomedical Engineering and Medical*

Physics, University of Keele, Stoke-on-Trent ST4 7QB, United Kingdom

ROBERT S. VAN DYCK, JR. (363), *Department of Physics, University of Washington, Seattle, Washington 98195*

C. R. VANE (299), *Physics Division, Oak Ridge National Laboratory, Oak Ridge, Tennessee 37831-6377*

JOHN R. WILLISON (437), *Stanford Research Systems, Inc., Sunnyvale, California 94089*

PREFACE

Since the publication in 1967 of "Atomic Sources and Detectors," Volumes 4A and 4B of this series, the field of atomic, molecular, and optical physics has seen exciting and explosive growth. Much of this expansion has been tied to the development of new sources, such as the laser, which have revolutionized many aspects of science, technology, and everyday life. This growth can be seen in the dramatic difference in content between the present volumes and the 1967 volumes. Not all techniques have changed however, and for those such as conventional electron sources, the earlier volumes still provide a useful resource to the research community. By carefully selecting the topics for the present volumes, Barry Dunning and Randy Hulet have provided us with a coherent description of the methods by which atomic, molecular, and optical physics is practiced today. We congratulate them on the completion of an important contribution to the scientific literature.

The present volume marks a turning point in this series. Beginning with this volume the series will be known as Experimental Methods in the Physical Sciences instead of Methods of Experimental Physics. The change recognizes the increasing multidisciplinary nature of science and technology. It permits us, for example, to extend the series into interesting areas of applied physics and technology. In that case, we hope such a volume can serve as an important resource to someone embarking on a program of applied research by clearly outlining the experimental methodology employed. We expect that such a volume would appear to researchers in industry, as well as scientists who have traditionally pursued more academic problems but wish to extend their research program into an applied area. We welcome the challenge of providing an important and useful series of volumes for all of those involved in today's broad research spectrum.

Robert J. Celotta
Thomas B. Lucatorto

METHODS OF EXPERIMENTAL PHYSICS

Editors-in-Chief

Robert Celotta and Thomas Lucatorto

Volume 1. Classical Methods
Edited by Immanuel Estermann

Volume 2. Electronic Methods, Second Edition (in two parts)
Edited by E. Bleuler and R. O. Haxby

Volume 3. Molecular Physics, Second Edition (in two parts)
Edited by Dudley Williams

Volume 4. Atomic and Electron Physics--Part A: Atomic Sources and Detectors; Part B: Free Atoms
Edited by Vernon W. Hughes and Howard L. Schultz

Volume 5. Nuclear Physics (in two parts)
Edited by Luke C. L. Yuan and Chien-Shiung Wu

Volume 6. Solid State Physics--Part A: Preparation, Structure, Mechanical and Thermal Properties; Part B: Electrical, Magnetic and Optical Properties
Edited by K. Lark-Horovitz and Vivian A. Johnson

Volume 7. Atomic and Electron Physics--Atomic Interactions (in two parts)
Edited by Benjamin Bederson and Wade L. Fite

Volume 8. Problems and Solutions for Students
Edited by L. Marton and W. F. Hornyak

Volume 9. Plasma Physics (in two parts)
Edited by Hans R. Griem and Ralph H. Lovberg

Volume 10. Physical Principles of Far-Infrared Radiation
By L. C. Robinson

Volume 11. Solid State Physics
Edited by R. V. Coleman

Volume 12. Astrophysics--Part A: Optical and Infrared Astronomy
Edited by N. Carleton

Part B: Radio Telescopes; Part C: Radio Observations
Edited by M. L. Meeks

Volume 13. Spectroscopy (in two parts)
Edited by Dudley Williams

Volume 14. Vacuum Physics and Technology
Edited by G. L. Weissler and R. W. Carlson

Volume 15. Quantum Electronics (in two parts)
Edited by C. L. Tang

Volume 16. Polymers--Part A: Molecular Structure and Dynamics; Part B: Crystal Structure and Morphology; Part C: Physical Properties
Edited by R. A. Fava

Volume 17. Accelerators in Atomic Physics
Edited by P. Richard

Volume 18. Fluid Dynamics (in two parts)
Edited by R. J. Emrich

Volume 19. Ultrasonics
Edited by Peter D. Edmonds

Volume 20. Biophysics
Edited by Gerald Ehrenstein and Harold Lecar

Volume 21. Solid State: Nuclear Methods
Edited by J. N. Mundy, S. J. Rothman, M. J. Fluss, and L. C. Smedskjaer

Volume 22. Solid State Physics: Surfaces
Edited by Robert L. Park and Max G. Lagally

Volume 23. Neutron Scattering (in three parts)
Edited by K. Sk"ld and D. L. Price

Volume 24. Geophysics--Part A: Laboratory Measurements; Part B: Field Measurements
Edited by C. G. Sammis and T. L. Henyey

Volume 25. Geometrical and Instrumental Optics
Edited by Daniel Malacara

Volume 26. Physical Optics and Light Measurements
Edited by Daniel Malacara

Volume 27. Scanning Tunneling Microscopy
Edited by Joseph Stroscio and William Kaiser

Volume 28. Statistical Methods for Physical Science
Edited by John L. Stanford and Stephen B. Vardeman

Volume 29. Atomic, Molecular, and Optical Physics--Part A: Charged Particles; Part B: Atoms and Molecules; Part C: Electromagnetic Radiation
Edited by F. B. Dunning and Randall G. Hulet

1. SPIN-POLARIZED ELECTRON SOURCES

Daniel T. Pierce

Electron Physics Group, National Institute of Standards and Technology,
Gaithersburg, Maryland

1.1 Introduction

Most polarized electron sources in use today are based on photoemission from negative-electron-affinity (NEA) semiconductor photocathodes, such as GaAs and related compounds, and thus these form the central focus of this chapter. It is worth noting at the outset that, in addition to producing spin-polarized beams, these photocathodes have other advantageous features, such as their high brightness, narrow energy spread of the emitted beam, and the possibility of modulating the beam intensity with an arbitrary time structure by controlling the photoexciting light. The experimental techniques presented in this chapter for electron guns with NEA photocathodes are also relevant for such guns in their numerous applications beyond those involving spin polarization.

Because there is no simple polarization filter for electrons equivalent to a calcite prism for light or a Stern–Gerlach magnet for atoms, a number of spin-dependent processes have been tried in attempts to produce beams of spin-polarized electrons [1,2]. Chief among these are photoionization of polarized Li atoms [3], the Fano effect in Rb [4] and Cs [5], field emission from W–EuS tips [6], photoemission from the ferromagnetic crystal EuO [7], scattering from an unpolarized target [1], chemi-ionization of optically oriented metastable He [8–11], and photoemission from NEA GaAs. For most applications, photoemission from NEA GaAs and related materials provides the most suitable source of polarized electrons. The source based on chemi-ionization of metastable He is competitive for applications which require high polarization in a continuous beam of moderate intensity and has the further advantage that ultrahigh vacuum is not required. Some characteristics of this source will be discussed in Section 1.6.

Photoemission of optically oriented electrons from NEA GaAs as a source of polarized electrons was proposed in the mid-1970s [12,13]. The feasibility of the GaAs-polarized electron source was demonstrated in spin-polarized photoemission experiments at the ETH—Zurich [14–17].

The first polarized electron guns were developed for low-energy condensed matter experiments at the National Institute of Standards and Technology (NIST) [18] and for high-energy physics experiments at the Stanford Linear Accelerator Center (SLAC) [19], where the GaAs source was a crucial part of the landmark parity violation experiment of Prescott et al. [20,21]. The condensed matter experiments required a low-energy, continuous source while the high-energy parity-violation experiment required a pulsed beam with high initial injection energy. Although there are some fundamental differences in source design, there are also a number of similarities. There have been a number of advances in the GaAs-type sources over the years which can be attributed in large part to the demanding requirements on polarized electron sources for accelerator applications. An account of this progress is summarized in the reports [22–25] on a series of workshops on polarized sources for accelerators.

This chapter attempts to distill the advances in GaAs-polarized electron source technology, to present important information and considerations for someone building such a source, and to compare its performance with that of other polarized electron sources. A source of spin-polarized electrons can be characterized by a number of parameters which allow one to determine how well it will meet the requirements of a particular application. Foremost among these is the polarization itself which we define as $P = (N\uparrow - N\downarrow)/(N\uparrow + N\downarrow)$, where $N\uparrow$ ($N\downarrow$) are the number of electrons with spins parallel (antiparallel) to a quantization direction. The ideal polarized electron source would produce a beam with the maximum polarization, $P = \pm 1$. In Section 1.2, the physics behind production of polarized electrons in photoemission from GaAs is reviewed and the progress toward obtaining a fully polarized electron beam is discussed. For a source producing an electron beam of current I, the quantity $P^2 I$ is a useful figure of merit when counting statistics are the chief source of experimental uncertainty. It is sometimes possible to trade off polarization and increase the current I, but the current may also be limited, for example, by space-charge or target damage considerations. The current available with a given light source is determined by the quantum efficiency which depends on the cathode material, how the surface is cleaned, and how it is activated, as presented in Section 1.3. An important consideration in the construction of a polarized electron gun, as discussed in Section 1.4, is the control of the incident radiation to determine the time structure of the electron beam intensity and, because many experiments involve measuring a small spin-dependent asymmetry, to reverse the sign of the electron spin polarization without affecting other beam parameters such as intensity, angle, or position. Electron optical properties of the beam of photoelectrons extracted from the GaAs photocathode, such as the beam brightness and emittance discussed in Section 1.5, set limits on the parameters of the beam which can be obtained at the target in a particular application. In Section 1.6, the main features of the GaAs source are summarized and compared with those of other polarized electron sources.

1.2 Background

Photoemission from NEA GaAs can be described in a particularly straightforward way by Spicer's three-step model: photoexcitation, transport, and escape [26,27]. The polarized electrons are generated in the photoexcitation process. The transport and escape strongly affect the quantum efficiency (i.e., the number of electrons emitted per incident photon) as well as the depolarization of the electrons. There are a number of experimental challenges in constructing such a spin-polarized electron source, but two paramount ones are optimizing the polarization and optimizing the emitted current. A brief description of the theory of the GaAs-polarized electron source which will form the framework for discussing the experimental approaches to obtaining the desired performance is presented.

1.2.1 Optical Spin Orientation

GaAs is a direct gap semiconductor with a minimum band separation, E_g, at Γ as in the $E(k)$ plot of the energy bands vs crystal momentum k shown in the left side of Figure 1. The conduction band is a twofold degenerate $s_{1/2}$ level. The spin–orbit interaction splits the sixfold degenerate p state of the valence band maximum into a fourfold degenerate $p_{3/2}$ level and a twofold degenerate $p_{1/2}$ level lying 0.34 eV lower in energy. The fourfold degenerate $p_{3/2}$ level consists of the heavy-hole band with $m_j = \pm\frac{3}{2}$ and the higher curvature light-hole band with $m_j = \pm\frac{1}{2}$. The transitions for circularly polarized light σ^+ (σ^-) between the m_j sublevels are shown by solid (dashed) lines on the right side of the figure. The selection rules require that $\Delta m_j = +1$ or $\Delta m_j = -1$ for σ^+ or σ^- light, respectively. The Clebsch–Gordan coefficients give the relative intensities of these transitions. Thus for σ^+ light, the theoretical polarization is

$$P_{th} = (1 - 3)/(1 + 3) = -0.5. \tag{1.1}$$

With increasing photon energy, transitions from the split-off $p_{1/2}$ level eventually contribute with a relative intensity of 2, and the polarization is reduced to zero.

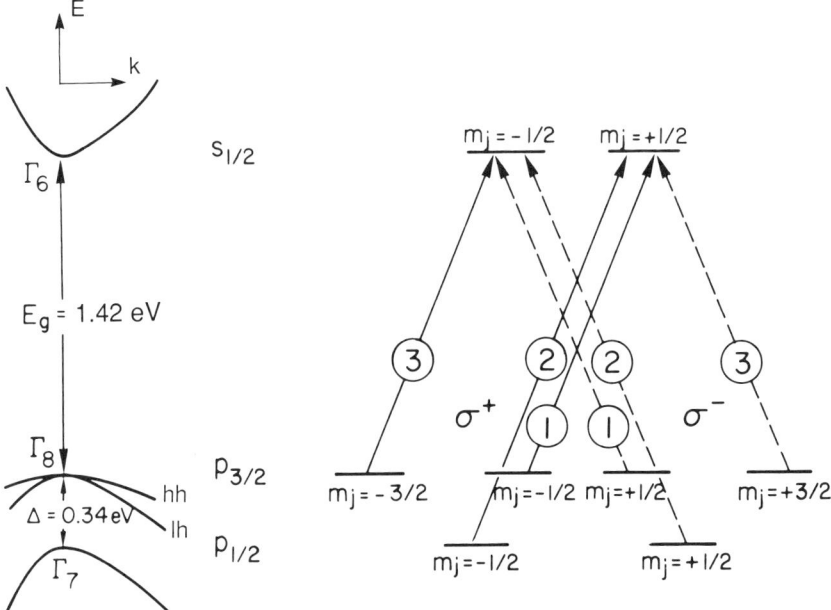

FIG. 1. The energy bands, $E(k)$, of GaAs near the center of the Brillouin zone are shown at the left of the figure. The room temperature bandgap is $E_g = 1.42$ eV, and the spin–orbit splitting of the valence bands is 0.34 eV. At the right, the allowed transitions between the m_j sublevels, for σ^+ and σ^- light are shown by the solid and dashed arrows, respectively. The circled numbers give the relative intensities. The degeneracy of the heavy-hole (hh) and light-hole (lh) bands at the valence band maximum limits the maximum polarization to -0.5 and $+0.5$ for σ^+ and σ^-, respectively. From Pierce and Meier [16].

1.2.2 Transport, Escape, and Depolarization

Ordinarily, the electrons excited to the conduction band minimum would be approximately 4 eV below the vacuum level and could not escape from the GaAs. By treating the surface of p-type GaAs with Cs and O_2, it is possible to lower the vacuum level at the surface below the energy of the conduction band minimum in the bulk to achieve the condition known as NEA [28]. This situation is illustrated in Figure 2. Instead of the escape depth being limited by the short mean-free path for inelastic scattering of hot electrons, the electrons from an NEA cathode are emitted from a depth determined by the diffusion length, L, which is on the order of 1 μm and comparable to the absorption length of the light, $1/\alpha$. The photoelectrons thermalize to the conduction band minimum in a time on

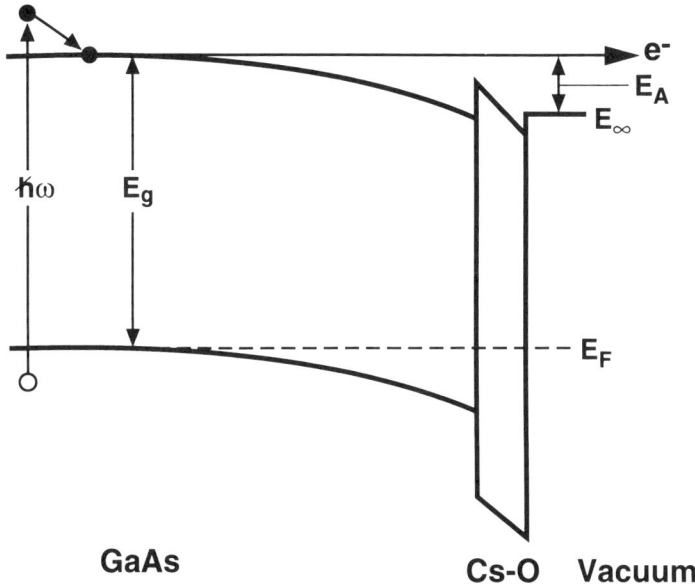

FIG. 2. The valence and conduction bands of p-type GaAs bend down at the surface. Activation with Cs and O_2 lowers the vacuum level E_∞ at the surface below the conduction band in the bulk, leading to an effective negative electron affinity, E_A. Electrons excited from the valence bands by photons of energy $\hbar\omega$ thermalize to the conduction band minimum and diffuse toward the surface, where they are accelerated in the band-bending region and escape into vacuum. From Pierce et al. [18].

the order of 10^{-12} sec and then diffuse toward the surface, where they are accelerated in the band-bending region, the width of which is determined by the doping level. When the electron reaches the surface, it can escape into vacuum or it may be reflected. A reflected electron may get turned around by scattering with phonons for subsequent attempts to escape at the surface. It may also recombine with a hole before it escapes. The probability that an electron which reaches the band-bending region is finally emitted is P_{esc}. For an NEA photocathode, the quantum efficiency or quantum yield, Y, is simply obtained in the three-step model in terms of these parameters as [26,27]

$$Y = P_{esc}/[1 + (\alpha L)^{-1}]. \quad (1.2)$$

This equation is valid for photon energies near E_g so that electrons thermalize into the Γ minimum rather than the higher-lying band minima [29]. For maximum yield, the light should be absorbed in a depth smaller than the diffusion length. The diffusion length can be increased from about 0.5–1 μm in a bulk GaAs crystal to 3 μm or more in epitaxially grown material with few defects. One has less control over the absorption coefficient α which depends on the electronic structure of GaAs. The factor which affects the yield most strongly is the escape probability P_{esc} which depends sensitively on the surface preparation, that is, the cleaning and activation.

The minority carrier lifetime, τ, of a photoexcited electron diffusing in GaAs is directly related to L through the diffusion constant D with $L = (D\tau)^{1/2}$. As the electrons diffuse to the surface, they can become depolarized by a number of mechanisms [30] which can be characterized by a spin relaxation time, τ_s. The polarization of the photoemitted electrons P can be related to the larger theoretical polarization P_{th} by

$$P = P_{th}[\alpha + (D\tau)^{-1/2}]/[\alpha + (DT)^{-1/2}], \qquad (1.3)$$

where $T = \tau_s\tau/(\tau_s + \tau)$ and approximations about the surface recombination velocity appropriate for an NEA cathode and a spin-independent escape probability have been assumed [18]. The equilibrium polarization of the photoexcited electrons inside the GaAs, P_L, as would be inferred from a photoluminescence measurement, is also less than P_{th} because electrons may undergo spin relaxation before they recombine. The polarization of photoemitted electrons has a direct relation to the equilibrium polarization, assuming no further depolarization in the escape through the activation layer,

$$P \approx [(\tau_s + \tau)/\tau_s]^{1/2} P_L. \qquad (1.4)$$

Photoemitted electrons spend less time in the sample and hence have less time to depolarize than electrons which remain in the sample until they recombine; this leads to a photoelectron polarization which is higher than the equilibrium polarization as has been shown experimentally [31].

1.2.3 Optimizing the Polarization

Optimizing the polarization would be synonymous with "increasing" the polarization if that could be accomplished without adversely affecting other source characteristics. From the above discussion, it should be possible to increase the polarization toward P_{th} by artificially constructing a cathode with an active semiconductor layer that is very thin, since the less time photoexcited electrons remain in the GaAs, the less time they have to depolarize. Indeed, the polarization of photoemitted electrons from molecular beam epitaxy (MBE)-grown GaAs layers 0.2 μm thick was 0.49, near the theoretical maximum, but for 1-μm-thick layers the polarization decreased to approximately 0.4 [32]. However, the higher polarization from the thinner active semiconductor layer was achieved at the sacrifice of quantum efficiency, since the light is absorbed over a distance much greater than the layer thickness. For thick GaAs cathodes, the measured polarizations range from roughly 0.25 to 0.4 depending on starting material, activation, and temperature. The causes of the differences in the reported polarizations are not entirely understood, although the difficulty in making accurate polarization measurements may also play a role.

The maximum polarization attainable from GaAs, $P_{th} = 0.5$, is a serious limitation. To obtain a higher P_{th} the degeneracy of the light-hole and heavy-hole bands at the valence band maximum must be lifted by reducing the symmetry of these states. This can be accomplished by: (1) choosing a material of lower crystal symmetry, such as a II–IV–V_2 chalcopyrite semiconductor like $CdSiAs_2$, (2) introducing strain in the emitting semiconductor layer, and (3) introducing periodic potential wells as in a GaAs–AlGaAs superlattice. The first measurements of spin-polarized photoelectrons from semiconductors with the chalcopyrite structure did not give high polarization [33]. Single crystal platelets of $CdSiAs_2$ [34] and $Zn(Ge_{0.7}Si_{0.3})As_2$ films grown by metal–organic chemical vapor deposition (MOCVD) [35] were successfully activated at or near NEA, but the measured polarizations did not exceed 0.2 even though $P_{th} = 1$. It is not known whether hybridization of the electron states or a shortcoming in the crystal preparation or activation is to blame for the large difference between theory and experiment for crystals with the chalcopyrite structure.

The first significant enhancement of the photoelectron spin polarization above 0.5 from an NEA photocathode was obtained using a strained $In_xGa_{1-x}As$ layer, with $x \approx 0.13$, grown on a GaAs substrate [36]. When the lattice constant of the substrate is less than the epilayer, as shown schematically in Figure 3a, there is a compressive biaxial strain in the plane of the layer and a tensile strain perpendicular to the layer. The lattice constant of GaAs is about 0.9% less than the $In_xGa_{1-x}As$ epilayer, resulting in a strain which causes the heavy-hole band to move up in energy and the light-hole band to move down as shown in Figure 3b. Theoretically, with σ^+ light of the proper photon energy, one has only the transition from $m_j = -\frac{3}{2}$ to $m_j = -\frac{1}{2}$, giving $P = -1$. The splitting δ of the light-hole and heavy-hole bands is proportional to the strain which is proportional to the In fraction x. For thin layers, the lattice mismatch is accommodated by elastic strain, but above a critical thickness, dislocations begin to relax the strain. Equilibrium thermodynamic arguments [37] predict a critical thickness on the order of 10 nm for InGaAs with $x = 0.13$. However, the strain is not significantly relaxed until a thickness

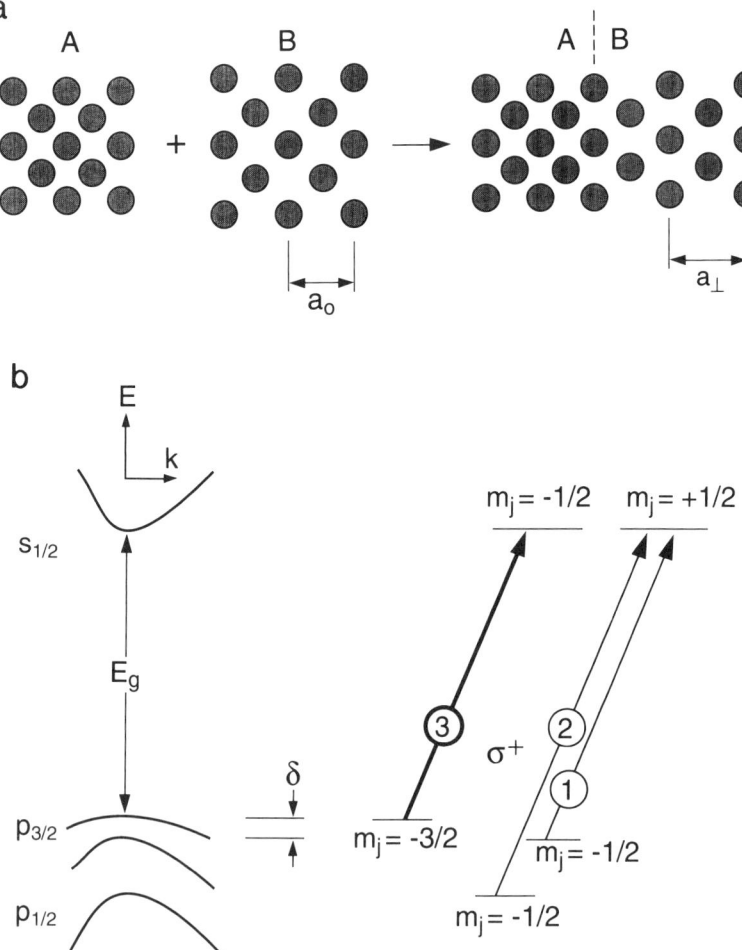

FIG. 3. (a) Schematic showing the growth of epilayer B on substrate A which has a smaller lattice constant. There is a compressive biaxial strain in the plane of the layer and a tensile strain perpendicular to the layer such that $a_\perp > a_0$. The pictured strain is about a factor of 25 greater than typical. (b) E vs. k band diagram showing how strain splits the heavy-hole and light-hole bands by an energy, δ, thus lifting the degeneracy. The allowed transitions for σ^+ light are shown on the right. Choosing the photon energy to select the transition shown by the heavy arrow gives $P = -1$.

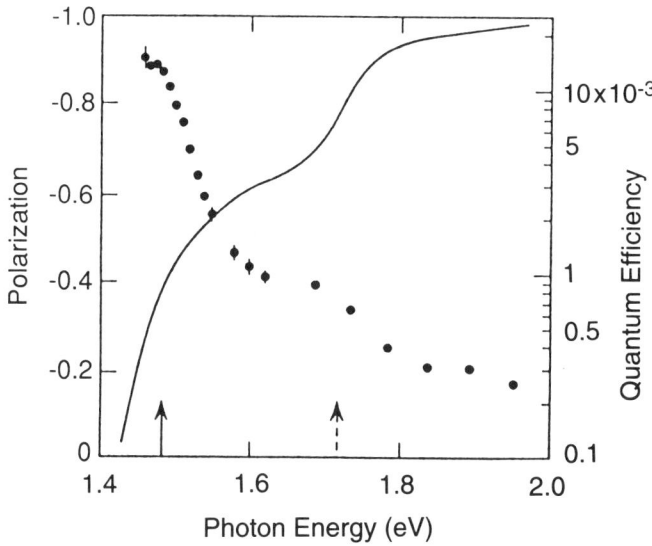

FIG. 4. Electron spin polarization (data points) and photocathode quantum efficiency (solid curve) as a function of excitation photon energy for GaAs grown on $GaAs_{1-x}P_x$ with $x = 0.28$. The calculated heavy-hole band gap is shown by the solid arrow, and the band gap of $GaAs_{1-x}P_x$ is shown by the dashed arrow. From Maruyama et al. [41].

approximately an order of magnitude larger is reached [38]. This result is very important, because a thicker layer leads to a higher quantum efficiency. Even though some of the light still passes through the active layer, the substrate has a larger band gap and does not contribute to the photocurrent until higher photon energies are reached.

Enhanced photoelectron spin polarization has also been observed in photoemission from strained GaAs layers grown on $GaAs_{1-x}P_x$ which has a smaller lattice constant than GaAs [39]. The dependence of spin polarization on strain has been investigated for this system [40,41]. Polarizations as high as 0.90 have been observed, and polarizations greater than 0.80 with a quantum efficiency greater than 0.1% have been achieved just above the heavy-hole band-gap energy indicated by the solid arrow in Figure 4 [41]. The dashed arrow shows the photon energy at which excitations in the GaAsP substrate begin to contribute and lead to an increase in quantum efficiency but a decrease in polarization.

A factor limiting the performance of strained-layer photocathodes is the limitation on the thickness required to prevent strain relaxation. Most of the incident light goes right through the active layer, and the quantum efficiency is low. A novel approach [42a] to get around this deficiency has been suggested: the cathode is engineered to include a distributed Bragg reflector between the GaAs substrate and the GaAsP layer, thus forming a Fabry–Perot cavity with the surface of the strained GaAs epilayer. By tuning the wavelength of resonant absorption, an increase in quantum efficiency of an order of magnitude may be achieved without apparent loss of electron spin polarization [42a]. Some caution is suggested, however, by subsequent measurements of a similar cathode structure which revealed reductions in polarization to nearly zero over a narrow wavelength range in the region of maximum quantum efficiency; this was attributed to optical anisotropies caused by a small anisotropy of the in-plane lattice strain [42b].

Another way to remove the degeneracy of the heavy- and light-hole bands is to confine the electrons in quantum wells in a superlattice. Early attempts to increase the polarization above 50% by using a superlattice were not successful. A significant enhancement of the photoelectron polarization to 71% was later achieved [43]. The key parameters are the fraction of Al in the AlGaAs and the thickness of the GaAs and AlGaAs layers to obtain a large enough splitting of the heavy- and light-hole bands compared with thermal energies. The overall thickness of the superlattice must be small enough to minimize depolarization within the structure. The successful results were obtained from a superlattice with repeats of a 3.11-nm $Al_{0.35}Ga_{0.65}As$ layer and a 1.98-nm GaAs layer for an overall thickness of 0.1 μm. For this structure the heavy-hole band is calculated to be 44 meV higher in energy than the light-hole band. The reported [43] quantum efficiency at the highest polarization was 2.7×10^{-6}. Further studies [44] of superlattice cathodes showed that the polarization can be increased by reducing the p doping within the structure to 1×10^{17} cm^{-3} and increasing the thickness of the top GaAs layer to 5 nm with a p doping of 5×10^{18} cm^{-3}. This yielded a polarization of $P \geq 0.7$ and a quantum efficiency of 0.02.

These specialized custom-engineered strained-layer or superlattice photocathodes will not be the choice for the average builder of a polarized electron source. Such specialized cathodes either are very costly to purchase commercially or require close collaboration with a group which can grow such materials. They are described here to show the range of possibilities that exist and to illustrate the beautiful results which have been obtained as a result of efforts to increase the electron spin polarization. Such photocathodes are being used in demanding high-energy physics experiments.

1.3 The Photocathode

The photocathode material and its preparation are key to the polarized electron source. The material can be as simple as a bulk wafer or as complex as the sophisticated structures engineered to obtain high polarization. The quantum efficiency depends on the diffusion length, a material property, and is also very significantly determined by how the surface is cleaned and activated. The quantum efficiency for a bulk crystal or thick epilayer can range from typical values in the neighborhood of 3% obtained in most research laboratories to around 30% which is not unusual for cathodes prepared commercially by proprietary processes. We present two approaches to preparing photocathodes for polarized electron sources: (1) a known, reliable method which gives cathodes with the lower quantum efficiency, but which are still quite adequate for most applications (we label these "adequate" cathodes), and (2) some considerations for obtaining the higher quantum efficiency cathodes (we label these "optimum" cathodes). High-quantum-efficiency cathodes that optimize the electron escape probability may be important in cases in which the incident light intensity must be limited or in which the thickness of the photocathode layer is constrained.

1.3.1 Material

A direct gap III–V semiconductor, such as GaAs or a related material, has the energy level scheme, shown in Figure 1, required to generate spin-polarized photoelectrons. In addition to GaAs, InGaAs [36], GaAsP [45,46], and AlGaAs [47] have been used. The Al and P concentrations in the last two materials were chosen to obtain a band gap appropriate for photoexcitation with a HeNe laser at 1.96 eV. The larger band gap leads to a larger negative electron affinity. This larger NEA has been reported to help achieve long cathode lifetimes [48], but the photoelectron energy distributions were also observed to narrow as the NEA decreased with time [46].

The (100), (110), and (111) faces of p-type GaAs can all be activated to NEA [49], and all have been shown to emit polarized photoelectrons [50]. Best results are obtained for crystals doped from 5×10^{18} to 2×10^{19} cm^{-3}. Wafers with (100) surface orientation are commonly used as substrates for epitaxy and are readily available. The (110) GaAs cleavage planes are perpendicular to the (100) surface. The (100) wafers, which are typically a few tenths of a millimeter thick, are easily cleaved into rectangular pieces by pressing a fine knife on the upper edge of the wafer.

The (100) surface is recommended when using a bulk (that is, sliced from a crystal) GaAs wafer for the photocathode material.

An alternative to the bulk water surface is to grow an additional epitaxial layer, for example, by liquid phase epitaxy, MBE, or MOCVD. Epitaxial layers have fewer defects, leading to diffusion lengths of 3 to 5 μm compared with 0.5 to 1 μm for the bulk wafer. Epitaxial material suitable for photocathodes is not ordinarily available and has to be specially prepared at significant cost.

1.3.2 Surface Cleaning

For optimum activation to NEA, the GaAs surface must be free from contaminants such as C and O. It is possible to get an atomically clean surface by cleaving a (110) crystal, but the cleaving apparatus and the crystal geometry can be cumbersome for source applications. Cleaved (110) surfaces have been tested [51]. There is indication that a lower polarization is obtained from the (110) face [50–52].

If the photocathode is an epitaxial layer grown by MBE, it is possible in the final growth step to lower the temperature of the crystal and expose it to the As_4 beam to grow a protective As layer on the order of 100 nm thick. This protective layer is easily removed by momentarily heating the sample to approximately 400°C a few times. Even with the As protective layer, the photocathode material can oxidize if left in the atmosphere and can degrade in a matter of hours in a particularly humid environment [53]. It is best to store As-capped samples under vacuum to minimize degradation.

The more typical photocathode material, a wafer which has been exposed extensively to atmosphere, requires a chemical cleaning and then heat cleaning in ultrahigh vacuum. The most widely used cleaning procedures employ a mixture of H_2SO_4, H_2O_2, and H_2O with composition ratios in the range of 8:1:1 to 3:1:1. A procedure [54] used at SLAC and in our work at NIST employs a 4:1:1 etch, followed by an etch in concentrated HF and then by a slow etch in a 1:1 solution of NaOH and H_2O_2. Details such as the quality of the rinse water and decanting methods are thought to be important. This procedure is described in Appendix A. The aim of the etch is to remove contaminants and form a thin oxide passivation layer before the material is exposed to the atmosphere. Ideally, this layer and any residual contaminants are volatile at temperatures well below temperatures at which As or Ga desorb from the surface.

If it is not possible to put the freshly cleaned photocathode material immediately under vacuum, the surface can be passivated by growing an anodic oxide. The anodization process involves running a current between the GaAs and a platinum cathode in a phosphoric acid solution as described in detail in Appendix B. The oxide grown is typically 50 to 100 nm thick. The oxide is readily removed by dipping in ammonium hydroxide for about 30 sec. There is some evidence that the anodization and subsequent stripping of the oxide leave a surface which is particularly well suited for the subsequent heat cleaning [53].

The freshly cleaned photocathode is mounted on a sample holder, typically an Mo block, and, ideally, is inserted into a prebaked ultrahigh vacuum chamber through a load lock. In this manner, contamination of the photocathode surface during bakeout is avoided. Nevertheless, adequate photocathodes can be prepared even when the freshly cleaned surface is inserted into the chamber and the chamber is pumped out and then baked. In this case, it is desirable to maintain the photocathode at a temperature approximately 50°C above the typical bakeout temperature of 150 to 200°C.

Once ultrahigh vacuum is attained, impurities at the surface can be desorbed by heating the photocathode to a temperature just below the highest congruent evaporation temperature for free evaporation which is near 660°C for GaAs (100) [55]. Above this temperature, the As evaporates preferentially and leaves behind Ga droplets which give the surface a foggy appearance when viewed with obliquely incident light. Since it is not possible to attach a thermocouple reliably to the GaAs itself, it is best to rely on a relative temperature measurement of an adjacent part and determine the optimum temperature for heat cleaning empirically. This is most easily measured by a using thermocouple mounted in the Mo block on which the GaAs is mounted. An infrared pyrometer can also provide a suitably reproducible measurement of the GaAs temperature.

The apparent temperature for heat cleaning GaAs that gives the best results will vary from one system to another depending on the cathode holder and cathode mounting. One method of determining the optimum heat cleaning temperature, if the cathode can be illuminated and viewed somewhat obliquely, is to heat it to successively higher temperatures, each time allowing it to cool and checking the surface for a frosty appearance. The cathode is sacrificed, but one then knows the correct temperature is 10 to 20°C below that at which the Ga droplets form. Another way to determine the best heat cleaning temperature is to activate the cathode with Cs and O each time after heat cleaning to successively higher temperatures. The quantum efficiency will improve as the heat cleaning temperature is raised to a certain point; above this temperature, the quantum efficiency that can be attained will drop, and one again has determined the optimum heat cleaning temperature. In using this second method, one must be aware that, even when the heat cleaning takes place at the optimum temperature, the subsequent activations are usually better than the first. This may be due to an additional reduction of surface contaminants when a cathode that has already been activated is heat cleaned again. Adequate photocathodes were obtained after heating at the predetermined temperature for as little as 5 min, the whole process of heating and cooling to near room temperature taking less than half an hour [18].

In contrast to the adequate photocathode preparation just described, the optimum photocathode preparation prevents formation of the refractory oxide Ga_2O_3 which is desorbed only at temperatures near the GaAs decomposition temperature. This is accomplished by avoiding any exposure of the chemically cleaned wafer to oxygen or by limiting the exposure to such a degree that less than a monolayer of oxide is formed [53]. In the absence of Ga_2O_3, the heat cleaning can take place at a lower temperature, and as a result there is less surface damage and a correspondingly improved quantum effciency.

Two approaches to cleaning the GaAs to achieve optimum photocathodes have been reported. In one [53], the chemically cleaned and anodized wafer was mounted with Pt clips and an In bond to the Mo sample holder. The anodization layer was stripped off by immersion in NH_4OH for 30 sec, followed by quick rinses in two beakers of deionized water. The cathode was then blown dry with N_2 and kept under N_2 as it was inserted in the vacuum load lock which was continuously purged with N_2. When this procedure was accomplished quickly, oxygen-free surfaces were obtained after heat cleaning at 500°C [53].

In a second approach [56], the final etch took place in an N_2-filled glove box using a 3 M solution of HCl in isopropyl alcohol (ethanol and water were also found to be satisfactory solvents [57]). Provision was made for transferring the sample to the load lock in the N_2 atmosphere to avoid contamination from air. The level of carbon impurities on the GaAs surface was typically 0.4 to 0.6 monolayer, determined to be from residual impurities in the atmosphere of the glove box or load lock. Heating such a surface in ultrahigh vacuum to 400°C reduced the level of carbon impurities to below the detection limit of 0.05 monolayer. It was further shown that when a heavy oxide was intentionally grown on the GaAs surface after introduction into the vacuum but before heat cleaning, the carbon formed a nonvolatile phase which remained on the surface even after heat cleaning to 620°C to remove the oxide [56]. This result is consistent with the general experience, using the usual chemical cleaning methods which also leave an oxide layer, that residual carbon cannot be removed by heating [58].

The key to high quantum efficiency is the avoidance of nonvolatile surface oxides. When this is achieved, the heat cleaning can take place at temperatures at which the surface stoichiometry is not disturbed by arsenic removal, and the low-temperature heat cleaning is adequate for removing all contaminants including carbon. The possibility of achieving carbon-free surfaces is an important result since there is a direct relationship between quantum efficiency and the residual carbon contamination [58].

1.3.3 Surface Activation

The vacuum level is lowered to achieve NEA by the application of Cs and an oxidant such as oxygen or fluorine (from NF_3). There are a number of ways the activation can take place, and a few of these will now be described. The activation is continuously monitored by measuring the photocurrent, for example, by biasing the anode structure and collecting the emitted electrons. It is worth making the effort to make sure that the light source used during activation can be set up in a reproducible way so that the behavior of one activation can be compared with another.

To monitor the photocurrent, a white light source is desirable because the higher photon energies in the spectrum allow the photocurrent to be monitored well before NEA is achieved. A laser at a photon energy sufficiently above the band-gap energy can also be used to monitor the activation. For comparison with data in the literature, in which photocathode sensitivities are often quoted in $\mu A/L$, it is convenient to monitor the activation with a tungsten light source operating at 2856 K which can be calibrated in lumens [55]. A quick check on the response of the cathode to red light is obtained by inserting a filter which transmits light at wavelengths longer than 715 nm. A well-activated cathode typically gives a white-light-to-red-light response ratio of 2 using a 2856 K tungsten light source, whereas this ratio will be higher for a poorly activated cathode. The white-to-red ratio clearly depends on the spectrum of the light source used.

Two sources of Cs are commonly used. The Cs can be obtained by passing a current through a well-outgassed cesium chromate channel as in a commercial Cs dispenser [59]. Alternatively, Cs metal (99.98% pure) is distilled into and sealed in a glass ampoule which can be inserted in a Cu pinch-off tube behind a stainless steel valve [60]. After bakeout, the ampoule is broken by squeezing the Cu tube. For deposition of Cs, the ampoule is maintained at 85 to 90°C. The stainless steel valve which controls the Cs flux is maintained at 150°C so that Cs passes through the open valve without sticking to it [18]. Oxygen or NF_3 is most often let into the cathode region through a leak valve. A heated, thin-walled Ag tube is permeable to oxygen and has also been used to control the oxygen flux [48]. The correct partial pressure of oxygen or NF_3 is determined from the photoresponse. For other pressure measurements, ion gauges with thoria-coated Ir filaments are used to avoid the CO production that takes place with a hot W filament in the presence of oxygen. However, during activation, it is best to turn off gauges, since any hot filament generates metastable excited oxygen which is more reactive on the GaAs surface and can form undesirable oxides [61].

The heat-treated GaAs surface should be cooled to about 20 to 40°C for activation. Since the Cs and O must have some mobility on the surface to achieve optimum NEA, much lower temperatures are not desirable. The Cs is applied first until the photocurrent reaches a maximum. At this point, one can proceed in a number of ways [55]. One technique is the "yo-yo" technique which is illustrated schematically in Figure 5. After reaching the photocurrent maximum with application of Cs, the Cs is turned off and oxygen is let in until the photocurrent is reduced to about $\frac{1}{3}$ of its previous peak value. The oxygen flow is then stopped, and the Cs is started until a new photocurrent peak is reached, higher than the previous one. After a number of cycles (from about 5 to 20 or more), there is little further increase in photocurrent, and the activation is ended with a slight overcesiation so the peak value decreases by about 10%; the photocurrent will recover a stable equilibrium value. A variation on this procedure is to maintain the Cs flux continuously during the activation and apply the oxygen on and off to peak the photoresponse. The bonding of oxygen to GaAs is greatly enhanced by the presence of a Cs layer; there is some evidence that better activations are obtained if there are two or more monolayers of Cs on the GaAs surface at the time of oxygen exposure [61]. As an alternative to the yo-yo procedure, after the first Cs peak is reached, the Cs can be left on and oxygen admitted concurrently, adjusting the oxygen flow to maximize the rate of increase of photocurrent [18,62]. When there is no further increase in photocurrent, first the oxygen and then the Cs is shut off.

A two-stage activation process has been found to produce about a 30–40% greater quantum efficiency on (100) GaAs surface [62,63]. In this method, the cathode is activated to optimize the sensitivity with a normal yo-yo process as described above and then heat cleaned at a lower temperature in the range 450–550°C. In this second heat cleaning, most of the Cs desorbs, but the oxygen remains. The cathode is then activated a second time to a new higher sensitivity. The higher sensitivity results from a larger NEA and decreased scattering in the activation layer [62].

A fundamental understanding of the Cs–O activation layer is still lacking. The stoichiometry and atomic arrangement of the layer are not known. The activation layer has been modeled as a heterojunction, dipole layer, or cluster system [55,61,62]. There is evidence for an interfacial barrier, shown schematically in Figure 2. Photoemission studies have shown that

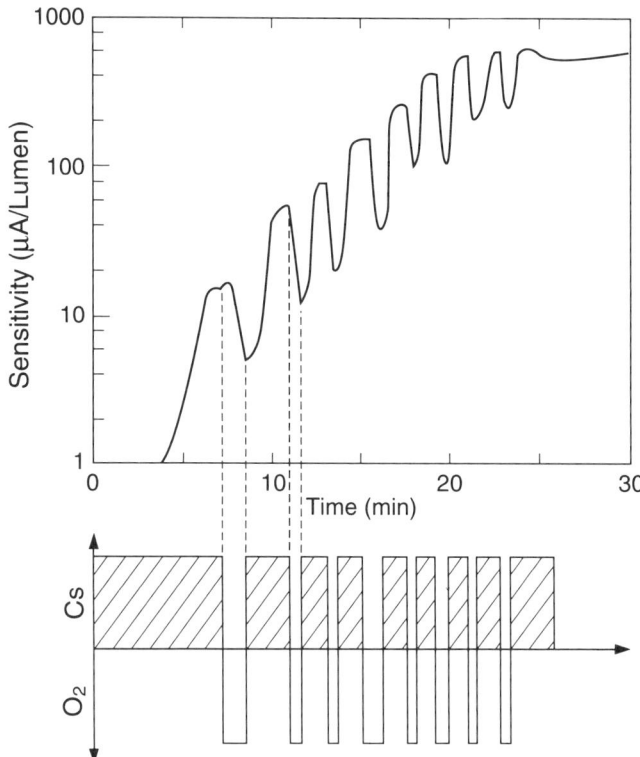

FIG. 5. Schematic of the yo-yo activation of GaAs with alternate cycles of Cs and O_2. A hypothetical but realistic photoresponse curve shows how the sensitivity increases on exposure to Cs and is decreased by O_2 exposure, the cycling of which is indicated at the bottom.

the best activations are achieved when the oxygen goes beneath the Cs to form a GaAs–O dipole layer [61]. This result is consistent with the two-stage activation process, the first stage of which forms a GaAs–O layer, Measurements of the work function at the surface of the activation layer, $E_\infty-E_F$ in Figure 2, showed that the work function continued to decrease with thickness beyond that thickness which gave optimum photoresponse [62]. The escape probability, P_{esc} of Equation (1.2), is increased by a lowering of the vacuum level, but it is decreased by scattering in the activation layer. Although much is not known about the Cs–O layer, it is found experimentally that the final Cs–O stoichiometry is very delicate; changes in either Cs or O of as little as 1–2% can decrease the photoresponse by a factor of 2 [62].

As an alternative to oxygen, NF_3 may be used. It decomposes on the Cs surface, leaving F bonded with the Cs. Because of the larger electronegativity of F compared with that of O, the use of NF_3 might be expected to produce a more stable photocathode. An activation procedure using NF_3 which is a modified yo-yo technique has been developed [35]. After the first peak in photocurrent with exposure to Cs has been obtained, the Cs is turned off and the NF_3 is turned on until a new maximum is reached. The NF_3 is left on and the Cs is turned on until the photocurrent is reduced to about $\frac{1}{3}$ of the previous maximum. The Cs is then turned off and the photocurrent is allowed to reach a new maximum. This process is repeated until there is little change, at which point the Cs is first turned off and then the NF_3 is slowly turned off as the photocurrent reaches a stable maximum. Substantially improved performance using NF_3 has been reported [22], but significant differences in quantum efficiency or lifetime were not observed by some others [48,64].

1.3.4 Quantum Efficiency

The quantum efficiency specifies the quality of the photocathode from the point of view of the current intensity which can be obtained with a given light source. For practical purposes, the definition of the quantum yield as the number of electrons emitted per incident photon is more useful than the yield per absorbed photon which is sometimes quoted. For a known incident light power, p, at wavelength λ, the measured photocurrent gives the yield according to

$$Y = 1.24 \, I \, (\mu A)/p \, (mW)\lambda \, (nm). \qquad (1.5)$$

Thus one obtains 6.45 μA per milliwatt of incident light power at $\lambda = 800$ nm from a GaAs photocathode which has a quantum efficiency of 1%. As a relative measure, some workers quote the yield of GaAs cathodes at the HeNe laser wavelength of 1.96 eV, but this yield can be over a factor of 2 higher than the yield for operation close to threshold. The yield at the operating photon energy is clearly the value of interest.

The yield as a function of photon energy is a useful diagnostic of the photocathode. A small monochromator, such as a 0.25-m Ebert grating monochromator, is adequate for the yield measurement as high resolution by optical standards is not required. A calibrated set of interference filters can also be used. Any of a number of different light sources can be used, such as a halogen lamp, a high-pressure Xe arc lamp, or a Zr arc lamp. At each photon energy, the photocurrent is compared with the incident light power which is easily measured by inserting a commercially available calibrated photodiode into the light beam after the monochromator [18].

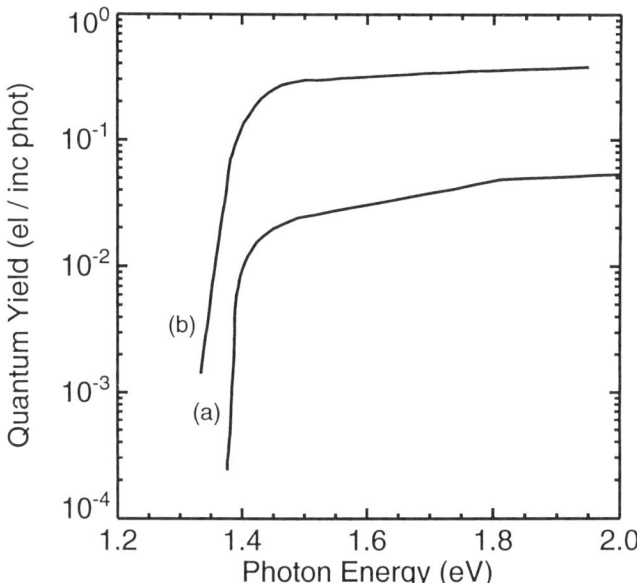

FIG. 6. The quantum efficiency curves of two NEA GaAs(100) reflection photocathodes are compared: (a) adequate photocathode [18] and (b) optimum photocathode [65].

A yield curve from what was described above as an adequate photocathode is compared in Figure 6 with a yield curve from a high-quantum-efficiency optimum photocathode. The adequate photocathode was made from a 5.6- × 10^{18}-cm^{-3} p-type GaAs (100) bulk wafer which was cleaned chemically and by heating [18]. The higher yield curve of Figure 6 was obtained from 7- × 10^{18}-cm^{-3} p-type GaAs (100) grown by vapor-phase epitaxy and activated by the two-stage process [65]. The sharp knee at threshold in the yield curves is characteristic of negative electron affinity. Fitting the yield curves using Equation (1.2) gives $L = 5$ μm and $P_{esc} = 0.55$ for the optimum photocathode [65] and $L = 0.4$ μm and $P_{esc} = 0.1$ for the adequate photocathode [18]. The magnitude of the yield is most sensitive to the escape probability which is much higher for the optimum photocathode presumably because of better cleaning and activation. The quantum efficiency of the optimum photocathode in Figure 6 is unusually high for a reflection photocathode as about 30% of the incident light is reflected. However, such a high-yield curve is typical for commercial transmission cathodes bonded to glass which combine antireflection coatings and the proper GaAs layer thickness to ensure near total absorption of the light [66].

1.3.5 Limitations of the Photocathode Response

When electron pulses are extracted from the photocathode, there are limitations on the minimum duration of the pulses. The photoexcitation process and the thermalization to the conduction band minimum take place rapidly in times on the order of 10^{-15} and 10^{-12} sec, respectively. Even for an arbitrarily short light pulse, the emitted electron distribution is spread in time because of the different transit times for electrons deep inside the active layer and for those near the surface to diffuse to the surface. The shortest electron pulses are obtained for the thinnest active layer, with an accompanying decrease in quantum efficiency. Pulses as short as 8 psec have been observed from an active photoemitting layer estimated to be 50 nm thick [67].

An apparent saturation of the amount of charge in a short, high-intensity electron pulse was observed to occur well below the space-charge limit of the electron gun [68]. The charge per pulse increased linearly with incident photon flux up to a point; beyond this point the charge increased approximately another factor of 2, but more slowly, saturating at a photon flux that was several times higher. This saturation of the photocurrent, that is the "charge limit," varied depending on the quality of activation of a particular cathode. The saturation photocurrent increased with increasing quantum efficiency, but saturated at the same photon flux. Thus the real limit is on the photon flux or light power density. For GaAs doped 2×10^{19} cm^{-3}, the charge per 2-nsec pulse increased linearly up to a power density of approximately 2 kW/cm^2. The nonlinear response is caused by the high light intensities which generate so many electrons that the surface states which determine the band bending become neutralized [69]. When this happens, the vacuum level is shifted higher in energy; this behavior is known as the surface photovoltage or photovoltaic effect. The shift in the vacuum level disappears, and the surface is restored by tunneling and thermionic emission of holes from the valence band into the surface states. For highly doped photocathode material, tunneling from holes into the surface states is the dominant mechanism that restores the surface [69]. Changing the doping of the photocathode from 5×10^{18} to 2×10^{19} cm^{-3} makes the band-bending region narrower, thereby increasing the tunneling probability and the rate at which the photocathode is restored.

In addition to the limit on peak light power density just described, one might expect there to be a limit on the average power density that a cathode could withstand. There have been reports of deterioration of the cathode response at average power densities on the order of 10 W/cm^2 and higher [23]. The mechanism affecting the photoresponse is not clear. It is not a surface photovoltage effect as in the case of high peak power. The heating of the GaAs surface, even assuming that all the light is absorbed at the surface, is calculated to be less than 10°C at this average power density [70]. Experimentally it is known that it is important to avoid contamination of the cathode from gases desorbed if the electron beam hits other surfaces. There needs to be further investigation into the maximum allowable average power density and the mechanism by which it affects the photocathode. As imprecise as these limits on the peak and average light power density are, they are important for estimating the performance of GaAs photocathodes, particularly the maximum brightness.

1.3.6 Photocathode Lifetime

The lifetime of the photocathode is defined as the time taken for the quantum efficiency to fall to $1/e$ of its initial value. Operating photocathode lifetimes vary from minutes to hundreds [48] or even thousands of hours [23]. The activation layer is delicately optimized and any changes in it can cause a decrease in quantum efficiency. One source of contamination is the residual gas in the vacuum chamber which can adsorb onto the photocathode surface and affect both the operating lifetime and the quiescent lifetime, which is analogous to the shelf life of a sealed-off phototube. Another factor limiting the operating lifetime is electron-stimulated desorption of atoms, molecules, or ions from any surfaces hit by the electron beam; desorbed neutrals may drift to the photocathode, but desorbed positive ions can be accelerated to the cathode surface.

An extremely good vacuum in the photocathode region, a pressure of $\leq 10^{-8}$Pa, is desirable. The usual ultrahigh-vacuum precautions regarding cleanliness and materials must be followed. An ion pump, possibly supplemented by a nonevaporable getter pump, can be used for pumping. The electron gun anode structure and subsequent electron optics must be carefully designed so that the electron beam does not strike their surfaces. If the polarized electron source must be attached to an apparatus with a poorer vacuum, differential pumping between the source and the apparatus can be employed [71].

The detailed mechanism at the photocathode surface which causes the decay in quantum efficiency is not well understood and, in fact, may be different in different situations. When the quantum efficiency has decreased, it is usually found that the photocathode is Cs deficient. Whether this is because an oxidant has adsorbed on the surface or because Cs has left the surface is not usually known. The photocathode efficiency, however, can be restored by addition of Cs. In sealed-off phototubes, there is a built-in excess of Cs which maintains cathode equilibrium and avoids a decrease in the quantum efficiency. In an ultrahigh-vacuum chamber, lifetimes are usually found to increase after a few activations, when the region around the photocathode becomes coated with Cs. In low-energy electron-gun applications, it is possible for the cathode to be continuously cesiated at a low rate to achieve very long lifetimes [72]. This treatment may not be possible in accelerator applications in which continuous cesiation may cause unwanted field emission from parts of the cathode structure at high voltage.

1.4 Incident Radiation

The photon energy of the photoexciting radiation should be within about 0.1 eV of the band-gap energy (much closer for strained layer cathodes). The incident photon intensity may be continuous or pulsed (subject to the limits on peak power density discussed above). Some examples of the many different kinds of lasers that have been used are an AlGaAs diode laser [18,73], an HeNe laser [46,73], a flash-pumped dye laser [19,48], and a YAG-pumped pulsed Ti:sapphire laser [74]. Because of the rapid photocurrent response time of GaAs photocathodes, it is possible to obtain current pulses with a wide variety of shapes by controlling the intensity of the incident radiation. For some accelerator applications, the photocurrent is sensed and a feedback system controls the light intensity during a pulse to obtain the required rectangular shape [48].

The light is circularly polarized using a linear polarizer, such as a Glan–Thompson prism, followed by a quarter-wave retarding element, such as a Pockels cell or quarter-wave plate as illustrated in Figure 7. When the quarter-wave voltage of the Pockels cell is reversed, the fast axis in Figure 7 becomes a slow axis and the photon helicity reverses from positive to negative. The degree of polarization can be measured using a photodiode and a second linear polarizer, crossed with the first, to detect the light passing through the Pockels cell. By measuring the maximum intensity I_{max} and minimum intensity I_{min} when the Pockels cell voltage is reversed, the degree of circular polarization P_{CP} can be found,

$$P_{CP} = 2(I_{max}I_{min})^{1/2}/(I_{max} + I_{min}), \qquad (1.6)$$

where the approximation has been made that the linear polarizers are perfect [75].

Very sensitive measurements of spin-dependent processes are achieved by detecting an experimental signal synchronously with the switching of the incident electron beam polarization. When small experimental asymmetries are to be measured, the electron beam polarization can be reversed

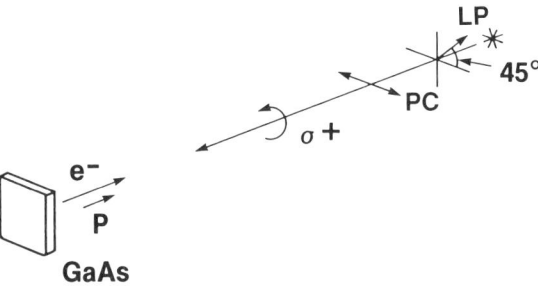

FIG. 7. The arrangement of optical elements to produce a particular electron spin polarization. Light from a source (∗) passes through a linear polarizer (LP) defining the plane of vibration of the electric vector at 45° to the fast axis (for the electric vector) of a Pockels cell (PC). The electric vector of the resultant circularly polarized light, as observed at a fixed point on the light axis, moves in the direction of the curved arrow. This is called σ^+ light and corresponds to the light angular momentum in the direction of light propagation. The spin polarization of electrons emitted from the GaAs is antiparallel to the incident light angular momentum.

rapidly and randomly on a pulse-to-pulse basis to minimize the affect of drift or possible switching of the beam polarization synchronously with other experimental parameters [20]. High sensitivity is achieved only if instrumental asymmetries, such as changes in the beam phase space or beam current when the polarization is reversed, are minimized. The instrumental asymmetry is defined as, $A = (I^+ - I^-)/(I^+ + I^-)$, where I^+ (I^-) is the current emitted for photoexcitation with σ^+ (σ^-) light. When extreme care is taken to align the Pockels cell, adjust the Pockels cell voltages, and compensate any birefringence of the vacuum windows, instrumental asymmetries as small as 6×10^{-6} can be achieved [48].

The energy spread of the electron beam from GaAs was observed to increase to as much as several electron volts when particular lasers were used to excite the photoelectrons [73]. This result was quite unexpected since most previous measurements found the energy spread ΔE in the range of 100 to 200 meV and under particular conditions as low as 30 meV [46,76,77]. This broadening was observed for excitation with intense HeNe lasers but not with diode lasers. It was attributed to very rapid fluctuations which result from interferences between the many modes of these lasers [73]. The apparently constant laser intensity actually consists of many rapid short pulses that give rise to intense bunches of emitted electrons° that result in an energy broadening [78]. Clearly, the laser used to photoexcite the electrons must be carefully selected when the energy spread of the beam is a consideration.

1.5 Operation of the Photocathode in an Electron Gun

Certain electron optical parameters of the photocathode determine what the optimum performance can be in a particular application. These are estimated for NEA GaAs and compared with measurements of actual GaAs photocathode performance and with the performance of other electron sources. Two examples are given of the wide variety of electron guns which have GaAs photocathodes.

1.5.1 Electron-Optical Considerations

The electron-optical phase-space volume of the emitted electron beam determines the fraction of the beam that can be accepted by the device to which the electron gun is coupled. A useful description of the beam is given by the product of the electron beam energy, E; the cross-sectional area, dA; and solid angle, $d\Omega$. The product is conserved according to the law of Helmholtz and Lagrange [79]. In the paraxial ray approximation this law leads to $E_1 A_1 \Omega_1 = E_2 A_2 \Omega_2$ for any two points, 1 and 2, along the beam, assuming conservation of current I. This conserved quantity is related to the concept of emittance which is used to describe electron beam quality. The emittance ε is defined as $1/\pi$ times the area in transverse phase space at a point along the beam. In the paraxial ray approximation, $\varepsilon = RR'$ for an axially symmetric beam, where R is the radius of the electron beam at the source or an image of the source and R' is the cone half-angle at that point. The quantity $\varepsilon E^{1/2}$ is sometimes referred to as the emittance invariant ε_{inv}, since $EA\Omega = (\pi \varepsilon_{inv})^2$. The emittance is readily generalized to nonaxially symmetric situations as well as to the relativistic case, in which the emittance invariant is written [80] $\varepsilon_{inv} = \beta \gamma \varepsilon$, where $\beta = v/c$, the electron velocity relative to the speed of light, and $\gamma = (1 - \beta^2)^{-1/2}$. Another quantity used to characterize the electron beam is the brightness, defined as the current per unit area per unit solid angle, $B = dI/dAd\Omega$. From the law of Helmholtz and Lagrange, it is seen that B/E is a conserved quantity and is known as the invariant brightness. In the relativistic limit the invariant brightness is written $B/\beta^2\gamma^2$. These relations show that the maximum current which can be incident at energy E onto a target in area dA and solid angle $d\Omega$ is determined by the invariant brightness B/E.

The emittance and brightness describe the GaAs photocathode for electron optical design purposes. Possible nonlinearities, aberrations, and space-charge effects can distort the emittance phase space and reduce the effective brightness in actual electron-optical systems. However, it is useful to estimate the photocathode brightness to obtain an upper limit.

In the absence of scattering in the band-bending region or at the surface, the angular spread of the photoemitted electrons is determined by the conservation of parallel wave vector and the ratio of the effective mass of an electron in the conduction band to the free-electron effective mass. For this ideal situation, Bell [28] calculated a minimum emission cone half-angle of $\alpha = 4°$ at room temperature. A larger value, $\alpha = 12°$, was measured [81] and attributed to roughness of the cathode surface. However, transverse energies of the photoelectrons of 40 to 100 meV corresponding to cone half-angles of approximately 25° to 30° have been measured and attributed to additional scattering, possibly in the activation layer [62,66,82]. The brightness can be written, $B = J/\pi\alpha^2$, where J is the current density. Using the measured $\alpha = 12°$ and an average electron emission energy of 0.2 eV, we arrive at an expression for the invariant brightness $B/E = 36J$ (A/cm^2-sr-eV), which would of course be decreased if a larger emission cone is assumed.

1.5.2 Performance

Some upper limits on the brightness can be estimated, taking the maximum average and peak light power densities of 10 W/cm^2 and 2 kW/cm^2, respectively, and a quantum efficiency of 0.3. The corresponding current densities are approximately 2 and 400 A/cm^2, from which we obtain values of the invariant brightness of 0.7×10^2 and 1.4×10^4 A/cm^2-sr-eV, respectively. In the pulsed mode, current densities of 180 A/cm^2 have been reported [83]. When a GaAs photocathode was used as the cathode in a scanning electron microscope, the electron beam brightness in the continuous mode at 3 keV was measured [81,84] to be 1×10^5 A/cm^2-sr. Measurements were made with the same average current in the pulsed mode with a duty cycle of 0.8%, leading to a pulsed brightness of 1.2×10^7 A/cm^2-sr. These measured brightness values can be compared with the 3-keV values of 2.1×10^5 and 4.2×10^7 A/cm^2-sr calculated from the continuous and pulsed invariant-brightness estimates, respectively.

Using the relativistic form of the invariant brightness, we calculate brightness values of 3.6×10^6 and 4.8×10^8 A/cm^2-sr at 100 keV, for the continuous and pulsed modes, respectively, from the measured brightness values at 3 keV. The 3-keV brightness values were not measured under optimum conditions, and further measurements are required to determine whether higher values can be attained [84]. These GaAs photocathode brightness values at 100 keV are compared in Table I with typical values [85] for LaB$_6$, a W hairpin filament, a pointed W filament, and heated field emission sources. The brightness of NEA GaAs in the continuous mode is about 10 times that of the W hairpin filament and approximately

TABLE I. NEA GaAs Compared with Sources of Unpolarized Electrons

Source	Brightness at 100 keV (A cm^{-2} sr^{-1})	ΔE FWHM (eV)
NEA GaAs		
Continuous	3.6×10^6	0.1–0.2
Pulsed	4.4×10^8	
LaB$_6$	7×10^6	1
W hairpin filament	5×10^5	0.7–2.4
Pointed W filament	2×10^6	2
Heated field emission	$10^7 – 10^8$	0.3

comparable to that of LaB$_6$. For pulsed applications, the brightness of NEA GaAs is much higher than that of LaB$_6$ and even higher than that of heated field emission sources.

The energy spread and vacuum requirements of the electron sources in Table I are quite varied. The LaB$_6$ and W cathodes operate at 1300–1500 and 2300–2500°C, respectively, leading to a fairly high energy spread of the emitted beams as noted in Table I. The electron affinity of a GaAs cathode can be adjusted to obtain an energy spread as narrow as 30 meV, but with much lower quantum efficiency. Values of $\Delta E = 0.1$–0.2 eV are estimated for GaAs operating in the high-brightness continuous mode. Somewhat larger values of ΔE can occur in systems with larger NEA or, in the case of reflection photocathodes, when the photon energy is significantly larger than the band gap and when nonthermalized electrons form a substantial part of the photoelectron energy distribution. There is evidence for further broadening in the pulsed mode [81], possible related to the broadening discussed in Section 1.4; additional investigation of this effect is needed. Thus, it may be possible to attain a factor of 2 or so lower ΔE values from NEA GaAs than from the heated field emitter; if attainable, such cathodes would be a significant improvement for low-energy electron microscopy applications. The field-emission and GaAs cathodes require ultrahigh vacuum, whereas the W and LaB$_6$ require only a moderate vacuum, a pressure of 10^{-3} Pa or lower.

The noise properties and stability of GaAs cathodes have been measured [84]. When a stabilization-feedback loop diminished the laser noise, the noise spectrum was found to be near the shot-noise limit. By measuring the emitted current on a spray aperture and controlling the laser power, the drift in a 3-hr period was less than 0.04%. The emission from the cathode as observed on a phosphor screen was found to be uniform without hot spots [81].

1.5.3 Examples of Polarized Electron Guns

Many electron-optical systems have been used with GaAs cathodes. The applications vary from a low-energy electron diffraction (LEED) gun [18] to a pulsed scanning electron microscope [84] and the injector for a linear accelerator [19]. For the low-energy applications, space-charge considerations dominate before brightness limits are reached. In higher energy applications, such as electron microscopy, guns are brightness limited. Electron guns employing GaAs photocathodes have generally used a simple diode configuration consisting of the photocathode and an anode. The triode configuration, in which a control electrode between the cathode and the anode forms a crossover, was ruled out for pulsed electron microscopy because of the space-charge-induced electron energy spread in the crossover [81].

The original SLAC GaAs-polarized electron gun [19] used for the parity-violation experiments [20,21] is shown in Figure 8. This gun, which has been a model for a number of later guns for accelerator applications, is very similar to thermionic guns used on SLAC, the main difference being the use of the GaAs photocathode instead of a thermionic emitter. The cathode geometry is shaped to optimize space-charge-limited operation, and the anode, as with the subsequent electron optics, is designed so as not to intercept any electrons which would desorb gases and limit the cathode lifetime as discussed in Section 1.3.6. The large insulator allows cathode operation at $-70\,\text{keV}$ as required for injection into the accelerator. Since the time of these experiments, it has been found that good results can be obtained without cooling the cathode or surrounding regions.

An example of a low-energy polarized electron gun is shown in Figure 9. For this type of application, the anode is simply a flat plate with a hole in it. When space charge is a limitation, it usually occurs when the beam is focused at low energy downstream in the electron optics. The anode is followed by a 90° spherical deflector which deflects the electron beam from the path of the incident laser radiation. The spherical deflector also acts as a focusing element. According to Barber's rule, the object (actually the virtual object which in this case is 3 mm behind the GaAs surface), the center of curvature of the deflector, and the image lie on a straight line. Two sets of deflection plates capable of changing the beam angle or laterally shifting the beam precede the acceleration to 1000 eV. Up to this point, labeled X in Figure 9, this polarized electron gun is fairly generic, while beyond this point it is designed to be a LEED gun as shown in the lower part of the figure. Alternatively, it could be designed to produce a lower energy and higher current beam as required for inverse photoemis-

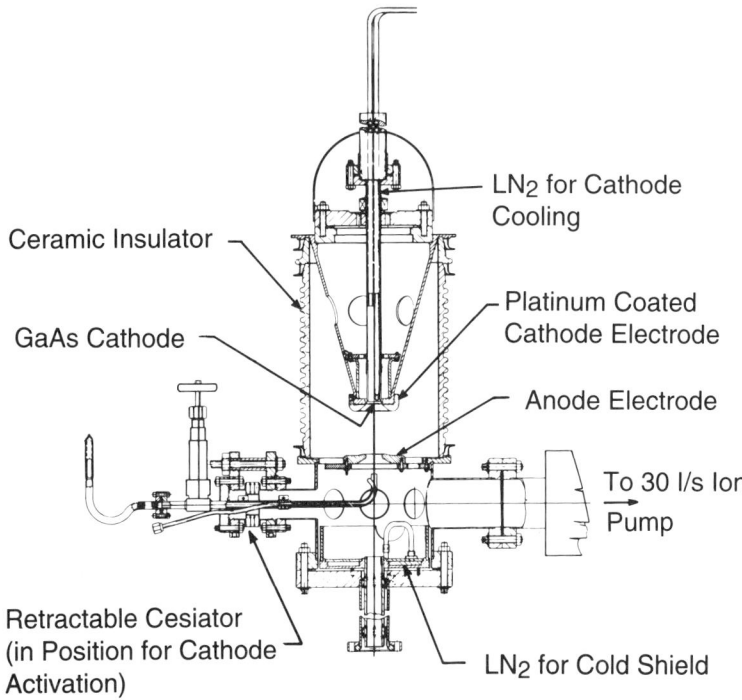

FIG. 8. Schematic of the initial SLAC GaAs polarized electron gun developed for the parity violation experiment. The cathode is at a potential of -70 kV, suitable for injection into the linear accelerator. From Sinclair [19].

sion [73]. The lens elements of this LEED gun were made of Cu, whereas those of the inverse photoemission gun [73] were made of Al and were coated on the inside with a graphite layer. In each case, care was taken to ensure that there was no line of sight from the electron beam to the insulators separating the electrodes. Molybdenum is a suitable material for the apertures.

1.5.4 Spin Rotation

The electron beam extracted from the GaAs photocathode is longitudinally polarized; that is, the axis of electron spin polarization is along the electron momentum. For most low-energy condensed matter experiments, a transverse polarization is desired. This objective is accomplished in the source shown in Figure 9 by the transverse electric field of the 90° electrostatic deflector which, at nonrelativistic energies, changes the beam

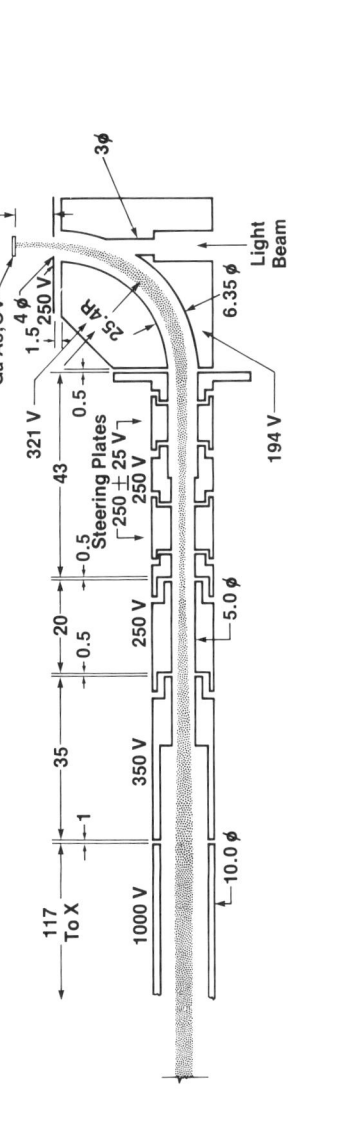
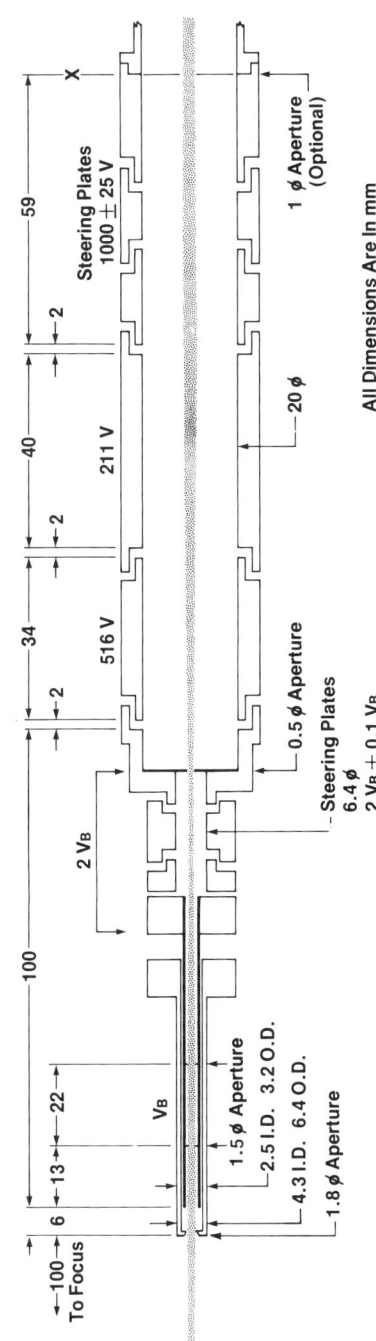

Fig. 9. Schematic of the low-energy polarized electron gun developed at NIST. The upper part of the figure is a generic polarized electron source. The light is incident through a hole in the 90° spherical deflector that deflects the beam to the electron optics that transport the electrons to the vacuum isolation valve. The lower part of the figure shows the electron optics which follow the valve and are specialized for a LEED gun. The calculated beam envelope is shown from the photocathode to the target. From Pierce et al. [18].

direction without changing the spin direction. At high energies, relativistic corrections are required [86]. For example, an electrostatic deflector with a bending angle of 107.7° is required to rotate the spin of a 100-keV beam by 90°. For a transverse magnetic field, the spin polarization direction follows the electron momentum in the nonrelativistic limit.

In most high-energy experiments, a longitudinal polarization is required at the scattering target. Magnetic fields in accelerators can cause the spin direction of the relativistic electrons to change during transport. It is desirable to be able to adjust the polarization direction of the beam relative to the electron momentum at the source in order to achieve the desired longitudinal polarization at the target for each energy. An elegant system, employing two electrostatic deflectors and two systems of solenoids each consisting of two double solenoids, that allows the selection of the spin direction of 100-keV electrons to be in any direction before injection into the accelerator has been described [87,88]. This spin rotator has the advantage that the electron-optical properties of the beam are otherwise unchanged.

1.6 Summary

A number of different polarized electron sources are compared in Table II, which is an updated version of previous comparisons [1,2,16,18]. There are significant changes in the entries for the NEA GaAs and flowing He afterglow sources, but there has not been further development of the other types of spin-polarized sources which are included for comparison. The optimum source for a particular application depends on a number of interrelated factors. For some, a minimum asymmetry on switching the electron beam polarization may be most important, whereas for others the highest polarization or the time structure of the beam is the overriding factor.

The developments in GaAs-type polarized electron sources are listed in the first part of Table II. For this type of source an extra column, the quantum efficiency, is included since the beam current is determined, within the limits discussed previously, by the incident radiation. Only published values of polarization and quantum efficiency are listed. The strained layer and superlattice cathodes are undergoing continuous improvement. Significant increases in the figure of merit $P^2 I$ can be expected, largely as a result of increases in the quantum efficiency without polarization reduction. The peak pulsed current listed is typical for SLAC [68]. The smallest energy spreads are attained by adjusting the vacuum level with a corresponding decrease in quantum efficiency. Energy spreads greater that 0.3 eV have been reported at higher photon energies, for

TABLE II. Comparison of Spin-Polarized Electron Sources

Method		P	Reversal of P	I_{dc}	I_{pulse} (el/pulse) (rep. rate)	ΔE (eV)	$EA\Omega$ (eV cm² sr)
Photoemission from NEA GaAs (100)			Optical				2.2×10^{-8}
Bulk GaAs	QE						
"Adequate" [18]	3×10^{-2}	0.25–0.4		20 µA/mW[a]			
"optimum"	3×10^{-1}			200 µA/mW[a]		0.03–0.3	
Thin (0.2 µm) GaAs [32]	1×10^{-2}	0.49			10^{11}/2 nsec 120 Hz		
GaAs/GaP$_x$As$_{1-x}$							
[40]	8×10^{-4}	0.83					
[41]	4×10^{-4}	0.90					
[41]	1×10^{-3}	0.80					
Superlattice [43]	3×10^{-6}	0.71					
Optically pumped flowing He afterglow [11]		0.80	Optical	1 µA		0.15–0.4	$<4.3 \times 10^{-2}$
		0.75		10 µA			
Photoemission from EuO [7]		0.61	Magnetic field	1 µA		2	1.8×10^{3}
Field emission from W-EuS [6]		0.85	Magnetic field	0.01 µA		0.1	10^{-11}
Fano effect							
Rb [4]		0.65	Optical		2.2×10^9/12 nsec 50 Hz	<500	1.1
Cs [5]		0.63				3	3.9
Photoionization polarized Li [3]		0.85	Magnetic field	0.01 µA	2.2×10^9/1.5 µsec 180 Hz	1500	<64

[a] The maximum current is determined by light power subject to limitations described in the text.

photocathodes such as GaAsP which have a larger NEA, and for some pathological cases mentioned in Sections 1.4 and 1.5.2. A beam energy spread of 0.1 to 0.2 eV is typical in ordinary operation. The energy-area-angle phase-space product for the GaAs source is calculated assuming an initial energy of 0.2 eV and an area corresponding to a 10-μm-diameter light spot which can be attained with ordinary optics. Using the emission cone half-angle of 12°, one calculates $EA\Omega = 2.2 \times 10^{-8}$ eV-cm^2-sr. This small value, which obviously depends on the assumed values, gives flexibility in the electron-optical design.

The optically pumped, flowing-He-afterglow polarized electron source has undergone further development which produced the significantly improved results listed in the second entry of Table II. This source also uses optical pumping, in this case of the metastable He(2^3S) in the flowing He afterglow generated by a microwave discharge. Using circularly polarized $2^3S \rightarrow 2^3P$ radiation one spin state is preferentially populated. When a target gas such as CO_2 is injected into the afterglow, spin conserving chemionization reactions take place, resulting in polarized electrons that can be extracted from the afterglow region and formed into a beam. Like the GaAs source, the spin polarization is conveniently reversed by reversing the circular polarization of the light. The absence of the need for ultrahigh vacuum is an advantage of the flowing He afterglow source for some applications. High polarizations of 0.80 and 0.75 at beam currents of 1 and 10 μA, respectively, have been achieved [11].

Cardman [89] compared various NEA GaAs sources and the flowing He afterglow source for three accelerator experiments: the low-current polarized target and higher current parity-violation experiments at the Continuous Electron Beam Accelerator Facility (CEBAF) and the Z^0 experiment at the Stanford Linear Collider (SLC). The high polarization that can be obtained at low currents from the flowing He afterglow source makes it competitive with the strained layer GaAs source for polarized target experiments at CEBAF [89]. The strained-layer GaAs/GaP$_x$As$_{1-x}$ source parameters are superior for the high-current CEBAF experiments and the pulsed SLC experiments.

In addition to its wide application as a source of spin-polarized electrons for particle, atomic, and condensed matter physics [90], the NEA GaAs photocathode makes possible the coupling of high-speed laser technology and electron-beam instrumentation in applications for which electron spin may be of no concern, such as time-resolved electron microscopy and spectroscopy. This feature, coupled with the high brightness of these photocathodes and small energy spread, presents many opportunities for fruitful application of NEA GaAs photocathodes.

Acknowledgments

Many helpful discussions with numerous colleagues at NIST and elsewhere are gratefully acknowledged. This work was supported in part by the Office of Naval Research.

Appendix A: Cleaning GaAs

The following procedure for cleaning GaAs (adapted from Pierce *et al.* [18]) is one of a number of procedures which have been reported. Important factors are thought to be using deionized water with a resistivity of >15 MΩ-cm, using electronic-grade chemicals and fresh etching solutions, and concluding etch steps (7,10,12) by flushing GaAs with deionized water without exposing it to air.

1. Ultrasonically clean four polyethylene beakers, Teflon tweezers, and a graduated cylinder in 1,1,1-trichloroethane, acetone, and methanol.
2. Prepare a 4 : 1 : 1 mixture of concentrated H_2SO_4, 30% H_2O_2, and H_2O by volume. Carefully add the H_2SO_4 to the H_2O_2 and H_2O to avoid heating above 80°C.
3. Prepare a 1 : 1 mixture of NaOH (1 M solution, 4 g NaOH to 100 ml H_2O) and H_2O_2 (0.76 M solution, 1 ml 30% H_2O_2 to 11.5 ml H_2O).
4. Ultrasonically clean the crystal at low power in trichloroethane for 3 min. Decant trichloroethane leaving the GaAs slightly covered. Add new trichloroethane and repeat this step two more times.
5. Decant the trichloroethane and rinse the crystal with methanol, decanting off the liquid leaving the GaAs slightly covered. Repeat this methanol rinse two more times. Ultrasonically clean the crystal at low power in methanol for 3 min. Decant the methanol leaving the crystal slightly covered. Add fresh methanol and repeat the ultrasonic step two more times.
6. Blow dry the crystal with filtered dry N_2.
7. Etch the crystal in the 4 : 1 : 1 mixture at 50°C for 3 min, face up. Agitate the solution to keep fresh etch at the surface.
8. Rinse the crystal in 10 changes of deionized water always keeping it covered by some water. Rinse the crystal in 6 changes of methanol, again always keeping the crystal covered with some of the liquid.
9. Blow dry the crystal with filtered dry N_2 and make a visual inspection at this point. If the crystal is not clean and shiny, start over with a new crystal.

10. Etch the crystal face up in undiluted (48%) HF, agitating the solution for 5 min at room temperature.
11. Rinse the crystal twice in deionized water.
12. Etch the crystal in 1 : 1 solution (from step 3) for 1 min at room temperature.
13. Rinse the crystal five times in deionized water keeping the crystal surface covered. Rinse five times in methanol again keeping the surface covered. Blow the crystal dry with filtered dry N_2 and install it in the vacuum system immediately.

Appendix B: Anodization of GaAs

The following procedure [91] is adapted from Schwartz *et al.* [92].

1. Prepare a 2.5 pH phosphoric acid (H_3PO_4) solution by adding two to three drops of 85% H_3PO_4 phosphoric acid to 800 ml deionized water.
2. Form a loop, about 2 cm in diameter, from 0.05-mm-diameter Pt wire for the cathode. Suspend it from the edge of a clean 100-ml beaker so it is about 1 cm from the bottom.
3. Sheath the ends of stainless steel tweezers with pure Al so that no metals other than Pt and Al contact the anodizing solution.
4. Add the anodizing solution to the beaker. Before using the sheathed tweezers to hold the GaAs, anodize the tips by applying -50 V to the Pt wire and wait until the current stabilizes. The disturbance of GaAs anodization is minimized by preanodizing the tweezers.
5. Remove the tweezers and scrape off a bit of the anodization for good electrical contact with the GaAs chip.
6. Hold the GaAs in the tweezers face up under the Pt loop in fresh anodizing solution. The GaAs can be transferred directly from the deionized water rinse of cleaning step 13 (Appendix A) to the anodizing solution.
7. Apply -40 to -50 V DC to the Pt wire to begin anodization. Monitor the current and continue anodization until it stabilizes at its minimum value for 1 min; bubbles will cease to form on the surface.
8. Remove the GaAs and rinse it in five changes of deionized water and methanol. Blow dry the GaAs with filtered dry N_2.
9. The anodization layer can be removed by agitating it face up in 30% NH_4OH for 30 sec. Rinse the GaAs with 10 changes of deionized water without exposing it to air. Blow dry the GaAs with N_2 and install it immediately in a vacuum system (preferably maintaining the GaAs under an N_2 atmosphere [53]).

References

1. J. Kessler, *Polarized Electrons*, Springer-Verlag, Berlin, 1985.
2. R. J. Celotta and D. T. Pierce, *Adv. At. Mol. Phys.* **16**, 101 (1980).
3. M. J. Alguard, J. E. Clendenin, R. D. Ehrlich, V. W. Hughes, J. S. Ladish, M. S. Lubell, K. P. Schüler, G. Baum, W. Raith, R. H. Miller, and W. Lysenko, *Nucl. Instrum. Methods* **163**, 29 (1979).
4. W. von Drachenfels, U. T. Koch, T. M. Müller, W. Paul, and H. R. Schaefer, *Nucl. Instrum. Methods* **140**, 47 (1977).
5. P. F. Wainwright, M. J. Alguard, G. Baum, and M. S. Lubell, *Rev. Sci. Instrum.* **49**, 571 (1978).
6. E. Kisker, G. Baum, A. H. Mahan, W. Raith, and B. Reihl, *Phys. Rev. B* **18**, 2256 (1978).
7. E. Garwin, F. Meier, D. T. Pierce, K. Sattler, and H. C. Siegmann, *Nucl. Instrum. Methods* **120**, 483 (1974).
8. P. J. Keliher, R. E. Gleason, and G. K. Walters, *Phys. Rev. A* **11**, 1279 (1975).
9. L. A. Hodge, F. B. Dunning, and G. K. Walters, *Rev. Sci. Instrum.* **50**, 1 (1979).
10. L. G. Gray, K. W. Giberson, C. Cheng, R. S. Keiffer, F. B. Dunning, and G. K. Walters, *Rev. Sci. Instrum.* **54**, 271 (1983).
11. G. H. Rutherford, J. M. Ratliff, J. G. Lynn, F. B. Dunning, and G. K. Walters, *Rev. Sci. Instrum.* **61**, 1460 (1990).
12. E. Garwin, D. T. Pierce, and H. C. Siegmann, *Helv. Phys. Acta* **47**, 393 (1974).
13. G. Lampel and C. Weisbuch, *Solid State Commun.* **16**, 877 (1975).
14. D. T. Pierce, F. Meier, and P. Zürcher, *Phys. Lett. A* **51A**, 465 (1975).
15. D. T. Pierce, F. Meier, and P. Zürcher, *Appl. Phys. Lett.* **26**, 670 (1975).
16. D. T. Pierce and F. Meier, *Phys. Rev. B* **13**, 5484 (1976).
17. D. T. Pierce, F. Meier, and H. C. Siegmann, U. S. Pat. 3,968,376 (1976).
18. D. T. Pierce, R. J. Celotta, G.-C. Wang, W. N. Unertl, A. Galejs, C. E. Kuyatt, and S. R. Mielczarek, *Rev. Sci. Instrum.* **51**, 478 (1980).
19. C. K. Sinclair, *AIP Conf. Proc.* **35**, 426 (1976).
20. C. Y. Prescott, W. B. Atwood, R. L. A. Cottrell, H. DeStaebler, E. L. Garwin, A. Gonidec, R. H. Miller, L. S. Rochester, T. Sato, F. J. Sherden, C. K. Sinclair, S. Stein, R. E. Taylor, J. E. Clendenin, V. W. Hughes, N. Sasao, K. P. Schüler, M. G. Borghini, K. Lübelsmeyer, and W. Jentschke, *Phys. Lett. B* **77B**, 347 (1978).
21. C. Y. Prescott, W. B. Atwood, R. L. A. Cottrell, H. DeStaebler, E. L. Garwin, A. Gonidec, R. H. Miller, L. S. Rochester, T. Sato, F. J. Sherden, C. K. Sinclair, S. Stein, R. E. Taylor, C. Young, J. E. Clendenin, V. W. Hughes, N. Sasao, K. P. Schüler, M. G. Borghini, K. Lübelsmeyer, and W. Jentschke, *Phys. Lett. B* **84B**, 524 (1979).
22. C. K. Sinclair, *Proc. Int. Symp. High Energy Spin Phys. 6th*, Marseille, *1984; J. Phys.* **46**, C2-669 (1985).
23. C. K. Sinclair, *Proc. Int. Symp. High Energy Spin Phys. 8th*, Minneapolis, *1988; AIP Conf. Proc.* **187**, 1412 (1989).
24. E. Reichert, *Proc. Int. Symp. High Energy Spin Phys. 9th*, Bonn, *1990*, Vol. 1, p. 303 (1991).
25. T. Nakanishi, *Proc. Int. Symp. High Energy Spin Phys. 10th*, Nagoya, *1992*. p. 279-290. (Universal Academy Press Inc., Tokyo).

26. W. E. Spicer, *Phys. Rev.* **112**, 114 (1958).
27. W. E. Spicer, *Appl. Phys.* **12**, 115 (1977).
28. R. L. Bell, "Negative Electron Affinity Devices." Oxford Univ. Press (Clarendon), Oxford, 1973.
29. L. W. James and J. L. Moll, *Phys. Rev.* **183**, 740 (1969).
30. G. Fishman and G. Lampel, *Phys. Rev. B* **16**, 820 (1977).
31. G. Lampel and M. Eminyan, *Proc. Int. Conf. Phys. Semiconduct. 15th, 1980; J. Phys. Soc. Jpn.* **49**, Suppl. A, 627 (1980).
32. T. Maruyama, R. Prepost, E. L. Garwin, C. K. Sinclair, B. Dunham, and S. Kalem, *Appl. Phys. Lett.* **55**, 1686 (1989).
33. P. Zürcher and F. Meier, *J. Appl. Phys.* **50**, 3687 (1979).
34. F. Meier, A. Vaterlaus, F. P. Baumgartner, M. Lux-Steiner, G. Doell, and E. Bucher, *Proc. Int. Symp. High Energy Spin Phys. 9th, Bonn, 1990*, Vol. 2, p. 25 (1991).
35. B. M. Dunham, Ph.D. Thesis, University of Illinois, Urbana (1993).
36. T. Maruyama, E. L. Garwin, G. H. Zapalac, J. S. Smith, and J. D. Walker, *Phys. Rev. Lett.* **66**, 2376 (1991).
37. J. W. Matthews and A. E. Blakeslee, *J. Cryst. Growth* **27**, 118 (1974).
38. P. J. Orders and B. F. Usher, *Appl. Phys. Lett.* **50**, 980 (1987).
39. T. Nakanishi, H. Aoyagi, H. Horinaka, Y. Kamiya, T. Kato, S. Nakamura, T. Saka, and M. Tsubata, *Phys. Lett. A* **158A**, 345 (1991).
40. H. Aoyagi, H. Horinaka, Y. Kamiya, T. Kato, T. Kosugoh, S. Nakamura, T. Nakanishi, S. Okumi, T. Saka, M. Tawada, and M. Tsubata, *Phys. Lett. A* **167A**, 415 (1992).
41. T. Maruyama, E. L. Garwin, R. Prepost, and G. H. Zapalac, *Phys. Rev. B* **46**, 4261 (1992).
42a. T. Saka, T. Kato, T. Nakanishi, M. Tsubata, K. Kishino, H. Horinaka, Y. Kamiya, S. Okumi, C. Takahashi, Y. Tanimoto, M. Tawada, K. Togawa, H. Aoyagi, and S. Nakamura, *Jpn. J. Appl. Phys.* **32**, L1837 (1993).
42b. J. C. Gröbli, D. Oberli, F. Meier, A. Dommann, Yu. Mamaev, A. Subashiev, Yu. Yashin, *Phys. Rev. Lett.* **74**, 2106 (1995).
43. T. Omori, Y. Kurihara, T. Nakanishi, H. Aoyagi, T. Baba, T. Furuya, K. Itoga, M. Mizuta, S. Nakamura, Y. Takeuchi, M. Tsubata, and M. Yoshioka, *Phys. Rev. Lett.* **67**, 3294 (1991).
44. T. Nakanishi, Nagoya University, private communication Feb. 15, 1994.
45. C. Conrath, T. Heindorff, A. Hermanni, N. Ludwig, and E. Reichert, *Appl. Phys.* **20**, 155 (1979); E. Reichert and K. Zähringer, *Appl. Phys. [Part] A* **A29**, 191 (1982).
46. J. Kirschner, H. P. Oepen, and H. Ibach, *Appl. Phys. [Part] A* **A30**, 177 (1983).
47. F. Ciccacci, S. F. Alvarado, and S. Valeri, *J. Appl. Phys.* **53**, 4395 (1982).
48. W. Hartmann, D. Conrath, W. Gasteyer, H. J. Gessinger, W. Heil, H. Kessler, L. Koch, E. Reichert, H. G. Andresen, T. Kettner, B. Wagner, J. Ahrens, J. Jethwa, and F. P. Schäfer, *Nucl. Instrum. Methods Sect. A* **286**, 1 (1990).
49. L. W. James, G. A. Antypas, J. Edgecumbe, R. L. Moon, and R. L. Bell, *J. Appl. Phys.* **42**, 4976 (1971).
50. S. F. Alvarado, F. Ciccacci, S. Valeri, M. Campagna, R. Feder, and H. Pleyer, *Z. Phys. B* **44**, 259 (1981).
51. B. Reihl, M. Erbudak, and D. M. Campbell, *Phys. Rev. B* **19**, 6358 (1979).
52. D. T. Pierce, G. C. Wang, and R. J. Celotta, *Appl. Phys. Lett.* **35**, 220 (1979).

53. C. J. Spindt, Ph.D. Thesis, Dept. of Applied Physics, Stanford University, Stanford, CA (1991).
54. I. Shiota, K. Motoya, T. Ohmi, N. Miyamoto, and J. Nishizawa, *J. Electrochem. Soc.* **124**, 155 (1977).
55. J. S. Escher, in *Semiconductors and Semimetals* (R. K. Willardson and A. C. Beer, eds.), Vol. 15. Academic Press, New York, 1981. p. 195.
56. Yu. G. Galitsyn, V. G. Mansurov, V. I. Poshevnev, and A. S. Terekhov, *Poverkhnost* Issue 10, 140 (1989).
57. Yu. G. Galitsyn, V. G. Mansurov, V. P. Opsheyenev, A. S. Terekhov, and L. G. Okorokova, *Poverkhnost* Issue 4, 147 (1989).
58. J. J. Uebbing, *J. Appl. Phys.* **41**, 802 (1976).
59. SAES Getters. Certain commercial instruments or materials are identified in this paper to clarify descriptions. In no case does such identification imply recommendation or endorsement by NIST.
60. W. Klein, *Rev. Sci. Instrum.* **42**, 1082 (1971).
61. C. Y. Su, W. E. Spicer, and I. Lindau, *J. Appl. Phys.* **54**, 1413 (1983); W. E. Spicer, Stanford University, Stanford, CA, private communication, September 1993.
62. D. C. Rodway and M. B. Allenson, *J. Phys. D* **19**, 1353 (1986).
63. B. J. Stocker, *Surf. Sci.* **47**, 501 (1975).
64. F. Ciccacci and G. Chiaia, *J. Vac. Sci. Technol., A* **9**, 2991 (1991).
65. G. H. Olsen, D. J. Szostak, T. J. Zamerowski, and M. Ettenberg, *J. Appl. Phys.* **48**, 1007 (1977).
66. V. Aebi, Intevac Corporation, private communication, September 1993.
67. C. C. Phillips, A. E. Hughes, and W. Sibbett, *J. Phys. D* **17**, 1713 (1984).
68. H. Tang, R. K. Alley, H. Aoyagi, J. E. Clendenin, J. C. Frisch, C. L. Garden, E. W. Hoyt, R. E. Kirby, L. A. Klaisner, A. V. Kulikov, C. Y. Prescott, P. J. Saez, D. C. Schultz, J. L. Turner, M. Woods, and M. S. Zolotorev, *Proc. Part. Accel. Conf.*, Washington, DC, *1993* Vol. 4, p. 3036 (1993).
69. A. Herrera-Gómez and W. E. Spicer, *Proc. SPIE—Int. Symp. Imaging Instrum.*, San Diego, *1993* (1993). p. 51.
70. M. Lax, *J. Appl. Phys.* **48**, 3919 (1977).
71. E. Mergl, E. Geisenhofer, and W. Nakel, *Rev. Sci. Instrum.* **62**, 2318 (1991).
72. F. C. Tang, M. S. Lubell, K. Rubin, A. Vasilakis, M. Eminyan, and J. Slevin, *Rev. Sci. Instrum.* **57**, 3004 (1986).
73. U. Kolac, M. Donath, K. Ertl, H. Liebl, and V. Dose, *Rev. Sci. Instrum.* **59**, 1933 (1988).
74. J. Frisch, R. Alley, M. Browne, and M. Woods, *Proc. Part. Accel. Conf.*, Washington, DC, *1993* Vol. 4, p. 3047 (1993).
75. U. Heinzmann, *J. Phys. E* **10**, 1001 (1977).
76. H.-J. Drouhin, C. Hermann, and G. Lampel, *Phys. Rev. B* **31**, 3872 (1985).
77. C. S. Feigerle, D. T. Pierce, A. Seiler, and R. J. Celotta, *Appl. Phys. Lett.* **44**, 866 (1984).
78. H.-J. Drouhin and P. Bréchet, *Appl. Phys. Lett.* **56**, 2152 (1990).
79. J. A. Simpson, in "Methods of Experimental Physics," Vol. 4A, (V. W. Hughes and H. L. Schultz, eds.) p. 124 Academic Press, New York, 1967.
80. J. D. Lawson, "The Physics of Charged-Particle Beams." Oxford Univ. Press (Clarendon), Oxford, 1977.
81. C. A. Sanford and N. C. MacDonald, *J. Vac. Sci. Technol., B* **8**, 1853 (1990);

C. A. Sanford, Ph.D. Thesis, Dept. Electrical Engineering, Cornell University, Ithaca, NY (1990).
82. D. J. Bradley, M. B. Allenson, and B. R. Holeman, *J. Phys. D* **10**, 111 (1977).
83. C. K. Sinclair and R. H. Miller, *IEEE Trans. Nucl. Sci.* **NS-28**, 2649 (1981).
84. C. A. Sanford and N. C. MacDonald, *J. Vac. Sci. Technol., B* **7**, 1903 (1989).
85. J. C. H. Spence, *Experimental High-Resolution Electron Microscopy*, p. 250. Oxford Univ. Press, Oxford, 1988.
86. V. Bargmann, L. Michel, and V. L. Telegdi, *Phys. Rev. Lett.* **2**, 435 (1959).
87. K.-H. Steffens, H. G. Andresen, J. Blume-Werry, F. Klein, K. Aulenbacher, and E. Reichert, *Nucl. Instrum. Methods A* **325**, 378 (1993).
88. D. A. Engwall, B. M. Dunham, L. S. Cardman, D. P. Heddle, and C. K. Sinclair, *Nucl. Instrum. Methods A* **324**, 409 (1993).
89. L. S. Cardman, *Nucl. Phys. A* **A546**, 317c (1992).
90. D. T. Pierce and R. J. Celotta, *in* "Optical Orientation" (F. Meier and B. P. Zakharchenya, eds.) p. 259, North-Holland Publ., Amsterdam, 1984.
91. T. Roder and E. Garwin, Stanford Linear Accelerator Center, Palo Alto, CA, private communication, 1979.
92. B. Schwartz, F. Ermanix, and M. H. Brastad, *J. Electrochem. Soc.* **123**, 1089 (1976).

2. POSITRON AND POSITRONIUM SOURCES

A. P. Mills, Jr.
AT&T Bell Laboratories, Murray Hill, New Jersey

2.1 Introduction

Positrons [1] are the antimatter partners of the electrons which are constituents of all matter. Positrons are positively charged and annihilate with electrons, producing γ rays. Many novel spectroscopic methods have been developed to study the properties of normal matter based on positron scattering or the observation of the annihilation photons. The application of such spectroscopies has, however, been hampered by lack of a simple, intense positron source. Nonetheless, it is now technically feasible to construct intense positron sources, and with their availability it will be finally possible to exploit the richness of positron interactions for the benefit of science and industry. The present chapter is devoted to the art and science of making positron sources, from the microcurie level as used by single investigators to the megacurie sources now proposed.

2.2 Positron Sources

Positrons are produced by certain β^+ decay processes and by e^+-e^- pair production via the electromagnetic interaction [2]. Note that positrons created by the weak decay of certain nuclei have traditionally been known as β^+ particles, whereas after they have slowed down we call them simply positrons, with the symbol e^+. Since positrons are created out of the vacuum in partnership with a second lepton, they appear with kinetic energies on the order of the positron rest energy, $mc^2 = 511$ keV. The positrons must be slowed to electron-volt energies or less to be useful for solid-state and atomic physics experiments. The present section concerns only the methods for producing the initial energetic positrons.

2.2.1 β Decay Sources

Positrons are created in β decay with a helicity equal to v/c due to the parity-violating weak interaction. Positron emission may compete with other channels such as α decay, K electron capture, β^- decay, fission, and isomeric decay. There are some 2700 known isotopes [3], roughly 272 of which are stable and 13 of which are naturally occurring radioactive isotopes. The rest, about 2400, are radioactive, and roughly half, or 1200, are neutron deficient and thus could be positron emitters provided there is sufficient energy in the β decay transition $(Z,A) \to (Z\text{-}1,A)$ to the nucleus with next-lower atomic number Z and the same atomic mass A. Only a few of the potential positron emitters have a favorable combination of (1) ease of production, (2) isotopic purity, (3) chemical and physical nature of the element, (4) positron yield, (5) positron energy, (6) accompanying prompt and background gamma rays, and (7) half-life for a particular use. The most common positron sources include $^{11}_{6}C$, $^{13}_{7}N$, $^{18}_{9}F$, $^{22}_{11}Na$, $^{58}_{27}Co$, $^{64}_{29}Cu$, and $^{68}_{32}Ge$. There are a few others, such as $^{26}_{13}Al$, $^{44}_{22}Ti$, $^{48}_{23}V$, $^{52}_{25}Mn$, $^{56}_{27}Co$, $^{65}_{30}Zn$, $^{72}_{34}Se$, $^{74}_{33}As$, $^{79}_{36}Kr$, $^{82}_{38}Sr$, and $^{126}_{53}I$, that are occasionally used or might be useful [4]. There are few useful β^+ emitters for $Z > 100$.

The easiest β decays to understand [5] are the mirror nuclei listed in Table I, for which the the atomic mass is $A = 2Z - 1$, where Z is the atomic number. Positron decay involves the change of a proton into a neutron with the emission of a positron and a neutrino. To the extent that the nuclear forces are invariant under rotations in isotopic spin space (i.e., neutrons and protons feel the same nuclear force) the nuclear part of the binding energy of two mirror nuclei is the same. The available energy is then simply the coulomb energy difference minus the neutron–proton mass difference times c^2. The nuclear part of the matrix element for β decay is also nearly unity, and the half-lives of the mirror nuclei provide information about the strength of the weak interaction [6]. As the mass increases, the decay energy increases and the half-life gets shorter, as can be seen from Figure 1. Figure 1a is a plot of the available decay energy versus the quantity $A^{2/3} - A^{-1/3}$ which should be proportional to the coulomb energy in a discrete proton nuclear model [5]. The straight line is fitted subject to the constraint that it intercept the vertical axis at -1.805 MeV $= (M_p - M_n - m_e)c^2$, where M_p, M_n, and m_e are the proton, neutron, and electron masses. Unfortunately, even using the latest data [3] the fit is no better than it was in Evans' version of the same plot from 1955 [5]. We conclude that details of the nuclear structure contribute ± 0.2 MeV to the energy. Similarly, the log–log plot of the half-lives versus atomic number in Figure 1b exhibits more scatter than one might hope.

The mirror isotopes can be made by the (p,n) reaction with relatively low-energy protons and with even lower threshold energies using the (d,n) reaction. Only $^{11}_{6}C$ and $^{13}_{7}N$ live long enough to be useful. The reaction $^{11}_{5}B(p,n)^{11}_{6}C$ has been used [7], and the reaction $^{12}_{6}C(d,2n)^{13}_{7}N$ is being contemplated as a convenient way to make positrons with a small accelerator [8]. Many other positron emitters may be made in the same manner as

TABLE I. β^+ Decay of the Mirror Nuclei, for which $A = 2Z - 1$

Z	Element	A	E_β (MeV)	β^+ yield (%)	Half-life (sec)
2	He	3	−1.003	0	Stable
3	Li	5		0	Particle unstable
4	Be	7	−0.160	0	No β^+
5	B	9	0.046	0	Decays to $p + 2\alpha$
6	C	11	0.960	99.8	1223
7	N	13	1.199	100	598
8	O	15	1.732	99.9	122
9	F	17	1.740	100	64.5
10	Ne	19	2.216	99.1	17.3
11	Na	21	2.525	94.9	22.47
12	Mg	23	3.037	91.4	11.3
13	Al	25	3.256	99.1	7.17
14	Si	27	3.787	99.8	4.13
15	P	29	3.922	98.6	4.1
16	S	31	4.373	98.8	2.6
17	Cl	33	4.561	98.3	2.51
18	A	35	4.943	98.3	1.78
19	K	37	5.127	98.0	1.23
20	Ca	39	5.502	100	0.86
21	Sc	41	5.473	100	0.596
22	Ti	43	5.839	100	0.49
23	V	45	6.103		Unknown
24	Cr	47			Unknown
25	Mn	49	6.694		Unknown
26	Fe	51			Unknown
27	Co	53	7.282	100	0.26
28	Ni	55			Unknown
29	Cu	57	7.46		0.18
30	Zn	59			Unknown
31	Ga	61			Unknown
32	Ge	63			Unknown

Note. The maximum energy available for emitting a positron is E_β. The fraction of the decays that result in β^+ emission is the β^+ yield.

shown in Table II. Of course, the higher Z nuclei may be excited only using higher energy protons or α particles to overcome the coulomb barrier. The most economical production of isotopes by this method takes advantage of the giga-electron-volt protons at the beam dump of an accelerator to create a spectrum of isotope products via spallation [9].

The simplest way to make a large positron source is by the (n,γ) reaction using thermal neutrons from a nuclear reactor. Table III gives some examples, with ^{64}Cu being the most famous [10].

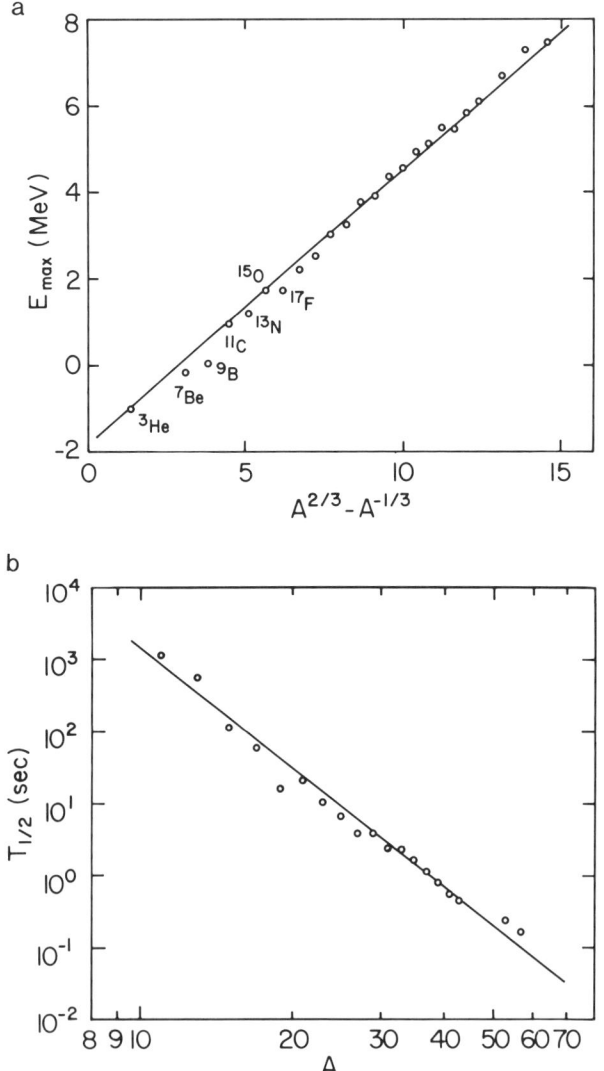

FIG. 1. (a) Maximum energy available for positron emission plotted vs a quantity that is proportional to the change in Coulomb energy for the mirror nuclei positron emitters. ^{3}He, ^{7}Be, and ^{9}B are included because they should follow the same energetics even though they do not emit positrons. (b) Half-life of the mirror nuclei vs the atomic mass A on a log–log plot.

TABLE II. β^+ Emitters Made by Energetic Charged Particle Nuclear Reactions

Isotope	Abundance (%)	Reaction	Product	E_β (MeV)	β^+ yield (%)	Half-life
^{11}B	80.2	(p,n)	^{11}C	0.960	99.8	20 min
^{12}C	98.9	(d,n)	^{13}N	1.199	100	10 min
^{18}O	0.2	(p,n)	^{18}F	0.64	97	110 min
^{20}Ne	90.5	(d,α)	^{18}F	0.64	97	110 min
^{24}Mg	79.0	(d,α)	^{22}Na	0.55	91	2.6 years
^{27}Al	100	$(p,2p3n)$	^{22}Na	0.55	91	2.6 years
^{26}Mg	11	(p,n)	^{26}Al	1.16	82	7×10^5 years
^{45}Sc	100	(α,n)	^{48}V	0.69	50	16 days
^{52}Cr	83.8	(p,n)	^{52}Mn	0.58	28	5.6 days
^{54}Fe	5.8	(d,n)	^{55}Co	1.51	77	18 hr
^{56}Fe	91.8	(p,n)	^{56}Co	1.46	19	79 days
^{56}Fe	91.8	(d,n)	^{57}Ni	0.85	40	36 hr
^{88}Sr	82.6	(p,n)	^{88}Y	0.76	0.2	108 days
^{95}Mo	15.9	(p,n)	^{95}Tc–m	0.68	0.3	61 days

Note. The first column is the target isotope; the second column is the natural isotopic abundance of the target. E_β is the maximum kinetic energy of the positrons.

Various β^+ emitters can also be made by (n,p) reactions using fast neutrons from a reactor. The advantage is that the product isotope is a different element and can be separated by chemical or physical means from the irradiated isotope to yield a source of high specific activity. The (n,p) reaction is useful only for producing an unstable even-A isotope that lies between two stable isotopes at $Z = \pm 1$. Although many isotopes satisfy this condition, the unstable isotope may mostly decay by K electron capture, and its lifetime may be too short or too long to be a good β^+ source. Often the abundance of the $(Z+1,A)$ isotope is very low, especially as Z becomes large. Also the probability for K capture increases with increasing A even for large available energies in the transition.

TABLE III. β^+ Isotopes Made by the (n,γ) Reaction Using Thermal Neutrons

σ_c (b)	Isotope	Abundance (%)	Product	E_β (MeV)	β^+ yield (%)	Half-life
4.6	^{58}Ni	68.3	^{59}Ni	0.051	1.5×10^{-5}	7.5×10^4 years
4.4	^{63}Cu	69.2	^{64}Cu	0.65	19	12.7 hr
0.78	^{64}Zn	48.6	^{65}Zn	0.325	1.46	244 days
4.7	^{78}Kr	0.36	^{79}Kr	0.61	7	35 hr

Note. σ_c is the thermal neutron capture cross-section in barns.

TABLE IV. β^+ Isotopes Made by the (n,p) Reaction Using Fast Neutrons

Isotope	Abundance (%)	Product	E_β (MeV)	β^+ yield (%)	Half-life
^{58}Ni	68.3	^{58}Co	0.474	15	72 days
^{64}Zn	48.6	^{64}Cu	0.65	90	12.7 hr
^{74}Se	0.9	^{74}As	1.53	31	18 days
^{80}Kr	2.3	^{80}Br	0.85	2.6	18 min
^{84}Sr	0.6	^{84}Rb	1.7	22	33 days
^{92}Mo	14.8	^{92}Nb	1	0.06	10 days
^{102}Pd	1.0	^{102}Rh	1.3	14	206 days
^{120}Te	0.1	^{120}Sb	1.6	44	16 min
^{124}Xe	0.1	^{124}I	2.1	25	4.2 days
^{126}Xe	0.1	^{126}I	1.1	1	13 days
^{130}Ba	0.1	^{130}Cs	1.97	40	30 min
^{132}Ba	0.1	^{132}Cs	0.40	1.5	6.5 days

Table IV gives several potential sources that could be made by (n,p) reactions. So far, only ^{58}Co has been made routinely for positron sources. Another possibility is ^{64}Cu which when made by ^{63}Cu$(n,\gamma)^{64}$Cu does not have a high specific activity because the thermal neutron cross-section is small. If, on the other hand, it were made by (n,p) from ^{64}Zn, the Zn might be vaporized to yield rather pure ^{64}Cu [11].

The energy valley [12] for the even-A isobars is double valued, with the even-Z–even-N nuclei lying on an energy parabola that is typically a few mega-electron volts lower than that of the odd–odd nuclei. Here, N is the neutron number of a nucleus. It may thus occur that an even–even nucleus has a rather small energy for β^+ decay (and therefore a long half-life) to an odd–odd nucleus that has a large energy for β^+ decay and therefore a large β^+ yield and a short half-life. Such even–even nuclei are thus candidates for mother–daughter β^+ sources. A few of the more promising possibilities are listed in Table V. ^{68}Ge is often used to obtain positrons without coincident γ rays or with a high polarization due to the large endpoint energy of the β spectrum. The odd-A isobars lie on a single energy parabola. Either the β^+ energies thus tend to be so small that the probability of positron emission is small or zero or the decay energy is so great that the lifetime is too short for the isotope to be useful. For this reason there are few useful odd-A positron sources.

The measurement of positron lifetimes in various materials is typically undertaken using rather weak sources of ^{22}Na which has a convenient 1.274-MeV γ ray that follows within a few picoseconds of the creation of the positron. The time interval between the γ ray and the annihilation

TABLE V. β^+ Mother–Daughter β^+ Sources

Mother	Half-life	Daughter	E_β (MeV)	β^+ yield (%)	Half-life
^{44}Ti	47 years	^{44}Sc	1.47	95	3.9 hr
^{68}Ge	288 days	^{68}Ga	1.9	90	68 min
^{72}Se	8.4 days	^{72}As	3.3	77	26 hr
^{82}Sr	25 days	^{82}Rb	3	96	1.2 min
^{100}Pd	3.6 days	^{100}Rh	2.61	5	21 hr
^{128}Ba	2.4 days	^{128}Cs	2.9	61	3.6 min
^{140}Nd	3.4 days	^{140}Pr	2.3	49	3 min
^{194}Hg	260 years	^{194}Au	1.49	3	39 hr

photons from the positron is measured by fast phototubes coupled to plastic or BaF scintillators. The source may be made *in situ* by direct irradiation of the sample in some special cases [13], it may be ion implanted into the sample [14], or it may be deposited onto the sample in liquid form and dried. Most often the source is deposited between two sheets of very thin material, Ni or Kapton foils, that contribute only a single well-known lifetime component to the measurement [15]. The ratio of the accidental coincidence rate to the true rate is equal to the source strength times a relevant time interval: if one wants a background rate that is less than 10^{-4} of the true rate when measuring a lifetime of about 0.37 nsec, the source strength must be less than 10 μCi of ^{22}Na.

In the preparation of large sources, the efficiency of the physical and chemical means of separation of the radioactive isotope from the parent material is important [16]. Large sources must be contained effectively, necessitating a covering window that can withstand exposure to vacuum [4,17].

Since positrons are created with a helicity, $h = v/c$, if no averaging over positron velocities occurs, a polarized source is obtained [18]. For maximum polarization, it is important to deposit the source on a low-Z material to reduce backscattering and to cover the source with a thicker window than ordinarily used to attenuate the low-momentum particles that detract from the polarization.

2.2.2 Pair-Production Sources

When relativistic electrons of energy T stop in a solid, most of their energy is lost to bremstrahlung radiation rather than to ionization. The bremsstrahlung photons are converted to lower energy relativistic electron–positron pairs. A shower of leptons and photons results, with the number of particles proportional to $\exp(x/l_{\text{rad}})$, where x is the distance through the solid in the direction of the initial particle and l_{rad} is a characteristic of the material known as the radiation length [19]. The shower reaches a maximum number of particles of order T/E_{crit}, where E_{crit} is a characteristic energy dependent on the Z of the solid. For a high-Z target like W, E_{crit} = 8 MeV and l_{rad} = 3 mm. A T = 100 MeV electron beam will thus yield about 10 positrons and 10 photons of roughly 3 MeV kinetic energy after a depth of about 7 mm. A suitable moderator can convert these to slow positrons to yield about 10^8 slow e^+ sec^{-1} per kilowatt of electrons [20]. Photons for pair production may also be obtained from the Cd(n,γ) reaction in a nuclear reactor [21]. In the severe radiation environment of a beam dump or a reactor, the positron moderation efficiency may be much less than ideal as we shall see in the next section.

2.3 Moderation

The key to the use of positrons for low-energy physics experiments is the fact that positrons with a few hundred kilo-electron volts of kinetic energy slow down to the point at which ionization is no longer significant in a time short compared with the annihilation lifetime in most materials. Thus β decay positrons may be implanted directly into a sample to study solid-state effects via the annihilation photons, as was done almost exclusively for the first half of the history of positron physics. Gradually conditions have been discovered that permit the slow positrons to be collected from the surface of solid moderators to form controlled-energy positron beams [22–24]. Slow positrons may be easily focused and transported for analyzing materials and for atomic physics studies. Possible solid moderators include metals such as W, Cu, and Ni and nonmetals such as diamond and the rare gas solids (Table VI).

The efficiency of moderation is given by a geometrical factor of order unity times the ratio of the positron diffusion length to the stopping length

TABLE VI. Characteristics of Positron Moderators

Material	ϕ_+ (eV)	E_{gap} (eV)	y_0 (%)	$E_{1/2}$ (keV)	ΔE (eV)
Al(100)	-0.16 ± 0.03	—	20	3	
Cu(111) + S	-0.78 ± 0.05	—	55	5	
Ni(100)	-1.30 ± 0.05	—	45	7	0.03
W(110)	-3.0	—	33	7	0.5
C(100)	-3.8 ± 0.1	5.3	67	3	0.4
Kr	$+2.00 \pm 0.05$	11.6	62	7	4
Ne	$+0.6 \pm 0.1$	21.4	70	10	10

and times the probability that diffusing positrons encountering the surface will be emitted. Thus far, the moderators are of two types: those that have a negative affinity for positrons and work by diffusion of thermal energy positrons in the solid and those that have a large band gap and a positive affinity for positrons and operate by the diffusion of hot positrons.

2.3.1 Negative Affinity Solids

Some metals [23,24] and diamond [25] have a negative affinity for positrons. Since the lowest energy unoccupied positron state is negligibly higher in energy than the highest energy occupied state, the positron affinity and the positron work function, denoted ϕ_+, are the same. The positron affinity for a solid is a sum of two contributions: (1) the chemical potential, μ_+, consisting of the positron's zero point energy, its average potential energy, and the correlation energy or attractive interaction with the electron gas and (2) the surface dipole potential Δ that contributes equally and oppositely to the electron and positron work functions [26],

$$\phi_\pm = -\mu_\pm \mp \Delta. \qquad (2.1)$$

Energetic positrons stop in a distance, λ, on the order of 15 μm in W. Thermal energy positrons can diffuse distances $L \approx 0.1$ μm before annihilation. Thus roughly 1% of the positrons from a radioactive source such as ^{22}Na that enter a W moderator will, following moderation, encounter its surface, where there is a $y \approx 30\%$ chance that they will escape into the vacuum because of their negative affinity. The efficiency for converting source positrons to slow positrons is $\varepsilon = gLy/\lambda$ and can be as much as 0.3% [27], depending on geometrical factors g.

The positron energy spectrum from a single-crystal negative-affinity moderator has a narrow peak (see, for example, Figure 2 [28]), which is nearly what one would expect from the thermal energy contributions alone. The peak energy may be identified with the positron-negative work function because the electron and positron work functions change by equal and opposite amounts for surfaces whose dipole potential is changed by absorbates as required by Equation (2.1) [29]. As expected from a negative-affinity emission mechanism, the positron angular distribution also has a narrow peak [30] composed of positrons of kinetic energy $-\phi_+$ directed perpendicular to the surface. The angular width of the narrow peak is roughly $\mathrm{atan}\sqrt{kT/-\phi^+}$ as expected from thermal effects. There is also a broad contribution in energy and momentum that may be associated with imperfect sample conditions. The total probability for positron emission is roughly proportional to $\sqrt{-\phi_+}$ [31].

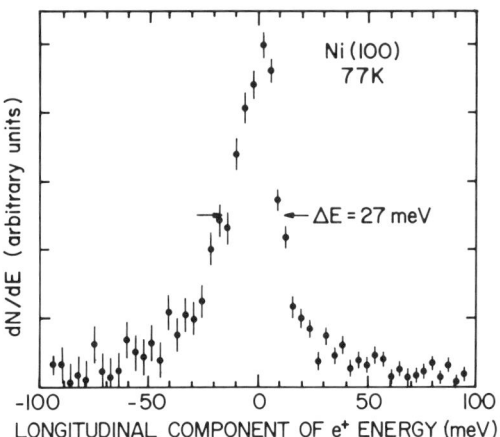

FIG. 2. Energy distribution of positrons from a Ni single crystal remoderator.

In order to obtain a long diffusion length, the moderator should be a single crystal. Its surface must be carefully prepared to preserve the negative affinity and to prevent scattering and trapping of the positrons in contaminant layers. It is important to remember that positrons can be trapped at vacancies and dislocations [32,33], and even a single crystal must be etched heavily after it is cut by spark erosion to prevent positron trapping. Various schemes have been devised for preparing moderators, from ion bombardment and annealing in ultrahigh vacuum to simple heat treatments followed by exposure to a reactive gas that passivates the surface. If moderators are prepared thin enough, they may operate in the transmission mode, with the energetic positrons entering one surface and slow positrons leaving from the other. Foils that are only a few thousand angstroms thick are difficult to prepare since deleterious dislocations form easily due to stress [34–36]. It is to be noted that the energy width of positrons from a thick moderator does not continue to narrow as the temperature becomes less than 100 K possibly because quantum reflection of the slow positrons at the surface prevent them from efficiently escaping [37]. However, a thin foil in which positrons encounter the surface significantly more often offers the possibility of a low-temperature positron source [38].

A popular form of moderator is made from polycrystalline tungsten foils. The foils are annealed in vacuum, exposed to air, and used as moderators without further treatment [39]. The efficiency of such moderators is not optimum, but they are easy to prepare especially because no heat treatment near a radioactive source is required. The W foil moderators are useful in a high radiation environment [20] because their surfaces are passivated by contaminants, and, already being somewhat polycrystalline, they are not so sensitive to defects.

2.3.2 Wide Band Gap Solids

The rare gas solids [40] are good moderators because after stopping in the solid, i.e., after the positron kinetic energy is no longer sufficient to cause ionization, the positrons lose energy only due to phonon emission, retaining several electron volts of kinetic energy while diffusing distances L of several micrometers. Upon encountering the rare gas solid surface, some of the hot positrons are able to escape into the vacuum despite their positive affinity, ϕ_+, for the solid if their kinetic energy projected perpendicular to the surface exceeds ϕ_+. A solid Ne moderator [41] has thus far shown the highest fast-positron-to-slow-positron conversion efficiency of any moderator. A Kr moderator is more easily prepared, since a condensed film can be prepared at higher temperatures and is nearly as efficient as solid Ne [42,43]. The ionic solids such as the alkali halides have large energy gaps too, but the diffusion length of hot positrons is not long enough to make them very efficient moderators [44,45] compared with the rare gas solids.

A moderator in which an electric field is used to significantly increase the positron diffusion length is a tantalizing concept because of the possibility of achieving 10% efficiency or more. So far diamond is the best candidate since it is an insulator and has a negative affinity. An increased slow positron yield for charged Ar and Kr moderator films [46] that may be partly due to the heating effect of the field keeping the hot positrons alive to diffuse farther in the solid has been reported.

2.3.3 Geometry

Ordinarily one would like a radioactive source to be close to a positron moderator so that the diameter of the slow positron source spot is a minimum. A thin moderator foil covering a source is optimum in this regard [36-39], but there may be other considerations. In particular, to maximize the polarization of a slow positron source one needs to sacrifice some of the count rate or brightness. Various source and moderator designs (see Figure 3) have been built or suggested, the venetian blind moderator [45], tungsten vanes [39], the cone moderator [47], the cylindrical and spherical sources [48], and the ribbon source [49] among others.

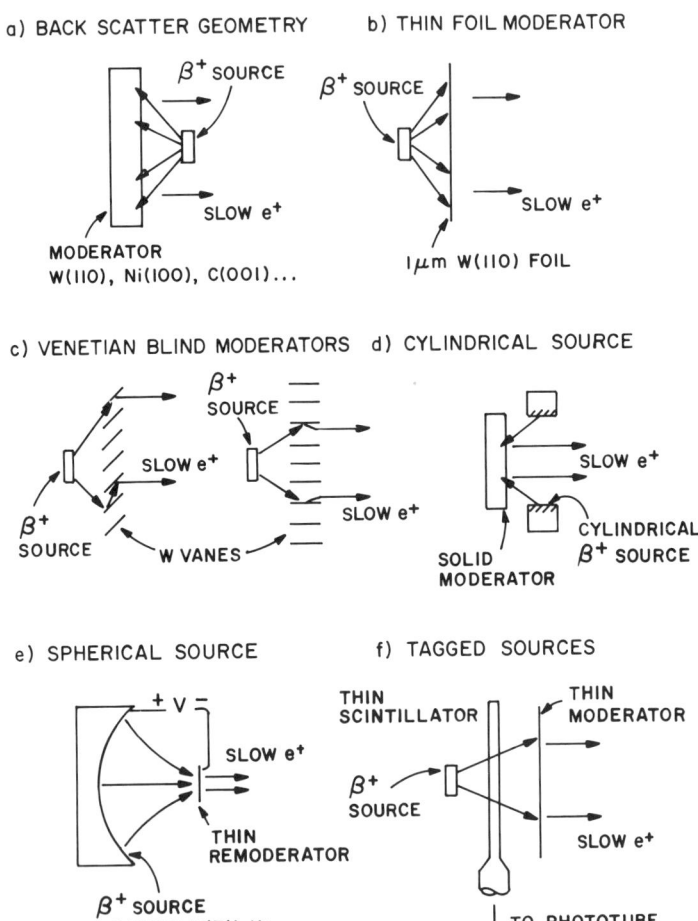

FIG. 3. Various possible geometries for radioactive positron sources and slow positron moderators.

2.4 Transport, Lenses, and Mirrors

The criteria for guiding positrons frequently differ from those for electron transport. In particular, because transmission is paramount, limiting apertures are typically avoided as they would reduce the beam intensity significantly. Furthermore, because positron interactions yield information about a material even if they are not focused to a small diameter spot, it is not always necessary to transport positrons with imaging optics.

There are many detailed descriptions of complete positron beams available that illustrate the techniques of positron transport [49–57].

2.4.1 Magnetic Transport

A constant solenoidal magnetic induction, B_0, provides a simple way to transport a positron beam from the slow positron moderator to the sample under study. Unwanted radiation may be filtered out by drifting the positrons across the field lines using a transverse electric field [49,58,59] to an off-axis position at which they pass through a hole in a W shield. The same principle may be used to velocity-analyze reemitted positrons from a sample. Positrons may also be captured and stored in a magnetic bottle or Penning trap.

Positrons may be focused in the presence of a background constant magnetic induction by means of rapidly varying magnetic lens fields. For example, if the positrons suddenly move into a region of increased induction, B_{lens}, the positrons will come to a focus spot smaller than the original beam diameter by a factor of $2B_0/B_{lens}$, where the constant solenoidal field B_0 is just sufficient to make the cyclotron orbit diameter comparable to the original source diameter. It is necessary for the longitudinal positron velocity to be large enough that the transition between fields is diabatic. If the transition between field regions is adiabatic, the beam diameter decreases only by the ratio $\sqrt{B_0/B_{lens}}$.

2.4.2 Electrostatic Transport

A good discussion of electrostatic positron optics design is given by Canter [60]. However, while electrostatic optics is thoroughly understood and highly developed, design and fabrication of a particular optics system can be time consuming. Electrostatic optics can take full advantage of "brightness enhancement" [61,62], discussed in the next section.

2.5 Remoderation

If energetic positrons from a radioactive source can be moderated to thermal energies with reasonable efficiency, they can be remoderated with high efficiency after acceleration to a few kilo-electron volts. The limiting quantity in the moderation process is the ratio of the positron diffusion length to the implantation depth, the probability for positron emission from a suitably prepared surface being about 50%. Since moderation reduces the momentum spread of a positron beam without appreciably affecting its spatial extent, moderation decreases the volume in phase space associated

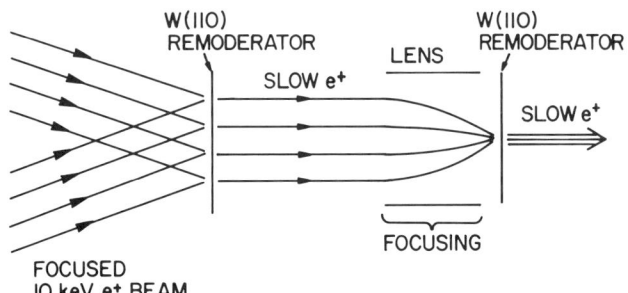

FIG. 4. The principle of brightness enhancement.

with the beam. Repeated stages of moderation, acceleration, and focusing (Figure 4) may thus be used to increase the brightness of a positron beam [61]. Similar phase-space compression may also be applied to ultracold neutrons, scintillator light using wavelength shifting light pipes, electron beams, and muon beams.

2.5.1 Microbeams

Liouville's theorem requires the constancy of the volume in phase space occupied by a swarm of particles under the influence of conservative forces. In the transverse subspace of a particle beam, the relevant quantities are the beam cross-sectional area $\pi d^2/4$ and the spread ΔE_T of the distribution of transverse energies. For a beam passing through various lens elements the product $\Omega = d^2 \Delta E_T$ cannot decrease, although the phase-space volume may appear to increase due to lens distortions. The minimum diameter of a converging beam of kinetic energy E_L is thus approximately $d \approx \alpha^{-1}\sqrt{\Omega/E_L}$, where α is the convergence half-angle of the beam. Given a moderated positron source of diameter d_0 and initial transverse energy spread ΔE_T the minimum spot size after the beam is accelerated to energy E_L is $d \approx d_0 \alpha^{-1} \sqrt{E_T/E_L}$. With $\alpha \approx 0.2$, $\Delta E_T \approx 0.25$ eV, and $E_L = 2.5$ keV (to ensure a high reemission probability from the moderator), the spot size is predicted to be $d \approx 0.05 \times d_0$. If a moderator is located at the focus, a moderated source of low-energy positrons of similar size will be obtained. Using conventional energy conserving optics and a centimeter-size source and moderator it is possible to obtain a microbeam with $d \approx 1$ μm, but only at the expense of a 10^6 reduction in intensity. In contrast, two stages of brightness enhancement have been used to focus positrons to a 3-μm-diameter spot with only a factor of 20 intensity loss [63]. The small positron spot size is the basis for the positron reemission microscope [64–66], the scanning positron microprobe [67], and the giant positron beam [68,69].

The ultimate in small positron beams would be associated with a diffraction-limited positron source. The latter would be a coherent source and would therefore have a negligible longitudinal energy spread and the minimum transverse energy spread consistent with the laws of quantum mechanics. The Airy disk of a coherent positron beam of kinetic energy E_L has a diameter

$$d = \frac{2.44\lambda}{\tan \alpha} = \frac{15 \text{ Å}}{\sqrt{E_L} \tan \alpha}, \qquad (2.2)$$

where $\lambda = h/p$ is the deBroglie wavelength of the positron and α is the half-angle of the converging beam. The minimum value the phase-space product $d^2 \Delta E_T$ may assume is thus

$$\Omega_0 = 2.98 h^2/m = 225 \text{ Å}^2 \text{ eV}. \qquad (2.3)$$

If a 3-μm spot of positrons were emerging from a moderator similar to that in Figure 2 the phase-space product would be $\Omega \approx 2.25 \times 10^7$. Since $\Omega/\Omega_0 \approx 10^5$, one could make a coherent positron beam by apertures with a loss of intensity by a factor of 10^5. One could imagine using a coherent positron beam for speckle microscopy, i.e., for studying the microstructure of nonperiodic surfaces by diffraction.

2.5.2 Large Sources

The half-life of radioactive sources and the penetration depth of β particles place an upper limit on the intensity per unit area of a positron source. For example, pure ^{64}Cu of a thickness equal to the positron range [70] of 28 cm^{2}g^{-1} would yield at most 10^{13} slow e^+ sec^{-1} cm^{-2}. If the isotope is made by the (n,γ) reaction (Table III) only 0.1% of the Cu would be radioactive given a flux of 5×10^{15} n cm^{-2} sec^{-1}. To obtain a source of 10^{13} slow e^+ sec^{-1} would require a 10^3-cm^{-2} source area. Such a large area source can be made generally useful only by some form of phase-space compression, i.e., brightness enhancement [61].

2.6 Trapping and Bunching

Capturing particles in a trap can be useful for cooling them, making various measurements, and releasing them at a propitious moment. In particular, a timed positron beam is useful for measuring positron lifetimes to detect defects in solids or to produce positronium in coincidence with a pulsed laser. From the dictates of Liouville's theorem we know that trapping a particle swarm in a certain region of phase space requires either expanding the volume occupied by the swarm in dimensions of phase space that are orthogonal to the trap or allowing nonconservative forces (friction) to operate. A simple example of a nondissipative trap is the capturing of positrons in a magnetic bottle by adding energy to their transverse degree of freedom after they have passed through the pinched field neck of the bottle [71]. Particles trapped in this manner are not permanently contained and will leak out at a rate inversely proportional to the increase in phase-space volume.

2.6.1 Gaseous Moderators

A successful way to dissipate positron kinetic energy is to allow the positrons to interact with a low-density gas under conditions in which not too many particles are lost due to positronium formation and annihilation. On the order of 10^6 or more positrons have been trapped using this approach in experiments to study the single-component positron plasma [72,73].

2.6.2 Resistive Traps

Transfer of positron kinetic energy via long-range coulomb forces to a resistive element can lead to cooling without loss of particles through annihilation or positronium formation. The most famous example is the Dehmelt trap [74] which has been used to measure the g-factors of the electron and positron with amazing accuracy [75]. The particles are caught in a Penning trap by allowing them to spread into a magnetron orbit that is much larger than the entrance hole into the trap. By the time the positrons or electrons get back to the hole and try to escape they have lost some kinetic energy due to coupling to a resistively damped mode of the trap. Because of the small entrance phase space, the Dehmelt–Penning trap will be efficient only if it can be combined with a bright positron beam [76]. Using a laser-cooled ion plasma to rapidly cool a positron beam, there being a potential to handle perhaps 10^{12} particles per second, has also been proposed [77].

2.6.3 RF Bunching

To obtain short bursts of positrons at a target, it is necessary to spread the positron energy distribution to compensate for the temporary increase in the spatial density of positrons. Periodic potentials applied to various electrodes can induce a time-dependent velocity distribution such that positrons from various parts of a beam will arrive simultaneously at a target [78]. Timing spreads of less than 100 psec are possible [79].

2.6.4 Harmonic Potential Bunching

Positrons captured in a magnetic bottle can be released in short bursts by appropriately timed spatial potential distribution [71]. According to Hulett's theorem, a harmonic potential of any time dependence will result in initially slow particles arriving simultaneously at the origin of the harmonic potential [80]. Positron bursts some 10 nsec long and containing about 10^5 particles have been produced in this way [81].

2.6.5 Debunching

A pulsed positron beam can yield a steady beam if the positrons are captured in a magnetic bottle and slowly released [82,83]. The beam brightness improves by this manipulation; i.e., the transverse energy spread and spatial extent of the beam can decrease if the release from the bottle is done properly.

2.6.6 Tagging

Individual positrons can be tagged through some nonannihilating interaction before they come to the target being studied. One method is to detect the secondary electrons they eject from a positron brightness enhancement remoderator [84]. Subnanosecond timing accuracy can be obtained [85]. Another method is to sense the positron charge as it passes through a Faraday cup without other interactions [86]. Meeting the signal-to-noise requirements will probably result in millisecond timing uncertainties.

2.7 Positronium Formation

Positronium is the lightest of the exotic hydrogen-like atoms made of a lepton and an antilepton [1,87]. The positronium atom provides an excellent testing ground for studying the relativistic bound-state two-body system and the quantum electrodynamic corrections to that system. Other interesting properties of positronium include the cross-sections for excitation and ionization by photons and electrons, the electron affinity, the properties of the positronium negative ion, and interactions of positronium with other atoms, molecules, and surfaces and in condensed matter. The positronium atom (Ps) in its ground state and first excited states and the negative ion have been produced in the laboratory by techniques that include Ps formation in a gas [88] or in a finely divided solid in which β^+

particles are stopping [89–92], slow-positron beam charge exchange in a thin foil or a gas target [93–95], and the interaction of slow positrons with the surface of a solid in vacuum [96,97].

2.7.1 Positronium Formation in Gases and Powders

Positronium can be formed by stopping energetic positrons from a radioactive source in a low-density gas [88]. Fast positrons that are scattering in a gas will slow down to a few electron volts of kinetic energy before capturing an electron. If the kinetic energy T of the positron is larger than E_{exc}, where E_{exc} in the lowest excitation energy of the stopping gas, inelastic collisions with the gas are the dominant interaction. When T is less than $E_i - E_{Ps}$, where E_i is the ionization energy of the gas atom or molecule and E_{Ps} is the binding energy of positronium, it is energetically impossible to form Ps. Also, for $T > E_i$, the kinetic energy of any Ps formed will exceed E_{Ps}, and the Ps has a high probability for dissociation in subsequent collisions. Thus, there is a narrow energy range, $E_{exc} \geq T \geq E_i - E_{Ps}$ or $E_i \geq T \geq E_i - E_{Ps}$, called the Ore gap [98], in which Ps formation is likely. A more realistic treatment includes the effect of the energy dependence of the cross-sections for positronium formation and breakup and for positron inelastic scattering [99]. In all but the simplest noble gases, such as Ar and Ne, the physics is even more complicated because the interactions of the positron with the electrons and ions in the fast positron's ionization trail or "spur" cannot be neglected. Thus, the actual yield depends on the gas density and impurities in a complicated manner [100]. Positronium is typically formed with probability on the order of 50% at gas pressures of 1 atm.

After the positronium is formed it thermalizes at a rate that depends on the target gas pressure and whether there are low-lying vibrational and rotational excited states of the target available. Collisions increase the Ps decay rate because it can be annihilated with a foreign electron, or the triplet and singlet substates can be perturbed if there are unpaired electrons on the target molecules. Collisions also perturb the Ps hyperfine interval proportionally to the gas density. Except in the case of helium, the hyperfine pressure shift is dominated by the van der Waals interaction which tends to pull the positron and electron apart. Collision effects are eliminated from measurements of the Ps lifetimes and hyperfine interval by extrapolation to zero gas density. The excited states are easily quenched by collisions and must be studied in vacuum or in a very-low-density environment.

Positronium has also been produced in solids and liquids [101–105], but the atoms are highly perturbed by their environment so that they are more useful as a material probe than for atomic measurements. Finely divided solids are also used to produce positronium. Nearly free Ps is formed in various oxide powders when irradiated by energetic positrons from a radioactive source [89]. This effect was explained [106] as being due to Ps forming within a powder grain and rapidly diffusing to the grain surface, where it escapes into the vacuum space between grains. A more careful analysis suggests that some of the Ps actually forms outside the grain surfaces with electrons liberated by the positron when it was slowing down [100]. Once outside the grains, the Ps can make many thousands of collisions with the grains during its lifetime and can come into thermal equilibrium with the powder [107]. Any precision measurement of Ps properties must be extrapolated to zero powder density to compensate for the effects of collisions [108]. The discovery that nearly free Ps can be formed in powders has been exploited in measurements of the annihilation decay rates and is the basis for one method for producing Ps in vacuum [109,110].

2.7.2 Positronium Formation in Vacuum

Frequently it is desirable to study Ps in vacuum, free of the perturbations introduced by gases or powders. Positronium formed near the surface of a layer of powder can escape into the vacuum [109,110], where it subsequently has no collisions. Although a thin powder layer makes Ps with kinetic energies on the order of 1 eV, Ps from deep inside a powder layer has been found to be thermalized [107] and is a possible source of 4-K Ps [111]. A similar powder layer has been used to make thermal muonium (the μ^+-e^- bound state) at room temperature [112,113]. It has been shown that beams of slow positrons can be used to form Ps atoms with ~1 eV of kinetic energy at the surface of a solid target in vacuum without the background radiation associated with radioactive sources [96]. Another important step in the development of vacuum Ps sources was the realization that thermal Ps can also be emitted from a surface [114–118].

Currently, a vacuum Ps source begins with a source of energetic positrons, either a radioactive source or a pair production source at an electron accelerator beam dump. A moderator is used to reduce the positron energy spread from mega-electron volt to electron volt widths. Once the brightness of the positron source has been increased by the moderator, the positrons are transported by magnetic fields or electrostatic lenses to a secondary target at which thermal Ps can be made with high efficiency. Ps is formed from positrons implanted into a sample at relatively high energies which then thermalize and diffuse to the sample surface [115]. The Ps observed at low temperatures is believed to form directly at the surface by capture of an electron and has a maximum kinetic energy equal to minus the Ps work function given by

$$-\phi_{Ps} = \tfrac{1}{2}R_\infty - \phi_- - \phi_+, \tag{2.4}$$

where ϕ_- and ϕ_+ are the electron and positron work functions of the metal and $\tfrac{1}{2}R_\infty$ is the positronium binding energy. For an Al(111) sample the direct Ps has a distribution in total energy and in momentum parallel to the surface consistent with the metal being left in a one-hold excited state due to the sudden removal of an electron from the solid [119]. The sharp cutoff in the distribution at energies above $-\phi_{Ps}$ corresponds to the Fermi energy and is clearly visible in the measurement reproduced in Figure 5 [120].

At elevated temperatures a different Ps formation mechanism occurs, and the fraction f of the incident positrons forming Ps has a temperature dependence characteristic of a thermally activated process [114–117]. The velocity distribution of the thermally activated Ps is consistent with the

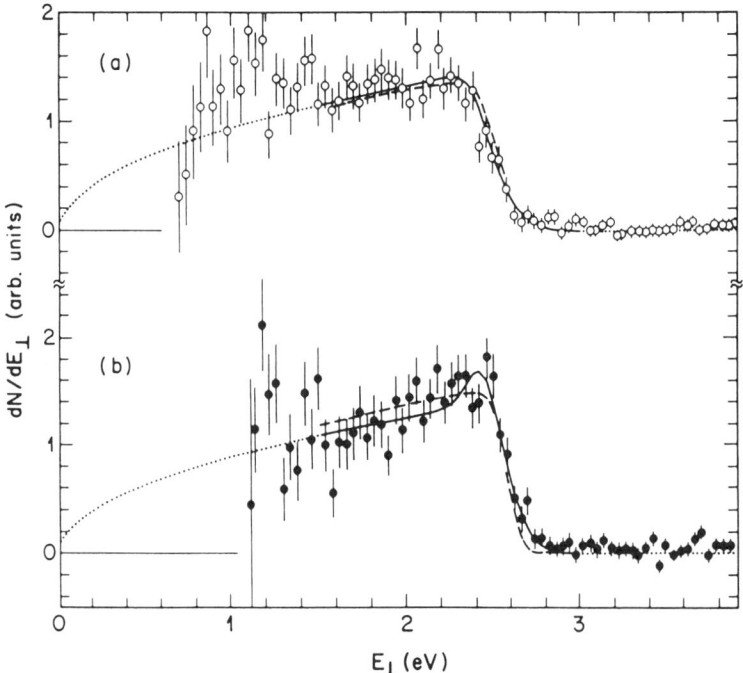

FIG. 5. Energy distribution in the normal direction of the work-function positronium from an Al(111) single-crystal surface.

Ps being thermally desorbed from the sample [118]. This phenomenon is explained by a model in which positrons bound in their "image" potential well at the surface are thermally desorbed as Ps when sufficient energy is supplied by thermal fluctuations. Thermal activation measurements have been analyzed using thermodynamic arguments to obtain the positron surface state binding energies, E_b, and estimates of the positronium formation rate constants.

The scenario for the trapping of positrons in surface states is as follows. Slow positrons penetrate the metal and quickly lose their energy by plasmon and phonon scattering. While some of the positrons are annihilated with electrons in the bulk crystal, most of them diffuse to the surface, where they encounter the surface dipole layer that may give the positron a negative work function. As the positron leaves the surface it sees an effective potential well due to a modified image potential at large distances from the metal surface and electron correlation at short distances. Either nonthermal free positrons or positronium will emerge or the positron may fall into the surface well as a result of inelastic collisions. Once in the well the positron can either be annihilated, principally into two γ rays, or escape from the surface as free Ps if sufficient energy is available from thermal fluctuations.

The fraction f of Ps produced from an incident positron is typically 0.5 at room temperature and increases to nearly unity at elevated temperatures. The temperature dependence of f fits an activation curve that can be described by a Ps formation rate of the form $z = z_0 \exp(-E_a/kT)$, where the activation energy E_a is dictated by energy balance

$$E_a = E_b + \phi_- - \tfrac{1}{2} R_\infty, \tag{2.5}$$

where $E_b > 0$ is the binding energy of the positron at the surface. If f_0 denotes the fraction of positrons that directly form Ps in the low-temperature limit and f_s denotes the fraction that becomes trapped in the surface well, then the fraction f_t of positrons that are thermally desorbed from the surface as free positronium will be

$$f_t = \frac{z f_s}{\gamma + z} = \frac{z_0 e^{-E_a/kT}}{\gamma + z_0 e^{-E_a/kT}}, \tag{2.6}$$

where γ is the two-photon annihilation rate of the positron surface state which has a unique lifetime characteristic of a low electron density [85]. The fraction f_t can be related to the experimentally determined high-temperature limit $f_\infty = f_0 + f_t$ at $T \to \infty$. If we assume that $\gamma \ll z_0$ and f_s is only a weak function of temperature, the Ps formation fraction f is

$$f = f_0 + f_t = \frac{f_0 + (z/\gamma) f_\infty}{1 + (z/\gamma)}. \tag{2.7}$$

If it is assumed that the surface positrons form a classical 2D gas, thermodynamic arguments [121] show that the factor z_0 is given by

$$z_0 = \frac{4kT}{h}(1 - \bar{r}), \qquad (2.8)$$

where $r(v_{\text{Ps}})$ is the reflection coefficient for a positronium atom of velocity v impinging on the surface and $\bar{r}$ is an average over a thermal distribution of velocities. Essentially, $(1 - \bar{r})$ is the analogue of the familiar emissivity of a body emitting thermal radiation.

The above model has been tested by measurements of the Ps formation probability as a function of temperature (Figure 6) and by time-of-flight measurements of the energy distribution of Ps thermally desorbed from Al(111) surfaces [122]. The thermally desorbed Ps energy spectrum from Al(111) is an exponentially decreasing function of energy and is consistent with Ps having a velocity-independent reflection coefficient. The Ps energy spectrum extends to zero perpendicular energy without diminution, and thus the low-energy tail of the Ps beam–Maxwellian distribution is of

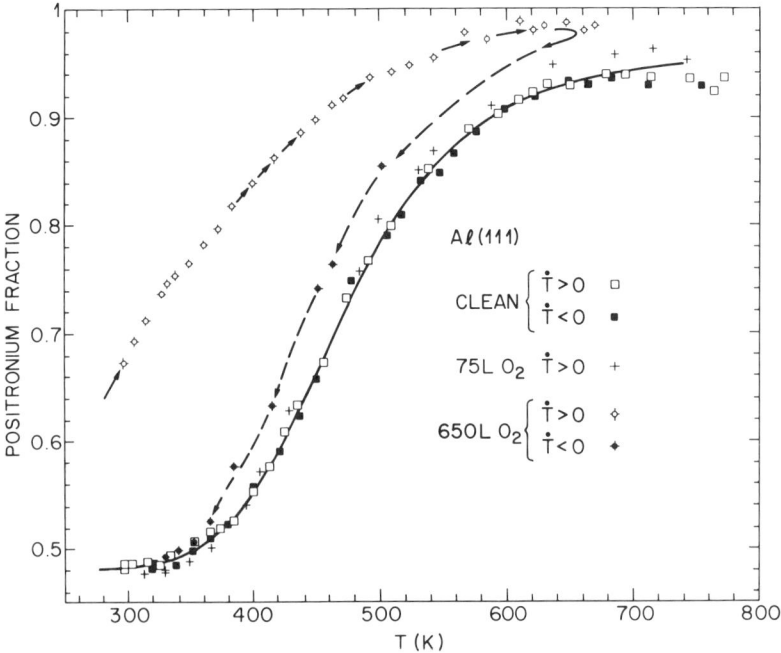

FIG. 6. Fraction of positrons forming positronium at an Al(111) surface vs sample temperature T.

sufficient intensity to be of use in precision experiments requiring low velocities. The activation energy for thermal desorption of Ps may be tuned by the addition of a partial monolayer of a material that will change the surface potential [117,123,124]. It is thus possible to cause Ps to be desorbed at temperatures below 300 K. The Ps energy spectrum of Al(111) with a partial layer of oxygen retains its beam–Maxwellian shape and has a significant intensity at temperatures as low as 80 K [125,126].

2.7.3 Cryogenic Positronium

Laser cooling techniques [127,128] could be applied to Ps [129,130] for reducing the transit time broadening in a CW laser measurement of the $1S$–$2S$ interval [131], for examining the quantum reflection of atoms with huge deBroglie wavelengths, and for more speculative possibilities such as the study of the Bose condensation of a dilute gas [132–135], the formation of a medium for γ-ray amplification via stimulated annihilation [136–140], and free-fall measurements on a particle–antiparticle system.

One possible choice for cooling is the $1S$–$2P$ Ly$_\alpha$ transition at 243 nm which has a natural lifetime of 3.2 nsec and a natural line width of 50 MHz. Because the atom has such a low mass, only 50 scattered photons are required to bring Ps to rest from room temperature. An atom with an initial velocity corresponding to a temperature of 100 K could be cooled in the 142-nsec lifetime of the triplet state. The velocity change that results from a single photon recoil gives a Doppler shift of 6 GHz, so that the atom recoils out of resonance with any monochromatic laser. Also, the initial Doppler width for 100-K atoms is ≈ 280 GHz. Conventional swept-frequency laser cooling [141] is thus impossible as would be cooling in an inhomogeneous magnetic field [142].

One solution to the problem would be to use "white light" cooling [143]. If a quasicontinuous spectrum of light is tuned to be on the red-wavelength side of the absorption line and if the atoms are illuminated in the "optical molasses" configuration [144], the atoms might be expected to cool down to a temperature, $T \approx 3.6$ mK, corresponding roughly to the recoil velocity of a single photon for the case of $1S$–$2P$ cooling.

2.7.4 Fast Positronium

Positrons passing through a thin foil pick up electrons to form positronium in various states and with a wide range of kinetic energies [145]. Monoenergetic positronium with kinetic energies of a few to several tens of electron volts may be made by charge exchange with a noble gas [94,95,146,147]. Ps$^-$ ions may be accelerated and photoionized to form a collimated beam of monoenergetic positronium with relativistic kinetic energies [148].

2.8 Summary and Conclusion

Sources of positrons and positronium have advanced rapidly to the point at which it is now possible to contemplate a wide range of experimental investigations similar to those using electrons and hydrogen atoms.

References

1. See, for example, A. T. Stewart and L. O. Roellig, eds., *Positron Annihilation*. Academic Press, New York, 1967; R. R. Hasiguti and K. Fujiwara, eds., *Positron Annihilation*. Japan Inst. Metals, Seendai, 1979; P. G. Coleman, S. C. Sharma, and L. M. Diana, eds., *Positron Annihilation*. North-Holland Publ., Amsterdam, 1982; W. Brandt and A. Dupasquier, eds., *Positron Solid-State Physics*. North-Holland Publ., Amsterdam, 1983; P. C. Jain, R. M. Singru, and K. P. Gopinathan, eds., *Positron Annihilation*. World Scientific, Singapore, 1985; Y. C. Jean, R. M. Lambrecht, and D. Horvath, *Positrons and Positronium, A Bibliography 1930–1984*. Elsevier, Amsterdam, 1988; E. H. Ottewitte and W. Kells, eds., *Intense Positron Beams*. World Scientific, Singapore, 1988; L. Dorikens-Vanpraet, M. Dorikens, and D. Seegers, eds., *Positron Annihilation*. World Scientific, Singapore, 1989; Y. C. Jean, ed., *Positron and Positronium Chemistry*. World Scientific, Singapore, 1990; A. P. Mills, Jr. and S. Chu, in *Quantum Electrodynamics* (T. Kinoshita, ed.), p. 774. World Scientific, Singapore, 1990; Z. Kajcsos and C. Szeles, eds., *Positron Annihilation. Mater. Sci. Forum*, Vols. 105–110, Trans Tech, Switzerland, 1992.
2. See, for example, R. D. Evans, *The Atomic Nucleus*. McGraw-Hill, New York, 1955; E. Segré, *Nuclei and Particles*, 2nd ed. Benjamin, Menlo Park, CA, 1977.
3. C. M. Lederer and V. S. Shirley, eds., *Table of Isotopes*, 7th ed. Wiley, New York, 1978.
4. E. A. Lorch, in *Positron Annihilation* (R. R. Hasiguti and K. Fujiwara, eds.), p. 403. Japan Inst. Metals, Sendai, 1979; K. G. Lynn and W. E. Frieze, in *Positron Scattering in Gases* (J. W. Humberston and M. R. C. McDowell, eds.), p. 165. Plenum, New York, 1984.
5. See R. D. Evans, *The Atomic Nucleus*, p. 35. McGraw-Hill, New York, 1955.
6. See, for example, A. Sirlin, *Nucl. Phys. B* **B71**, 29 (1974); *Rev. Mod. Phys.* **50**, 573 (1978); *Ann. N. Y. Acad. Sci.* **461**, 521 (1986); W. J. Marciano and A. Sirlin, *Phys. Rev. Lett.* **56**, 22 (1986); A. Sirlin and R. Zucchini, *ibid.* **57**, 1994 (1986); A. Sirlin, *ibid.* **72**, 1786 (1994).
7. T. S. Stein, W. E. Kauppila, and L. O. Roellig, *Phys. Lett. A* **51A**, 327 (1975).
8. R. Xie, M. Petkov, D. Becker, K. F. Courter, F. M. Jacobsen, K. G. Lynn, R. Mills, and L. O. Roellig, *Nucl. Instrum. Methods Phys. Res. Sect. B* **93**, 98 (1994).

9. R. Michel, B. Dittrich, U. Herpers, F. Peiffer, T. Schiffmann, P. Cloth, P. Dragovitsch, and D. Filges, *Analyst* **114**, 287 (1989). It is interesting to note that the proposed post cold war production of tritium via GeV proton bombardment of LiAl targets would also make large amounts of ^{22}Na [I. Goodwin, *Phys. Today*, May, p. 53 (1993)].
10. See, for example, K. G. Lynn, A. P. Mills, Jr., L. O. Roellig, and M. Weber, in *Electronic and Atomic Collisions* (D. C. Lorents, W. E. Meyerhof, and J. R. Peterson, eds.), p. 227. Elsevier, Amsterdam, 1986; ^{79}Kr produced by (n,γ) on ^{78}Kr would be a possible advantageous material for a large positron source [A. P. Mills, Jr., *Nucl. Sci. Eng.* **110**, 165–167 (1992)]; a scheme for producing ^{126}I by two successive (n,γ) reactions on ^{124}Xe [M. Skalsey and J. Van House, *Nucl. Instrum. Methods Phys. Res. Sect.* B **30**, 211–216 (1988)] may not be feasible due to the large thermal neutron cross section for the product ^{126}I [B. L. Brown, private communication, 1993].
11. ^{74}Se(n,p)^{74}As [M. Deschuyter and J. Hoste, *Radiochim. Acta* **7**, 90 (1967)] is also a possible route to a strong source.
12. R. D. Evans, *The Atomic Nucleus*, pp. 284–289. McGraw-Hill, New York, 1955.
13. H. L. Weisberg and S. Berko, *Phys. Rev.* **154**, 249 (1967); H. L. Weisberg, Ph.D. Thesis, Brandeis University, Waltham, MA (1965), available from University Microfilms International, Ann Arbor, MI, UMI Catalog No. 514449; J. D. McGervey, B. S. Chandrasekhar, and J. W. Blue, in *Positron Annihilation*. (P. G. Coleman, S. C. Sharma, and L. M. Diana, eds.), p. 877. North-Holland Publ., Amsterdam, 1982.
14. M. J. Fluss and L. C. Smedskjaer, in *Positron Annihilation* (R. R. Hasiguti and K. Fujiwara, eds.), p. 407. Japan Inst. Metals, Sendai, 1979.
15. I. K. MacKenzie, in *Positron Solid-State Physics* (W. Brandt and A. Dupasquier, eds.), p. 196. North-Holland Publ., Amsterdam, 1983; L. C. Smedskjaer and M. J. Fluss, *Methods Exp. Phys.* **21**, 77 (1983); A. Saoucha, N. J. Pedersen, and M. Eldrup, *Mater. Sci. Forum* **105–110**, 1971 (1992).
16. K. G. Lynn and J. H. Hurst, private communication (1986); K. G. Lynn, ^{22}*Na Source Production and Chemical Separation*, Seminar. University of Michigan, Ann Arbor, 1986; T. D. Steiger, J. Stehr, H. C. Griffin, J. H. Rogers, M. Skalsey, and J. Van House, *Nucl. Instrum. Methods Phys. Res., Sect. A* **299**, 255 (1990).
17. G. R. Massoumi, P. J. Schultz, W. N. Lennard, and J. Ociepa, *Nucl. Instrum. Methods Phys. Res., Sect. B* **30**, 592 (1988); L. D. Hulett, J. M. Dale, and S. Pendyala, *Surf. Interface Anal.* **2**, 204 (1980).
18. T. D. Lee and C. N. Yang, *Phys. Rev.* **105**, 1671 (1957); J. D. Jackson, S. B. Treiman, and H. W. Wyld, *ibid.* **106**, 517 (1957); S. S. Hanna and R. S. Preston, *ibid.* **106**, 1363 (1957); see A. Rich, in *Positron Studies of Solids, Surfaces and Atoms* (A. P. Mills, Jr., W. S. Crane, and K. F. Canter, eds.), p. 177. World Scientific, Singapore, 1986.
19. R. B. Leighton, *Principles of Modern Physics*, pp. 690–697. McGraw-Hill, New York, 1959; see also Particle Data Group, *Rev. Mod. Phys.* **56**, S1, S48–S53 (1984).
20. R. H. Howell, R. A. Alvarez, and M. Stanek, *Appl. Phys. Lett.* **40**, 751 (1982).
21. B. Krusche and K. Schreckenbach, *Nucl. Instrum. Methods Phys. Res., Sect. A* **295**, 155 (1990); W. Triftshäuser, G. Kögel, K. Schreckenbach, and B. Krusche, *Helv. Phys. Acta* **63**, 378 (1990).

22. W. H. Cherry, Ph.D. Thesis, Princeton University, Princeton, NJ (1958), available from University Microfilms International, Ann Arbor, MI; D. E. Groce, D. G. Costello, J. W. McGowan, and D. F. Herring, *Bull. Am. Phys. Soc.* [2] **13**, 1397 (1968); D. G. Costello, D. E. Groce, D. F. Herring, and J. W. McGowan, *Phys. Rev. B* **5**, 1433 (1972).
23. A. P. Mills, Jr., in *Positron Solid-State Physics* (W. Brandt and A. Dupasquier, eds.), p. 432. North-Holland Publ., Amsterdam, 1983.
24. P. J. Schultz and K. G. Lynn, *Rev. Mod. Phys.* **60**, 701 (1988).
25. G. R. Brandes, A. P. Mills, Jr., and D. M. Zuckerman, *Mater. Sci. Forum* **105-110**, 1363 (1992).
26. B. Y. Tong, *Phys. Rev. B* **5**, 1436 (1972).
27. A. Vehanen, K. G. Lynn, P. J. Schultz, and M. Eldrup, *Appl. Phys.* **A32**, 163.
28. E. M. Gullikson and A. P. Mills, Jr., *Phys. Rev. B* **36**, 8777 (1987).
29. C. A. Murray, A. P. Mills, Jr., and J. E. Rowe, *Surf. Sci.* **100**, 161 (1980).
30. C. A. Murray and A. P. Mills, Jr., *Solid State Commun.* **34**, 789 (1980); D. A. Fischer, K. G. Lynn, and D. W. Gidley, *Phys. Rev. B* **33**, 4479 (1986).
31. E. M. Gullikson, A. P. Mills, Jr., and C. A. Murray, *Phys. Rev. B* **38**, 1705 (1988).
32. S. Berko and J. C. Erskine, *Phys. Rev. Lett.* **19**, 307 (1967).
33. I. K. MacKenzie, T. L. Khoo, A. B. McDonald, and B. T. A. McKee, *Phys. Rev. Lett.* **19**, 946 (1967).
34. E. Gramsch, J. Throwe, and K. G. Lynn, *Appl. Phys. Lett.* **51**, 1862 (1987).
35. G. R. Brandes, K. F. Canter, and A. P. Mills, Jr., *Phys. Rev. B* **43**, 10103 (1991); G. R. Brandes, Ph.D. Thesis, Brandeis University, Waltham, MA (1989), available from University Microfilms, Ann Arbor, MI.
36. N. Zafar, J. Chevallier, F. M. Jacobson, M. Charlton, and G. Laricchia, *Appl. Phys.* [*Part*] *A* **A47**, 409 (1988); N. Zafar, J. Chevallier, G. Laricchia, M. Charlton, and F. M. Jacobson, in *Positron Annihilation* (L. Dorikens-Vanpraet, M. Dorikens, and D. Seegers, ed.), p. 595. World Scientific, Singapore, 1989; M. R. Poulson, M. Charlton, J. Chevallier, B. I. Deutsch, and F. M. Jacobson, *ibid.*, p. 597.
37. D. T. Britton, P. A. Huttunen, J. Makinen, E. Soininen, and A. Vehanen, *Phys. Rev. Lett.* **62**, 2413 (1989).
38. K. G. Lynn, private communication, 1993.
39. J. M. Dale, L. D. Hulett, and S. Pendyala, *Surf. Interface Anal.* **2**, 472 (1980); R. S. Brusa, M. Duarte Naia, E. Galanetto, P. Scardi, and A. Zecca, *Mater. Sci. Forum* **105-110**, 1849 (1992).
40. E. M. Gullikson and A. P. Mills, Jr., *Phys. Rev. Lett.* **57**, 376 (1986).
41. A. P. Mills, Jr. and E. M. Gullikson, *Appl. Phys. Lett.* **49**, 1121 (1986).
42. T. Grund, K. Maier, and A. Seeger, *Mater. Sci. Forum* **105-110**, 1879 (1992).
43. A. P. Mills, Jr., S. S. Voris, Jr., and T. S. Andrew, *J. Appl. Phys.* **76**, 2556 (1994).
44. A. P. Mills, Jr. and W. S. Crane, *Phys. Rev. Lett.* **53**, 2165 (1984).
45. K. F. Canter, P. G. Coleman, T. C. Griffith, and G. R. Heyland, *J. Phys. B* **5**, L167 (1972).
46. J. P. Merrison, M. Charlton, B. I. Deutsch, and L. V. Jorgensen, *J. Phys.: Condens. Matter* **4**, L207 (1992). Other schemes have been suggested: see C. D. Beling, R. I. Simpson, M. Charlton, F. M. Jacobsen, and T. C. Griffith, *Appl. Phys.* **A41**, (1986).
47. K. G. Lynn, E. Gramsch, S. G. Usmar, and P. Sferlazzo, *Appl. Phys. Lett.* **55**, 87 (1989).

48. R. S. Brusa, R. Grisenti, S. Oss, A. Zecca, and A. Dupasquier, *Rev. Sci. Instrum.* **56,** 1531 (1985); A. Zecca and R. S. Brusa, *Mater. Sci. Forum* **105–110,** 2021 (1992).
49. A. P. Mills, Jr., *Appl. Phys. Lett.* **35,** 427 (1979); **37,** 667 (1980).
50. Y. Ito, M. Hirose, I. Kanazawa, O. Sueoka, and S. Takamura, *Mater. Sci. Forum* **105–110,** 1893 (1992).
51. K. G. Lynn and H. Lutz, *Rev. Sci. Instrum.* **51,** 977 (1980).
52. J. Lahtinen et al., *Nucl. Instrum. Methods Phys. Res., Sect. B* **17,** 73 (1986).
53. P. J. Schultz, *Nucl. Instrum. Methods Phys. Res., Sect. B* **30,** 924 (1988).
54. C. Lei, D. Mehl, A. R. Koyman, F. Gotwald, M. Jibaly, and A. Weiss, *Rev. Sci. Instrum.* **60,** 3656 (1989).
55. W. Bauer et al., in *Positron Annihilation* (L. Dorikens-Vanpraet, M. Dorikens and D. Seegers, eds.), p. 579. World Scientific, Singapore, 1989.
56. L. D. Hulett, Jr., T. A. Lewis, D. L. Donohue, and S. Pendyala, in *Positron Annihilation* (L. Dorikens-Vanpraet, M. Dorikens, and D. Seegers, eds.), p. 586. World Scientific, Singapore, 1989.
57. H. Huomo et al., in *Positron Annihilation* (L. Dorikens-Vanpraet, M. Dorikens, and D. Seegers, eds.), p. 603. World Scientific, Singapore, 1989.
58. A. Rose and H. Iams, *Proc. IRE* **27,** 547 (1939); W. L. Barr and W. A. Perkins, *Rev. Sci. Instrum.* **37,** 1354 (1966).
59. S. M. Hutchins, P. G. Coleman, R. J. Stone, and R. N. West, *J. Phys. E* **19,** 282 (1986).
60. K. F. Canter, in *Positron Studies of Solids, Surfaces, and Atoms* (A. P. Mills, Jr., W. S. Crane, and K. F. Canter, eds.), p. 102. World Scientific, Singapore, 1966; K. F. Canter, P. H. Lippel, W. S. Crane, and A. P. Mills, Jr., *ibid.*, p. 199; K. F. Canter, P. H. Lippel, and D. T. Nguyen, *ibid.*, p. 207.
61. A. P. Mills, Jr., *Appl. Phys.* **23,** 189 (1980) U.S. Pat. 4,365,160 Dec 21, 1982.
62. K. F. Canter and A. P. Mills, Jr., *Can. J. Phys.* **60,** 551 (1982).
63. K. F. Canter, G. R. Brandes, T. Roach, and A. P. Mills, Jr., in *Positrons at Metal Surfaces* (A. Ishii, ed.), 1993. p. 133. Trans Tech Publications Aedermannsdorf.
64. L. D. Hulett, Jr., J. M. Dale, and S. Pendyala, *Mater. Sci. Forum* **2,** 133 (1984).
65. J. Van House and A. Rich, *Phys. Rev. Lett.* **61,** 488 (1989).
66. G. R. Brandes, K. F. Canter, and A. P. Mills, Jr., *Phys. Rev. Lett.* **61,** 492 (1989).
67. A. P. Mills, Jr., *Science* **218,** 335 (1982).
68. A. P. Mills, Jr., *Comments Solid State Phys.* **10,** 173 (1982).
69. L. D. Hulett, Jr. et al., *Workshop Report on the Application of Positron Spectroscopy to Materials Science.* (Office of Basic Energy Sciences, Washington, DC, 1993.
70. R. D. Evans, *The Atomic Nucleus*, p. 628. McGraw-Hill, New York, 1955.
71. A. P. Mills, Jr., *Appl. Phys.* **22,** 273 (1980).
72. C. M. Surko, A. Passner, M. Leventhal, and F. J. Wysocki, in *Positron Annihilation* (L. Dorikens-Vanpraet, M. Dorikens, and D. Seegers, eds.), p. 161. World Scientific, Singapore, 1989; T. E. Cowan, B. R. Beck, J. H. Hartley, R. H. Howell, R. R. Rohatgi, J. Fajans, and R. Gopalan, *Hyperfine Interact.* **76,** 135 (1993).
73. R. S. Conti, B. Ghaffari, and T. D. Steiger, *Nucl. Instrum. Methods Phys. Res., Sect. A* **299,** 420 (1990); *Hyperfine Interact.* **76,** 127 (1993).

74. P. B. Schwinberg, R. S. Van Dyck, and H. G. Dehmelt, *Phys. Lett.* A **81A**, 119 (1981).
75. R. S. Van Dyck, P. B. Schwinberg, and H. G. Dehmelt, *Phys. Rev. Lett.* **59**, 26 (1987); P. B. Schwinberg, in *Quantum Electrodynamics* (T. Kinoshita, ed.), p. 332. World Scientific, Singapore, 1990.
76. L. Haarsma, K. Abdullah, and G. Gabrielse, *Hyperfine Interact.* **76**, 143 (1993).
77. D. J. Wineland, C. S. Weimer, and J. J. Bollinger, *Hyperfine Interact.* **76**, 115 (1993).
78. D. Schödelbauer, P. Sperr, G. Kögel, and W. Triftshäuser, *Nucl. Instrum. Methods, Sect. B* **34**, 258 (1988); P. Willutzki, P. Sperr, D. T. Britton, G. Kögel, R. Steindl, and W. Triftshäuser, *Mater. Sci. Forum* **105-110**, 2009 (1992).
79. R. Suzuki, Y. Kobayashi, T. Mikado, H. Ohgaki, M. Chiwaki, T. Yamazaki, and T. Tomimasu, *Jpn. J. Appl. Phys.* **30**, L532 (1991); R. Suzuki, Y. Kobayashi, T. Mikado, H. Ohgaki, M. Chiwaki, T. Yamazaki, and T. Tomimaso, *Mater. Sci. Forum* **105-110**, 1993 (1992).
80. L. D. Hulett, D. L. Donohue, and T. A. Lewis, *Rev. Sci. Instrum.* **62**, 2131 (1991).
81. A. P. Mills, Jr., E. D. Shaw, R. J. Chichester, and D. M. Zuckerman, *Rev. Sci. Instrum.* **60**, 825 (1989).
82. Y. Ito, O. Sueoka, M. Hirose, M. Hasagawa, S. Takamura, T. Hyodo, and Y. Tabata, in *Positron Annihilation* (L. Dorikens-Vanpraet, M. Dorikens, and D. Seegers, p. 583. World Scientific, Singapore, 1989.
83. T. Akahane, T. Chiba, N. Shiotani, S. Tanigawa, T. Mikado, R. Suzuki, M. Chiwaki, T. Yamazaki, and T. Tomimasu, in *Positron Annihilation* (L. Dorikens-Vanpraet, M. Dorikens, and D. Seegers, eds.), p. 592. World Scientific, Singapore, 1989; *Appl. Phys.* **AA51**, 146 (1990).
84. J. Van House, A. Rich, and P. W. Zitzewitz, in *Positron Annihilation* (P. C. Jain, R. M. Singru, and K. P. Gopinathan, eds.), p. 992. World Scientific, Singapore, 1985.
85. K. G. Lynn, W. E. Frieze, and P. J. Schultz, *Phys. Rev. Lett.* **52**, 1137 (1984).
86. A. P. Mills, Jr., in *Positron Scattering in Gases* (J. W. Humberston and M. R. C. McDowell, eds.), p. 121. Plenum, New York, 1984.
87. J. A. Wheeler, *Ann. N. Y. Acad. Sci.* **48**, 219 (1946); V. W. Hughes, in *Atomic Physics* (J. J. Smith, ed.) Plenum, New York, 1972; in *Adventures in Experimental Physics* (B. Maglic, ed.), Vol. 4, pp. 64-127. World Science Education, Princeton, NJ, 1975; M. A. Stroscio, *Phys. Lett. C* **22C**, 215 (1975); T. C. Griffith and G. R. Heyland, *Nature (London)* **269**, 109 (1977; S. Berko, K. F. Canter, and A. P. Mills, Jr., in *Progress in Atomic Spectroscopy* (W. Hanle and H. Kleinpoppen, eds.), Part B, p. 1427. Plenum, New York, 1979; G. T. Bodwin and D. R. Yennie, *Phys. Lett. C* **43C**, 267 (1978); S. Berko and H. N. Pendleton, *Annu. Rev. Nucl. Part. Sci.* **30**, 543 (1980); A. Rich, *Rev. Mod. Phys.* **53**, 127 (1981); D. W. Gidley, A. Rich, and P. W. Zitzewitz, in *Positron Annihilation* (P. G. Coleman, S. C. Scharma, and L. M. Diana, eds.), p. 11. North-Holland Publ., Amsterdam, 1982; J. W. Humberston and M. R. C. McDowell, eds., *Positron Scattering in Gases*. Plenum, New York, 1984; V. W. Hughes, in *Precision Measurements and Fundamental Constants II* (B. N. Taylor and W. D. Phillips, eds.), Spec. Publ. 617, p. 237. National Bureau of Standards, Washington, DC, 1984;

W. E. Kaupilla, T. S. Stein, and J. M. Wadehra, *Positron (Electron)-Gas Scattering*. World Scientific, Singapore, 1985; M. Charlton, *Rep. Prog. Phys.* **48,** 737 (1985); A. P. Mills, Jr., K. F. Canter, and W. S. Crane, eds., *Positron Studies of Solids, Surfaces, and Atoms*. World Scientific, Singapore, 1986; A. P. Mills, Jr., in *The Spectrum of Atomic Hydrogen: Advances* (G. W. Series, ed.), p. 447. World Scientific, Singapore, 1988; R. J. Drachman, ed., *Annihilation in Gases and Galaxies*, NASA Conf. Publ. 3058, NASA, Greenbelt, MD, 1990.
88. M. Deutsch, *Phys. Rev.* **82,** 455 (1951).
89. R. Paulin and G. Ambrosino, *J. Phys. (Orsay, Fr.)* **29,** 263 (1968).
90. H. Morinaga, *Phys. Lett. A* **68A,** 105 (1978).
91. H. Morinaga and Y. Matsuoka, *Phys. Lett. A* **71A,** 103 (1979).
92. U. Zimmerman, F. Stucki, and F. Heinrich, *Phys. Lett. A* **74A,** 346 (1979).
93. A. P. Mills, Jr. and W. S. Crane, *Phys. Rev. A* **31,** 593 (1985).
94. M. H. Weber, S. Taug, S. Berlio, B. L. Brown, K. F. Canter, K. G. Lynn, A. P. Mills, Jr., L. O. Roellig, and A. J. Viescas, *Phys. Rev. Lett.* **61,** 2542 (1988).
95. B. L. Brown, in *Positron Annihilation* (P. C. Jain, R. M. Singru, and K. P. Gopinathan, eds.), p. 328. World Scientific, Singapore, 1985.
96. K. F. Canter, A. P. Mills, Jr., and S. Berko, *Phys. Rev. Lett.* **33,** 7 (1974).
97. K. F. Canter, A. P. Mills, Jr., and S. Berko, *Phys. Rev. Lett.* **34,** 177 (1975).
98. A. Ore, *Univ. Bergen, Arbok, Naturvitensk. Rekke* No. 9 (1949).
99. D. M. Schrader and R. E. Svetic, *Can. J. Phys.* **60,** 517 (1982).
100. O. E. Mogensen, *J. Chem. Phys.* **60,** 998 (1974).
101. R. L. Garwin, *Phys. Rev.* **91,** 1571 (1953).
102. M. Dresden, *Phys. Rev.* **93,** 1413 (1954).
103. L. A. Page and M. Heinberg, *Phys. Rev.* **102,** 1545 (1956).
104. W. Brandt, G. Cussot, and R. Paulin, *Phys. Rev. Lett.* **23,** 522 (1969).
105. A. L. Greenberger, A. P. Mills, Jr., A. Thompson, and S. Berko, *Phys. Lett A* **32A,** 72 (1970).
106. W. Brandt and R. Paulin, *Phys. Rev. Lett.* **21,** 193 (1968).
107. T. Chang, M. Xu, and X. Zeng, *Phys. Lett A* **126A,** 189 (1987).
108. D. W. Gidley, K. A. Marko, and A. Rich, *Phys. Rev. Lett.* **36,** 395 (1976).
109. S. M. Curry and A. L. Schawlow, *Phys. Lett. A* **37A,** 5 (1971).
110. D. W. Gidley and P. W. Zitzewitz, *Phys. Lett. A* **69A,** 97 (1978).
111. A. P. Mills, Jr., E. D. Shaw, R. J. Chichester, and D. M. Zuckerman, *Phys. Rev. B* **40,** 2045 (1989).
112. G. A. Beer, G. M. Marshall, G. R. Mason, A. Olin, Z. Gelbant, K. R. Kendall, T. Bowen, P. G. Halverson, A. E. Pifer, C. A. Fry, J. B. Warren, and A. R. Kunselman *Phys. Rev. Lett.* **57,** 671 (1986).
113. S. Chu, A. P. Mills, Jr., A. G. Yodh, K. Nagamine, Y. Miyake, and T. Kuga, *Phys. Rev. Lett.* **60,** 101 (1988).
114. A. P. Mills, Jr., P. M. Platzman, and B. L. Brown, *Phys. Rev. Lett.* **41,** 1076 (1978).
115. A. P. Mills, Jr., *Phys. Rev. Lett.* **41,** 1828 (1978).
116. K. G. Lynn, *Phys. Rev. Lett.* **43,** 391, 803 (1978).
117. A. P. Mills, Jr., *Solid State Commun.* **31,** 623 (1979).
118. A. P. Mills, Jr. and L. Pfeiffer, *Phys. Rev. Lett.* **43,** 1961 (1979).
119. A. P. Mills, Jr., L. Pfeiffer, and P. M. Platzman, *Phys. Rev. Lett.* **51,** 1085 (1983).

120. A. P. Mills, Jr., E. D. Shaw, R. J. Chichester, and D. M. Zuckerman, *Phys. Rev. B* **40**, 8616 (1989).
121. S. Chu, A. P. Mills, Jr., and C. A. Murray, *Phys. Rev. B* **23**, 2060 (1981).
122. A. P. Mills, Jr. and L. Pfeiffer, *Phys. Rev. B* **32**, 53 (1985).
123. K. G. Lynn, *Phys. Rev. Lett.* **44**, 1330 (1980).
124. D. W. Gidley, A. R. Köyman, and T. W. Capehart, *Phys. Rev. B* **37**, 2465 (1988).
125. A. P. Mills, Jr., E. D. Shaw, M. Leventhal, P. M. Platzman, R. J. Chichester, D. M. Zuckerman, T. Martin, R. Bruinsma, and R. R. Lee, *Phys. Rev. Lett.* **66**, 735 (1991).
126. A. P. Mills, Jr., E. D. Shaw, M. Leventhal, R. J. Chichester, and D. M. Zuckerman, *Phys. Rev. B* **44**, 5791 (1991).
127. See, for example, the Special Issue on *The Mechanical Effects of Light*, edited by P. Meystre and S. Stenholm, *J. Opt. Soc. Am. B* **2** (1985).
128. See also the Special Issue on *Laser Cooling and Trapping*, edited by S. Chu and C. Weiman, *J. Opt. Soc. Am. B* **6** (1989).
129. E. P. Liang and C. D. Dermer, *Opt. Commun.* **65**, 419 (1988).
130. K. Danzmann, M. S. Fee, and S. Chu, *Laser Spectrosc.* **9**.
131. S. Chu, A. P. Mills, Jr., and J. L. Hall, *Phys. Rev. Lett.* **52**, 1689 (1984); M. S. Fee, A. P. Mills, Jr., S. Chu, E. D. Shaw, K. Danzmann, R. J. Chichester, and D. M. Zuckerman, *ibid.* **70**, 1397 (1993); M. S. Fee, S. Chu, A. P. Mills, Jr., E. D. Shaw, R. J. Chichester, D. M. Zuckerman, and K. Danzmann, *Phys. Rev. A* **48**, 192 (1993).
132. A. Loeb and S. Eliexer, *Laser Part. Beams* **4**, 3 (1986).
133. P. M. Platzman, in *Positron Studies of Solids, Surfaces, and Atoms* (A. P. Mills, Jr., W. S. Crane, and K. F. Canter, eds.), p. 84. World Scientific, Singapore, 1986.
134. A. P. Mills, Jr., P. M. Platzman, S. Berko, K. F. Canter, K. G. Lynn, and L. O. Roellig, *Bull. Am. Phys. Soc.* [2] **34**, 588 (1989).
135. P. M. Platzman and A. P. Mills, Jr., *Phys. Rev. B* **49**, 454 (1994).
136. C. M. Varma, *Nature (London)* **267**, 686 (1977).
137. M. Bertolotti and C. Sibilia, *Appl. Phys.* **19**, 127 (1979).
138. R. Ramaty, J. M. McKinley, and F. C. Jones, *Astrophys. J.* **256**, 238 (1982).
139. A. Loeb and S. Eliezer, *Laser Part. Beams*, **4**, 577 (1986).
140. G. Kurizki and A. Friedman, *Phys. Rev. A* **38**, 512 (1988).
141. W. Ertmet, R. Blatt, J. L. Hall, and M. Zhu, *Phys. Rev. Lett.* **54**, 996 (1985).
142. J. V. Prodan, A. Mignall, W. D. Phillips, I. So, H. Metcalf, and J. Dalibard, *Phys. Rev. Lett.* **54**, 992 (1985).
143. J. Hoffnagle, *Opt. Lett.* **13**, 102 (1988).
144. S. Chu, L. Hollberg, J. E. Bjorkholm, A. Cable, and A. Ashkin, *Phys. Rev. Lett.* **55**, 48 (1985).
145. A. P. Mills, Jr. and W. S. Crane, *Phys. Rev. A* **31**, 593 (1985).
146. R. Khatri, K. G. Lynn, A. P. Mills, Jr. and L. O. Roellig, *Mater. Sci. Forum* **105–110**, 1915 (1992).
147. N. Zafar, G. Laricchia, M. Charlton, and T. C. Griffith, *Mater. Sci. Forum* **105–110**, 2017 (1992).
148. A. P. Mills, Jr., *Phys. Rev. Lett.* **46**, 717 (1981); **50**, 671 (1983); *Hyperfine Interact.* **44**, 107 (1988).

3. SOURCES OF LOW-CHARGE-STATE POSITIVE-ION BEAMS[1]

G. D. Alton

Oak Ridge National Laboratory, Oak Ridge, Tennessee

3.1 Introduction

Ion beams are used pervasively in many areas of fundamental and applied research and for a growing number of industrial applications, thus emphasizing the importance of the ion source to modern science and technology. Ion source development has been, historically, driven by the needs of the basic and applied research communities and, in more recent years, by the needs of the industrial communities for sources with improved performance attributes, including operational reliability, lifetime, beam quality (emittance), and intensity, for a growing number of applications. These applications include basic and applied research, isotope separation, mass spectroscopy, fusion energy, inertial confinement, and radiation therapy, as well as a growing and diverse number of industrial applications such as ion beam lithography, semiconducting material doping, ion beam deposition, modification of material surfaces (e.g., conductivity, wear, and corrosive resistance), and probe beams for the important analytical fields of secondary ion mass spectrometry (SIMS), accelerator mass spectrometry (AMS), Rutherford backscattering spectroscopy (RBS), proton-induced X-ray excitation (PIXE), nuclear reaction analysis (NRA), and elastic recoil detection analysis (ERDA). All of these applications are dependent on the use of ion beams and, consequently, on the technologies for their production. Such sources of ions constitute an important set of technologies, each of which may vary widely in complexity, depending on the method required to produce the desired ion species and beam intensity.

Due to the many types of sources and variations within a particular category, a comprehensive description of all source types will not be possible in this chapter. Since most applications involve the use of low-

---

[1] Research was sponsored by the U.S. Department of Energy under Contract No. DE-AC05-84OR21400 with Martin Marietta Energy Systems, Inc.

charge-state ion sources, multiply charged sources, such as the electron cyclotron resonance (ECR) and electron bombardment ion source (EBIS), will not be included. However, because the ECR technology can be used effectively to produce low-charge-state ion beams, sources based on this technology will be briefly described. Since the performance of a particular source type can best be measured by the results obtained from its use in a particular field, the reader is encouraged to consult references contained in previously published review articles [1-3], books [4,5], and conference proceedings which deal with ion sources and their applications.

3.1.1 Ion Source Selection Considerations

The probability for ion formation depends on the atomic structure properties of the particular atom or molecule. Several physical and physiochemical mechanisms can be employed to effect positive ion formation, and, therefore, a wide variety of ion sources have been developed that use one or sometimes a combination of mechanisms for producing ions. The mechanism utilized for positive ion formation may be rather general in scope in terms of species capability, as is the case for ionization through electron impact, or rather selective, as is the case for surface ionization. The ion species capability of a source may be design limited. For example, to provide a wide range of ion species, the source must be equipped with a means of supplying adequate vapor flow of the material in question to the discharge; that is, the source must be equipped with either a vaporization oven or a high-voltage sputter probe, as well as a gas-inlet-metering valve system. Many source concepts and designs are not commensurate with ovens or probes and, therefore, are species limited because they must operate with a restricted number of high-vapor-pressure liquid or gaseous-feed materials.

The selection of an ion source for a particular application should be made with due consideration of factors such as species and intensity capabilities, beam quality (emittance and brightness), ionization efficiency, material use efficiency, reliability, ease of operation, maintenance, duty cycle, and source lifetime. These factors obviously are not of equal importance, and the ion source selection process should be based on the most desirable combination of characteristics for the particular application. No truly universal source that can meet all application requirements exists.

3.1.2 Ion Source and Transport System Figures of Merit

Liouville's Theorem. Liouville's theorem states that the motion of a group of particles under the action of conservative force fields is such that the local number density in the six-dimensional phase space hypervolume $xyzp_xp_yp_z$ is a conserved quality for which x, y, and z are position coordinates and p_x, p_y, and p_z, their respective canonical components of momenta. If the transverse components of motion of a group of particles are mutually independent in configuration space, they are also independent in the orthogonal phase space planes (x,p_x), (y,p_y), and (z,p_z), and the corresponding phase space areas are separately conserved. The transverse phase space areas are proportional to the emittances ε of the beam which are, in turn, also conserved. The quality of an ion beam is usually measured in terms of the emittance ε and brightness B which are, in turn, figures of merit for the ion source from which the ion beam was generated. Both ε and B are related to Liouville's theorem.

Emittance. For ion beam transport, the components of phase space transverse to the direction of motion of the beam are usually the most important. If the transverse components of moton of a group of particles are independent in configuration space, the motion of the particles in the orthogonal planes (x,p_x), (y,p_y), and (z,p_z) will be uncoupled, and, therefore, the phase spaces associated with each of these planes will be separately conserved. These conserved areas of phase space are referred to the emittance ε of the ion beam in the respective direction.

The conserved components of transverse phase space have areas given by

$$d^2A_x = dxp_x; \quad d^2A_y = dyp_y. \tag{3.1}$$

For the case in which the component of momentum along the z direction of the beam p_z is approximately constant

$$d^2A_x = dxp_x = dxM\beta\gamma c \tan\theta_{xz}; \quad d^2A_y = dyp_y = dyM\beta\gamma c \tan\theta_{yz}, \tag{3.2}$$

where $\beta = v/c$ and $\gamma = 1/\sqrt{1-(v/c)^2}$. In Expressions (3.2), M is the mass of the particle of velocity v, c is the velocity of light, and θ_{xz} and θ_{yz} are the angles that the respective x and y transverse components of the particle's motion make with the component of motion along the z axis. In the small angle approximation,

$$\tan\theta_{xz} \approx \theta_{xz} = \frac{dx}{dz} = x' \tag{3.3}$$

and

$$\tan\theta_{yz} \approx \theta_{yz} = \frac{dy}{dz} = y' \tag{3.4}$$

so that

$$d^2A_x \cong M\beta\gamma c\,dxdx' \quad \text{and} \quad d^2A_y \cong M\beta\gamma c\,dydy' \tag{3.5}$$

for the relativistic case. For the nonrelativistic case,

$$d^2A_x \cong \sqrt{2ME}\,dxdx' \quad \text{and} \quad d^2A_y \cong \sqrt{2ME}\,dydy', \quad (3.6)$$

where E is the energy of the ion beam. The emittance ε of the ion beam is proportional to the transverse phase space and thus is also a conserved quantity. The following E- and v-dependent definitions have been adopted historically for the definitions of normalized emittance ε_{nx} and ε_{ny},

$$\varepsilon_{nx} \approx \pi \left(\iint dxdx'/\pi \right) \sqrt{E} \quad \text{and} \quad \varepsilon_{ny} \approx \pi \left(\iint dydy'/\pi \right) \sqrt{E} \quad (3.7)$$

or

$$\varepsilon_{nx} = \pi \left(\iint dxdx'/\pi \right) \beta\gamma \quad \text{and} \quad \varepsilon_{ny} \approx \pi \left(\iint dydy'/\pi \right) \beta\gamma, \quad (3.8)$$

where the integrations are performed over the emittance contour which contains a specified fraction of the beam (10%, 20%, etc.). The normalized emittances as defined by Equation (3.7) are often measured in units of π mm · mrad (MeV)$^{1/2}$ while the units of emittance represented by Equation (3.8) are usually expressed in π mm · mrad.

Both mechanical [6,7] and electrostatic deflection-type devices [8,9] have been developed for the measurement of emittance. Figure 1 displays the stepping-motor-controlled emittance detector unit described in Alton

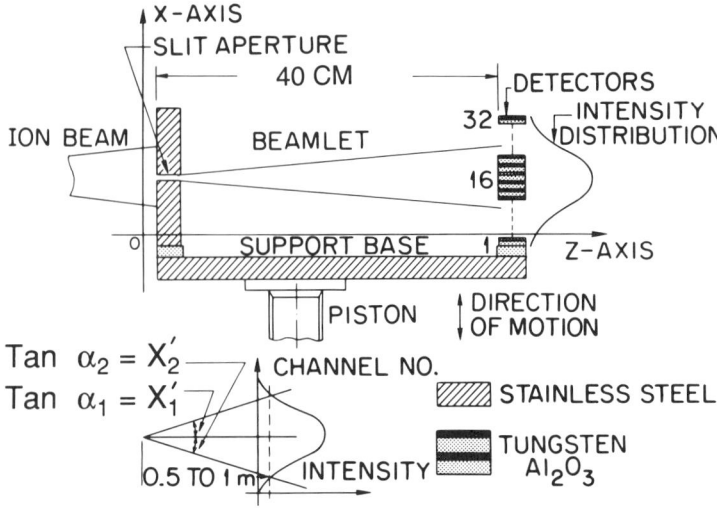

FIG. 1. Schematic drawing of the emittance measurement detector unit.

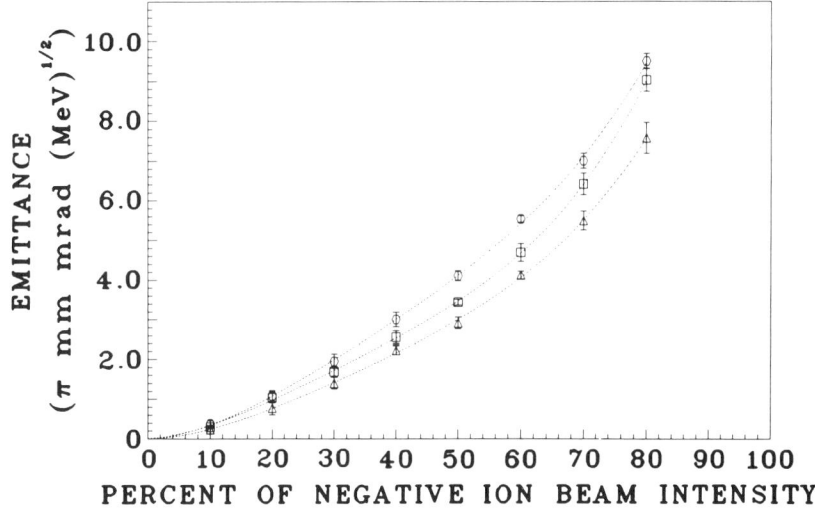

FIG. 2. Normalized emittance, ε_n, versus percentage ion beam intensity for ($\triangle$) ^{28}Si$^-$ (51 μA), ($\square$) ^{58}Ni$^-$ (13 μA), and ($\bigcirc$) ^{197}Au$^-$ (41 μA), illustrating a species-dependent effect. Ion energy, 20 keV [10]. Source, cesium sputter.

and McConnell [7]. Figure 2 displays normalized emittance versus percentage of the ion beam contained within a particular contour for Au$^-$, Ni$^-$, and Si$^-$ [10].

Brightness. Another figure of merit often used for evaluating ion beams is the brightness B. Brightness is defined in terms of the ion current d^2I per unit area dS per unit solid angle $d\Omega$ or

$$B = \frac{d^2I}{dS\,d\Omega}. \quad (3.9)$$

In terms of normalized brightness, Equation (3.9) can be shown to be equivalent to [11]

$$B_n = \frac{2d^2I}{\varepsilon_{nx}\varepsilon_{ny}}. \quad (3.10)$$

Acceptance. The term acceptance A_n refers to the beam transport system and is complementary to the transverse phase space or emittance ε. The acceptance of a device is defined as the phase-space area containing all particles that can be transmitted through the device without impediment. In order to transmit an ion beam through a system of optical components (lenses, magnets, electrostatic analyzers, drift spaces, etc.) without loss, the beam must be transformed into the acceptance phase-space A_n of each individual element from the source to the target. The two-dimensional acceptance phase space for the respective orthogonal x and y directions can be expressed through the following relations.

$$A_{nx} = \pi \left(\iint dx dx'/\pi \right) \sqrt{E}$$

and (3.11)

$$A_{ny} = \pi \left(\iint dy dy'/\pi \right) \sqrt{E},$$

or

$$A_{nx} = \pi \left[\iint dx dx'/\pi \right] \beta \gamma$$

and (3.12)

$$A_{ny} = \pi \left[\iint dy dy'/\pi \right] \beta \gamma,$$

where the units of normalized acceptance are again measured in π mm·mrad (MeV)$^{1/2}$ in Equations (3.11) and π mm·mrad in Equations (3.12). If the acceptance of the device is less than the emittance of the beam, then only that portion of the beam will be transported for which $\varepsilon_{nx} \leq A_{nx}$ and $\varepsilon_{ny} \leq A_{ny}$. In general, in order to transport a beam through a system, it is necessary to incorporate properly positioned and sized lenses, steerers, magnets, etc. to transform the beam emittance to properly match into the acceptance A_n of the system.

3.2 Arc Plasma Discharge Ion Sources

The predominant method for producing a highly ionized medium from which intense beams of positive ions can be extracted is creation of a magnetically directed or confined electrical (plasma) discharge supported wholly or in part by a gaseous vapor containing the material of interest. This method is universal in that any species can be ionized and, therefore, is the method most often used in the design of versatile heavy ion sources for the production of low-charge-state, intense, ion beams. In simple plasma sources, the discharge is initiated and sustained by accelerating electrons that are thermonically or secondarily emitted from a directly (hot) or indirectly (cold) heated cathode to energies above the threshold for ionization of the gaseous vapor. Alternatively, the electrons may be accelerated by high-frequency alternating electric fields generated by RF or ECR techniques. The discharges are most generally directed or confined by a magnetic field—usually oriented parallel to the direction of electron acceleration. Ions may be extracted in a direction parallel to the discharge axis or perpendicular to the direction of the discharge. The former method is often referred to as an axial or end-extraction geometry, and extraction perpendicular to the discharge axis is usually referred to as a side or radial-extraction geometry. The end-extraction geometry is compatible with circular extraction apertures, whereas the side-extraction geometry permits extraction from the length of the plasma column using slit-type apertures. Figures 3a and 3b illustrate the two commonly used discharge configurations.

3.2.1 Elementary Physical Processes in a Plasma Discharge

The ensemble of complex physical and chemical processes, which are in dynamic equilibrium in stable discharges containing chemically active elements, makes detailed analysis and understanding of such media difficult. Most investigations involve noble gases and, therefore, avoid the additional complexities present in more typical discharges, which may involve several different chemically active elements, simultaneously. Regardless of the complexity of the situation, it is instructive to consider some of the basic physical processes which are operative in plasma discharges.

Thermionic Emission. The discharge in hot-filament sources is initiated and sustained primarily by thermionically emitted electrons from the heated filament which are accelerated to energies sufficiently high to produce ionization of the medium. The emission current density is given by the well-known Dushman–Richardson equation,

$$j = A(1 - r)T^2 e^{-\phi/kT}, \tag{3.13}$$

where A is a constant with value $A = 120$ A/cm^2, r is the reflection coefficient of the electron at the metal surface, T is the absolute temperature, ϕ is the work function of the metal, and k is Boltzmann's constant; the term $(1 - r)$ in Equation (3.13) is often defined as the emissivity of the metal ε. From this equation, we can readily see that for equivalent current densities, high-work-function metals must be heated to higher temperatures than those with lower work functions. In practice, the emission current may not reach the value predicted by Equation (3.13) because of space-charge effects. However, theoretical values can be achieved if the cathode-to-anode potential is sufficiently high so that electrons are extracted as fast as they are emitted. In the case of a plasma discharge situation, the theoretical value can also be achieved when positive ions

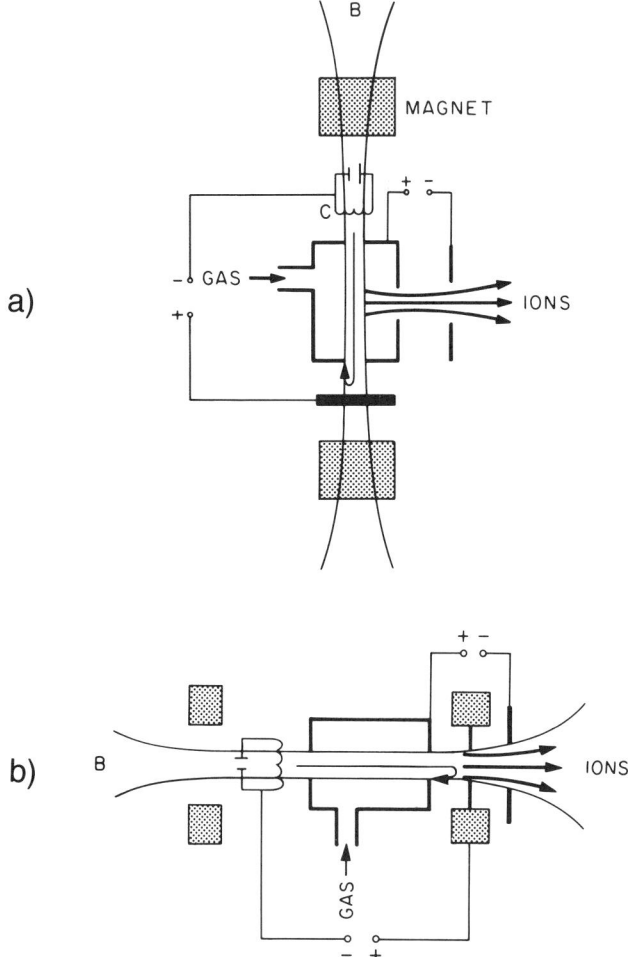

FIG. 3. Commonly used ion-source-discharge configurations for low-charge-state, positive ion production: (a) side- or radial-extraction geometry and (b) end-extraction geometry.

arriving at the filament negate space-charge effects that would otherwise limit electron emission from the filament. Space-charge effects are always present in plasma discharge ion sources.

Ionization. Ionization in an arc discharge is accomplished by introducing a small amount of gas or vapor into the discharge chamber at flow rates high enough to establish an adequate pressure for maintaining a stable discharge but low enough to allow acceleration of electrons to an energy sufficient to ionize a fraction of the vapor. In this way, a medium with plasma-like properties is produced.

The probability of creating a positive particle in a discharge of arc current density j_a is given by

$$P = 1 - e^{-\sigma_i j_a l/e\bar{v}_0}, \qquad (3.14)$$

where σ_i is the cross-section for ionization, l is the average distance traveled by a neutral atom of mean velocity $\bar{v}_0$ in the discharge, and e is the charge on the electron. For a neutral density of n_0, the rate of ion generation in terms of current density j_+ can be expressed as

$$j_+ \cong q \frac{n_0 \bar{v}_0}{4} P(\sigma_i, j_a, l, \bar{v}_0). \qquad (3.15)$$

Stable Discharge Criteria. In order to maintain a stable discharge, the relationship between electron j_e and an ion current density, j_+, flowing across the sheath formed between the cathode and the plasma boundary must satisfy the following inequality,

$$j_e \leq \gamma \left(\frac{M}{m_e}\right)^{1/2} j_+, \qquad (3.16)$$

where M and m_e are the masses of the positive ions and electrons, respectively, and γ is the correction factor which varies with the state of the cathode and has values between $\frac{1}{2}$ and $\frac{2}{3}$ [12]. Equation (3.16) is referred to as the Langmuir criterion for a stable discharge [13].

The Langmuir criterion provides an upper limit for the electron current density emitted by the filament. When the ratio of positive ions arriving to electrons leaving the filament falls below the limit imposed by Equation (3.16), the discharge will become unstable and eventually go out. The rate of generation of positive ions is directly proportional to pressure, and, therefore, there is a critical pressure below which the condition imposed by Equation (3.16) cannot be satisfied.

Minimum Flow Rate, Pressure, and Neutral Density Requirements. In the following derivation, reference is made to Figure 4. After steady-state conditions have been reached, the rate of ion generation is just balanced by the rate at which ions leave the plasma column. For a density, n_+, the ion current density arriving at the sheath edge is

$$j_+ = qn_+(kT_e/M)^{1/2}. \qquad (3.17)$$

If we assume that the field penetrates to a distance, δ, into the plasma and that losses occur with equal probability at each end of the plasma

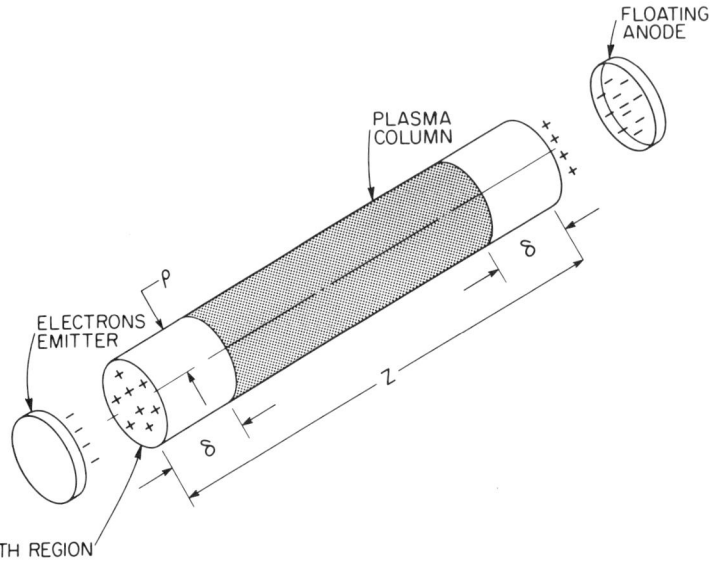

FIG. 4. Schematic diagram of a plasma-discharge column of length Z and diameter ρ with field penetration δ at the ends.

column, then the fractional loss at the ends of the column $j_{+\|}(\text{loss})$ is given by

$$j_{+\|}(\text{loss}) = \frac{2\delta}{Z} qn_+ \left(\frac{kT_e}{M}\right)^{1/2}. \tag{3.18}$$

In plasmas directed along a magnetic field, the velocity v_r of diffusion across the field (radial direction) is given by

$$v_r = D_\perp \frac{\nabla n_+}{n_+}, \tag{3.19}$$

where n_+ is the plasma density and $D_\perp$ is the diffusion coefficient across the field, which is approximately given by

$$D_\perp \cong \frac{1}{16} \frac{kT_e}{eB} \tag{3.20}$$

for a turbulent plasma, according to Bohm et al. [14]. In the expression for $D_\perp$, k is Boltzmann's constant, T_e is the electron temperature of the plasma, and B is the magnetic flux density. This solution to the diffusion equation results in an exponential decrease in the ion density given by

$$n_+ = n_{0+} e^{-r/\rho}, \tag{3.21}$$

ARC PLASMA DISCHARGE ION SOURCES   79

where n_{0+} is the value of the plasma density at the center of the plasma column and ρ is a characteristic distance in which the density drops to $1/e$ of its value at the center of the column. The fractional rate of loss of ions across the magnetic field through diffusion processes is given by

$$j_{+\perp}(\text{loss}) \cong 4qn_+ \frac{\rho}{Z}\left(\frac{kT_e}{M}\right)^{1/2}. \tag{3.22}$$

By equating the rates of ion loss for a plasma column, such as shown in Figure 4, to the rate of ion generation and using the Langmuir stability criterion, expressions for the minimum flow rate f, the minimum neutral density n_0, and the minimum pressure $P_{\min}$ necessary to maintain a stable discharge can be derived [2]. These expressions are given by

$$f \geq \frac{4}{Z\gamma}\left(\rho + \frac{\delta}{2}\right)\left(\frac{m_e}{M}\right)^{1/2}\frac{j_a}{eP} \tag{3.23}$$

$$n_0 \geq \frac{16}{Z\gamma}\left(\rho + \frac{\delta}{2}\right)\left(\frac{\pi m_e}{8kT_0}\right)^{1/2}\frac{j_a}{eP} \tag{3.24}$$

$$P_{\min} \geq \frac{8}{Z\gamma}\left(\rho + \frac{\delta}{2}\right)\left(\frac{\pi m_e kT_0}{2}\right)^{1/2}\frac{j_a}{eP}, \tag{3.25}$$

where $\bar{v}_0 = [(8/\pi)(kT_0/M)]^{1/2}$ has been used for the average thermal velocity of the neutral atoms of temperature T_0 and the arc current density j_a has been equated to j_e. The previous expressions, although not general because of the specific geometry considered, illustrate the dependencies of the quantities f, n_0, and $P_{\min}$ on certain physical or source operational parameters.

Minimum pressures for stable discharge generally lie between 10^{-4} and 10^{-2} Torr for hot filament ion sources. However, the high-frequency (RF) source generally requires pressures between 10^{-2} and 10^{-1} Torr for discharge initiation.

3.2.2 Ion Extraction and Space-Charge-Limited Flow: Analytical Approximations

Poisson's equation for the potential distribution ϕ in the presence of space-charge density ρ is

$$\nabla^2\phi = -\rho/\varepsilon_0, \tag{3.26}$$

where ε_0 is the permittivity of free space. Space-charge effects are always present in high-intensity sources, the influence of which can be estimated by solution to an appropriate form of Equation (3.26).

Ion Extraction from Solid Emitters: Analytical Approximations. Space-charge flow solutions to Poisson's equation have been made by Langmuir for rectilinear flow between parallel plates [15] and by Langmuir and Blodgett for flow between concentric cylinders [16] and spheres [17]. Methods of solutions for the cylindrical and spherical geometries were first given by Langmuir and Compton [18]. These solutions are appropriate for emission from a solid surface. The following analytical solutions to Equation (3.26) are for ionic flow between solid electrodes in the respective geometries.

Equation (3.26) can be solved analytically for the ion current I flowing in a planar diode (parallel plate geometry) configuration which results in the following expression [15]:

$$I = \frac{4}{9}\varepsilon_0 \left(\frac{2q}{M}\right)^{1/2} \frac{V^{3/2}}{d^2} A_c. \qquad (3.27)$$

In terms of perveance P, the solution is

$$P = \frac{I}{V^{3/2}} = \frac{4}{9}\varepsilon_0 \left(\frac{2q}{M}\right)^{1/2} \frac{A_c}{d^2}, \qquad (3.28)$$

where V is the potential difference, d is the spacing between the anode and the cathode, q is the charge on the ion, A_c is the area of the cathode, M is the mass of the ion, and ε_0 is the permittivity of free space.

In general, analytic solutions to Equation (3.26) cannot be obtained for the concentric cylindrical and spherical electrode configurations, and, therefore, approximate solutions must be made. For rectilinear current flow in the spherical geometry, the solution to Poisson's equation is taken to be

$$I = \frac{16\pi}{9} \left(\frac{2q}{M}\right)^{1/2} \varepsilon_0 \frac{V^{3/2}}{\alpha^2}, \qquad (3.29)$$

where $\alpha(r_a/r_c)$ is a dimensionless parameter which is a function of the ratio of the anode radius, r_a, to the cathode radius, r_c [17]. It can be shown that the parameter α has the series solution

$$\alpha = \gamma - 0.3\gamma^2 + 0.075\gamma^3 \ldots, \qquad (3.30)$$

with $\gamma = \ln r_a/r_c$ [18]. In terms of perveance P, Equation (3.29) becomes

$$P = \frac{I}{V^{3/2}} = \frac{16\pi}{9} \left(\frac{2q}{M}\right)^{1/2} \frac{\varepsilon_0}{\alpha^2} \quad \text{(spherical geometry)}. \qquad (3.31)$$

Similarly, an expression for the ion current flowing between two concentric cylinders can be obtained,

$$I = \frac{8\pi}{9}\left(\frac{2q}{M}\right)^{1/2}\varepsilon_0\frac{V^{3/2}l}{\beta^2 r_a}, \tag{3.32}$$

where l is the length of the cylindrical system [16]. The perveance for this geometry is

$$P = \frac{I}{V^{3/2}} = \frac{8\pi}{9}\left(\frac{2q}{M}\right)^{1/2}\varepsilon_0\frac{l}{\beta^2 r_a} \quad \text{(cylindrical geometry)}. \tag{3.33}$$

The function $\beta(r_a/r_c)$ has the solution

$$\beta = \gamma - \frac{2}{5}\gamma^2 + \frac{11}{120}\gamma^3 \ldots, \tag{3.34}$$

where again $\gamma = \ln r_a/r_c$. Values for α, β, and γ have been tabulated as a function of r_a/r_c and r_c/r_a by Langmuir and Compton [18].

Ion Extraction from a Plasma: Analytical Approximations. Because of similarities between the solid electron and the ion emission plasma boundaries, several investigators have proposed that the same analytical procedure can be applied to the case of ion emission from a curved plasma boundary. Several elementary analyses have been made of the ion optics of the extraction region of plasma sources based on the Langmuir–Blodgett theoretical formulations for current flow between cylindrical and spherical electrodes or through simple apertures as treated by Pierce [19] and by Brewer [20]. For example, Expressions (3.29) and (3.32) have been applied to two-electrode extraction geometry to obtain the space-charge-limited flow between the curved sector segments of solid spherical and those of solid cylindrical electrodes. The cathode is taken to be r_c, and the anode radius r_a is calculated from the difference between r_c and d, the anode spacing. Application of the procedure to the case of emission from a spherical sector boundary of aperture radius a yields expressions for the perveance P,

$$P = \frac{I}{V^{3/2}} = \frac{4}{9}\varepsilon_0\left(\frac{2q}{M}\right)^{1/2}\pi\frac{a^2}{d^2}\left(1 - 1.6\frac{d}{r_c}\right)$$

(spherical emission boundary) (3.35)

and

$$P = \frac{I}{V^{3/2}} = \frac{8}{9}\varepsilon_0\left(\frac{2q}{M}\right)^{1/2}\frac{al}{d^2}\left(1 - 0.8\frac{d}{r_c}\right)$$

(cylindrical emission boundary), (3.36)

for the case of a cylindrical sector emission boundary of width $2a$. In both cases, it is assumed that $r_c \gg d$. The expansion has been carried to order d/r_c.

If we define the perveance P in terms of the perveance for the parallel plate arrangement P_p for the two types of extraction apertures, then

$$P = P_p\left(1 - 1.6\frac{d}{r_c}\right) \quad \text{(circular aperture)} \quad (3.37)$$

and

$$P = P_p\left(1 - 0.8\frac{d}{r_c}\right) \quad \text{(slit aperture).} \quad (3.38)$$

Whenever the radius of curvature of the plasma is positive, i.e., $P < P_p$, the beam is convergent. Conversely, whenever the radius of the plasma is negative, i.e., $P > P_p$, the beam is divergent. (The curvature of the plasma is defined to be positive (negative) if the center of curvature is on the extractor (discharge) side of the plasma boundary.) Thus, we see that space-charge-limited currents may vary, depending on the source discharge parameters.

The previous analysis does not take into consideration the presence of the aperture in the second electrode which alters the particle trajectories because of transverse electric field components. In the following analysis, reference is made to the two-electrode system illustrated in Figure 5. The effect of this aperture may be analyzed in terms of the well-known

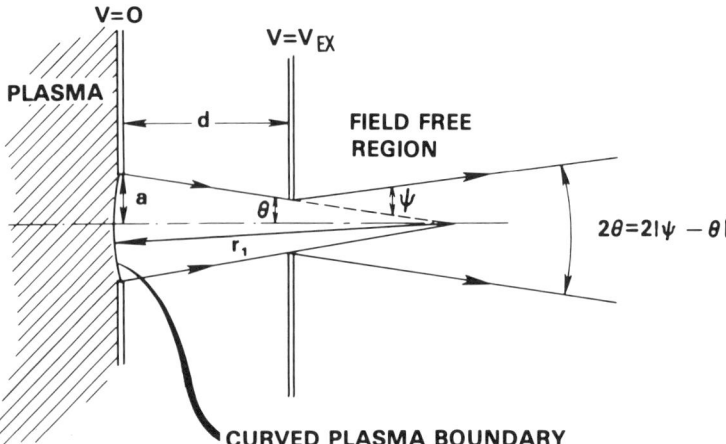

FIG. 5. Simplified ion optics in the ion extraction region of a two-electrode system.

Davisson–Calbick [21] thin-lens formula, taking into account modification of the vacuum field by the presence of space charge ($V \to \frac{4}{3}V$) (see the analysis made by Green [22]).

The angular divergence ψ for a circular aperture of radius a is found to be

$$\psi \cong \frac{a}{3d} \quad \text{(circular aperture)}, \tag{3.39}$$

and for a slit aperture of width $2a$ we have

$$\psi \cong \frac{2a}{3d} \quad \text{(slit aperture)}. \tag{3.40}$$

Assuming the fact that the initial angle is convergent, $\rho \cong a/r_c$, and making substitutions for r_c in Equations (3.37) and (3.38), the following expressions are obtained:

$$\theta \cong \frac{5}{8}\frac{a}{d}\left(1 - \frac{P}{P_p}\right) \quad \text{(circular aperture)} \tag{3.41}$$

and

$$\theta \cong \frac{5}{4}\frac{a}{d}\left(1 - \frac{P}{P_p}\right) \quad \text{(slit aperture)}. \tag{3.42}$$

The final angle, after passing through the second aperture (assuming that a, in Equations (3.39) and (3.40), is the beam half-width in the second aperture), is

$$\omega = \theta - \psi = \frac{1}{2}\frac{a}{d}\left(1 - 1.67\frac{P}{P_c}\right) \quad \text{(circular aperture)} \tag{3.43}$$

and

$$\omega = \theta - \psi - 1.41\frac{a}{d}\left(1 - 1.47\frac{P}{P_c}\right) \quad \text{(slit aperture)}. \tag{3.44}$$

These expressions, even though very approximate, illustrate how the angular divergence is affected through changes in the plasma boundary curvature which are, in turn, affected through changes of the discharge parameters.

The addition of a third electrode between the ground and the extraction electrodes which is maintained at a negative v^- potential relative to the ground will also alter the final angular divergence. This electrode permits a change in perveance without changing the energy of the ion beam, adds an additional controllable optical element in the extraction system, and prevents electrons generated by the ion beam and residual background gas from backstreaming to the source, thus reducing space-charge neutralization of the beam. Figure 6 shows schematically a three-element circular-aperture electrode system which has been studied by Cooper *et al.* [23] by computational analyses. Experimentally, Coupland *et al.* [24] have evaluated a three-element circular-aperture electrode system similar to the configuration of Figure 6. The most critical parameter was found to be the ratio of the radius of the extraction aperture r_1 to the first electrode gap d_1, or $S = r_1/d_1$, with the highest current densities obtained at $S \leq 0.5$. The optimum beam divergence for the system was found to be $\omega = \pm 1.2°$. The measured perveance of the system at minimum beam divergence agrees well with values predicted with the simple model based on the Langmuir–Blodgett formula for the spherical-electrode system. The increase in divergence due to the introduction of a third electrode can be expressed as

$$\Delta\omega = \frac{1}{3}\frac{r_b}{d_1}\left\{1 + \frac{d_1}{d_2}\left[2 - \left(\left|\frac{v_-}{v_+}\right| + 1\right)^{-3/4} + \left(\left|\frac{v_-}{v_+}\right| + 1\right)^{3/4}\right]\right\}, \quad (3.45)$$

where r_b is the beam diameter through the electrodes, d_1 is the first gap, and d_2 is the second gap.

Optimum Perveance. The forms of the expressions suggest an optimum perveance for minimum angular divergence, i.e.,

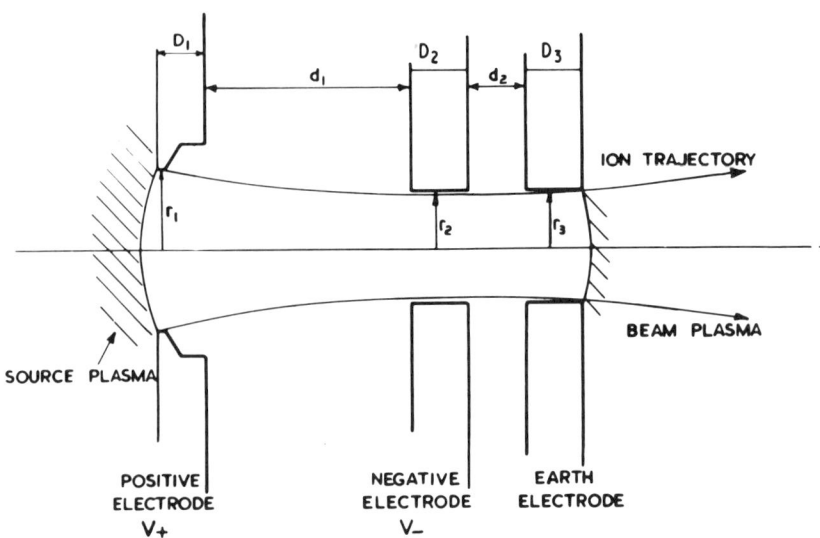

FIG. 6. Schematic drawing of a three-electrode ion extraction system.

$$P \cong \frac{P_\text{p}}{1.67} \quad \text{(circular aperture)} \tag{3.46}$$

$$P \cong \frac{P_\text{p}}{1.47} \quad \text{(slit aperture)} \tag{3.47}$$

Although minimum divergences ($\omega \cong 0$) may be desirable in order to transport the beam through limited acceptance beam transport systems, this mode of operation is not compatible with maximum resolution at which sharp virtual objects are desirable. Therefore, it is desirable to operate the source with angular divergence whenever high resolution is required. (This is a consequence of Liouville's theorem which is discussed in Section 3.1.2.)

Several experimental measurements which verify these previous qualitative predictions have been made [24–26]. Figure 7 illustrates the dependence of angular divergence ω on perveance and the extraction gap as experimentally determined by Chavet and Bernas [25], Coupland et al. [24], and Raiko [26].

3.2.3 Computational Simulation of Ion Extraction from a Plasma

Considerable advances have been made toward appropriate computer modeling of the plasma–vacuum interface ion extraction problem. Ion optics and extraction electrode optimization have been made considerably easier by the development of sophisticated computer codes which can calculate, in a self-consistent manner, ion beam extraction from realistic plasmas through real electrode geometries. Several programs have been developed for these applications, including those described in Whealton et al. [27–31].

In general, these codes involve the iterative solution to either the Poisson equation (Equation (3.26)) [29–31] or the Poisson–Vlasov equations [27,28]. In the former case, space-charge-limited solutions can be numerically estimated by solving Poisson's equation for extraction from a quasineutral plasma at potential ϕ_p with respective ions and electron populations of n_i and n_e. The equation appropriate for this scenario can be expressed by

$$\nabla^2 \phi_0 = \frac{-e}{\varepsilon_0}(qn_\text{i} - n_\text{e}), \tag{3.48}$$

where q is the charge on the ion. The electron velocity v distribution function is usually assumed to be a Boltzmann function of the form

$$f(v) = A_0 \exp\left[-\left(\frac{1}{2}m_\text{e}v^2 + q\phi_\text{p}\right)\right] \Big/ 2kT_\text{e}, \tag{3.49}$$

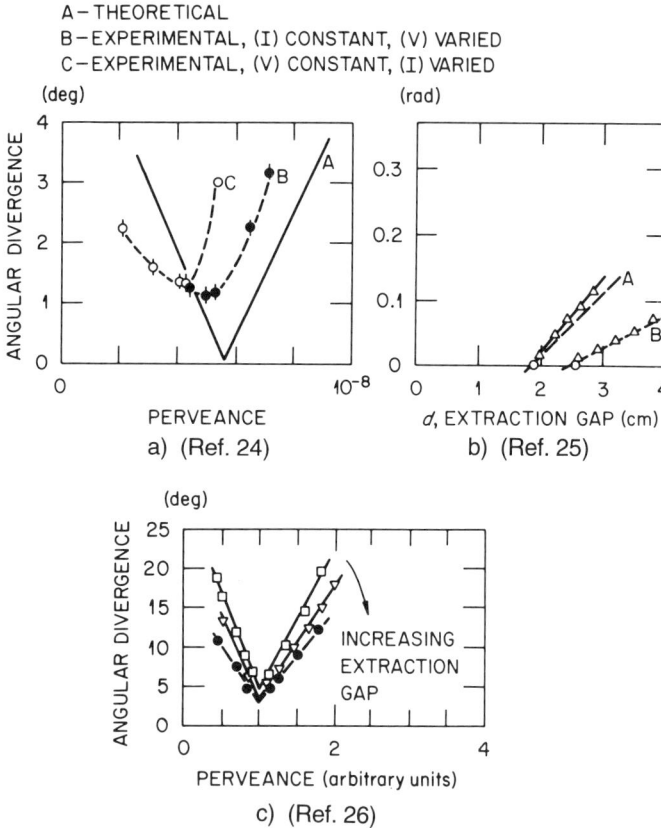

FIG. 7. Experimentally determined ion beam angular divergences versus perveance during extraction from high-density-plasma positive ion sources: (a) from Coupland et al. [24], (b) Chavet and Bernas from [25], and (c) from Raiko [26].

where A is a normalization constant, k is Boltzmann's constant, and T_e is the electron temperature of the plasma. The normalization constant A_0 is related to the particle density n through the requirement that $n = \iiint_{-\infty}^{\infty} f(\vec{v})d\vec{v}$ which leads to

$$A_0 = n \left(\frac{M}{2\pi k T_e}\right)^{3/2}. \quad (3.50)$$

The ion temperature T_i is usually taken to meet the Bohm energy criterion [32] for the ion temperature T_i given by

$$E_i = 2kT_i \cong kT_e. \quad (3.51)$$

ARC PLASMA DISCHARGE ION SOURCES 87

Thus, according to Bohm, the ions must reach the plasma sheath with energy $T_i \geq \tfrac{1}{2} T_e$; ion energies are typically found to be a few electron volts for hot cathode plasma discharge ion sources.

When Equation (3.49) is integrated over the complete velocity spectrum, then we find an expression for n_e given by

$$n_e = n_0 \exp[e\phi/kT_e]. \qquad (3.52)$$

The one-dimensional form of Equation (3.48) can be expressed by substituting for n_e (Equation (3.52)) and $qn_i = n_0$, leading to

$$\nabla^2 \phi = -\frac{e}{\varepsilon_0}(qn_i - n_e) = \frac{en_0}{\varepsilon_0}\{(\exp[e\phi/kT_e] - 1)\}$$
$$\cong \frac{en_0}{\varepsilon_0}\left[e\phi/kT_e + \frac{1}{2}(e\phi/kT_e)^2 + \cdots\right], \qquad (3.53)$$

where the exponential has been expanded in a Taylor series. Dropping higher order terms, the one-dimensional form of Poisson's equation (Equation (3.48)) reduces to

$$\frac{d^2\phi}{dz^2} \cong (n_0 e^2/\varepsilon_0 k T_e)\phi. \qquad (3.54)$$

Substituting $\lambda_d = (\varepsilon_0 k T_e/n_0 e^2)^{1/2}$ into Equation (3.54), we obtain the solution:

$$\phi = \phi_0 \exp[-|z|/\lambda_d]. \qquad (3.55)$$

λ_d is referred to as the Debye length and is a measure of the shielding distance or thickness of the sheath which surrounds the plasma boundary. If the dimensions of the plasma volume are much larger than the Debye length, then the externally applied potentials are shielded so as to protect the inner plasma particles from influences by the electric fields, thus preserving the quasineutrality requirement of a plasma.

Poisson Representation. To examine the exact behavior of $\phi(z)$ in the sheath, the one-dimensional form of Poisson's equation is assumed. At the plane $z = 0$, ions are assumed to enter the sheath region from the bulk plasma with drift velocity v_0. This drift velocity is needed to account for the loss of ions to the wall from the region in which they were created by ionization. For simplicity, we assume $T_i = 0$, so that all ions have the same velocity v_0 at $z = 0$ and that a steady state, in a collisionless sheath region, is formed in which the potential ϕ decreases monotonically with z.

The velocity of the ions after acceleration through the potential $\phi(z)$ can be obtained from the conservation of energy relation, or

$$\frac{1}{2}Mv^2 = \frac{1}{2}Mv_0^2 - e\phi(z)$$

$$v = \left(v_0^2 - \frac{2e\phi}{M}\right)^{1/2}. \tag{3.56}$$

From the equation of continuity, the ion density n_i can be related in terms of the plasma density n_0 in the main plasma:

$$n_0 v_0 = n_i(z) v(z) \tag{3.57}$$

so that

$$n_i(z) = n_0 \left(1 - \frac{2e\phi}{Mv_0^2}\right)^{-1/2}.$$

In the steady state, the electrons will follow the Boltzmann relation (Equation (3.52)) closely or

$$n_e(z) = n_0 \exp[e\phi/kT_e].$$

Poisson's equation then becomes

$$\nabla^2 \phi = \frac{e}{\varepsilon_0}(n_e - n_i) = en_0\left[\exp\left(\frac{e\phi}{kT_e}\right) - \left(1 - \frac{2e\phi}{Mv_0^2}\right)^{-1/2}\right]. \tag{3.58}$$

This equation can be simplified by making the following substitutions:

$$\chi \equiv -\frac{e\phi}{kT_e}; \quad \xi \equiv \frac{z}{\lambda_d} = z\left(\frac{n_0 e^2}{\varepsilon_0 kT_e}\right)^{1/2}; \quad \mathcal{M} \equiv \frac{v_0}{(kT_e/M)^{1/2}}. \tag{3.59}$$

Equation (3.58) then can be written

$$\chi'' = \left(1 + \frac{2\chi}{\mathcal{M}^2}\right)^{-1/2} - e^{-\chi}, \tag{3.60}$$

where the prime denotes $d/d\xi$. This is a nonlinear equation for a plane sheath, which has an acceptable solution only if $\mathcal{M}$ is large enough.

Equation (3.60) can be integrated once by multiplying both sides by χ',

$$\int_0^\xi \chi' \chi'' \, d\xi' = \int_0^\xi \left(1 + \frac{2\chi}{\mathcal{M}^2}\right)^{-1/2} \chi' \, d\xi' - \int_0^\xi e^{-\chi} \chi' \, d\xi', \tag{3.61}$$

where ξ' is a dummy variable. Since $\chi = 0$ at $\xi = 0$, the integrations yield

$$\frac{1}{2}(\chi'^2 - \chi_0'^2) = \mathcal{M}^2\left[\left(1 + \frac{2\chi}{\mathcal{M}^2}\right)^{1/2} - 1\right] + e^{-\chi} - 1. \tag{3.62}$$

A second numerical integration is required to find χ, for which the right-hand side of Equation (3.62) must be positive for all χ.
By expanding the right-hand terms in a Taylor series,

$$\mathcal{M}^2\left[1 + \frac{\chi}{\mathcal{M}^2} - \frac{1}{2}\frac{\chi^2}{\mathcal{M}^4} + \cdots - 1\right] + 1 - \chi + \frac{1}{2}\chi^2 + \cdots - 1 > 0$$

$$= \frac{1}{2}\chi^2\left(-\frac{1}{\mathcal{M}^2} + 1\right) > 0 \quad \mathcal{M}^2 > 1 \quad \text{or} \quad v_0 > (kT_e/M)^{1/2}. \quad (3.63)$$

The inequality represented by Equation (3.63) is referred to as the Bohm criterion for stable sheath formation [32].

Poisson–Vlasov Representation. Basic mathematical formulation of the problem in terms of the Poisson–Vlasov equations for extraction of ions from a collisionless plasma requires the solution of the following equations,

$$\left.\begin{array}{l}\nabla^2\phi = \int f_i dv_i - e^{-\phi} \\ v \cdot \nabla_r f - \dfrac{q}{M}\nabla_r\phi \cdot \nabla^v f_i = 0\end{array}\right\} z > z_0, \quad (3.64)$$

where ϕ and f are, respectively, the potential and ion distribution functions within the plasma. The boundary conditions are taken to be

$$\phi = \phi_0 \quad f = f_0 \quad \text{(Dirichlet)} \quad (3.65)$$

and

$$\nabla_{r\perp}\phi = 0; \quad \nabla_{v\perp}f = 0 \quad \text{(Neumann)}.$$

The Dirichlet boundary conditions are easily specified, and a solution to the extraction problem may be obtained wherever a solution to the source boundary value problem is achieved. The source plasma boundary is traditionally related through the one-dimensional sheath potential distribution described by the following equation:

$$\frac{a^2}{2}\frac{d^2\phi(z)}{dz^2} = \int_0^z \frac{g(y)\,dy}{\sqrt{[\phi(z) - \phi(y)]}} - e^{\phi(z)}, \quad z \le z_0. \quad (3.66)$$

In this equation, $a = \sqrt{2(\lambda_d/\lambda)}$, where λ_d is the Debye length and λ is the mean free path for ionization, usually given in units of Debye lengths for the plasma, g is a source function for ionization, and z is an arbitrary axial distance from the center of the plasma in Debye lengths. The equations are appropriate for descriptions of the ion extraction problem as long as the Debye length is short in relation to electrode dimensions.

The previous equations and their adequacy for properly representing the extraction problem have been critiqued in Whealton [28]. Such programs are very useful in designing low-aberration electrode systems, estimating the emittances of ion sources, and determining optimum perveances for source operation. Figure 8 illustrates a stable solution for ion extraction from a cylindrical-geometry electrode configuration as calculated by use of the program described in Whealton et al. [27].

Beam transport is affected by the angular divergence of the beam during extraction from a plasma boundary. Therefore, optimum perveance operational conditions should be met in order to optimally transport beams through the beam transport system. Computational simulation of the effect is illustrated in Figure 9, which displays the angular divergence θ_{rms} as a

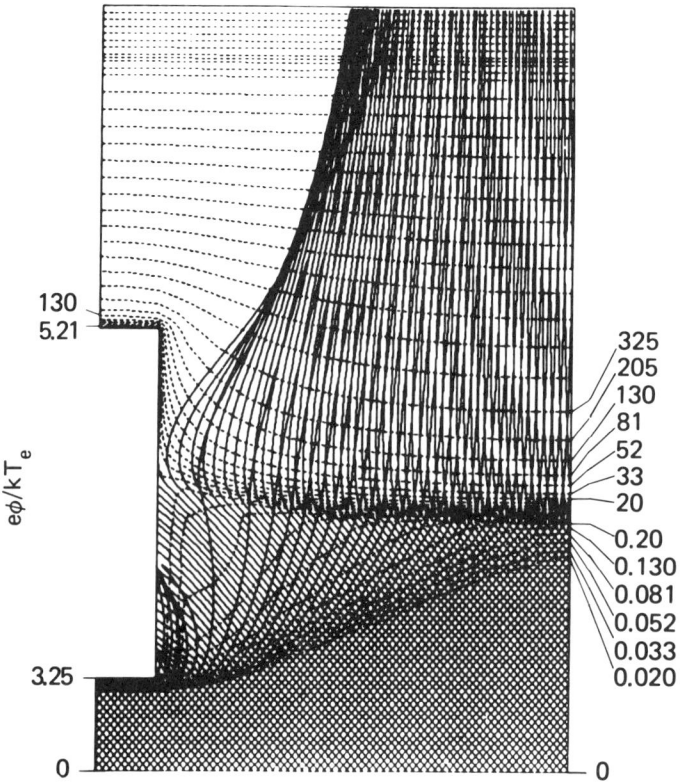

FIG. 8. Computational analyses of the ion-extraction region near a high-density-plasma boundary for a circular-aperture extraction electrode system, computed using the code described in Whealton et al. [27].

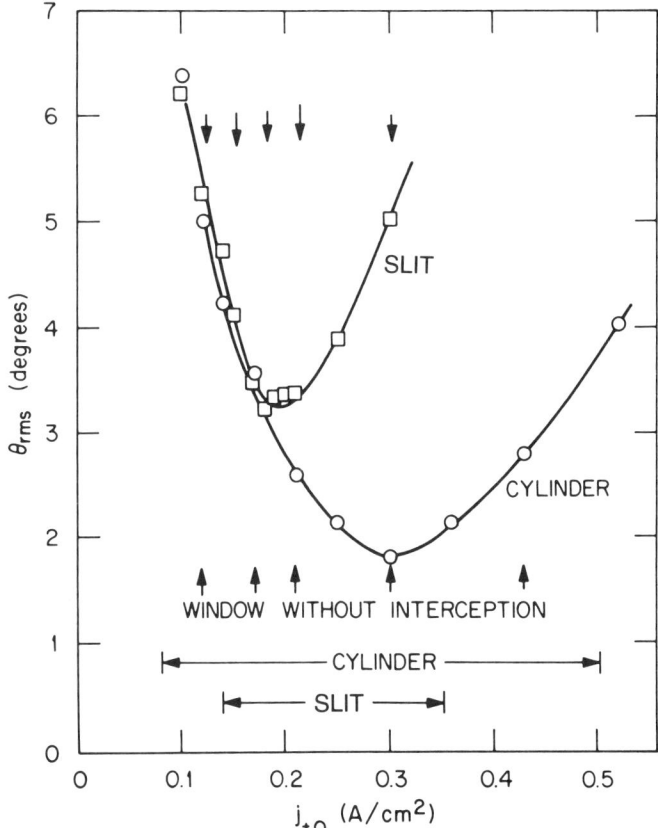

FIG. 9. Computationally determined ion beam angular divergence θ_{rms} versus ion extraction current j_+ for slit and cylindrical apertures. From Whealton et al. [27].

function of ion current density j_{+0} extracted from both slit and circular aperture sources [27].

3.2.4 Plasma Sheath Formation, Plasma Potential, External Field Penetration, and Ion Optical Effects in the Plasma Boundary

Sheath Formation. Under space-charge-limited flow conditions with no plasma boundary present, the electron current between the cathode and the anode approximately follows the familiar Langmuir space-charge-limited flow formula for a planar diode [15],

$$j_e = \frac{4}{9} \varepsilon_0 \left(\frac{2e}{m_e}\right)^{1/2} \frac{V^{3/2}}{d^2}, \quad (3.67)$$

where e is the electronic charge, m_e is the electronic mass, V is the potential drop between the cathode and the anode, and d is the electrode spacing. The sheath that forms around a volume of plasma has a greater thickness, z, than that formed when only positive ions are present [33]. This is due to the fact that the electrons reduce the space charge in the sheath so that a greater distance is required to adsorb the potential drop. The actual sheath thickness according to Bohm [33] is given by

$$z = \beta \frac{2}{3} \left(\frac{\varepsilon_0}{\pi j_+}\right)^{1/2} \left(\frac{2q}{M}\right)^{1/4} V^{3/4}. \quad (3.68)$$

Plasma Potential. After plasma ignition, a steady-state balance between electron and ion losses from the plasma to the walls of the chamber is established. Immediately following ignition, the plasma potential is zero. However, since electrons have much higher thermal velocities than ions, they leave the plasma at a faster rate so that, in general, the net potential of the plasma reaches a positive value. Since the plasma is positive relative to the chamber walls, the potential near the plasma chamber walls will begin to vary over a distance of several Debye lengths; this region, which must exist at the plasma chamber walls, is referred to as the plasma sheath, and the thickness of the layer is referred to as the sheath thickness. The function of the sheath is to provide a barrier so that the more mobile species, usually electrons, are confined electrostatically. The height of the potential barrier adjusts itself dynamically so that the net flow of negatively charged particles with energies sufficiently high to exceed the barrier height is in precise balance with the net flow of positively charged particles striking the wall. Because of the difference in the mobilities of electrons and positive ions, the plasma potential V_p is slightly positive with respect to the walls of the ionization chamber of the source and usually is on the order of 5 to 20 V. At the exit boundary, a sheath is formed between the plasma surface area and the extraction electrode which is composed of ions extracted from the plasma boundary.

External Field Penetration. Extraction of ions from a plasma boundary can be accomplished by applying a potential, V_{ex}, between the exit aperture and an extraction electrode that is negative with respect to the plasma potential V_p. The space-charge-limited ion current density for a planar plasma boundary can be approximated by

$$j_+ = \frac{4}{9}\varepsilon_0 \left(\frac{2q}{M}\right)^{1/2} \frac{V^{3/2}}{z^2}. \quad (3.69)$$

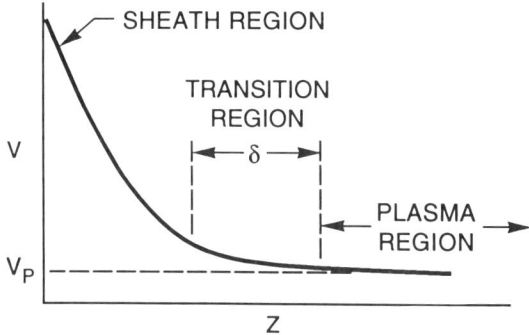

FIG. 10. Potential variation near the plasma sheath.

Equation (3.69) assumes that the field terminates abruptly at the plasma boundary. However, if we assume, as in the case for electromagnetic waves interacting with a plasma [34], that static fields penetrate into the plasma, dropping off exponentially to a depth, δ_{plasma}, given by

$$\delta_{plasma} \cong c/\omega_p, \qquad (3.70)$$

where $\omega_p = (n_e e^2/\varepsilon_0 m_e)^{1/2}$, ω_p is the plasma frequency, n_e is the plasma density, m_e is the electronic mass, and e is the electronic charge. For laboratory-observed plasma densities corresponding to $n_e \approx 10^{12} \rightarrow 10^{14}/$ cm^3, the field penetration depth will be on the order of 5 to 0.5 mm. Because of the penetration of the field into the plasma, positive ions at depth δ_{plasma} are accelerated toward the cathode. (A schematic of a plasma discharge, illustrating penetration of the external field to a depth, δ, is shown in Figure 4.) Figure 10 illustrates schematically the potential distribution in the sheath region.

Ion Optical Effects in the Plasma Boundary. The ion generation and extraction system may contribute considerably to the final beam quality, and thus careful consideration should be given to its selection and design. For example, the ion production and extraction system may introduce energy spreads associated with the fact that the ions are generated at differing equipotentials within the ionization volume; the effect is illustrated in Figure 11b. The problem is reduced by increasing the plasma density so that the external field is terminated more abruptly (Figure 11a). Thus, ions may be generated from sources with ion densities that vary from the very low to the fluid surfaces of high-density plasma-discharge sources.

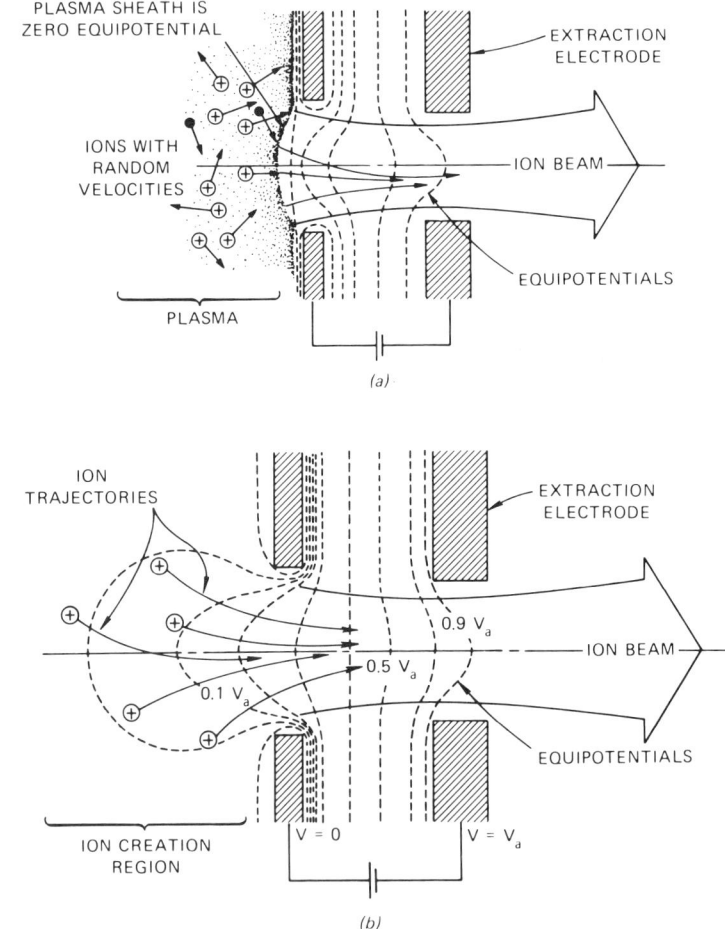

FIG. 11. Schematic diagram of ion extraction from a source operated in modes: (a) high plasma density and (b) low plasma density, illustrating how ion energy spreads can be induced by the mode of source operation or by the design of the ion generation extraction regions of the source.

The ion extraction boundary (meniscus) may assume a variety of shapes, depending on the plasma density n_e, variations in plasma density, and the potential V_{ex}. For fixed ion source parameters and the low extraction voltage V_{ex}, the plasma may assume a convex shape. As V_{ex} is gradually increased, more positive ions are extracted and the boundary recedes until it becomes flat and finally concave at appropriately high values of V_{ex}. These effects are schematically illustrated in Figure 12. Since the shape of the extraction plasma boundary varies with extraction potential V_{ex} and source parameters (arc current, arc voltage, vapor flow rate, and magnetic field), the boundary acts as the first and most important lens in the system and thus plays an important role in the final angular divergence of the ion beam, as well as in beam quality.

3.2.5 Ion Extraction: Ion Beam Intensity

Ion Beam Intensity. The ion current density which can be extracted from a plasma boundary generally exhibits a $V_{ex}^{3/2}$ dependence [35–37]. Current density versus extraction voltage V_{ex} has the form

$$j = C(q/M)^{1/2} V_{ex}^{3/2}/z^2, \qquad (3.71)$$

where the constant C depends on the source geometry, q is the charge, M is the mass of the ion, and z is the distance between the extraction electrode and the plasma boundary.

For space-charge-limited flow to be in effect, the rate of arrival of ions dn_+/dt at the plane of the extraction surface dA must equal or exceed the value obtained by solution of the appropriate form of Poisson's equation (Equation (3.26)), given by

$$\frac{dn_+}{dt} = n_+ \left(\frac{2kT_i}{M}\right)^{1/2} dA = n_+ \left(\frac{kT_i}{M}\right)^{1/2} dA = \frac{4}{9}\varepsilon_0 \left(\frac{2q}{M}\right)^{1/2} \frac{V_{ex}^{3/2}}{z^2} dA \qquad (3.72)$$

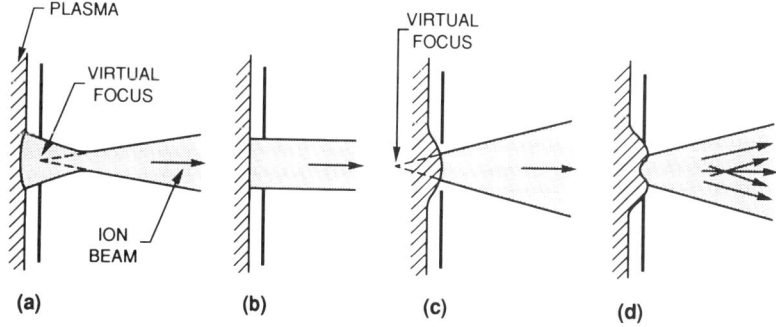

FIG. 12. Illustration of possible plasma boundary curvatures during ion extraction that may be effected by changes in plasma density or ion extraction parameters: (a) concave meniscus, (b) flat meniscus, (c) convex meniscus, and (d) complex curvature that leads to beam aberrations.

for a planar-geometry extraction system with electrode spacing z, where n_+ is the number density of positive ions within the plasma. Thus, the position of the plasma boundary (meniscus) relative to the extraction electrode in the source changes with ion temperature so that the equality represented by Equation (3.72) holds. For example, an increase in plasma density n_+ or electron temperature T_e causes the plasma boundary to move to a position of higher field strength, that is, closer to the extraction electrode. The meniscus shape may be concave, flat, or convex. A low plasma density or low electron temperature plasma meniscus moves toward the interior of the ion source, and, therefore, the field strength is reduced.

It should be pointed out that many species of ions reflecting the relative populations of elements which are present in the discharge will be generated in the plasma because the ions have different masses and may exist in a variety of charge states; the space-charge-limited current will also reflect the ion mass and charge state population within the discharge. The presence of other charge states and masses can substantially reduce the beam intensity of the desired species. This is, in general, true, even if an elemental feed material is utilized because of contaminants from the ion source materials of construction due to sputter erosion and surface contaminants and especially true if the ion is generated from a complex molecular feed material or if an auxiliary discharge support gas is used. Magnetically analyzed Calutron ion current density data from various feed materials versus atomic number of the desired species, as shown in Figure 13, illustrate this point. We note that atomic or simple molecular compounds yield the highest current densities of the desired species while the current densities are lower for more complex feed materials.

Because of the capability of adjusting the source parameters for optimum production of the desired component, some marginal species differentiation is possible. As a rule, the arc discharge voltage should be maintained at approximately $5I_i$, where I_i is the first ionization potential of the desired species. This setting should roughly correspond to the maximum in the cross-section versus electron energy for single ionization.

Ionization Efficiency and Material Utilization Efficiency. The ionization efficiency η of an ion source is defined as the ratio of the flux of charged particles j_+ to the sum of the flux of charged particles j_+ and neutral particles j_0 flowing across the ion extraction boundary or

$$\eta = \frac{j_+}{j_+ + j_0}. \qquad (3.73)$$

The positive ion current density is given by the formula of Bohm *et al.* [14] and is

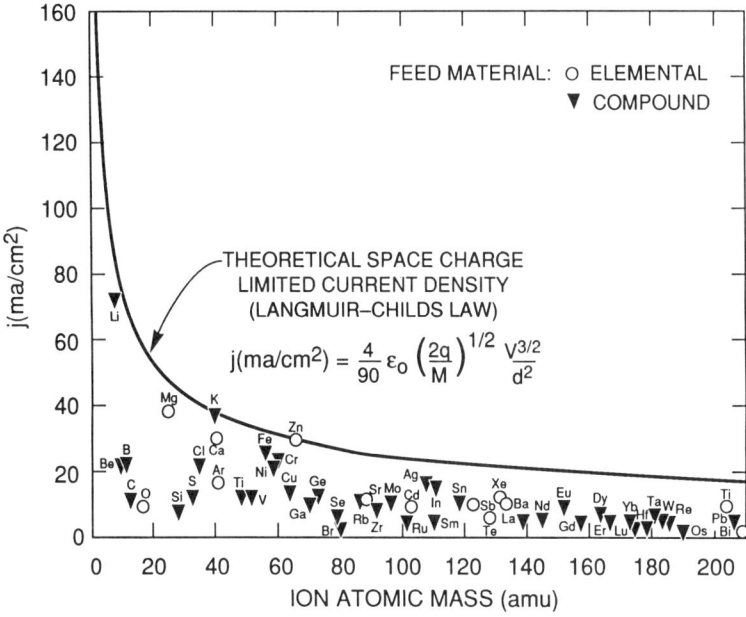

FIG. 13. Magnetically analyzed ion current density data as a function of atomic mass for the Calutron ion source. Ion currents represent maximum values observed for the respective species. Feed material: (○) element and (▼) compound.

$$j_+ = 0.4 \, qn_+ \left(\frac{kT_e}{M}\right)^{1/2}, \quad (3.74)$$

where k is Boltzmann's constant and T_e is the electron temperature of the plasma. The equivalent current density of neutrals flowing across the sheath is given by the well-known formula from the kinetic theory of gases,

$$j_0 = q \frac{n_0 \bar{v}_0}{4} = qn_0 \left(\frac{kT_0}{2\pi M}\right)^{1/2}, \quad (3.75)$$

where T_0 is the temperature of the neutrals of mass M. Substituting Expressions (3.74) and (3.75) into Equation (3.73), we obtain

$$\eta = \left[1 + 0.7 \frac{n_0}{n^+}\left(\frac{T_0}{T_e}\right)^{1/2}\right]^{-1}. \quad (3.76)$$

Ion source performance measurements are very often made immediately after extraction and, therefore, include all ion species generated by the discharge. In such measurements, the natural tendency is to adjust all ion source parameters to achieve a maximum extracted ion current without regard to beam quality, space-charge effects, or the spectrum of charges or masses that make up the extracted beam. Therefore, measurements made under these conditions are of limited value. The most important component in the spectrum of ion beams generated by the ion source is the desired ion beam, which must be optimally transmitted through the beam-handling equipment. The ion source parameters and other optical components should be adjusted so as to maximize the current through an aperture at some point after mass analysis. The material efficiency measured in this way may be expressed as

$$\eta = 4\frac{I}{qn_0v_0A} = \left(\frac{2\pi M}{kT_0}\right)^{1/2}\frac{I}{qn_0A}, \quad (3.77)$$

where I is the analyzed current of the desired species of charge q and mass M and where A is the area of the ion source extraction aperture.

3.2.6 Breakdown Voltage in a Vacuum

As a matter of practical importance in the design of an electrode system, the value of d should be chosen so that sparking is minimized and voltage breakdown does not occur. Coupland et al. [24] and Hamilton [38] have shown that the maximum voltage V_B that can be applied across a gap, d, is given by

$$V_B \cong 6 \times 10^4 d \text{ (cm)}^{1/2} \quad (V). \quad (3.78)$$

In sources from which condensable vapors are emitted, additional sparking and perhaps voltage breakdowns may result due to deposits on the extraction electrode system before reaching the voltage predicted by Equation (3.78).

3.2.7 Theoretical Estimation of the Hot Cathode Lifetime

In heavy-ion sources which utilize hot cathodes, the cathode lifetime is more often than not the limiting factor which determines the sustained length of source operation. The ion generation mechanism which produces sputtering of the cathode and various chemical reactions which the cathode may be subjected to erode the cathode away. In most cases, the principle mechanism which eventually wears down the filament or cathode to the point of breakage or arc extinction in the case of cold cathode sources is physical sputtering. However, high concentrations of oxygen, halogens, and other electronegative elements severely reduce the lifetime of filaments made of chemically active metals such as Ta and W.

A formula which relates cathode lifetime to the physical properties of the cathode materials by using the previously discussed elementary concepts in conjunction with the applicable Sigmund [39] or modified Sigmund [40] formula for sputtering can readily be derived. In this analysis, the wear mechanism is assumed to be solely due to physical sputtering. The amount of material from the cathode removed per unit of time depends on the rate of positive ion bombardment and the sputter ratio according to the following relationship:

$$\frac{dN_0}{dt} = \frac{dN^+}{dt} S \quad \text{(atoms/sec)}. \tag{3.79}$$

Singly charged ions bombard the cathode with energy T_c equal to the sum of the energy received across the plasma sheath plus the energy of arrival at the sheath edge, T_i.

$$T_c = T_a + T_p + T_i,$$

where

$$T_a = qV_a; \quad T_i \cong \frac{1}{2} T_e; \quad T_p \cong qV_p. \tag{3.80}$$

In the previous expression, V_a is the discharge voltage, T_e is the electron temperature of the plasma (~2 eV), and V_p is the plasma potential (typically 5–20 V). The range of potential differences over which most hot or cold cathode discharges operate is $30 \leq V_a \leq 1000$ V. The sputtering ratio S can be approximated by the Sigmund formula [39] for low-energy sputtering,

$$S \text{ (atoms/ion)} = \frac{3}{4\pi^2} \frac{\alpha \lambda T_c}{U_0} \cong \frac{3}{4\pi^2} \frac{\alpha \lambda}{U_0} qV_a, \tag{3.81}$$

where α is a function of the ratio of target to bombarding ion mass M_2/M_1 which has been numerically calculated by Sigmund and experimentally extracted from sputtering data compiled by Andersen and Bay [41]. λ is given by

$$\lambda = \frac{4M_1M_2}{(M_1 + M_2)^2}, \tag{3.82}$$

and U_0 is the heat of sublimation of the cathode material.

The arc current density j_a is the sum of the ion and electron current densities,

$$j_a = j_e + j_+. \tag{3.83}$$

Using the Langmuir criterion for a stable discharge [13], we obtain an expression for j_+ in terms of j_a,

$$j_+ = \frac{j_a}{1 + \gamma\sqrt{(M_1/m_e)}} \simeq \frac{j_a}{\gamma\sqrt{M/m_e}}. \tag{3.84}$$

Thus,

$$\frac{dN^+}{dt} \cong j_+ A = \frac{I_a}{\gamma\sqrt{(M/m_e)}}, \tag{3.85}$$

where A is the area of the cathode.
The rate of loss per unit time is given by

$$\frac{dN_0}{dt} \cong -\frac{3}{4}\frac{\alpha\lambda}{\pi^2}\frac{1}{\gamma\sqrt{(M/m_e)}}\frac{q}{\rho_0}\frac{I_a V_a}{U_0}, \tag{3.86}$$

so that

$$t \leq \frac{4}{3}\frac{\pi^2 U_0 \rho_0}{\alpha\lambda q l_a V_a M_2}\gamma\sqrt{(M/m_e)}\int_0^{\tau_0} d\tau, \tag{3.87}$$

where ρ_0 is the density of the cathode material and $d\tau$ is the differential volume of the material being sputtered [1].

Figure 14 shows the theoretical upper limit of a few equidiameter cathode materials subjected to uniform argon bombardment, predicted from Equation (3.87) for the indicated source and filament parameters. The theory does not take into account the influence of the neutral particles which continually strike the surface and reflect or impede the sputtering process. As is known but neglected here, a fraction of the cathode material may subsequently become ionized and accelerate back to the cathode which would tend to increase the sputtering in this case. The equation appears to predict lifetimes which agree reasonably well with those experienced in practice. In reality, the filament probably never or rarely wears completely through by sputtering, but nearly always breaks due to melting as a result of ohmic heating after having been worn thin by ion erosion. Because of the several factors which determine the length of cathode lifetime, lifetimes vary widely with nominal values for T_a and W filaments, typically ranging between 50 and 100 hr, depending on the discharge parameters and the types and concentrations of chemically active materials in the discharge.

The choices of filament materials are very limited, and no ideal material exists. (The ideal material should have a high melting temperature, have a very low sputtering ratio, and be chemically inert.) The filament cross-sectional area, length, and resistivity at the operational temperature determine the total resistance of the filament which, in turn, specifies the minimum power required to drive the filament to the desired temperature.

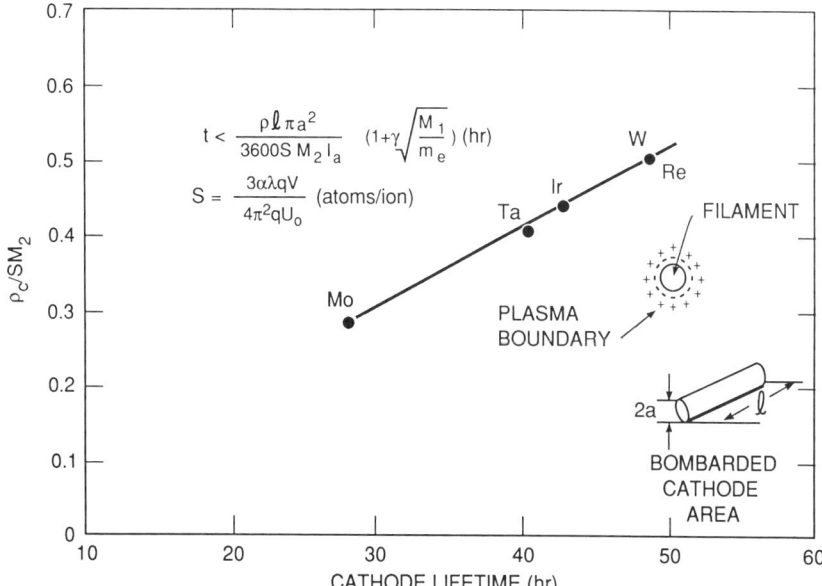

FIG. 14. Theoretical hot-cathode lifetime, limited by physical sputtering with $\gamma = 0.33$, $a = 1$ mm, $M_1 = 40$ amu, $I = 2.54$ cm, $I_a = 2$ A, and $V_2 = 100$ V.

Because of the different resistivities for the various cathode materials, the filament diameters will be different for equal resistance values so that the actual lifetimes will be affected. For example, the resistivity of tantalum is ~1.75 that of tungsten so that a tantalum filament is increased in diameter relative to that of a tungsten filament for equivalent resistances, and thus the relative lifetime is increased.

3.2.8 Types of Arc Plasma Discharge Source

The Duoplasmatron. The duoplasmatron source has been the subject of further development since its introduction by von Ardenne [42]. A schematic of the extraction region of the source is shown in Figure 15.

The plasma discharge is usually generated by applying a potential between a cathode and an anode which are separated by an intermediate electrode. The cathode may be a hot tungsten or tantalum filament hollow cathode or low-work-function emitter. A strong axial inhomogeneous magnetic field (3–10 kG) maintained between an intermediate ferromagnetic electrode and the anode concentrates the discharge near the extraction aperture in the anode region by the action of the field. As a consequence,

102 SOURCES OF LOW-CHARGE-STATE POSITIVE-ION BEAMS

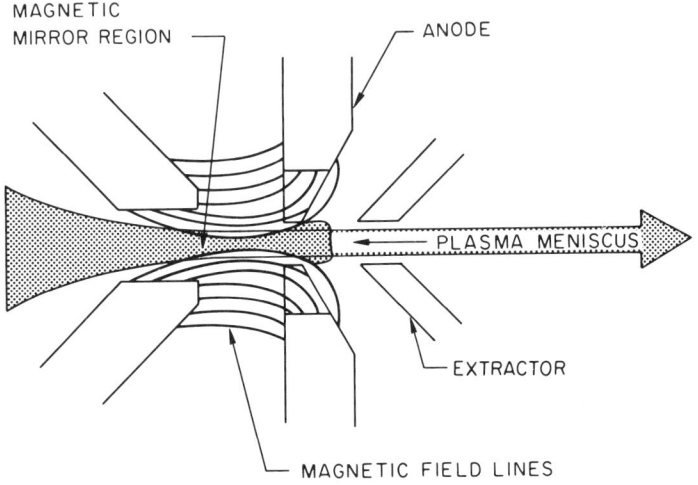

FIG. 15. Illustration of the ion extraction region of a duoplasmatron ion source.

the plasma discharge is characterized by two distinct regions: a high-pressure region between the cathode and the intermediate electrode and a lower-pressure region between the intermediate electrode and the anode. Figure 16 displays the potential distribution within the plasma discharge of the source. The plasma in the anode region attains densities on the order of $n_e = 1 \times 10^{14}/\mathrm{cm}^3$, from which ion beams are extracted. The source is characterized by high efficiency (greater than 80% for hydrogen) and high current densities and is best suited for the generation of ion beams from noncorrosive gaseous feed materials.

A number of descriptions of the emissive properties of various geometry sources can be found in the literature [43–57]. Because of the high plasma densities present near the anode, most sources are equipped with a plasma expansion cup which allows the plasma to expand and thereby cool [51]. This technique increases the area over which the beam can be extracted and transmitted through the system. The characteristics of the source plasma have been studied by a number of research groups, including Morgan et al. [51–53]. Figure 17 displays a long-lifetime duoplasmatron which utilizes an LaB_6 filament [55]. The incorporation of the LaB_6 filament for H^+ (8 mA) and He^+ (10 mA) beam generation extends the lifetime from 150 hr to several months over that for a tantalum filament. A single-aperture source has been designed for use as a high-current (≥ 200 mA) deuteron source for the generation of high-energy neutron beams for use in cancer therapy. The source can produce intense beams from gaseous feed material covering a range of intensities, 1–100 mA in CW mode, and

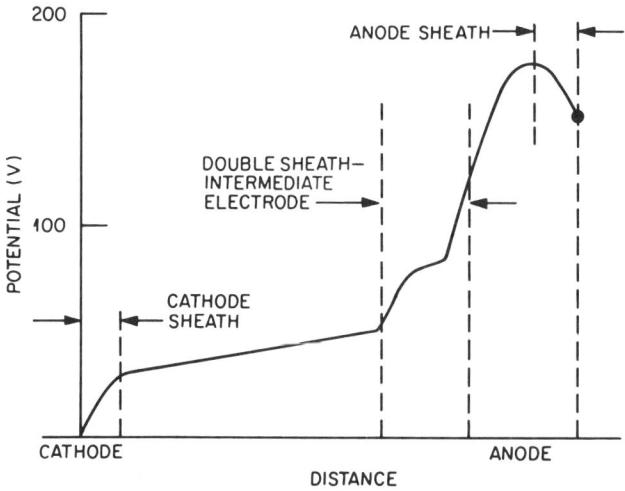

FIG. 16. Potential distribution within the plasma discharge of a duoplasmatron ion source.

has been used to produce peak beam intensities in pulsed mode up to ~20 A. The source is used widely in SIMS microprobe applications, for which high beam densities must be imaged to spot sizes in the range from less than 1 up to 100 μm [58].

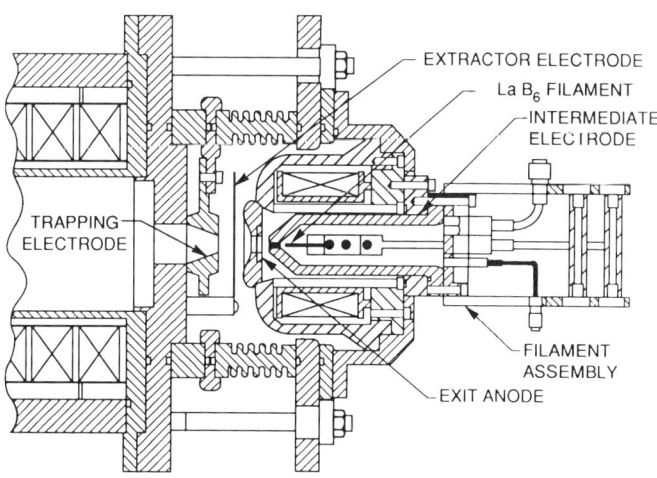

FIG. 17. Illustration of a long-lifetime duoplasmatron source which uses an LaB_6 filament [55].

Although the source is used primarily for noncorrosive gaseous materials, versions that can be used with low vapor pressure and corrosive materials have been constructed [49,50]. In the source described in Masic *et al.* [49], the required feed vapor is introduced in the expansion cup while the primary discharge is sustained with helium or argon. The source described in Illigen *et al.* [50] has been designed for multiply charged ion beams and is fitted with a ring-shaped vaporization oven between the intermediate electrode and the anode for processing solid materials.

The Duopigatron Source. The duopigatron source is a duoplasmatron source to which has been added a reflex discharge section between the intermediate electrode and the anode structure. The concept of the source is illustrated in Figure 18. The modification, first proposed by Demirkhanov *et al.* [59], improves the ionization efficiencies of the source and permits the use of multiple-aperture extraction electron systems because of the large-area plasma region which can be formed by expansion of the plasma flowing from the intermediate structure into the reflex discharge region of the source [59–61]. Ion beam intensities of several amperes have been realized with this source geometry. Because of the high intensity and broad beam capabilities of the duopigatron, the source has been

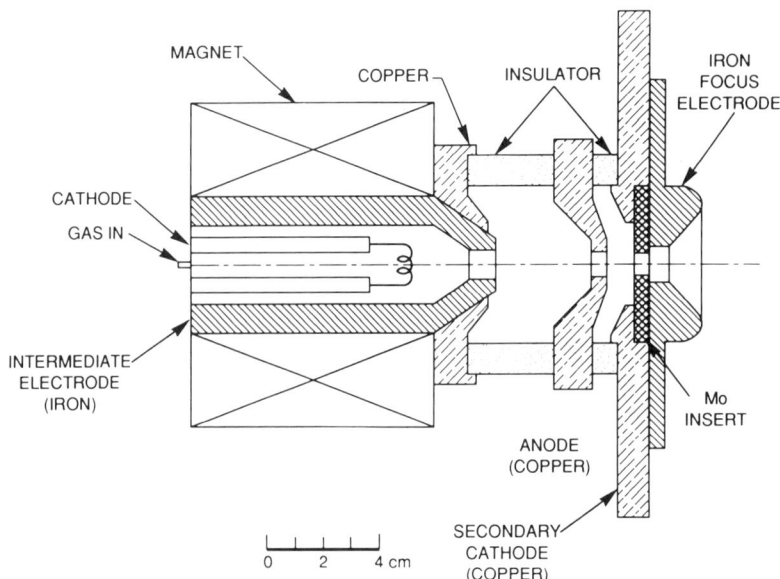

FIG. 18. Illustration of the duopigatron source.

developed by a number of groups for use in plasma heating experiments, including those cited in Stirling *et al.* [60–62].

A single-aperture source has been developed for the generation of high beam intensities (≥ 200 mA) of mixtures of hydrogen and deuterium gases for use in neutron generators [63]. Coaxial forms of the source have also been developed, which are designed to produce high-intensity, uniformly distributed H^+ beams [64].

A single-aperture source has been modified to produce metal ions [65]. Beams are formed of the metal of interest by placing inserts on the extraction aperture/electron reflection electrode. The material from the electrode is sputtered into the discharge, where it is ionized and extracted. Beam intensities of ~ 50 μA for singly charged ions are typical with useful intensities of low-charge-state, multiply charged ions also present in the spectrum. This technique overcomes some of the species limitations of the source which has otherwise been restricted to gaseous or high-vapor-pressure feed materials. A single-aperture form of the source has also been modified for the generation of Li^+ ion beams at densities exceeding 15 mA/cm^2 [66].

The duopigatron source geometry allows independent control of the primary and reflex discharges. For example, an inert gas such as Ar can be introduced to support the cathode discharge while a very corrosive material such as oxygen can be introduced into the reflex discharge region for the generation of O^+ ion beams. The argon serves as a buffer gas which helps to reduce back-diffusion of the O_2 into the cathode region, where it can severely reduce the lifetime of the hot filament. By utilizing Re filaments and argon support gas in the cathode region, the lifetime of these sources has been extended from 2 or 3 hr, which is a typical lifetime for chemically active filaments such as tantalum, to greater than 25 hr. This concept has been incorporated by Shubaly *et al.* to produce total current densities of 160 mA/cm^2 and O^+ ion beams exceeding 140 mA from a multiple-aperture source [67]. This source has also been incorporated into an ion implanter for ion implantation into silicon (separation by ion implantation into silicon (SIMOX)) [68]. This source type has also been developed for solid materials [69]. This source is equipped with an oven which feeds vaporous materials into the reflex discharge region of the source. The source oven, which can reach temperatures up to 1100°C, greatly extends the range of species that can be processed in this source type. Ion beam densities of 16, 24, 27, and 8 mA/cm^2, respectively, are typical of Li^+, P^+, Ca^+, and Bi^+ extracted from a linear array of three 5-mm-diameter circular apertures.

Hollow-Cathode Ion Sources. The hollow-cathode ion source was developed for use in electromagnetic isotope separation processes [70].

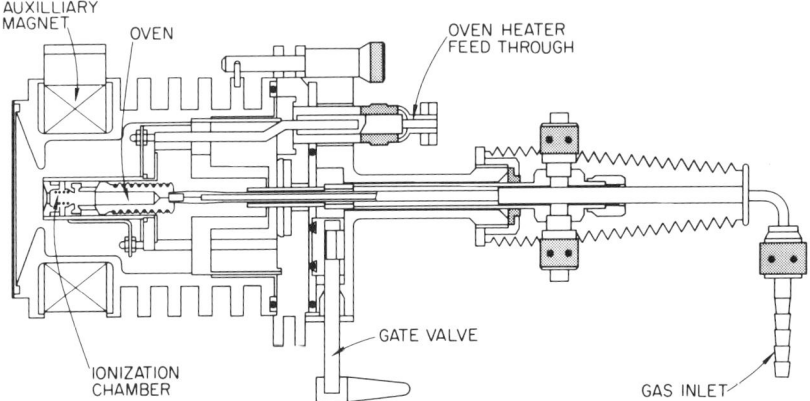

FIG. 19. Schematic drawing of a hollow cathode similar to the source described in Sidenius [70].

The hollow-cathode ion source relies on emission of electrons from a thin wall tube that may be heated directly or self-heated to maintain the discharge. The magnetic field produced by the cathode heating filament is canceled by the windings of the auxiliary magnet coil. The oven temperature may be varied between 200 and 2000°C by regulating the oven's position in the innermost tube of the source. Analyzed beams of 400 μA for argon and 150 μA for lead have been observed using a 0.5-mm-diameter exit aperture. The source is illustrated in Figure 19, while the ionization chamber of the source is shown in Figure 20.

Cold-Cathode Penning Discharge Ion Sources. The reflex discharge originated by Penning [71], often referred to as a PIG (Penning ionization gauge) source, has been used for many years for generating singly and low, multiply charged ion beams. The discharge mechanism is very complex and involves avalanche or cascade ionization processes. Electrons born within the electrode system are accelerated toward the anode, but cannot strike it because their radial motion is constrained by the magnetic field. They, therefore, execute oscillatory motion between the cathodes and the anode, making collisions with the residual gas, which, in turn, liberates other electrons, which are also accelerated. Ions are extracted through an aperture in the end of the cathode (end or axial extraction) or through an aperture in the anode (side or radial extraction). The latter geometry is compatible with slit apertures. Some sources employ hot cathodes that emit electrons thermionically to initiate the discharge while others use cold cathodes for sources in which electrons are emitted princi-

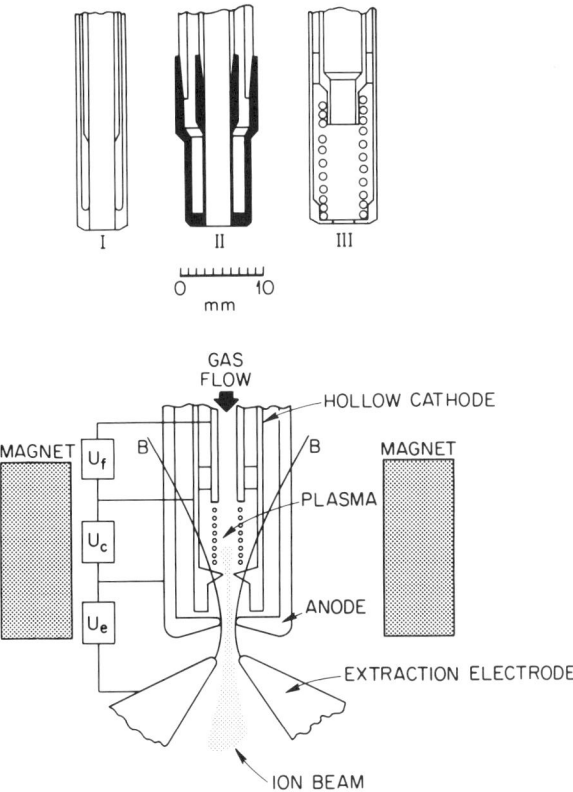

FIG. 20. Schematic diagram of the ionization chamber of the hollow-cathode source [70].

pally by secondary processes. Several types of discharges may occur in the cold-cathode source, depending on the pressure during operation. For most ion source applications, the discharge is characterized by modes of operation involving relatively low voltages and high pressures (10^{-4}–10^{-3} Torr), which produces predominantly low-charge-state ions. For multiply charged ion production, the pressure of operation must be significantly lower because charge exchange at typical discharge pressures depletes higher-charge-state components produced in the ionization processes. Figure 21 displays an axial-geometry cold-cathode Penning source. The Penning discharge ion source consists of two cathodes placed at the ends of a cylindrical hollow anode. Electrons, produced in the discharge, oscillate between the cathodes and are constrained from moving to the anode cylinder by means of a magnetic field directed parallel to the axis of the

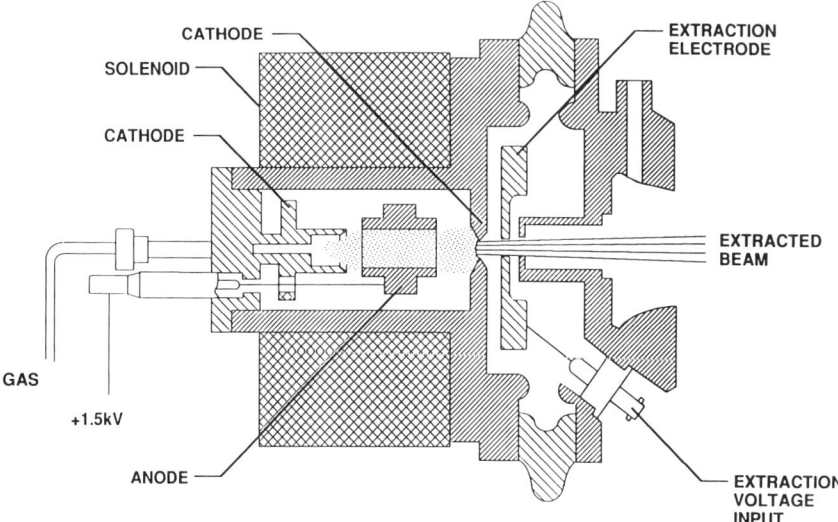

FIG. 21. Schematic representation of an axial-geometry, cold-cathode, Penning-discharge source.

anode. Initiation of the discharge is accomplished by introducing a small amount of gas or vapor into the discharge chamber at flow rates high enough to establish an adequate pressure in the anode–cathode.

In the cold-cathode source, the cathode potential is adjusted to values typically between 600 and 2000 V. After initiation, the discharge is aided by secondary electrons emitted as a result of positive ion bombardment of the cathode when the source is operated in the low-power mode. In high-power sources, the cathodes reach thermionic emission temperatures and thus are self-heated; the discharge, then, is sustained by both thermally emitted and secondarily emitted electrons. Because of the relatively high voltages required to maintain the discharge compared with the hot-cathode version, the cold-cathode Penning source is often used for production of multiply charged heavy ion beams. The radial extraction geometry is widely used for multiply charged heavy ion beam production for cyclotron applications and for heavy ion synchroton applications [72–77]. Higher charge states presumbly result from longer containment times than those in the axial extraction geometry source. The cold-cathode Penning source is used primarily to produce ions from gaseous materials, but its range of capability may be extended by direct vaporization, internal chemical synthesis, or both [73–75] or the use of the sputtering technique [72,75,77]. The ionization chamber and cathode/anode regions of an axial-geometry

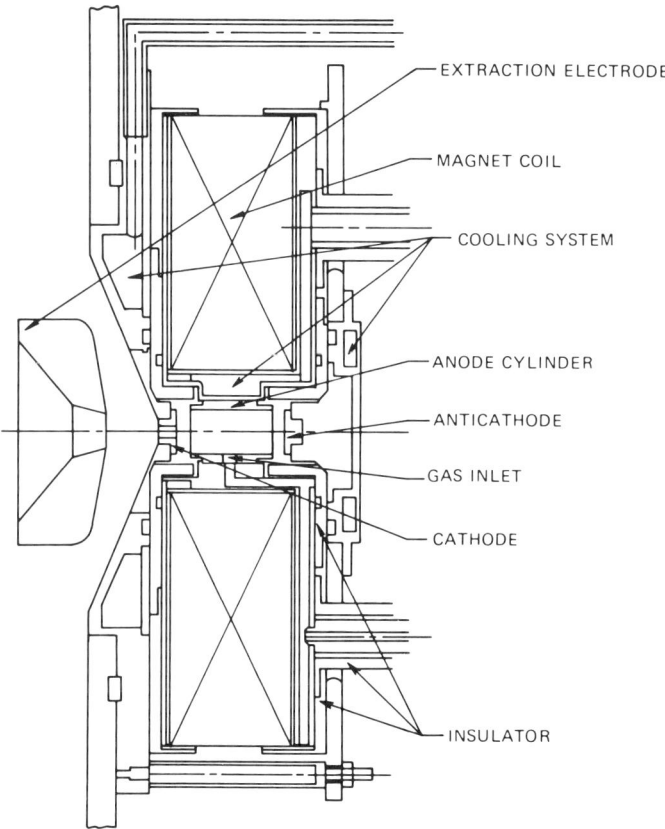

FIG. 22. Schematic drawing of a low-power-consumption, cold-cathode, Penning-discharge ion source [78].

source designed for low power consumption and multiply charged ion production [78] are displayed in Figure 22.

Hot-Cathode Penning Discharge Sources. The hot-cathode PIG source uses a directly or indirectly heated filament as one of the cathodes. The operational range of the discharge voltage for these sources is typically between 50 and 150 V. A wide variety of materials may be processed in the ion source by external feed of gaseous materials, by direct evaporation of solids in either of two ovens having a temperature range of 600–1200°C, or by halogenation of heated oxides or elemental materials. Typically, ion currents of a few to several hundred microamperes can be extracted from laboratory-scale sources of this type. The hot-cathode source is principally

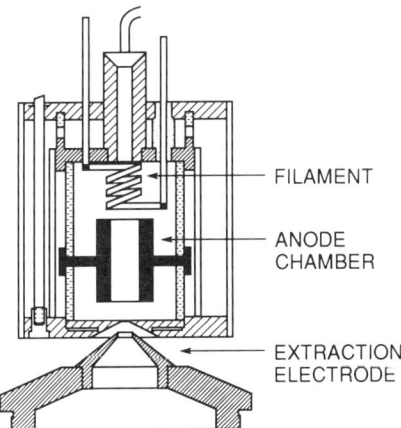

FIG. 23. Ionization chamber structure of a Nielsen-type, hot-cathode, Penning-discharge ion source [79].

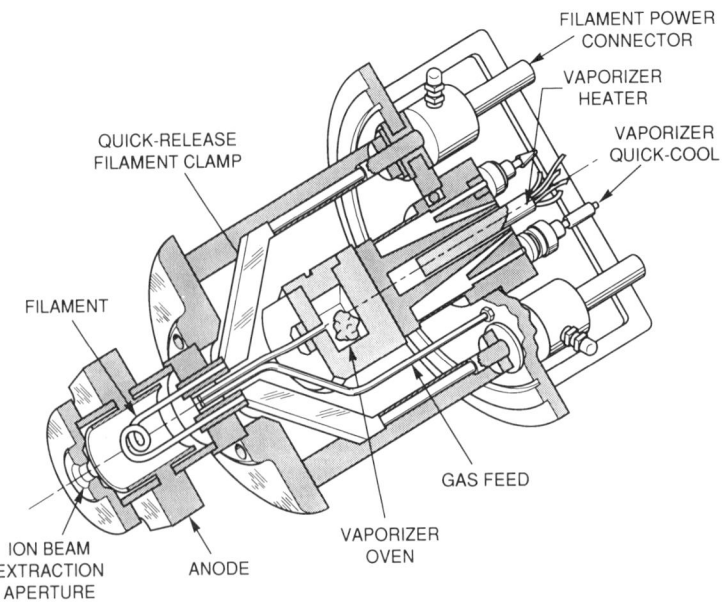

FIG. 24. Schematic drawing of the high-intensity, hot-cathode, Penning-discharge source of O'Connor et al. [81].

used for low-charge-state ion production because of the limited lifetimes at higher power discharge operation. The discharge chamber of the Nielsen-type hot-cathode source described in Nielsen [79] is illustrated schematically in Figure 23. An improved version of the source is described in Ma *et al.* [80].

High-intensity, universal forms of the hot-cathode Penning discharge source have also been developed. An example of a high-intensity source developed for ion implantation applications is shown in Figure 24 [81]. The operating parameters and performance of the source for generating ion beams of the group III B and V B elements are displayed in Table I. The source is equipped with an oven with a controllable temperature range from ~300 to ~1200°C and is used principally for processing group III B and V B elements for ion implantation doping of semiconducting materials. The source is used in conjunction with a tandem electrostatic accelerator for high-energy ion implantation applications, and, therefore, the initially positive ion beams must be converted to negative ion beams through charge-exchange processes.

Freeman-Type Sources. The Freeman source [82], shown in Figure 25, was developed for use with an electromagnetic isotope separator and, therefore, has the capability of producing a wide variety of ion beams. One of the most important attributes of the source is the quiescent, high-frequency-oscillation-free characteristic of ion beams extracted from the source which are highly desirable for high-resolution isotope separation applications. The plasma from which the ion beam is extracted must be free of high-frequency oscillations or "hash" in order to accomplish this objective. Such oscillations modulate the ion beam; the consequent

TABLE I. Typical Ion Source Operating Conditions for the High-Intensity Hot-Cathode PIG Source [81]

Filament current (A)	170
Arc current (A)	2.5 to 8
Arc voltage (V)	60
Solenoid current (A)	3
Extraction voltage (kV)	35
Extraction current (mA)	20 to 30
Emission area (cm^2)	0.30
Current density (mA/cm^2)	65–100
Lifetime (hr)	80–100
^{31}P$^+$ (mA)	>10
^{11}B$^+$ (mA)	>4
Positive-to-negative ion conversion efficiencies (%)	20–25 for ^{11}B$^+$; >50 for ^{31}P$^+$ or ^{75}As$^+$

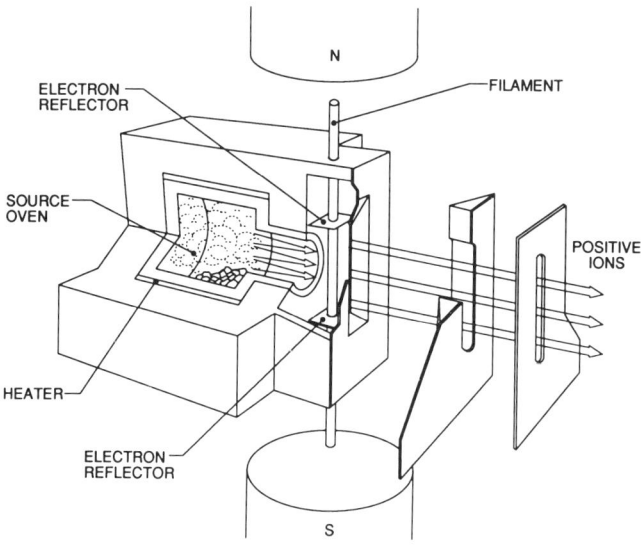

FIG. 25. Schematic drawing of the Freeman (slit-aperture) ion source.

changes in space-charge compensation cause the quality of the focus at the image position of the isotope separator to be degraded. The temperature range of the source oven lies between 550 and 1100°C. The source oven is also equipped with an inlet for introducing halogenating agents such as CCl_4, CF_4, or ClF_3. Ion beams of the platinum or palladium group of metals may also be produced by mounting a sputtering probe made of the material in the rear of the discharge chamber. The cathode is usually a tantalum or tungsten rod, often with a flat emitting surface, mounted 3 mm behind a slit aperture. An auxiliary magnetic field of 0–150 G is maintained parallel to the axis of the cathode. The source produces analyzed currents up to a few milliamperes for most elemental materials. This particular source type is widely employed as the principal source used in medium-to-high-intensity ion implantation systems [83,84].

Calutron-Type Sources. The Calutron source is a side-extraction, hot-cathode source with a slit-extraction aperture and was originally designed to operate within the main magnetic field volume of the large-scale Calutron isotope separators [85]. Several smaller versions of the source have been designed and constructed for use outside the main magnetic field and, therefore, use small auxiliary magnets to direct the discharge. A schematic drawing of the source is shown in Figure 26. The discharge is effected by acceleration of electrons thermionically emitted by a hot

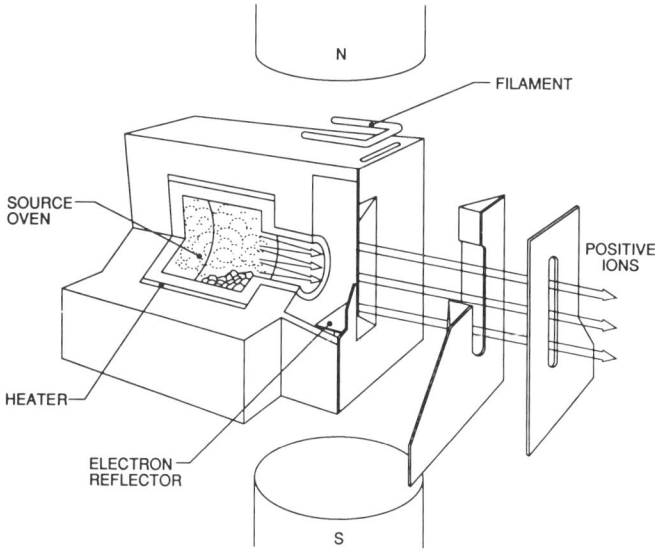

FIG. 26. A schematic drawing of a Calutron, slit-aperture, high-intensity ion source.

filament located at one end of the discharge chamber. The electron beam is collimated parallel to the magnetic field. This arrangement permits use of a nonreactive primary discharge support gas such as Ar which acts as a differential buffer against corrosive feed gases. Inert support gases are fed near the filament while corrosive feed gases are fed independently into the ionization chamber of the source, thus extending the lifetime of the source. The source is versatile and may be used with gases, elemental molecular solids, or internal chemical synthesis. The controllable temperature range of the source is ~300–1200°C. Ion beams up to 200 mA have been obtained from a source with an extraction aperture of 4.8×127 mm^2. The material use efficiency from the large-scale source typically ranges from 5% to more than 30%. (Average current densities versus atomic masses which have been extracted from the source are shown in Figure 13.) The source described in Yabe [86] and developed for use in ion implantation is, in essence, a Calutron source. The source generates current densities of ~20 mA/cm^2 and has a lifetime of greater than 90 hr.

The Calutron/Bernas/Nier configuration [87] of the source is almost identical to the Calutron source, except that, in this version, the filament is mounted within the discharge instead of outside the discharge as generally found in Calutron sources. Thus, the hot cathode is exposed to the

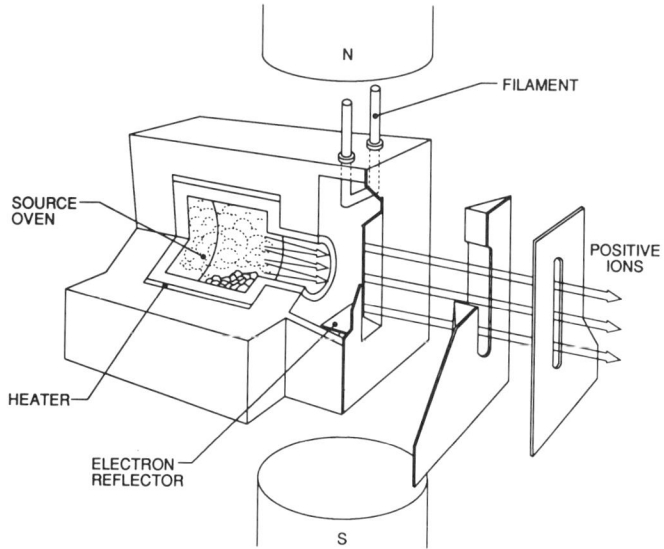

FIG. 27. A schematic drawing of a Calutron/Bernas/Nier, slit-aperture, high-intensity ion source.

corrosive lifetimes of all components in the discharge. The lifetime of the cathode, in principle, should be less than the lifetime of the Calutron cathode. This source is also used in ion implantation applications [84,88,89]. This source configuration is illustrated schematically in Figure 27.

3.2.9 Sources Based on the Use of Multicusp, Magnetic-Field Plasma Confinement Techniques

A multicusp, magnetic field provides a simple, but effective, means for reducing electron and ion losses at vacuum chamber walls. Cusp-type magnetic-field boundaries can be formed by a so-called "picket" fence array of parallel conductors with currents flowing in opposite directions [90]. The wires may be placed around the circumference of a cylindrical vacuum chamber or in a planar array on a flat vacuum chamber. When the cusp fields are strong enough so that the electron precession radii are small compared with the cusp-to-cusp distance, most of the electrons will be reflected back into the plasma volume; only those electrons which are moving along the loss cone or cusp direction can escape. Arrays of permanent magnets (e.g., $SmCo_5$, NdFeB, and AlNiCo) are often used to simulate the picket fence plasma confinement current-carrying wire

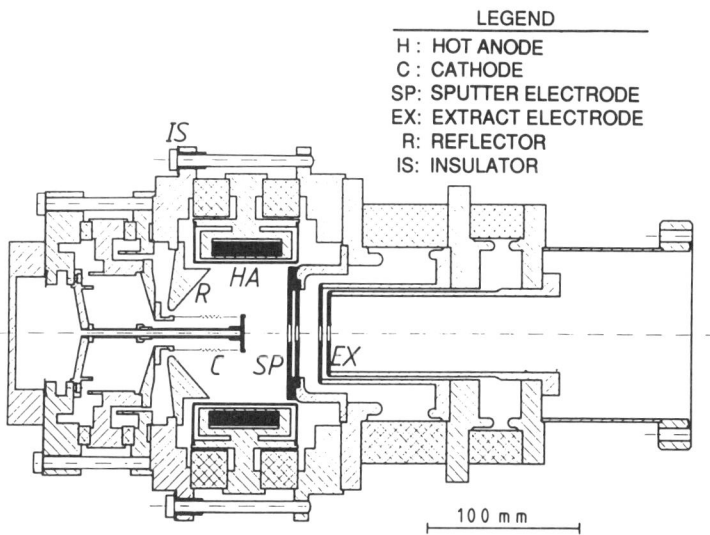

FIG. 28. Schematic drawing of the cold- or hot-reflex-discharge ion source equipped with provisions for sputtering the cathode material [91].

arrangement. Various configurations have been utilized effectively to confine the plasmas in a number of different types of ion sources.

Types of Multicusp, Magnetic-Field Sources. The multicusp, magnetic-field plasma confinement technique is being used at an increasing frequency in both positive and negative ion sources. A few examples will be provided to illustrate the varieties of applications to which this technique may be applied. A versatile, high-intensity ion source, developed for a variety of applications [91], is shown schematically in Figure 28. The CHORDIS (cold and hot reflex discharge ion source) is a modular source which is designed to produce milliampere beam intensities from a wide variety of metal, gaseous, and volatile compound feed materials [91,92]. The source can be operated either in DC or pulsed mode. The discharge is effected by hot tantalum cathodes maintained at the discharge potential. The source is equipped with an oven for processing elemental or compound materials which have vapor pressures between 10^{-4} and 10^{-2} Torr at temperatures between ~300 and 1200°C and a sputter cathode for metal elements. Ions are usually extracted from the source with a three-electrode, multiple-aperture extraction system; the source may also be operated with a single-slit or single-circular-aperture extraction system. Table II lists beam intensities and source operational parameters for a few elements that have been provided from the source.

TABLE II. Performance of the CHORDIS Ion Source [91]

Species	Ion beam intensity (mA)	V_{ex} (kV)	Number of apertures	Emission area (cm^2)
H	80	25	13	1.0
He	120	47	7	0.5
Li	59	30	7	2.0
N	52	25	7	2.7
Ne	53	40	7	0.9
Al	2.4	20	1	0.8
Ar	42	50	7	1.3
Kr	86	45	1	2.7
Xe	71	50	7	2.0
I	28	31	7	1.4
Bi	37	30	7	2.0

3.3 Radiofrequency (RF) Discharge Ion Sources

The RF discharge is maintained by application of high-frequency (HF) electromagnetic fields. In these discharges, the major mechanism by which the power is coupled into the plasma is collisions of charged particles with the background gas. Discharges of this type can be categorized by the means by which the discharge is effected. The two principal means by which the RF plasma is excited and maintained are capacitively coupled or "E" discharges and inductively coupled or "H" discharges. E discharges are capacitive electrodeless discharges excited by an electric field. Capacitively coupled discharges are effected by the electrodes within the discharge. H discharges are eddy (or ringlike) electrodeless discharges, excited by an alternating magnetic field. H discharges are generally referred to as inductively coupled discharges; for some applications, such as those in multicusp field sources, the coil producing the magnetic field may also be in the plasma. "Hybrid" discharges, in which the plasma is maintained by an H discharge and an RF-biased electrode within the discharge, can also be utilized. The RF discharge can also be generated by traveling waves. For this coupling scheme, the electromagnetic waves are introduced into the plasma chamber through a wave guide.

HF discharges can also be classified by the relative magnitudes of the applied frequency ω and the electron and ion plasma frequencies ω_{pe} and ω_{pi}, respectively, defined through the following relation:

$$\omega_{pj}^2 = \frac{n_j e^2}{m_j \varepsilon_0}. \tag{3.88}$$

RADIOFREQUENCY (RF) DISCHARGE ION SOURCES   117

In the low-frequency regime ($\omega \ll \omega_{pi} \ll \omega_{pe}$) both ion and electron motion is modulated by the field. In the high-frequency regime ($\omega_{pi} \ll \omega \ll \omega_{pe}$) ions cannot follow the alternating field; the ion distribution remains fixed in time and the energy is transferred to the electron population. Thus, high-frequency RF discharges are desirable since low-frequency discharges deliteriously affect the emittance of ion beams extracted from the discharge.

Ionization, effected by an alternating electric field (AC), differs in many respects from that produced by a steady-state (DC) field. During the periodic variation of the field, the charge carriers are not swept out of the discharge region toward the walls or the acceleration electrodes, where they can be captured. The reduction in charge loss makes possible a slow increase in ionization with rather weak fields, which can lead to a self-sustained discharge. The secondary processes that occur at the electrode boundaries no longer are important for sustaining the discharge unless the phase of the field is correct for acceleration. As a consequence, the dramatic electrode wear, attributable to sputtering associated with steady-state discharges, is not present, and high-frequency (RF) discharge ion sources have much longer lifetimes than those using a steady-state plasma discharge.

3.3.1 Elementary Physical Processes in a High-Frequency Discharge

In the case of a continuous discharge, electrons are accelerated by a steady-state electric field. At sufficiently low pressures, the electrons will pass through a gas without undergoing ionizing collisions—provided that the mean free path λ is greater than the spacing between the electrodes d. In this case, the discharge can be sustained only by secondary electron emission followed by acceleration. At sufficiently high pressures, the mean free path λ will be small compared with the spacing of the electrodes d, and the primary electrons will make numerous ionizing collisions and liberate secondary electrons that also contribute to the maintenance of the discharge. The latter process is a much more effective means of producing ions and stabilizing the discharge. Some of the important parameters that characterize a high-frequency (RF) discharge, oscillating under the influence of an electric field, $E = E_0 e^{i\omega t}$, are given below [93]: (1) The pressure p and, consequently, the mean free path λ and collision frequency $\nu = \nu/\lambda$ of the electron, where ν is the velocity of the electron; (2) the frequency f of the alternating field $f = \omega/2\pi$ and the wavelength of the alternating field λ_E; and (3) the electrode spacing d and radius of the discharge volume r.

Types of Discharge. If $\lambda > d$ and r, the electrons strike the walls more often than they collide with the gaseous vapors, and secondary wall effects dominate. Such processes generally occur at very low pressures ($p \leq 10^{-3}$ Torr). At medium or high pressures ($p \leq 10^{-3}$–10^{-1} Torr) and with low-frequency fields such that $\lambda < d$ and r and $f \leq v$, the electrons undergo collisions during the time required for reversal of the alternating field and drift in-phase with the field. The electrons that collide with the vaporous material absorb energy from the electric field. The collisional processes cause the electrons to change phase, and thus they gain energy only as a consequence of collisions. Electrons that gain additional momentum from the field and lose it during collisions with atoms or molecules in the discharge move with drift velocity v_d. In a medium in which the frictional losses are proportional to the velocity of the electrons v_d, the electron cloud experiences a force given by

$$m\frac{dv_d}{dt} = e\vec{E}_0 \sin \omega t - mvv_d, \qquad (3.89)$$

which can be readily integrated to give

$$\vec{v}_d = \frac{e\vec{E}_0 \sin(\omega t - \phi)}{m(v^2 + \omega^2)^{1/2}}. \qquad (3.90)$$

In the previous expression, E_0 is the amplitude of the electric field which oscillates at frequency ω, ϕ is the phase angle, and v is the collision frequency of the electrons in the medium. Integration of Equation (3.90) results in the average motion of an electron in the medium,

$$\bar{x} = \frac{e\vec{E}_0}{m\omega} \frac{\cos(\omega t - \phi)}{(v^2 + \omega^2)^{1/2}} + \text{const}, \qquad (3.91)$$

which occurs about a fixed position in the medium. The electron current density can be expressed as

$$j_e = n_0 e v_d = \frac{n_0 e^2}{m} \vec{E} \frac{\sin(\omega t - \phi)}{(v^2 + \omega^2)^{1/2}}, \qquad (3.92)$$

where n_0 is the number density of neutral atoms or molecules per unit volume. The energy per unit volume P gained by the electrons then can be readily determined from the product $P = j\vec{E}$ or

$$P = \frac{n_0 e^2 \vec{E}_0^2 (\cos \phi - \cos 2\omega t)}{2m(v^2 + \omega^2)^{1/2}}, \qquad (3.93)$$

which, when averaged over time, gives the mean value of energy gained by the electrons:

$$\overline{P} = \frac{n_0 e^2}{2m} E_0^2 \frac{v}{v^2 + \omega^2}. \qquad (3.94)$$

If the electric field is represented by $\vec{E} = \vec{E}_0 e^{i\omega t}$, then the medium possesses complex conductivity given by

$$\sigma = \frac{1}{m} \frac{n_0 e^2}{(i\omega + v)}, \qquad (3.95)$$

with real part σ_r reaching a maximum whenever $\omega = v$ is given by

$$\sigma_r = \frac{n_0 e^2}{m} \frac{v}{v^2 + \omega^2}. \qquad (3.96)$$

3.3.2 Types of High-Frequency RF Ion Source

The radiofrequency discharge is usually affected by an RF oscillator with frequencies in the range of 20–100 MHz. An axial magnetic field is usually employed to direct the discharge, and the ions are extracted through a canal in one end of the chamber. The source generally requires an order of magnitude higher gas pressure (10^{-2}–10^{-1} Torr) than conventional hot-cathode discharge sources. The RF power may be capacitatively or inductively coupled to the plasma. A capacitatively coupled source is shown in Figure 29. The vessel is made of quartz or glass to reduce recombination at the inside surface of the vessel. Wall recombination includes not only neutralization of ions by electrons, but also recombination of neutral atoms and ions. A large steady potential difference is maintained between the extraction electrode and the anode. The degree of ionization is observed to increase monotonically with the high-frequency power applied. Typical RF discharges require between 30 and 700 W of power, depending on the source type. In certain sources, a steady magnetic field is applied parallel to the steady-state electric field or transverse to it. The greatest increase of ion density is achieved with a transverse field whose magnitude corresponds to the electron cyclotron frequency. RF power may be inductively coupled to the plasma by placing a coil around the discharge vessel. In this case, the gas electrons are accelerated by the magnetically induced rotational electric field. An additional DC magnetic field can be used to constrict the plasma to the central portion of the discharge chamber.

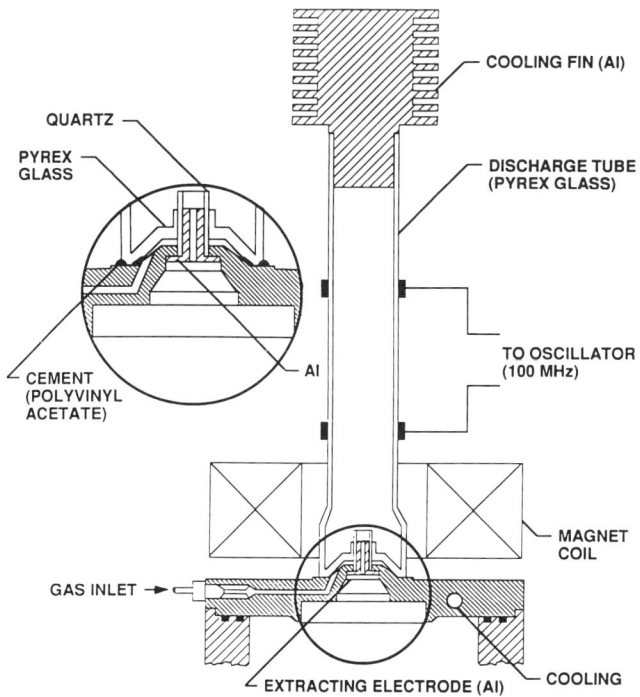

FIG. 29. Illustration of a capacitatively coupled RF ion source.

These sources are particularly useful for the production of beams of high proton content (~70%) in comparison with H_2^+ ions and are also suitable for heavier ions. Many sources, including those described in Morozov et al. [94–97], have been described in the literature. An early review of high-frequency sources is given in Blanc and Degeilh [98], which includes a tabulation of the physical details of several sources. The pressure and intensity characteristics of parameters of the RF discharge are described in Misra and Gupta [99]. Various extraction geometries have been investigated to optimize beam shape, source efficiency, and gas consumption. For microfocused beam applications, the potential of this source appears to be limited due to the high-energy spread (30–500 eV) of the extracted ions, which is about one order of magnitude higher than that in the duoplasmatron and more than two orders higher than that in the surface-ionization source [100].

Proton currents of the order of 100 mA have been obtained in pulsed-mode operation [101]. The source is generally used for simple gases, but

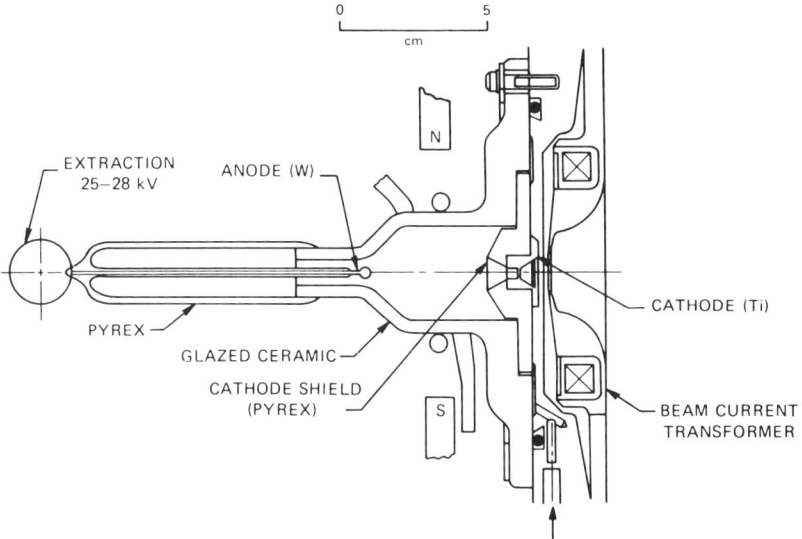

FIG. 30. Schematic representation of a high-intensity RF ion source for protons [101].

versions that can be used to produce beams for more complex gases and solids have been reported. A schematic of the source described in Norbeck and York [102] for the production of proton beams is shown in Figure 30. While the RF source is clearly better suited for processing gaseous-feed materials, a few sources have been developed; such as the sources described in Norbeck and York [102] and Saito et al. [103], which use sputtering of the exit canal made of the metal from which the ion species is produced. However, the useful lifetimes of RF sources which use sputtering or gaseous molecular compounds may be reduced from ≥ 50 hr to a few hours by coating of the quartz vacuum envelope with sputter deposits or chemical reaction and dissociation products. The ratio of H^+ to H_2^+ and H_3^+ is also known to be seriously affected by deposits on the quartz tube.

3.3.3 Types of Multicusp-Field RF Source

Compact, single-aperture, RF-driven, multicusp-field-type sources which are capable of generating nearly pure atomic ion beams of both positive and negative ions have been developed. In these sources, the plasma is ignited by means of a coil immersed in the plasma discharge.

A magnetic filter is used to achieve high-purity N^+ ion beams. The magnetic filter serves to eliminate high-energy electrons from the extraction region, which results in an increase in the fraction of N^+ ions extracted. Nitrogen ion implantation is used industrially to increase the surface hardness and wear resistance of metals. The implantation process also has the effect of producing much smoother surfaces than is possible with untreated material, resulting in less friction for contacting surfaces such as ball bearings. Nitrogen ion implantation is usually carried out at energies of 10–400 keV and dose levels of 10^{16}–10^{18} ions/cm^2. A pure beam of N^+ ions is desirable for this application [104]. Sources of this type can also be used to generate relatively pure atomic ion beams from diatomic molecules such as H_2, O_2, and N_2, as well as from the noble gases (He, Ne, Ar, Kr, and Xe) and a variety of other species. The source described in Reference [104] is schematically illustrated in Figure 31. This source requires high RF power for pulsed-mode operation; beam intensities up to 1000 mA/cm^2 for H^+ and 500 mA/cm^2 for N^+, Ar^+, Kr^+, and Xe^+ have been extracted from the source. A compact RF source equipped with a magnetic filter field has been developed for generation of high-purity beams of O^+ for SIMOX applications [105].

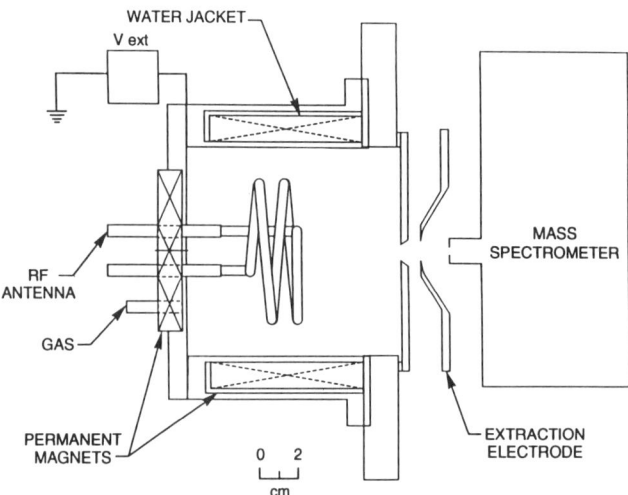

FIG. 31. Schematic diagram of the multicusp magnetic-field RF ion source of Leung *et al.* [104].

3.3.4 Types of Wave-Guide RF Sources

RF-discharge ion sources which utilize traveling waves have been developed. In these sources, the discharge forms in a waveguide-like structure as a result of a propagating electromagnetic wave excited at one end of the discharge by a wave-launching mechanism. This type of source has been reviewed in Moisan and Zakrzewski [106].

3.4 Electron–Cyclotron Resonance and Microwave Ion Sources

During the past several years, multiply charged ion sources based on the ECR principle have been developed. The ECR source technology has advanced steadily over the years, driven by needs for high-charge-state beams for high-energy applications. The ECR ion source was first used by Geller et al. for multiply charged ion beam generation [107,108], and the technology has continually grown. ECR ion source technology for multiply charged ion generation has rapidly advanced due to developments at several laboratories [109–111]. Reviews for high-charge-state ECR ion sources are given in Lyneis and Antaya [112] and in another chapter in this volume [113]. The principles of operation and history of ECR source developments for multiply charged ion beam generation are discussed in Jongen and Lyneis [114]; scaling laws for high-charge-state ion wave generation have been postulated by Geller [115]. A new ECR concept for multiply charged ion beam generation has also been proposed [116]. Taylor [117] reviews high-current, low-charge-state ECR ion sources. Computational studies have been utilized to design an ECR ion source based on a new magnetic-field geometry for multiply charged ion beam generation. The properties of ECR plasmas generated by low-frequency microwave power have been measured by a number of investigators, including Amemiya et al. [118] and Popov [119]. A major advantage of this source concept is that the bombarding and ionizing electron current is generated from the plasma itself, and there are no emissive electrodes which can erode as a result of ion bombardment and thus limit the lifetime of continuous operation.

3.4.1 Principles of an ECR and Microwave Ion Source

An ECR and microwave ion source comprises a multimode cavity that serves as the plasma generation and containment cavity. Magnetic mirror coils are situated at the inlet and ion-extraction ends of the cavity to confine the plasma in the axial direction. Multicusp, magnetic fields are usually used to assist in containing the plasma in the radial direction. A high-frequency cavity is positioned on the inlet side of the source, and high-frequency power of several gigahertz is used to produce a plasma from the gaseous-feed material that is metered into the source. The operating pressure in this so-called first stage of the source is $\sim 10^{-3}$ Torr. After generation, the plasma drifts down the axial magnetic field gradient into a second cavity of the source, where the electrons are resonantly excited by the high-frequency field at a frequency

$$\omega_{RF} \cong \omega_c = \frac{eB}{m_e}, \tag{3.97}$$

where ω_{RF} is the excitation frequency, ω_c is the electron cyclotron frequency, e is the magnitude of the electronic charge, B is the magnetic flux density, and m_e is the mass of the electron. Wherever the resonance condition of Equation (3.97) is satisfied, the electrons are stochastically heated, a small fraction can actually attain energies of a few to several kilo-electron volts, and they are thus capable of removing tightly bound electrons from heavy ions. The pressure in this so-called second stage of the ECR source is $\sim 10^{-6}$ Torr.

Ionization occurs principally by single-electron removal with contributions from multiple-electron removal and inner shell vacancy creation, which may result in Auger electron ejection. Single-electron-loss processes are believed to dominate so that multiple collisions are necessary to produce highly charged ions. Thus, it is necessary to make the product $n_e \tau_i$ as large as possible, where n_e is the electron density and τ_i is the containment time of the ion beam. Power can be coupled into the plasma until the plasma density reaches the critical value n_c, at which time the electron plasma ω_{pe}, excitation ω_{RF}, and electron cyclotron ω_c frequencies are equal, i.e., $\omega_{RF} = \omega_c = \omega_{pe}$. The plasma frequency is related to the critical plasma density through the following expression,

$$\omega_{pe} = \left(\frac{n_c e^2}{\varepsilon_0 m_e}\right)^{1/2}; \tag{3.98}$$

therefore, $n_c \propto \omega_{RF}^2$. However, for a fixed magnetic field design configuration, ω_{RF} and thus n_c are limited by the attainable magnetic field B in the multimode cavity of the source.

To increase the number of highly charged ions, multiple-stage sources (usually two stages) have been developed so that the ion containment time τ_i can be increased to values compatible with those required for production of higher-charge-state ions. Ion–ion collision recombination processes tend to lower the charge-state distribution at higher operating pressures, and, therefore, differential pumping is used so that lower pressures can be maintained from stage to stage. Ions diffuse along the axial field gradient and flow from stage to stage until they are extracted. The RF power generator usually has the capability of producing a few hundred watts to a few kilowatts of power. The power may be pulsed with typical widths of tens to hundreds of milliseconds at variable repetition rates or be continuous.

Electrons at the resonance frequency are excited essentially in a direction perpendicular to the axial magnetic field so that their energy perpendicular to the field is much greater than that parallel to the field, or $T_\perp \gg T_\parallel$. Because of the direction of excitation and the magnetic-mirror geometry, the electrons are very effectively trapped, and, therefore, it is possible to bombard ions continually with several amperes of electron current during their diffusion along the field. Ions arriving at the outlet end of the source diffuse through an aperture into a field region where they are extracted. Total current densities of a few amperes per square centimeter have been extracted from the single-stage source. While the source is perhaps the most efficient and best method for producing CW beams of highly charged ions for injection into high-energy acceleration devices, e.g., the RFQ linear accelerator and cyclotron, low-charge-state, single-stage ECR sources have also been developed for use in producing beams of rare isotopic elements [120,121].

According to Equation (3.98) the maximum plasma density or critical density that can be achieved under ECR conditions is limited by the plasma frequency ω_{pe}. In order to avoid cutoff at ω_{pe}, the power source frequency ω_{RF} should be chosen such that $\omega_{RF} > \omega_c$. However, the density limitation can be overcome by operation in the so-called "overdense" mode [122,123]. This means that a low-cost 2.45-GHz power source can be used to achieve densities far exceeding the theoretical ECR limit of $n_e = 7.45 \times 10^{10}/cm^3$. For example, densities of $n_e \geq 3 \times 10^{13}/cm^3$ have been achieved using a 2.45-GHz magnetron power source [124]. The magnetic field necessary to excite an "overdense" plasma, however, must be increased over that required to meet the ECR condition.

3.4.2 Low-Charge-State ECR and Microwave Ion Sources

Several low-charge-state microwave ion sources which operate in the overdense mode have been developed. Multiple-aperture sources have been developed for fusion-reactor-related research, modification of material surfaces, and SIMOX applications [125], as well as slit-type sources such as those described in Sakudo et al. [126], which are suitable for use

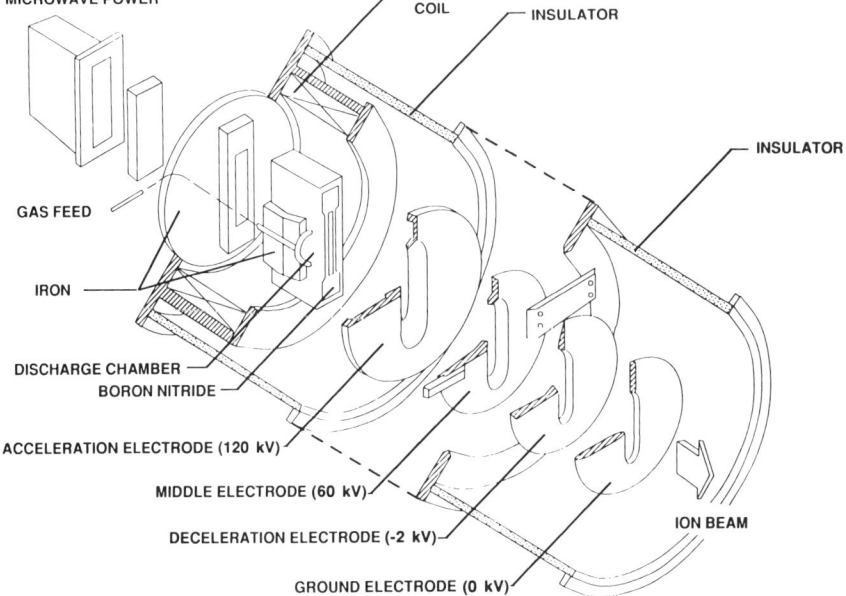

FIG. 32. Schematic drawing of the high-intensity, slit-aperture, microwave source described in Sakudo et al. [126].

in ion implantation or electromagnetic isotope separator applications. A schematic representation of the slit-type source of Sakudo et al. [126], developed for ion implantation applications, is shown in Figure 32. In this source, the discharge is initiated and sustained by microwave-induced excitation and ionization of the gaseous feed material. The frequency of the microwave generator is 2.45 GHz. Waveguide techniques are used to optimally transmit and couple several hundred watts of microwave power to the discharge volume, which is effective in producing a highly ionized plasma. A dielectric plate is utilized as a seal against atmospheric pressure. Analyzed 10-mA beams of P^+ from an exit aperture of 2×40 mm^2 have been observed. Because of the nature of the plasma-generation technique, the source has a very long lifetime. The principal problem which limits lifetime is backstreaming electrons which strike the dielectric plate located in the rear of the source and ultimately cause vacuum leaks. A comparison of the performance of this source with those of selected Freeman, Calutron, Calutron-Bernas, and microwave sources is made in Table III.

Oxygen is particularly deleterious to hot-cathode sources, the use of which may reduce the lifetime to a very few hours due to the chemical reactions between the oxygen feed gas and the hot Ta or W filaments traditionally used in such sources. The microwave or ECR ion source has an advantage over most conventional sources in that it lacks a filament, the component which limits the ultimate functional lifetime of the source. Thus, stability of operation with reactive gases can be easily achieved due to the microwave or ECR ion source's nonfilament structure while simultaneously they have the benefit of long lifetime. A few sources have been successfully developed for corrosive feed materials processing such as oxygen [127–129]. O^+ beams are important in the semiconducting materials industry for use in SIMOX technology. These sources have the capability of producing O^+ currents exceeding 100 mA as required for use in commercial ion-implantation devices. Figure 33 displays a source developed for this application which generates O^+ beams exceeding 138 mA at current densities exceeding 150 mA/cm².

A multicusp-field source which uses microwave plasma cathodes to supply electrons for ignition of the primary plasma in the cusp-field region of the source has been developed [130]. The plasma in the cathode region of the source is ignited by three rod coaxial-type antennae located on the back plate of the source which are driven at 2.45 GHz under ECR conditions. Total beams of Ar (exceeding 230 mA) and O (exceeding 130 mA) have been extracted from the source with a multiple-aperture emission diameter of 115 mm. The lifetime of the source, which is determined essentially by the lifetime of the antenna used to produce the microwave plasma, exceeds 100 hr.

TABLE III. Comparison of the Performances of Calutron, Calutron/Bernas, and Sakudo Ion Implantation Sources

	Calutron[a]		Calutron/Bernas[b]		Freeman[b]	Sakudo[c]
Ion species	B^+	P^+	B^+	P^+	B^+	P^+
Source gas	BCl_3	P^0	BF_3	P^0	BF_3	P^0
Extraction area (cm²)	3.2	3.2	1.2	1.2	0.8	0.8
Arc voltage (V)	100	80	100	89	100	NA
Arc current (A)	4.5	3.0	5.5	3.0	4.5	NA
Microwave power (W)	NA	NA	NA	NA	NA	300
Extraction voltage (kV)	35	35	20	20	35	30
Current density (mA/cm²)	65–100	65–100	44	35	44	50
Gas flow (atm cm³/min)	10–12	1–2	7	1	~7	0.1–0.5
Lifetime (hr)	60–80	60–100	40–50	43	40–50	≥100

[a] Oak Ridge National Laboratory, TN
[b] Eaton Corporation, Beverly, MA [84].
[c] Sakudo microwave.

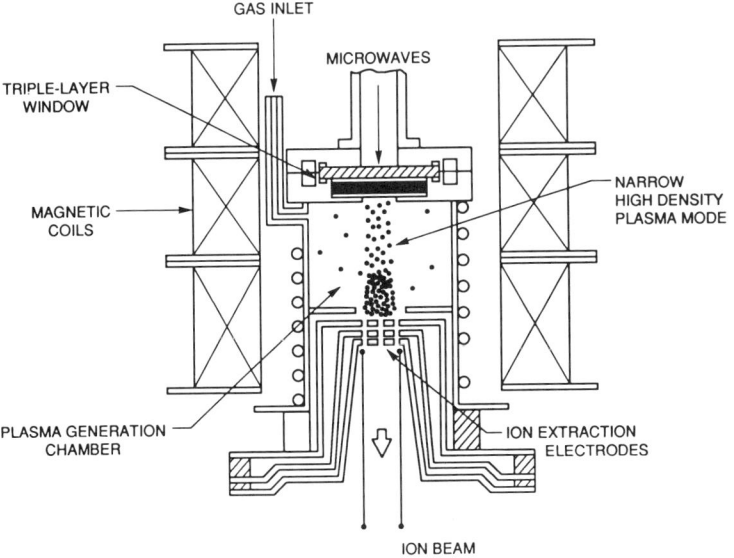

FIG. 33. Schematic diagram of the high-intensity, O^+, microwave source described in Torii et al. [127].

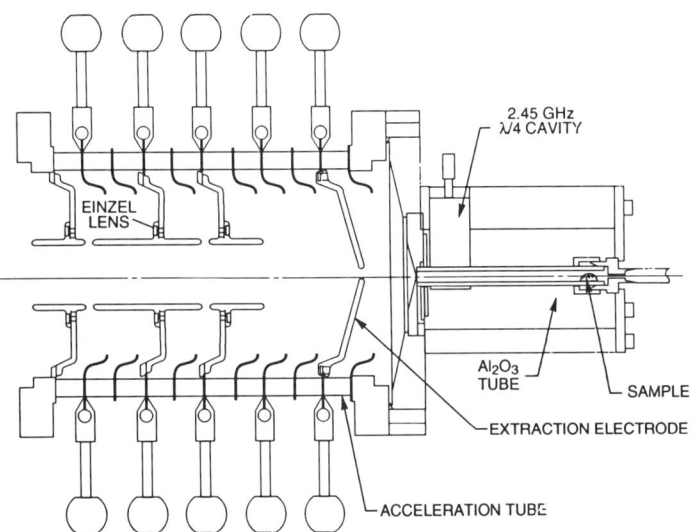

FIG. 34. Schematic diagram of a microwave ion source similar to the source described in Walther et al. [132].

A low-emittance, high-intensity ECR (microwave) source operating with a 2.45-GHz magnetron has been developed for producing high-proton-ratio H$^+$ beams for postacceleration with an RFQ injector [131]. Proton fractions up to 90% and total extracted beam densities up to 350 mA/cm^2 have been extracted from a 4-mm-diameter single-aperture source. The rms emittance of the source for a total beam intensity of 25 mA is reported to be 1.55 π mm · mrad (MeV)$^{1/2}$.

The compact microwave source, illustrated schematically in Figure 34, uses a quartz tube [132]; the plasma is produced with a 2.45-GHz magnetron power supply which couples power through the quartz tube. Both solid materials and gases have been generated with the source. The source produces hydrogen ion densities of 200 mA/cm^2 with an 80% proton fraction.

3.5 Vacuum-Arc Ion Sources

A simple and practical method of producing ions from solid materials, often used in mass spectrometry applications, employs a periodic low- or high-voltage-spark discharge between two conducting or semiconducting electrodes in a vacuum. A book which describes the theory and applications for both low- and high-voltage vacuum arcs has been written [133]. The results of early studies of both low- and high-voltage-spark discharges have been reported by Dempster [134] with primary emphasis on the low-voltage type. Additional information has been accumulated as a result of studies of low-voltage discharges by Venkatasubramanian and Duckworth [135] and by Wilson and Jamba [136]. The high-voltage-spark discharge has also been investigated by Saludze and Plyutto [137]; these authors have also reviewed previous work on the subject. Sources of each type are illustrated schematically in Figure 35.

3.5.1 The Low-Voltage Vacuum-Arc Discharge

The low-voltage discharge is produced by mechanically separating two current-carrying electrodes initially in contact, one of which contains the material of interest. During the mechanical interrupt, the contact area decreases and, consequently, the resistance of the interface increases, leading to a localized heating of a small volume of the material. The temperature of the material quickly rises to a high value, which results in vaporization of the electrodes. As the contacts are further separated and the resistance increases, the potential difference increases through the vapor, producing a highly ionized plasma if the potential difference is greater than the first ionization potential of the vaporized material. The

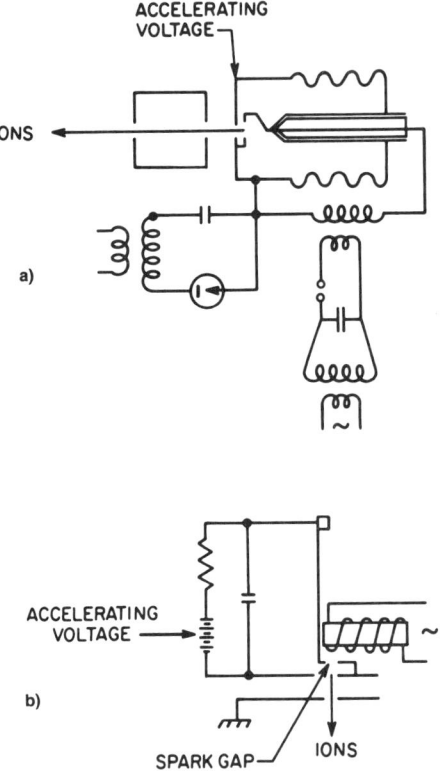

FIG. 35. Illustrations of spark-discharge ionization sources: (a) high-voltage spark-discharge arrangement and (b) low-voltage spark-discharge arrangement.

discharge may sustain itself for a long period of time if the correct conditions prevail. Such discharges principally produce singly charged ions.

3.5.2 The Triggered, High-Voltage, Vacuum-Arc Discharge

The high-voltage breakdown or so-called cold-cathode arc usually occurs on nonrefractory metals such as iron and copper. The cathode phenomena of this type of arc are different from those of the thermionic arc, in that the cathode "spot" consists of a considerable number of active areas which move about on the metallic surface. These active areas are often, in fact, complexes of emission sites that are in a continuous and rapid state of flux. In contrast with the thermionic arc, which is "explained" by the well-known equation for thermionic emission, there has been little general agreement on the physical mechanism responsible for the observed cold-cathode spots. On application of a high voltage to the cathode, the vapor necessary to support the arc is supplied by the multiplicity of highly mobile "cathode spots," which move about randomly over the negative electrode. The current density in each of these small spots can be very high, often on the order of 10^6 A/cm^2. Metallic vapor jets with velocities of up to 1000 m/sec originate at the cathode spots. In these jets, which are the principal source of vapor in the vacuum arc, one atom of metal can be removed for about every 10 electrons emitted.

The high-voltage discharge is initiated by applying a high voltage across a gap between two electrodes that exceeds the vacuum breakdown voltage. Because of the complexities involved in initiation and sustenance, a comprehensive theoretical basis for the breakdown vacuum arc discharge has not been formulated to date. However, the initiation mechanism can be explained in terms of field emission theory; the current density from a field emitting protrusion or needle can be represented by the well-known Fowler–Nordheim relation [138].

The current density from a field emitting tip as a function of electric field at the tip is given by the Fowler–Nordheim equation:

$$J = \frac{1.541 \times 10^{-2} E^2}{\varphi t^2(y)} \exp\left[\frac{-6.831 \times 10^9 \varphi^{3/2} v(y)}{E}\right] \quad (\text{A/m}^2)$$

$$y = \frac{3.795 \times 10^{-3} \sqrt{E}}{\varphi},$$

(3.99)

where J is the current density, E is the electric field strength, and φ is the work function of the metal. The expressions $v(y)$ and $t(y)$ are slowly varying, tabulated functions of work function and electric field and are usually regarded as constants.

The high-voltage-spark discharge, usually effected by applying 10–50 kV across the gap, produces a very highly ionized plasma at high densities (10^{15}–10^{16} ions/cm^3) with relatively high electron temperatures ($T = 10^{5\circ}$K). Such high-temperature plasmas produce a spectrum of highly ionized particles with charge states up to +4 or +5. Positive ion current densities of $j_+ = qn_+ (kT_i/M)^{1/2}$ of $10 \rightarrow 100$ A/cm^2 can be generated. Metals differ widely in their high-voltage breakdown characteristics in vacuum. This difference is determined by a number of parameters: the bulk properties of the pure metal itself; the degree of oxide formation; the maximum temperature permitted by the melting point and vapor pressure; the impurity content; and the microgeometry of the surface (smoothness–roughness).

Such sources have the advantage that ions can be produced from almost any element from which electrodes can be fabricated. Because of the nature of the discharge, electrode lifetime is very short and the discharge duration is also variable, having typically low duty cycles. The ion energy spread in the low-voltage source is significantly lower than that produced in the high-voltage source [134,135].

3.5.3 Types of Vacuum-Arc-Discharge Ion Source

Wilson and Jamba [136] have utilized a low-voltage, low-frequency-spark source to produce ions of the group I, II, III, V, and VI elements. Two electrodes of the desired element are employed, and the discharge is initiated by striking the arc at a third central electrode of either the same element or carbon using 50- to 60-Hz interrupted DC voltage of 45 V. Their data indicate that high-conductivity elements (such as Al, Cu, Ag, and Au) produce large C^+ components of current when they are used with carbon electrodes. The unwanted C^+ ion current can be eliminated by choosing identical electrodes containing the element of interest. Techniques for large-area (5–7 cm^2) emitting surfaces in high-voltage sources have also been developed, and ion current densities of 5–15 A/cm^2 and total ion currents of 50–200 A of H^+ and D^+ have been achieved in microsecond pulses [138].

The high-voltage-type source produces multiply charged ions from a triggered high-voltage-spark source as evidenced by the results of Zwally et al. [139]. Such sources have also been used extensively in high-resolution secondary-ion mass spectrometry [140,141].

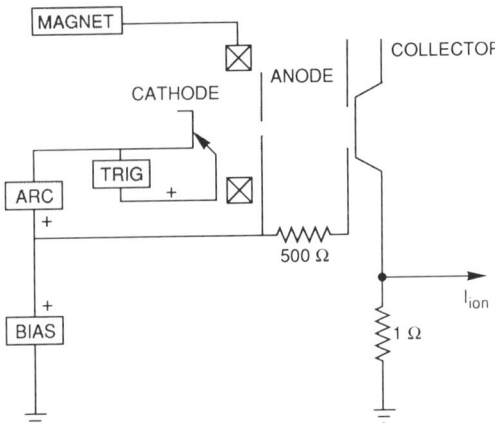

FIG. 36. Schematic representation of the trigger circuitry for the vacuum-arc sources described in Brown et al. [142–144].

Series of high-current-density, metal-vapor, vacuum-arc-type sources have been developed by Brown et al. [142–144]. The arc power supply for this source is a simple, low-impedance (~0.5 Ω) line with a pulse length of ~300 μsec, the line is charged to several hundred volts with a small DC power supply, and the output is connected to the anode and cathode terminals of the source. A schematic of the trigger circuitry is shown in Figure 36. When the triggering pulse is applied to the trigger electrode, the cathode–anode circuit is closed by the plasma and the arc proceeds throughout the duration of the pulse. The trigger pulse is generated by the discharge of a 0.1-μF capacitor of a few kilovolts (up to 5 kV) and is switched with a thyraton, through a step-up-isolation transformer, the output of which is connected trigger-to-cathode. The trigger voltage is connected positive with respect to the cathode, attains values between 10 and 20 kV, and lasts several microseconds. The ion source repetition rate is limited by the charging times for the arc and trigger circuitries and by average power constraints on various components. Typical repetition rates of a few pulses per second up to a maximum repetition rate of ~10 Hz are achieved. The usual impetus for triggering is from an annular

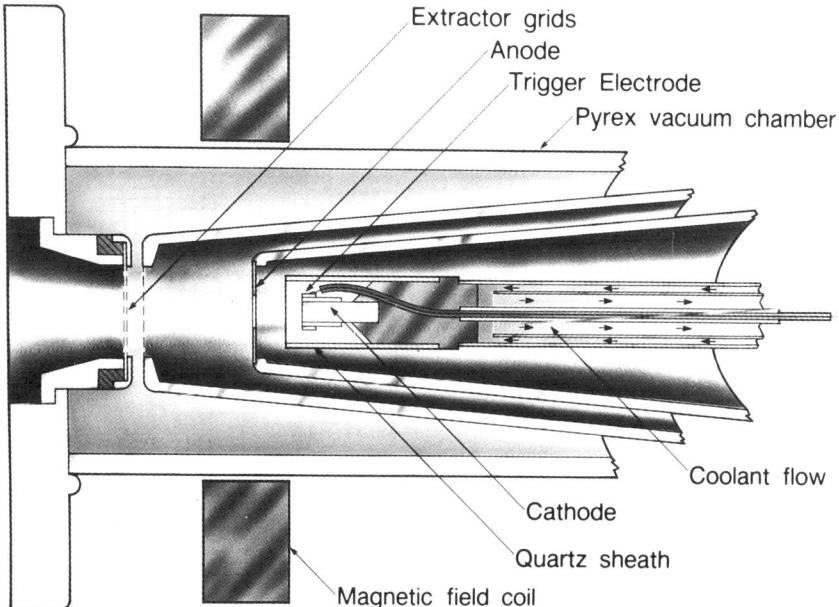

FIG. 37. Schematic diagram of a high-intensity, vacuum-arc source described in Brown et al. [143].

ring which surrounds the cathode. For hard materials such as niobium, >100,000 triggerings can be effected before failure. For softer materials, the number of triggerings possible before failure is considerably lower. The long pulse length (≥ 250 μsec) and low repetition rate (a few Hertz) of the source make it well suited for synchroton accelerator basic research applications; the source is also used for implantation and modification to material surfaces. The source has demonstrated intensities for many metal ions of several hundred milliamperes when the ions are accelerated with extraction voltages up to 100 kV. A schematic of a single-electrode, vacuum-arc source [143] is shown in Figure 37. A high-frequency-spark (1 MHz) source which produces C^{+3} and C^{+4} beams with an rms variation from pulse to pulse of less than 10% has been developed [145].

3.6 Sources Based on Field Ionization and Field Evaporation

3.6.1 Field Ionization

Theory of Field Ionization. Field electron emission refers to the transfer of electrons from the surface of a metal into the vacuum as a result of the action of very high electric fields at the surface ($\sim 10^7$–10^8 V/cm), which lowers the potential barrier so that electrons can leak out of the metal. Ionization of atoms or molecules adsorbed at or near a surface at high potential can also occur wherever the field polarity is reversed, and thus the technique can be used to form positive ions. The field ionization process usually requires electric fields on the order of 10^8 V/cm. To achieve surface fields of such high values at moderate potential differences, small-diameter spherically tipped or thin, hollow, needle-like electrodes are usually employed. The basic mechanisms of the two processes are essentially the same and involve the tunneling of electrons from the metal into the vacuum (field emission) or from the atom or molecule into the surface (field ionization). The tunneling mechanism can be explained only by quantum mechanical theory and has no classical analog. Analytical expressions for the probability of field emission have been derived by Fowler and Nordheim [138] for an abrupt potential step at the metal surface that neglects the image potential term $-e^2/4x$, where x is the distance of the electron from the surface. More rigorous theoretical calculations that include the image term have been made by Nordheim [146]. Summaries of work in field emission have been given by Good and Mueller [147], and a textbook on the subject has been written by Gomer [148].

The problem of field ionization can be treated in an analogous manner, but is slightly more difficult because the image potential term now must include both nuclear and electronic terms, as well as polarization effects.

An atom near a surface in a strong electric field is strongly polarized as a result of the superposition of the field and image effects. The potential barrier is lowered by the superposition of the strong electric field E and the induced image charges of the electron and nucleus V_{im}. The one-dimensional potential energy for the atom is, therefore, the sum of the applied and induced image potentials or

$$V(x) = -eEx + V_{im}, \qquad (3.100)$$

where e is the electronic charge and x is the distance of the atom from the surface. For the transfer process to occur, the electric field must raise the potential energy of the atomic electron at least to the Fermi level of the metal. The critical distance for which tunneling can occur is given approximately by

$$x_c \cong \frac{(eV_i - \phi)}{|E|}, \qquad (3.101)$$

where eV_i is the first ionization potential of the atom.

The probability of barrier penetration or tunneling, P_i, and hence ionization can be estimated by applying the Wentzel–Kramers–Brillouin (WKB) quantum theory approximation, provided realistic potential energy curves $V(x)$ are available. For the case in which the image potential in Equation (3.100) is assumed to be a coulomb potential of the form

$$V_{im} = \frac{Ze^2}{x}, \qquad (3.102)$$

where Z is the effective nuclear charge, P_i becomes

$$P_i \cong \exp\left\{-6.8 \times 10^7 \frac{(eV_i)^{3/2}}{|E|}\left(1 - 7.6 \times 10^{-4} Z^{1/2} \frac{E^{1/2}}{eV_i}\right)^{1/2}\right\}. \qquad (3.103)$$

The transition rate of barrier penetration is equal to the product of P_i and the frequency of arrival of electrons at the barrier, given by

$$P_i = v_e P_t, \qquad (3.104)$$

where

$$v_e = \tfrac{1}{2} v_e / r_e, \qquad (3.105)$$

so that the probability of ionization per unit time P_i is approximately given by $P_i = v_e P_t$. In Equation (3.105) v_e is the velocity of the electron and r_e is the effective orbital radius of the electron in an s state; for example, the frequency of arrival of the electron at the barrier is on the order of 10^{15}–$10^{16}/s$.

The time τ (lifetime) during which the transition takes place is given by

$$\tau = (v_e P_t)^{-1}. \tag{3.106}$$

The field ionization current can be calculated in principle by multiplying the arrival rate of atoms at the surface of the ionizer, dn_0/dt, and the probability of ionization, P_i. However, the mechanism of generation depends in a complex way on the ambient temperature, field strength, and polarizability of the atom or molecule. These factors affect the voltage–current characteristics of the GFIS which depend in a complex way on the temperature of the ionizer needle, the field strength, and the gas particle density in the vicinity of the tip of the needle.

The ion current field emitter is a complex function of the electric field and temperature of the ionizer needle and can best be approximated by consideration of certain limiting cases for the conditions which exist during operation of the sources. We shall consider the ion-field and high-field cases for purposes of illustration.

The Ion Current Low-Field Ionization. At low fields, for which the total rate of ionization is small compared with the arrival rate of atoms or molecules, the current is determined by the equilibrium number of particles near the ionizing tip multiplied by P_i, the probability of ionization factor (Equation (104)). For the case in which $T \neq T_t$, the steady-state or equilibrium population density of fully thermally accommodated atoms or molecules in the immediate vicinity of the needle n_t can be expressed as

$$n_t = n_0 \left(\frac{T}{T_t}\right)^{1/2} \exp[-V(\vec{E})/kT], \tag{3.107}$$

where n_0 is the concentration of atoms or molecules far from the needle tip and T and T_t are, respectively, the temperatures of the gas far from and at the tip of the ionizer.

Under the assumption that the field ionization process does not appreciably affect the number density of particles n_t, the total ion current can be estimated by integrating over the volume which extends from the emitter tip of radius r_t to the transition region which defines the boundary between n_t and n_0, or

$$i = 2\pi n_0 \int_{r_t + x_c}^{\infty} P_i (T/T_t)^{1/2} \exp[-V(r)/kT] r^2 \, dr. \tag{3.108}$$

For cases in which $-V(\vec{E}) \geq \frac{3}{2}kT$, the polarization energy $-V(r) = -V(\vec{E}) = \frac{1}{2}\alpha|\vec{E}|^2$ is dominant, and α is the polarizibillity of the atom or molecule, the current can be approximated from Equation (3.108) by the following relation:

SOURCES BASED ON FIELD IONIZATION AND FIELD EVAPORATION 137

$$i \simeq 2\pi n_0 r_t^2 x_c (T/T_t)^{1/2} P_i \exp[-\tfrac{1}{2}\alpha \vec{E}^2/kT]. \qquad (3.109)$$

Equation (3.109) is valid for cases in which the ionization rate is low compared with the arrival rate of particles striking the tip of the needle.

We now turn to supply-limited (high field) ionization. At very high electric fields, $\vec{E}$, all particles, approaching the tip are assumed to be ionized before reaching the tip. For this case, the current is determined by the supply function. Molecules passing near the ionizing tip are attracted by polarization forces [147]. For the case in which the apex of the ionizer is spherical with radius r_t, the field at distance r can be expressed by

$$\vec{E}(r) = \vec{E}(r_t)\left(\frac{r_t}{r}\right)^2. \qquad (3.110)$$

The potential energy of a polarizable particle with a permanent dipole moment in this central force field is then the sum of the potential energies given by

$$-V(\vec{E}) = p\vec{E} + \frac{1}{2}\alpha \vec{E}^2, \qquad (3.111)$$

where α is the polarizability of the atom or molecule and p is the dipole moment of the particle. In the one-dimensional centrifugal force field approximation, the potential energy can be expressed by

$$V'(\vec{E}) = V(\vec{E}) + \frac{M}{2}v^2\frac{\rho^2}{r^2} = V(\vec{E}) + \frac{3}{2}kT\frac{\rho^2}{r^2}, \qquad (3.112)$$

where ρ is the distance of closest approach to the apex of the ionizer if the field $\vec{E}$ were zero and v is the velocity of the particle of mass M far from the tip of the needle. When the particle strikes the needle tip at grazing incidence, from conservation of energy, the total energy of the particle E_T, taken as $\tfrac{3}{2}kT$, can be equated to the potential energy of the particle as expressed in Equation (3.112) so that E_T becomes

$$E_T = \tfrac{3}{2}kT = V(\vec{E}_0) + \tfrac{3}{2}kT\left(\frac{\rho}{r_t}\right)^2. \qquad (3.113)$$

From Equation (3.113), we can derive an expression for the effective tip surface area,

$$\sigma = \left(\frac{\rho}{r_t}\right)^2 = 1 - \frac{2}{3}V(\vec{E}_0)/kT, \qquad (3.114)$$

where $-V(\vec{E}_0)$ is a positive quantity. The effective area A_{eff} of the ionizer is increased by a factor of σ over the geometric area A_t or

138   SOURCES OF LOW-CHARGE-STATE POSITIVE-ION BEAMS

$$A_{\text{eff}} < A_t \sigma = 2\pi r_t^2 \sigma = 2\pi r_t^2 \left(1 - \frac{2}{3}\frac{V(\vec{E}_0)}{kT}\right). \quad (3.115)$$

The arrival rate of particles at the tip of the needle is enhanced by polarization forces by a factor of $\sigma(\Phi)$, where Φ is defined as $\Phi = \frac{1}{2}\alpha|\vec{E}|^2/kT$. For simple geometries, the enhancement factor $\sigma(\Phi)$ may be calculated analytically. Southon [149] derived an expression for a sphere, given by

$$\sigma(\Phi) \simeq \sqrt{\pi\Phi} \quad \text{for} \quad \Phi > 2, \quad (3.116)$$

whereas van Eekelen [150] derived an expression for the enhancement factor $\sigma(\Phi)$ for a hyperboloid which can be expressed according to

$$\sigma(\Phi) = \tfrac{1}{4}(\Phi + 2.7\Phi^{2/3} + 2.7\Phi^{1/3} + 1). \quad (3.117)$$

At very high fields, all particles arriving at the ionizer tip are ionized so the current is limited only by the rate of arrival. Since polarization forces attract atoms in the vicinity of the ionizer, the rate of arrival is faster than that associated with thermal motion. At very high fields, for which the ionization probability tends to be unity, the ion current is determined by the number of particles which strike the ionization zone which surrounds the emitter.

The high-field ion current i from the tip of a hemispherical ionizer of radius r_t and area $A = 2\pi r_t^2$ in the central field approximation for polarization forces is

$$i = \frac{n_0}{4}qvA_t = \frac{n_0}{4}qv2\pi r_t^2 \sigma. \quad (3.118)$$

By substituting $n_0 = P/kT$ and $v = \sqrt{(8/\pi)(kT/M)}$, where v is the velocity of the particle of mass M and charge q and P is the pressure into Equation (3.118), the equation becomes

$$i = 2\pi r_t^2 \frac{qP}{(2\pi MkT)^{1/2}}\sigma. \quad (3.119)$$

Since $-V(\vec{E}_0)$ is a positive quantity, the rate of arrival due to polarization forces is increased by the factor σ over that associated with thermal motion alone. The quantity σ may have values from 10 to 100 whenever the particles do not have permanent dipole moments.

Field Ionization Sources. The method of field ionization has been applied in mass spectrometry ion sources, in field ion microscopy, and as a source of intense ion beams for general use. Two source concepts that produce cylindrically symmetric ion beams are illustrated in Figures 38a and 38b. The hemispherically tipped ionizer (Figure 38a) produces ions from atoms that pass within the critical distance of the tip. At high

SOURCES BASED ON FIELD IONIZATION AND FIELD EVAPORATION 139

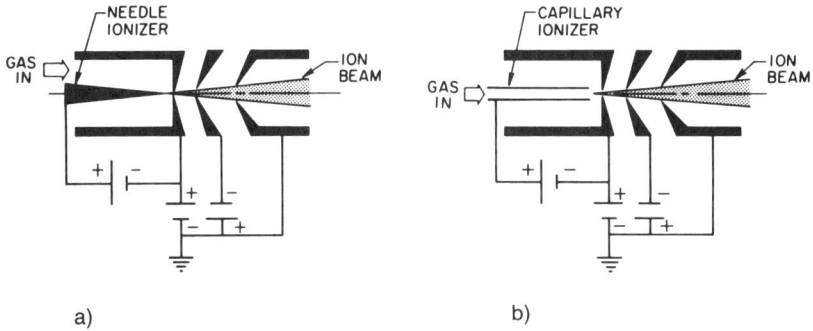

FIG. 38. Illustration of two types of field-ionization sources: (a) needle ionizer and (b) capillary ionizer.

pressures, the un-ionized particles in the path of the accelerated ion beam may act as scattering centers and thus degrade the beam quality. An attractive method, shown in Figure 38b, reduces this effect and thus may be more desirable. It consists of a thin capillary tube through which the gaseous or vaporous material to be ionized can be fed. The method was implemented by Hendricks [151] and later by a number of groups for microfocused beam applications [152–154]. The needle-type H_2 gas field ionizer (GFIS) source described in Allan *et al.* [155] is shown in Figure 39. In MacKenzie and Smith [156] there is a compilation of work related to the performance of several field-ionization sources.

3.6.2 Field Evaporation

Theory of Field Evaporation. A potential route to increased-brightness beams is through the use of field-ionization or liquid-metal (field-evaporation) ion sources. The liquid-metal ion source (LMIS) has the advantage of higher beam intensities and longer lifetimes and avoids the gas loading and scattering problems associated with the field-ionization source. In addition, the LMIS is simple in structure, is easy to operate, has a very long lifetime, and consumes very little power. Microfocused ion beams are now used in a wide range of fields. These include such diverse applications as microfabrication, ion implantation, materials characterization, and space propulsion. The extremely high brightnesses of the central cores of beams extracted from field-ionization and liquid-metal ion sources make them primary choices for such applications. As a consequence, the LMIS has been studied by many groups. A comprehensive bibliography, which contains ~1100 references to both field-ionization and liquid-metal ion sources and their applications, as well as several

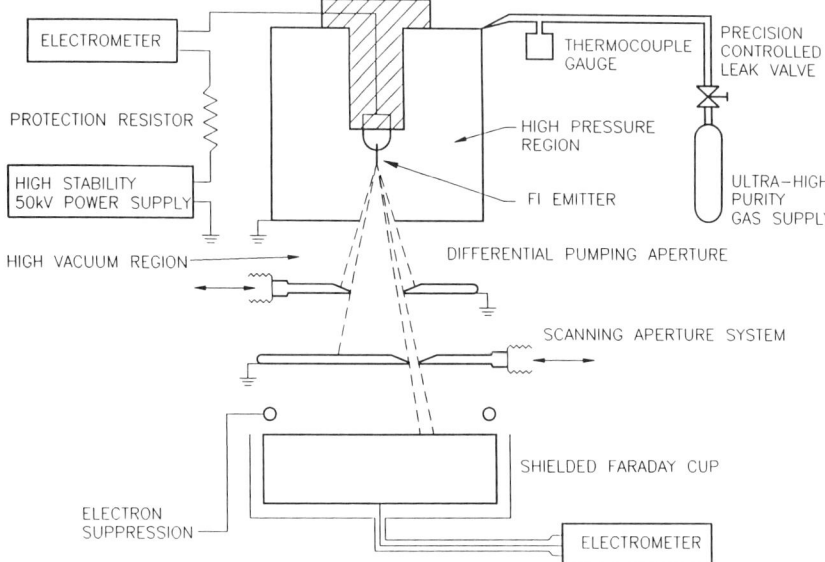

FIG. 39. Schematical representation of the room-temperature H_2 GFIS, equipped with an Ir needle-type ionizer, described in Allan et al. [155]; the experimental apparatus for measuring the angular divergence of ion beams generated in the source is also displayed.

review articles on these source types, has been compiled by MacKenzie and Smith [156], while a textbook which deals with the physics and technology of the LMIS with emphasis on the use of microfocused beams for microscopy and analysis, micromachining and deposition, microcircuit lithography, and ion implantation has been written by Prewett and Mair [157]. A review article which emphasizes the use of the LMIS for microfocused beam applications has been written by Orloff [158].

Although the precise ion formation mechanism in the LMIS is still the subject of controversy, generation of ion beams in this source is quite simple. A strong electric field, applied between a capillary or needle wetted with molten metal at the anode potential and an extraction aperture, pulls the metal up into a modified Taylor cone (apex angle, 98.6°) with a rounded apex [159]. The two principle types of liquid-metal ion sources are shown schematically in Figure 40. If the field is sufficiently strong (10^8 V/cm) at the surface of the molten cone, ion desorption will occur. The process whereby ions are removed from the tip region of the molten surface is often referred to as field evaporation. It is highly likely that field-ionization

SOURCES BASED ON FIELD IONIZATION AND FIELD EVAPORATION   141

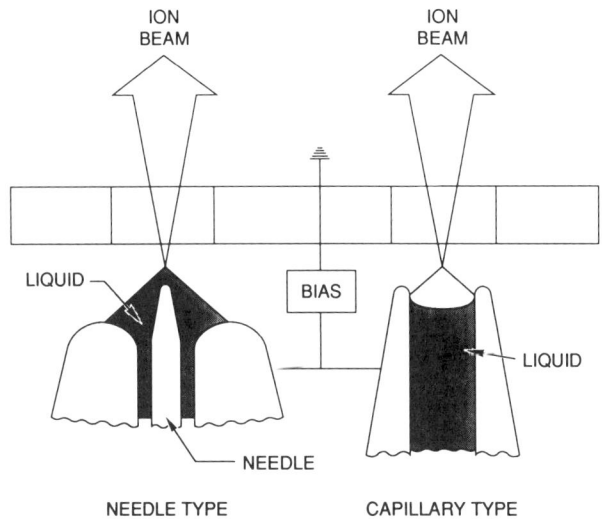

FIG. 40. Illustration of two principle types of liquid-metal ion sources: needle ionizer and capillary ionizer.

and field-evaporation processes both contribute to the ionziation process. Simulation studies of extractions from a needle-type source have been conducted using the computer code described in Whealton et al. [27] and shown in Figure 41. Aberrational effects are evident in the phase-space diagrams as displayed in Figure 42.

Because the field-ionization model has been discussed previously, only a brief description of the field-evaporation model is given here. For further details of models used to describe the process, the reader is referred to the review article by Marriot and Riviere [160] and the more comprehensive field-evaporation model described by Forbes [161]. Although the "image-hump" model has certainly not been confirmed, it is usually used to explain the field-evaporation–ionization process for the high-temperature case. The model views the field-evaporation process as the thermal escape of an ion over a one-dimensional barrier. The ion current density j which can be extracted through a barrier height of ΔH can be expressed as

$$j = qNA(\omega) \exp\left\{-\frac{\Delta H}{kT}\right\}, \qquad (3.120)$$

where q is the charge on the ion, N is the number of active sites per unit area, $A(\omega)$ is a factor which depends on the vibration frequency of the atoms in the potential well, k is Boltzmann's constant, and T is the absolute

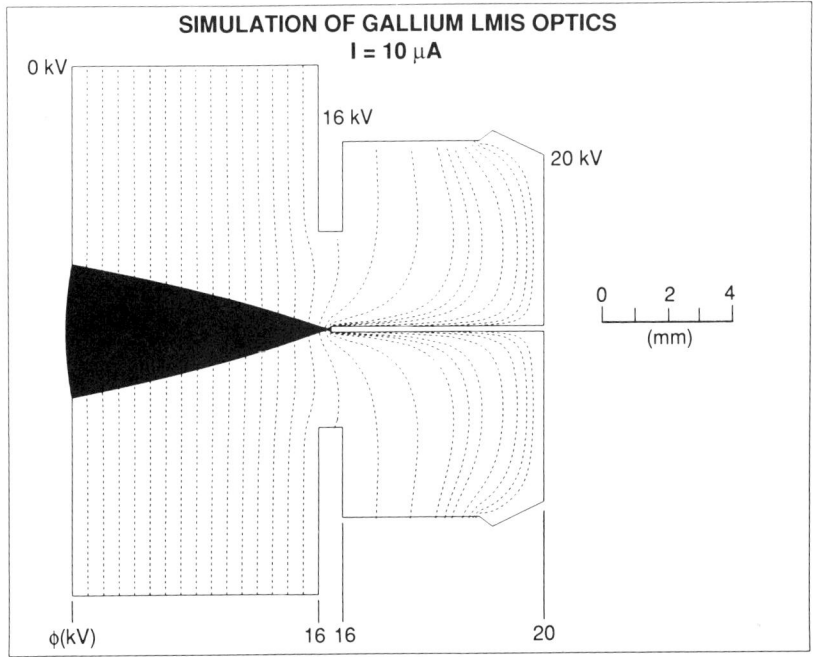

FIG. 41. Simulation of extraction from a Ga LMIS with the computer code described in Whealton *et al.* [27]. Ion beam intensity, 10 μA. Extraction voltage, 4 kV.

temperature of the surface. ΔH can be expressed according to the following simple relationship,

$$\Delta H = H_{ad} + V_{im} + I_q - \Phi - \left(\frac{q|\vec{E}|}{4\pi\varepsilon_0}\right)^{1/2}, \quad (3.121)$$

where H_{ad} is the heat of adsorption, V_{im} is the image potential energy between an ion and an image charge given by Equation (3.121). I_q is the ionization energy required to ionize the atoms to charge state q, Φ is the work function of the emission surface, $|\vec{E}|$ is the electric field strength at the surface, and ε_0 is the permittivity of free space. The last term in Equation (3.121) is attributable to the Schottky effect which lowers the escape barrier. In this way, beam intensities of a few to several tens of microamperes can be formed. The remarkable feature of this source type is that the area of ion emission is extremely small—having a radius of curvature of a few angstroms [162,163]. Thus, the LMIS is, in principle, the brightest of existing positive ion sources. The central cores of beams

SOURCES BASED ON FIELD IONIZATION AND FIELD EVAPORATION 143

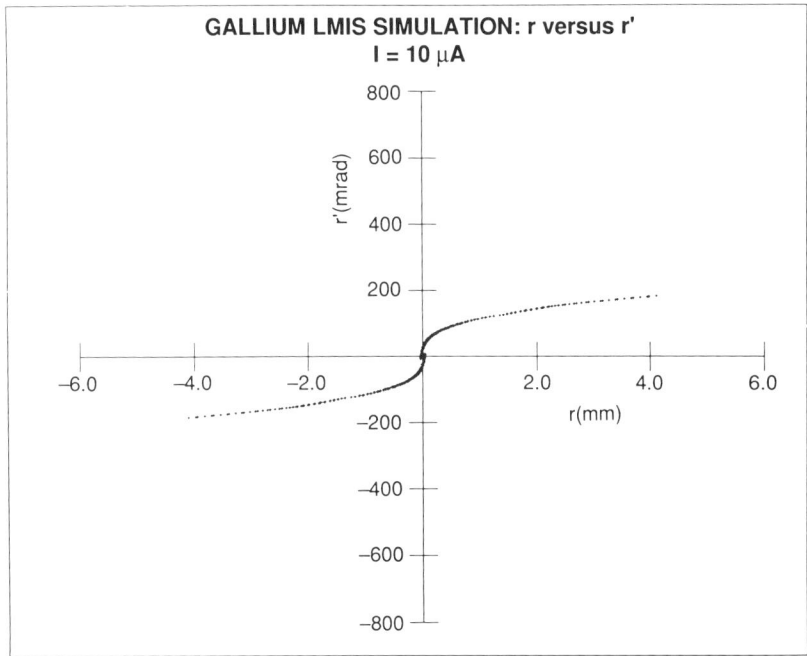

FIG. 42. Computed r' versus r emittance diagram for the Ga$^+$ ion beam shown in Figure 41. The aberrational effects associated with space charge and spherical aberrations, which increased during extraction, are obvious. The computer code described in Whealton et al. [27] was used in the computations.

extracted from this source are estimated to have brightnesses ~10^5 times those of conventional gaseous discharge ion sources [155]. The liquid-metal ion source offers a means for generating beams with central brightnesses of approximately 1×10^{16} amp/sr · m^2 [164–166]. LMIS beams can be focused into submicrometer images with current densities of a few A/cm^2 and thus offer a means for direct writing during ion implantation doping of semiconducting materials and for a number of ion beam lithography applications, including mask repair. However, in order to utilize the source for submicrometer imaging applications (e.g., microprobe surface analysis, ion implantation, microfabrication, and microetching), the angular divergence and, consequently, ion beam intensity must be severely limited. This is a consequence of the highly divergent characteristic of ion beams from this source type and the difficulty of accelerating and focusing beams without severely distorting their phase spaces through spherical aberrational effects (such effects increase as θ_0^3, where θ_0 is the half-angular divergence into the optical device). The half-angular divergence of LMIS beams ranges from a few degrees at low beam intensities to a few tens of degrees at higher intensities [167,168].

Inherent in LMIS beams are space-charge effects which cause increases in the angular and energy spreads; these, in turn, lead to increases of the phase space of the ion beam through spherical and chromatic aberrations. Energy spread attributable to space-charge influences near the region of ion emission may be caused by collective and individual particle coulomb–coulomb interactions (Boersch effect) [169]. Theoretical treatment of the coulomb-collision-dominated problem in a high-density laminar-flow electron beam predicts an energy spread, ΔE, proportional to the square root of the ion current I or $\Delta E \propto I^{1/2}$ [170]. For the case of a high-density, highly divergent beam dominated by coulomb–coulomb interactions, an energy spread, ΔE, proportional to $I^{2/3}$ or $\Delta E \propto I^{2/3}$ [171] is theoretically predicted. (Energy spreads, ΔE, in the LMIS are found to follow, approximately, the $I^{1/3}$-to-$I^{2/3}$ relation and have widths of several electron volts [167,171–173].) The energy spread ΔE associated with collective space-charge effects increases linearly with ion current of $\Delta E \propto I$. The resultant energy distribution associated with these effects is superposed on other distributions intrinsic to the source which are independent of ion current (e.g., energy spreads attributable to thermal and ion-formation processes). All additively affect the emittance through chromatic aberrational and dispersive effects in the extraction postacceleration, lensing, and momentum-analysis systems used in the experiments.

Types of Field Evaporation (Liquid-Metal) Ion Source. Liquid-metal ion sources usually rely on the wicking of a molten elemental or alloy metal along the outer surface of a metal needle [170–178], along the inner surface of a metal capillary tube [179,180], or through a sintered matrix [181–184]. Many examples of each of these source types have been developed as exemplified in MacKenzie and Smith [156]. The source may be composed of single or multiple emitters. The ion currents achievable from the LMIS increase linearly with the number of emitters [183,184]. A schematic representation of a needle-type LMIS equipped with provisions for inserting or withdrawing the needle in the molten liquid metal is shown in Figure 43 [185].

The currents which can be achieved with this source type range from a few to several hundred microamperes. The current which can be achieved can be enhanced by use of multiple emitters such as those described in Ishikawa *et al.* [183,184]. Clampitt [180] has reported high-brightness beams of Li^+, Cs^+, Sn^+, Ga^+, and Hg^+ from single emitters with beam intensities ranging from 500 to 700 μA. Current densities from these sources reached densities in excess of 10^4 A/cm² at extraction volt-

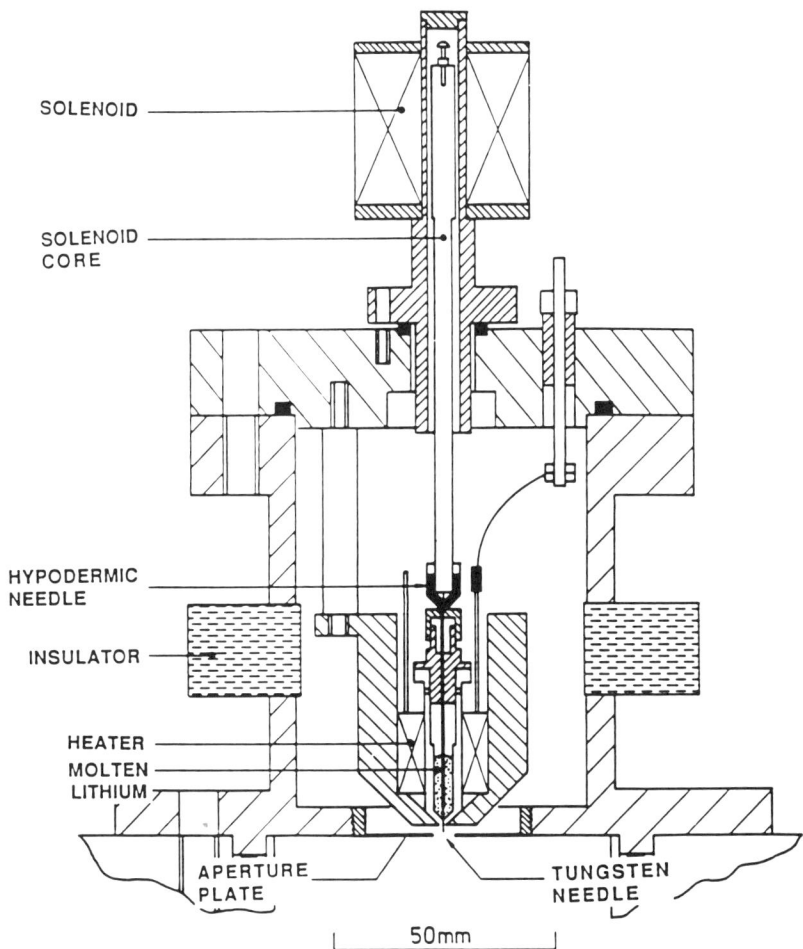

FIG. 43. Schematic drawing of a needle-type Li⁺ LMIS equipped with provisions for remote wetting or rewetting the needle [185].

ages of 3–10 kV. The operating characteristic of a liquid-metal source, therefore, is a strong function of the characteristics of the particular metal or alloy along the surface of the metal (tungsten) needle or capillary tube. Wagner [186] has characterized the metal flow along the surface of the needle or capillary in an LMIS with a hydrodynamic model which balances pressures due to surface tension and electrostatic forces at the end of the liquid cone. A number of sources which are predicated on ion beam generation from elemental metals have been developed [187–191]. The

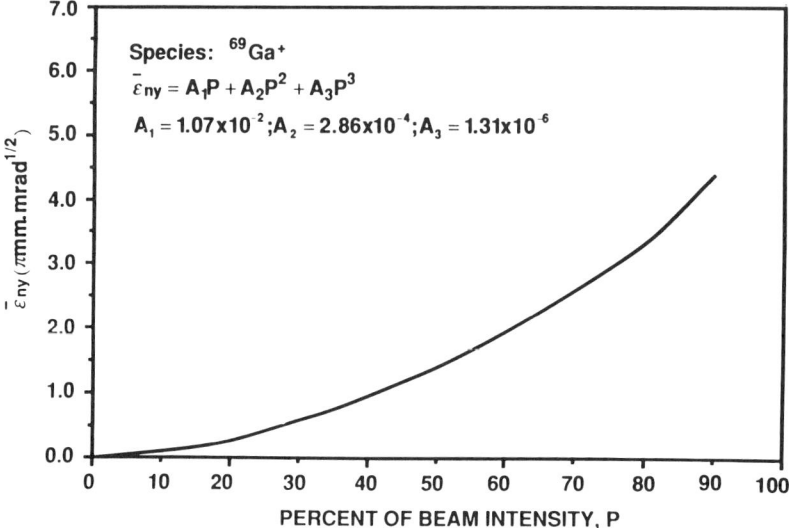

FIG. 44. Emittance as a function of the percentage of the beam contained within a particular emittance contour for a Ga LMIS [166,174,175].

TABLE IV. A Partial Listing of Metals and Alloys for Use in Liquid Metal Ion Sources

Species	Liquid element or alloy	Reference
Be	Au–Be, Au–Be, Au–Si–Be, Ga–Si–Be	187–190
B	Pt–B, Pd–Ni–B, Pt–Ni–B	187, 187, 191
Al	Al	192
Si	Au–Si, Ga–Si–Be	187, 191, 190
Ni	Pd–Ni–B	187
Cu	Cu	193
Zn	Zn	194
Ga	Ga	195
Ge	B–Pt–Au–Ge	196
As	Pd–Ni–B–As, As–Sn–Pb, As–Pt	187, 196, 197
Rb	Rb	198
Pd	Pd–Ni–B	187
In	In	195
Sn	Sn, As–Sn–Pb	193, 196
Sb	Sb–Pb–Au	191, 197
Cs	Cs	193
Pt	Pt–B	187, 193
Au	Au	199
Pb	Pb, Sb–Pb–Au	193, 196
Bi	Bi	200
U	U	201

versatility of the technique has been significantly extended by using the low-melting (eutectic) property of alloys. By using low-melting-point alloys, the LMIS can be used to generate beams of most of the metallic elements. The versatility of this source type in terms of species is illustrated in Table IV, which references a number of liquid-metal ion sources that have been developed to generate a variety of ion beams from both elemental and eutectic alloy materials [187–201]. A more comprehensive listing of field-emission and liquid-metal ion sources is given in MacKenzie and Smith [156]. Figure 44 displays the measured emittance of a gallium LMIS as a function of the percentage of the beam contained within a particular phase-space contour as measured by Alton and Read [166,174,175].

3.7 Surface Ionization and Thermal Emitter Ion Sources

3.7.1 Theory of Surface Ionization

Positive ion beams of several elements can be produced by surface ionization. An atom of low ionization potential (eV_i) can be ionized by contact with a surface of high work function that is hot enough to thermally evaporate ions. In this process, a valence electron of the adsorbed atom or molecule is lost to the surface upon evaporation or desorption of the positive ion. The most easily ionized materials for which the technique can be effected are ions of the alkali metals (Cs, Rb, K, Na, and Li). The alkaline-earth, rare-earth, and transuranic elements can also be formed as positive ions through the process, but with lower efficiencies. The elements that are most difficult to ionize are the elements with high heats of vaporization or high ionization potentials.

If we supply enough energy to ionize the atom, transferring the electron to the metal in the case of electropositive atom adsorption in the process, then the atomic and ionic potential curves are separated by amount $(eV_i - \phi)$, where eV_i is the first ionization potential of the electropositive atom and ϕ is the work function of the surface. If enough energy is supplied to evaporate the atom or ion from the hot surface, the probability for arrival at a position far from the metal in a given state depends on the magnitude of $(eV_i - \phi)$. For thermodynamic-equilibrium processes, the ratio of ions to neutrals that leave an ideal surface can be predicted from Langmuir–Saha surface-ionization theory. The probability of positive ion formation P_i, leaving a heated surface for the degree of ionization, is given by

$$P_i = \frac{\omega_+}{\omega_0}\left(\frac{1 - r_+}{1 - r_0}\right)\exp\left(\frac{\phi - eV_i}{kT}\right)\left[1 + \frac{\omega_+}{\omega_0}\left(\frac{1 - r_+}{1 - r_0}\right)\exp\left(\frac{\phi - eV_i}{kT}\right)\right]^{-1},$$
(3.122)

where r_+ and r_0 are the reflection coefficients of the positive and neutral particles at the surface, respectively, ω_+ and ω_0 are statistical weighting factors, T is the absolute temperature, and k is Boltzmann's constant. Optimum ionization efficiencies are obtained for high-work-function materials and low-ionization-potential atomic species. For elements for which $eV_i > \phi$, the process is much less efficient. For example, the work function for clean tungsten is about 4.6 eV, and the ionization potential for indium is 5.8 eV. Thus, in this case, the exponential term $(\phi - eV_i)$ in the Langmuir–Saha relation is negative, and, therefore, the probability of ionization is low. A particular technique that helps to improve the ionization efficiency is to incorporate an oxygen spray that is directed onto the ionization surface. This increases the work function of the emitting surface and, hence, the efficiency of ionization. The work function of a particular material may vary significantly with crystallographic orientation of the surface. A plot of the efficiency of ionization of a number of elements as a function of the ionization potential eV_i for the high-work-function iridium surface is shown in Figure 45. It is seen that the efficiency is relatively high for some atoms, but is low (10^{-2} or 10^{-3}) for atoms of In, Ca, Al, Ga, and Tl. The incident particles which are not ionized are evaporated as neutral atoms. Surface-ionization sources have been used widely to generate Cs^+ ion beams for sputtering and effecting low-work-function surfaces, from which negative ion yields can be enhanced for basic and applied research and SIMS applications.

3.7.2 Types of Surface-Ionization Sources

Ion sources based on the surface-ionization principle are generally characterized by a high degree of ion beam purity, low energy spread, and limited range of species capability. The ionization efficiency can be very high or low, depending on the ion–substrate combinations as evidenced by the Langmuir–Saha relationship (Equation (3.122)). However, the energy spread of the ion beam is characteristically very low and is on the order of thermal energies, $\sim 2kT$ ($\ll 1$ eV). Ion current densities of alkali metals ranging from 10^{-3} to above 10^{-2} and densities 10^{-6} to $\geq 10^{-5}$ A/cm^2 of the group IIIB elements have been produced by surface ionization.

Several types of surface-ionization sources have been used and are characterized by the means by which the atomic vapor is fed onto the ionizing surface and the method used to extract the ions (see, for example, Figure 46). Numbers of sources such as those described in Harrison *et al.* [202–213] have been developed. The most highly developed, and in many respects the simplest type of surface-ionization source, is illustrated in Figure 46a. The ionizer is a porous body made of sintered tungsten.

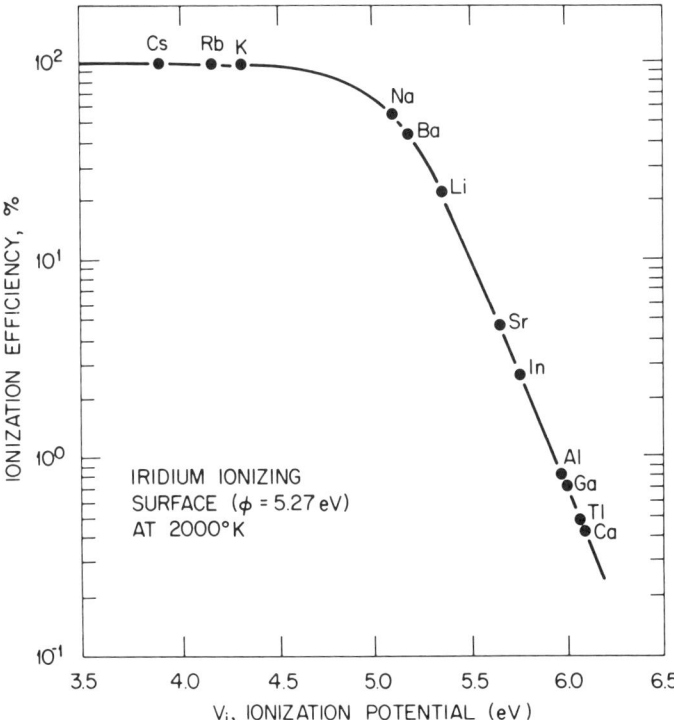

FIG. 45. Surface-ionization efficiencies of the group I, II, and III elements, evaporated from hot iridium metal, calculated from Eq. (3.122).

The atomic vapor is fed from the atomic oven through a sealed tube, and then the atoms are diffused through the porous tungsten ionizer, where they are ionized. The porous tungsten ionizer has a density ~80% of that of solid tungsten. At the front surface, a substantial fraction of the metal is ionized and extracted with an acceleration–deceleration type of electrode system. This type of ion source has also been highly developed to form ion-propulsion devices [207]. It is especially suitable for the alkali metals (Cs, Rb, K, and Na), which exhibit relatively high vapor pressures and relatively low critical temperatures—the temperature required to evaporate the ion. The properties of a porous ionizer may change at the high temperatures required for ionization of those metals that have high critical temperatures because of recrystallization processes that tend to change the porosity of the ionizer. Figure 47 illustrates the critical temperature effect for the ionization of cesium in a source of this type [213]. The

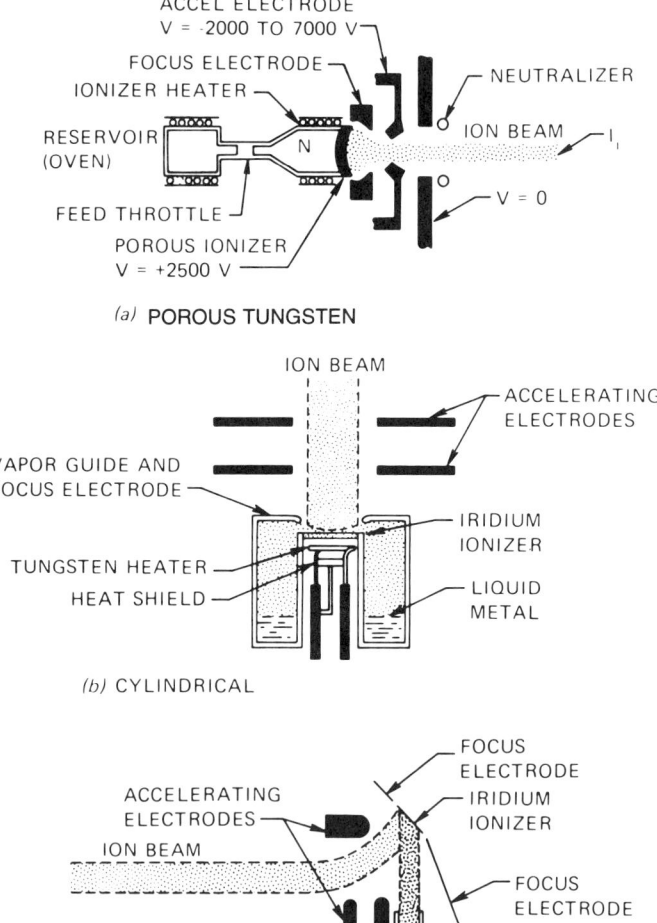

FIG. 46. Illustration of (a) porous-tungsten, (b) cylindrical, and (c) curvilinear-cesium-surface ionization sources.

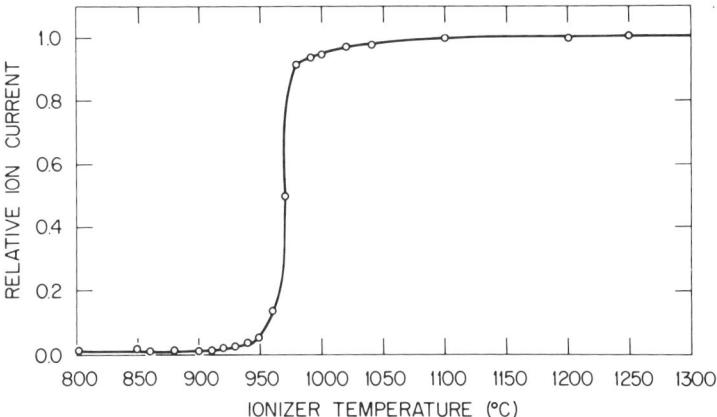

FIG. 47. Illustration of the critical temperature effect for ionization of cesium with a porous-tungsten ionizer [213].

perveance of the source and, consequently, cesium ion current is found to depend on the porosity of the tungsten ionizer. Cesium ion current versus extraction voltage, for various cesium oven temperatures, is shown in Figures 48 and 49 for ionizers with porosities $\rho = 0.7\, \rho_0$ and $\rho = 0.8\, \rho_0$, respectively, where ρ_0 is the theoretical density of tungsten; these data were derived from the source described in Alton [213]. Examples of surface ionization sources which use porous tungsten ionizers that have been developed for generation of Li^+, Na^+, and In^+ ions are reviewed with several other types of sources in Daley et al. [203,212].

An alternative to porous-tungsten-ionizer-type sources is to directly ionize a volatile vapor as it comes in contact with a hot, high-work-function surface [see, for example, Figures 46b and 46c of References 202–213]. An example of this source type is shown in Figure 50 [214]. The source utilizes a spherical-geometry ionizer which is designed to focus a Cs beam through a small emission aperture. The ionizer is customarily operated at a fixed temperature of $\sim 1100°C$ so that the positive ion beam intensity is governed by the rate of arrival of neutral vapor at the ionizer surface and the rate of evaporation of the ions formed on the ionizer surface. The number of neutral particles dN striking the ionizer per unit area dA and per unit time dt is given by the familiar relation

$$\frac{dN}{dAdt} = \frac{n\bar{v}}{4}, \qquad (3.123)$$

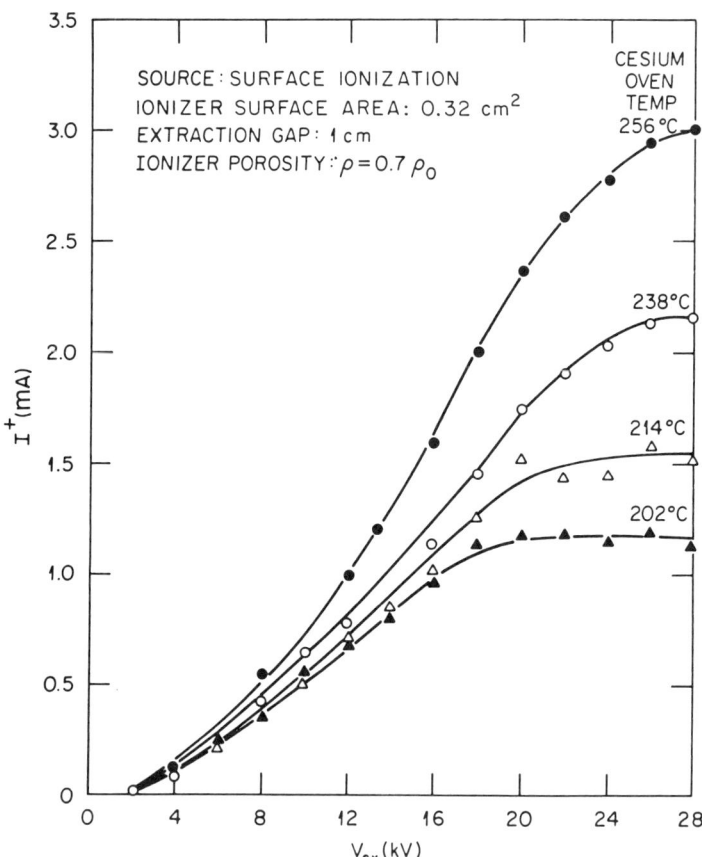

FIG. 48. Cesium ion beam intensity I^+ versus extraction V_{ex} for a cesium–porous-tungsten ion source. Porosity, $\rho = 0.7\,\rho_0$ [213].

where $\bar{v}$ is the average velocity of a cesium atom within the cesium reservoir/ionizer tube chamber at temperature T and n is the number of particles per unit volume within the chamber.

Positive ion production in this source is governed by the rate at which cesium atoms strike the ionizer surface. This rate is governed by the vapor pressure of cesium in the reservoir of the cesium oven. The cesium ion beam intensity is governed by the flow rate of cesium vapor, which is determined by the temperature of the cesium oven. The ion optics of the source are displayed in Figure 51, while comparisons between the

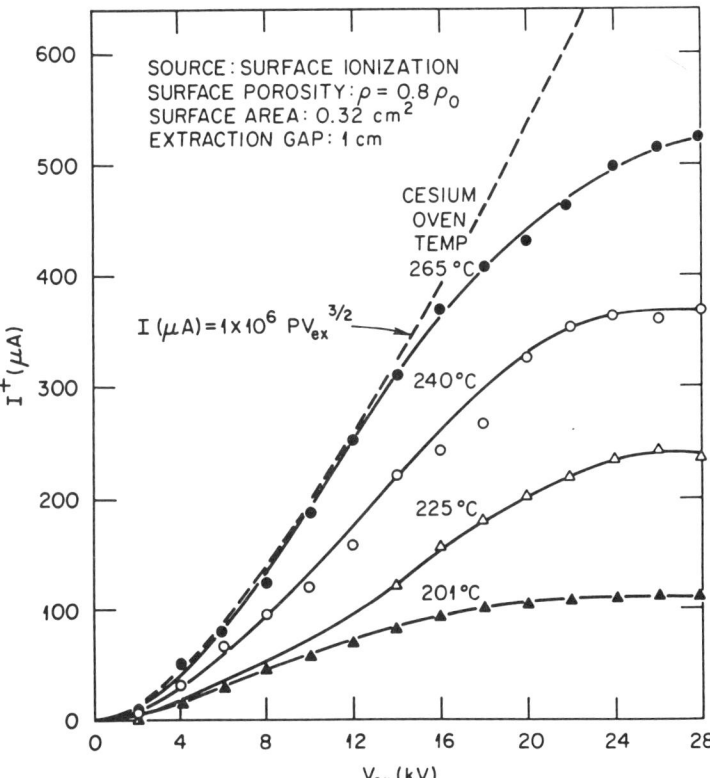

FIG. 49. Cesium ion beam intensity I^+ versus extraction voltage V_{ex} for a porous-tungsten-type cesium-surface-ionization source. Porosity, $\rho = 0.8\,\rho_0$ [213].

theoretical space-charge-limited cesium ion current versus extraction voltage and experimentally measured values are shown in Figure 52.

Ions of Al, Ga, and Tl require high ionizer temperatures, and the ionization process generally exhibits low efficiencies. A solid ionizer is desirable for these materials with the neutral vapor directed away from the target. A curvilinear-type source has been developed by Jamba [206], adapted from the original work of Kino [207]. In this source, the atoms to be ionized are directed from an atomic oven onto the ionizer surface, which is resistively heated. The ions are extracted in a curvilinear path by a relatively complex arrangement of electrodes. In this way, the neutral atoms evaporated from the ionizer can be directed toward a cold surface. The source was designed especially to produce a rectangular beam with

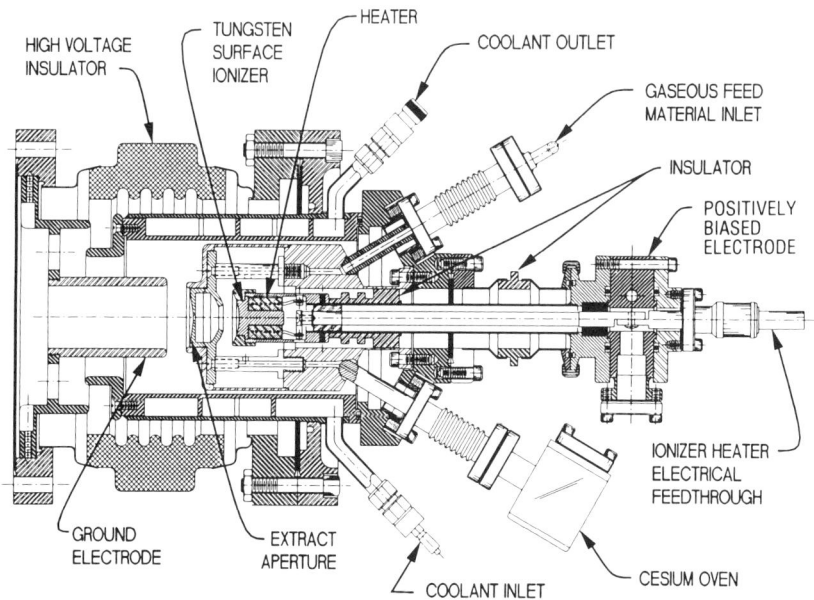

POSITIVE SURFACE IONIZATION SOURCE

FIG. 50. Schematic drawing of the self-extraction, spherical-geometry, positive- or negative-surface-ionization source. For positive ion generation, the ionizer is chosen to be Mo, Ta, or W, whereas, for negative ion generation, the ionizer is chosen to be LaB_6 [214].

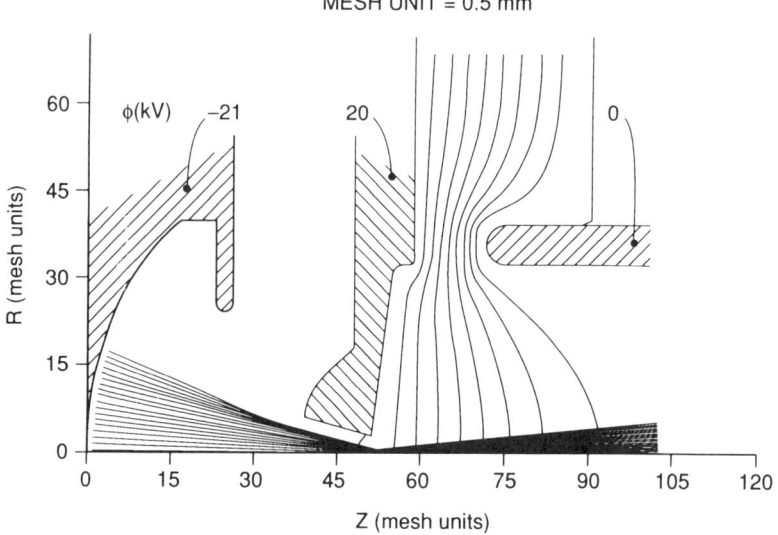

FIG. 51. Ion optics of the spherical-geometry surface ionization source for the space-charge-limited flow of Cs^+ [214].

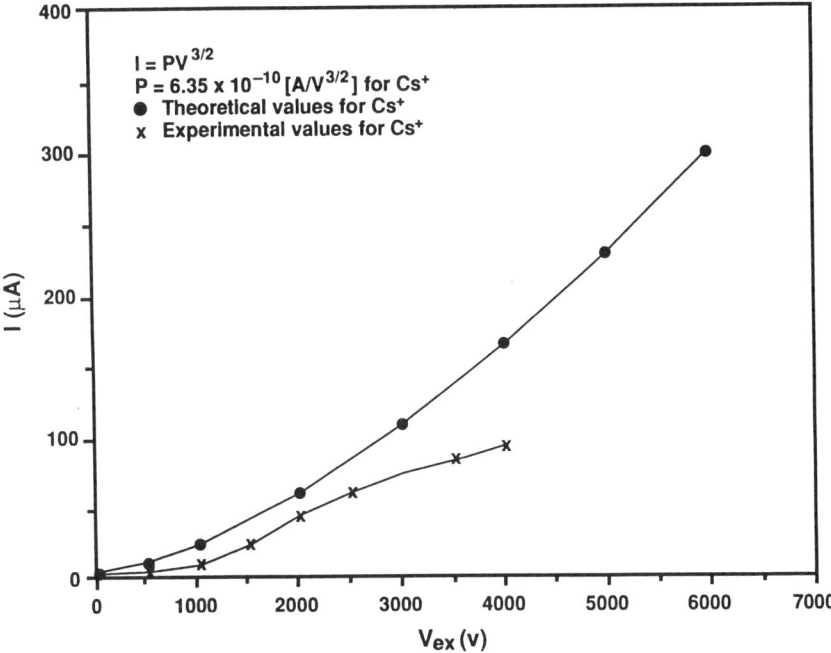

FIG. 52. Comparison of observed as well as theoretical and predicted space-charge-beam-limited Cs^+ intensities as a function of ionizer voltage [214].

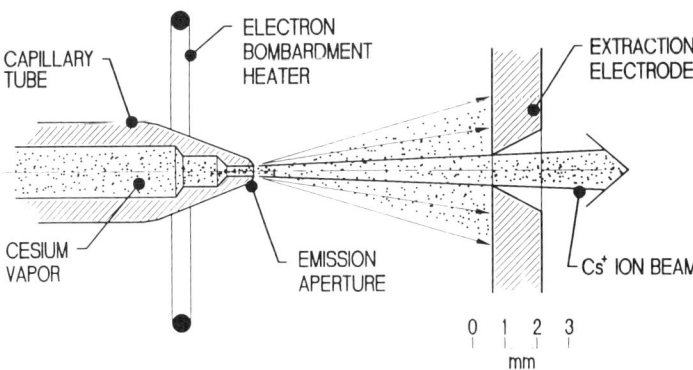

FIG. 53. Schematic representation of the capillary-type cesium-surface-ionization source Liebl and Senftinger [216].

parallel ion trajectories and with uniform current density. The Tl^+ surface-ionization source described in Dyer and Robertson [215] uses an electron-bombarded Ir ionizer to ionize Tl vapor, which impinges on the surface of the ionizer from an independently heated reservoir.

Raiko *et al.* have used surface ionization for the production of very-high-intensity beams of potassium and rubidium in an electromagnetic isotope separator with a nickel ionizing surface that has a high work function (5.03 eV) [204]. In the source, the alkali vapor from an oven is passed between a heated nickel surface and the front face of the ionization chamber and extracted through an aperture.

A capillary-type cesium ion source, which is especially well suited for microfocused beam applications such as micro-SIMS, has been developed [216]. The source, shown in Figure 53, uses a heated tantalum tube (~1100°C) through which cesium vapor is transported from an independently heated reservoir. Cesium vapor emitted from the hot tantalum capillary (0.1 mm) is surface ionized, and the resulting Cs^+ ions are extracted by impression of a high-electric field between the capillary tip and the electrode at ground potential. The brightness of the source is estimated to be 80 $A/cm^2 \cdot sr$ at a beam energy of 4 kV. This correlates to an ion emission density from the orifice of ~380 mA/cm^2. The source produces

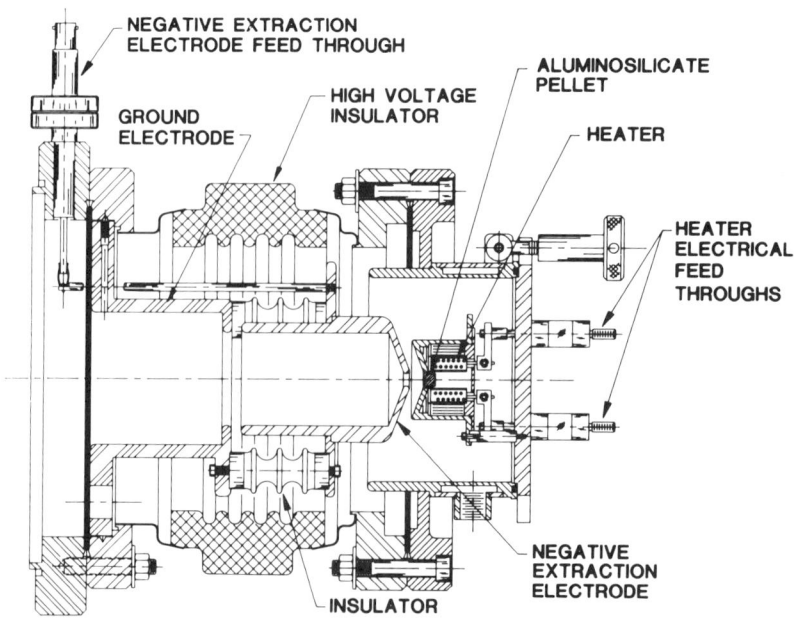

FIG. 54. Aluminosilicate ion source for the generation of Li^+, Na^+, K^+, Rb^+, and Cs^+ ion beams [225].

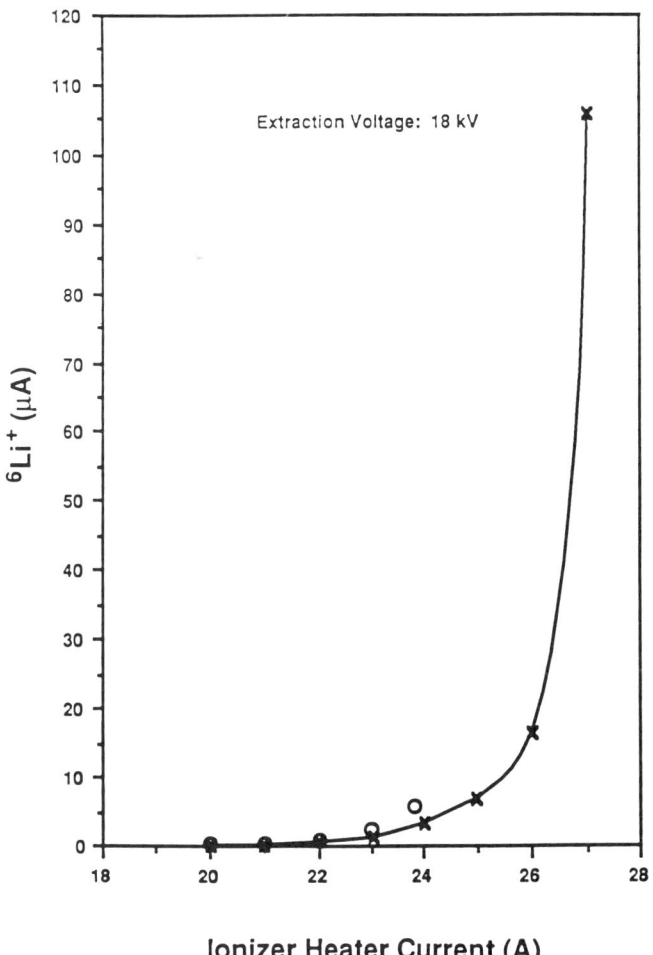

Ionizer Heater Current (A)

FIG. 55. Li$^+$ ion beam intensity versus ionizer heater current [225].

Cs$^+$ ion beams of 30 μA within a solid angle of $\sim$ 0.5 sr for periods of time exceeding six weeks.

3.7.3 Thermal Emitter Ion Sources

The sources described earlier are of the conventional type, in which surface ionization is effected by a hot metal surface. An alternative method for producing high currents of alkali metal ions is by heating certain chemical compounds. This is one of the oldest methods of producing such ion beams [217–223]. For lithium ions, β-eucryptite (lithium aluminum silicate) can be used. This is a naturally occurring compound, but for ion source requirements, it is normally synthesized by heating alumina and silica with lithium carbonate at around 1450°C,

$$Li_2CO_3 + 2SiO_2 + Al_2O_3 \rightarrow Li_2O, Al_2O_3, 2SiO_2 + CO_2. \quad (3.124)$$

The compound can be melted onto a high-work-function material such as platinum, which is then used as the ion-emitting surface. β-Eucryptite Li$^+$ currents of several milliamperes have been achieved from ions using a low extraction potential (9 kV) [218].

Other alkali metals can be ionized in a similar way using analogous and other aluminosilicate compounds. A source which produces 100-μA cesium beams for up to 50 hr from a cesium oxide–alumina–ferric oxide mixture has been developed [220]. Crystalline aluminosilicate compounds containing group IA elements (Li, Na, K, Rb, and Cs) will emit singly charged ions of these elements when heated to $\sim 1000°C$ [223,224]. Sources based on this principle can be very simple as evidenced by the Li$^+$ source shown in Figure 54 [225]. The ion beam intensity versus heater current is shown in Figure 55. For low beam intensities (low operational temperatures), sources of this type, formed from thin layers of the aluminosilicate material coated on the emission surface, can run for 100 to 200 hr before depletion of the particular group IA element. The lifetimes of Cs$^+$ sources have been extended to greater than 2000 hr by forming the aluminosilicate in a pellet and then drifting the cesium ions forward to the emission surface by placing a bias voltage across the pellet [226]. This technique can be applied as a means of increasing the lifetime of aluminosilicate pellets containing other members of the group IA elements, as well. A schematic of the Cs$^+$ ion source based on this principle is shown in Figure 56.

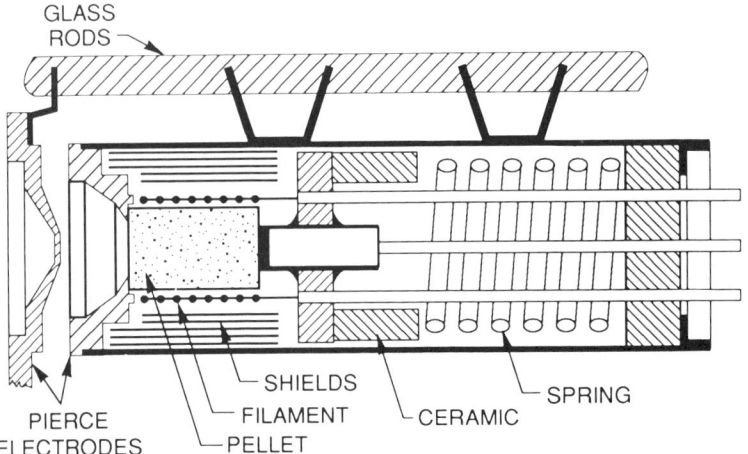

FIG. 56. Schematic illustration of a long-lifetime aluminosilicate ion source with provisions for drifting the Cs$^+$ to the extraction surface of the source [226].

3.8 Thermal Ionization Sources

3.8.1 Theory of Thermal Ionization

At high temperatures, collisions between gas particles may produce ionization provided that their relative energies exceed the first ionization of the atoms or molecules that make up the gas. For the case of a monoatomic gas, a fraction of the total number of gas particles will be in various stages of ionization at thermal equilibrium. Ionization equilibrium is a particular case of chemical equilibrium corresponding to a series of ionization reactions, symbolically written as

$$A_0 = A_1 + e; \quad A_1 = A_2 + e, \quad (3.125)$$

where A_0 denotes a neutral atom, A_1 and A_2 denote, respectively, singly and doubly ionized atoms, and e is the electron removed in the collision. For such reactions, the law of mass action can be used to derive an expression for the ionization efficiency as a function of temperature and pressure [227]. The ionization efficiency for an atom with the first ionization potential I_p can be expressed as

$$\eta = \left[1 + p \frac{g_0}{2g_i} \left(\frac{2\pi\hbar^2}{m}\right)^{3/2} \frac{\exp(I_p/kT)}{(kT)^{5/2}}\right]^{-1/2}, \quad (3.126)$$

where m is the mass of the electron, T is the temperature, I_p is the first ionization potential of the atom, g_i gives the respective statistical weights of the electron, atom, or ions, and $\hbar = h/2\pi$, where h is Planck's constant. For electrons $g = 2$ and for atoms or ions $g = (2L + 1)(2S + 1)$, where L and S are the orbital and spin angular momentum quantum numbers of the atom or ion.

This expression determines the degree of ionization as a function of pressure and temperature. In general, the process of thermal ionization is not a practical means of producing ions in the laboratory, but accounts for ionization processes in the stars. However, the temperature need not be exceedingly high for a reasonable degree of ionization for easily ionized elements. Figure 57 illustrates the degree to which atomic hydrogen, mercury, calcium, and cesium are ionized as a function of temperature at $p = 1 \times 10^{-3}$ and 1×10^{-2} Torr. The first reported applications of this technique are described in Beyer et al. [228] and Johnson et al. [229]. Such sources are relatively efficient for elements with ionization potentials of $I_p \leq {\sim}7$ eV.

3.8.2 Types of Thermal Ionization Sources

Metallic ions may also be produced by heating metal oxides or halides to very high temperatures in a high-work-function crucible. A source of

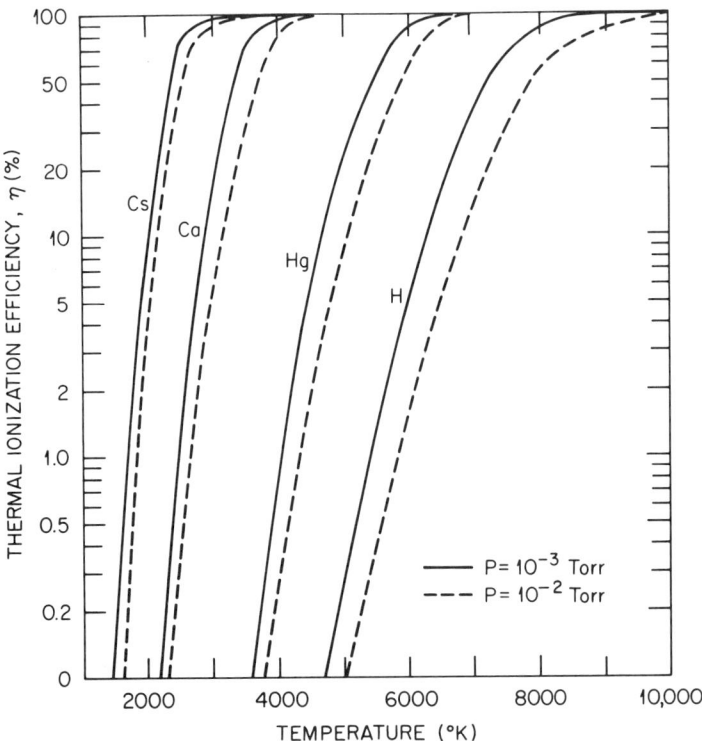

FIG. 57. Thermal ionization efficiencies η for Cs, Ca, Hg, and H, calculated from Eq. (3.126).

this type which functions by heating a tungsten crucible containing one of the halides of the rare-earth metals to ~2800°C has been developed [229]. Thermal dissociation of the halide releases elemental rare-earth atoms that have many collisions with the crucible walls before being emitted. A schematic diagram of the source is displayed in Figure 58. The source has been used to generate a wide variety of rare-earth species, with material utilization efficiencies greater than 70% recorded for europium and samarium. Table V compiles ionization efficiencies for a number of elements, including Ca, Sr, Nb, and La; all of the lanthanides; and U, Pu, and Cm of the actinides as measured by use of the thermal ionization source described in Johnson et al. [229]. This source type can be used for a variety of elements, as indicated by the results described in Johnson et al. [229] and Kirchner [230]. Because of the high thermal temperatures involved, perhaps both surface and thermal ionization processes may be involved in the production of ions from this source. Because of its chemical

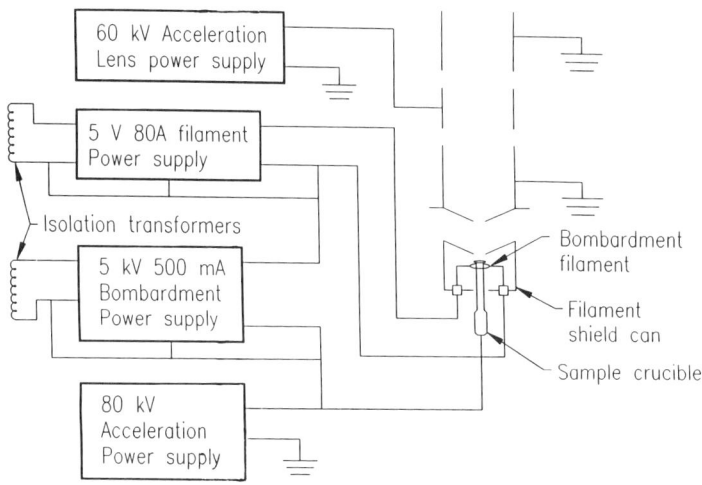

FIG. 58. Schematic drawing of the power supply arrangement for the thermal ionization source described in Johnson et al. [229].

TABLE V. Thermal Ionization Efficiencies for Various Elements[a]

Sample		Yield data (% of initial charge)			Number of runs
Element no.	Form	Low	High	Average	
20	CaO		13.2	13.2	1
38	$SrCO_3$	20.4	74.0	44.1	9
39	Y_2O_3	4.6	27.5	16.0	2
41	Nb_2O_5	0.12	15.0	2.31	72
57	La_2O_3	10.0	41.0	24.6	6
58	CeO_2	—	36.0	36.0	1
59	Pr_6O_{11}	21.4	40.3	33.0	3
60	Nd_2O_3	19.9	61.0	42.9	6
61	Pm_2O_3	50.5	93.5	75.3	3
62	Sm_2O_3	57.0	71.5	65.9	10
63	Eu_2O_3	13.1	73.0	45.8	105
64	Gd_2O_3	18.0	28.0	23.0	2
65	Tb_2O_3	17.4	51.8	28.8	7
66	Dy_2O_3	33.3	51.8	43.8	3
67	Ho_2O_3	26.6	44.4	32.5	4
68	Er_2O_3	28.1	38.1	32.6	3
69	Tm_2O_3	20.6	37.6	32.4	5
70	Yb_2O_3	8.4	23.0	14.9	4
71	Lu_2O_3	14.5	39.2	19.5	8
92	U_3O_8	10.8	20.2	15.1	11
94	PuO_2	7.9	24.4	16.2	2
96	CmO_2	1.0	14.8	8.8	10

[a] From Ref. 229.

selectivity and efficiency for ionization, the source has been utilized for ionization of short-lived nuclei created "on line" by neutron activation or for light ion transmutations of target materials [231].

Acknowledgments

The author expresses his gratitude to Ms. Jeanette McBride for typing and to C. Havener for helping edit the manuscript.

References

1. G. D. Alton, *Nucl. Instrum. Methods* **189**, 15 (1981).
2. G. D. Alton, in *Applied Atomic Collisions Physics* (H. W. Massey, ed.), Vol. 4, Chapter 2. Academic Press, New York, 1983.
3. G. D. Alton, *Nucl. Instrum. Methods* **B73**, 221 (1993).
4. A. T. Forrester, *Large Beams: Fundamentals of Generation and Propagation*. Wiley, New York, 1988.
5. I. G. Brown, ed., *The Physics and Technology of Ion Sources*. Wiley, New York, 1989.
6. G. Doucas, H. R. Mck. Hyder, and A. B. Knox, *Nucl. Instrum. Methods* **124**, 11 (1975).
7. G. D. Alton and J. W. McConnell, *Nucl. Instrum. Methods* **A268**, 445 (1983).
8. J. H. Billen, *Rev. Sci. Instrum.* **46**, 33 (1975).
9. P. W. Allison, J. B. Sherman, and D. B. Holtkamp, *IEEE Trans. Nucl. Sci.* **NS-30**(4), 2204 (1983).
10. G. D. Alton and R. W. Sayer, *Phys. Rev. B* **41**, 1770 (1990).
11. T. R. Walsh, *J. Nucl. Energy Part C* **4**, 53 (1962).
12. D. Bohm, in *Characteristics of Electrical Discharges in Gases* (A. Guthrie and R. K. Wakerling, eds.), Chapter 4, p. 87. McGraw-Hill, New York, 1949.
13. I. Langmuir, *Phys. Rev.* **33**, 954 (1929).
14. D. Bohm, E. H. S. Burhop, and H. S. W. Massey, in *Characteristics of Electrical Discharges in Gases* (A. Guthrie and R. K. Wakerling, eds.), Chapter 2, p. 65. McGraw-Hill, New York, 1949.
15. I. Langmuir, *Phys. Rev.* **2**, 450 (1913).
16. I. Langmuir and K. R. Blodgett, *Phys. Rev.* **22**, 347 (1923).
17. I. Langmuir and K. R. Blodgett, *Phys. Rev.* **24**, 49 (1924).
18. I. Langmuir and K. Compton, *Rev. Mod. Phys.* **13**, 191 (1931).
19. J. R. Pierce, *J. Appl. Phys.* **11**, 548 (1940).
20. G. R. Brewer, *IEEE Spectrum* **2**, 65 (1965); in *Physics and Technology of Ion Motors* (F. E. Marble and J. Surugue, eds.), p. 255. Gordon & Breach, London, 1966.
21. C. J. Davisson and C. J. Calbick, *Phys. Rev.* **38**, 585 (1931).
22. T. S. Green, *Rep. Prog. Phys.* **37**, 1257 (1974).
23. W. S. Cooper, K. H. Berkner, and R. V. Pyle, *Nucl. Fusion* **12**, 163 (1972); *Proc. Int. Conf. Ion Sources, 2nd*, Vienna, *1972*, p. 264 (1973).
24. J. R. Coupland, T. S. Green, D. P. Hammond, and A. C. Riviere, *Rev. Sci. Instrum.* **44**, 1258 (1973).
25. I. Chavet and R. Bernas, *Nucl. Instrum. Methods* **47**, 77 (1967).
26. V. I. Raiko, *Kernenergie* **10**, 89 (1967).

27. J. H. Whealton, R. W. Gaffey, and E. F. Jaeger, *Appl. Phys. Lett.* **36**, 91 (1980).
28. J. H. Whealton, *Nucl. Instrum. Methods* **189**, 55 (1981).
29. R. Becker and W. B. Hermannsfeldt, *Rev. Sci. Instrum.* **63**, 2756 (1992).
30. P. Spädke, *Rev. Sci. Instrum.* **63**, 2647 (1992).
31. J. Boers, *PBGUNS, Thunderbird Simulations.* Garland, TX.
32. D. Bohm, in *Characteristics of Electrical Discharges in Gases* (A. Guthrie and R. K. Wakerling, eds.), Chapter 3, p. 82. McGraw-Hill, New York, 1949.
33. D. Bohm, in *Characteristics of Electrical Discharges in Gases* (A. Guthrie and R. K. Wakerling, eds.), Chapter 4, p. 91. McGraw-Hill, New York, 1949.
34. J. D. Jackson, *Classical Electrodynamics*, Chapter 7. Wiley, New York, 1962.
35. W. L. Rautenbach, *Nucl. Instrum. Methods* **12**, 196 (1961).
36. O. Almen and K. O. Nielsen, *Nucl. Instrum. Methods* **1**, 302 (1957).
37. I. Chavet, Ph.D. Thesis, Univ. Paris (1965).
38. G. W. Hamilton, *Proc. Symp. Ion Sources Form. Ion Beams, 1971,* Rep. BNL-50310, p. 171 (1971).
39. P. Sigmund, *Phys. Rev.* **184**, 383 (1969).
40. G. D. Alton, D. H. Olive, M. L. Pinkham, and J. R. Olive, *Radiat. Eff. Defects Solids* **126**, 331 (1993).
41. H. H. Andersen and H. L. Bay, "Sputtering Yield Measurements" ch. 4, p. 145 in *Surface Interaction, Sputtering and Related Phenomena* (R. Behrish, W. Heiland, W. Poschanrieder, P. Staib, and H. Verbeek, eds.). Gordon & Breach, London, 1973.
42. M. von Ardenne, *Tabellen der Elektronenphysik, Ionenphysik und Übermikroskopie.* VEB Dtsch. Verlag Wiss., Berlin, 1956.
43. C. D. Moak, H. E. Banta, J. N. Thurston, J. W. Johnson, and R. F. King, *Rev. Sci. Instrum.* **30**, 694 (1959).
44. H. Fröhliche, *Nucleonic* **1**, 183 (1959).
45. H. Winter, *Proc. Int. Conf. Electromagn. Isot. Sep.*, Marburg, **BMBW-FB-K70-28**, 447 (1970).
46. R. A. Demirkhanov, H. Fröhliche, V. V. Kursanov, and T. T. Gutkin, *Brookhaven Natl. Lab.* [*Rep.*] *BNL* **BNL 767**, 218 (1962).
47. J. Kistemaker, P. K. Rol, and J. Politiek, *Nucl. Instrum. Methods* **38**, 1 (1965).
48. C. Lejeune, *Nucl. Instrum. Methods* **116**, 417 (1976).
49. R. Masic, R. J. Warnecke, and J. M. Sautter, *Nucl. Instrum. Methods* **71**, 339 (1969).
50. J. Illigen, R. K. Kirchner, and J. Schulte, *IEEE Trans. Nucl. Sci.* **NS-19**, 35 (1971).
51. O. B. Morgan, G. G. Kelley, and R. C. Davis, *Rev. Sci. Instrum.* **38**, 467 (1967).
52. N. B. Brooks, P. H. Rose, A. B. Wittkower, and R. P. Bastide, *Rev. Sci. Instrum.* **15**, 894 (1964).
53. J. H. Ormrod, *Proc. Symp. Ion Sources Form. Ion Beams, 1971,* Rep. BNL-50310, p. 151 (1971).
54. G. Gautherin, C. Lejeune, F. Prangere, and A. Septier, *Plasma Phys.* **11**, 397 (1969).
55. G. Stover and F. Zajae, *Proc. 1989 Part. Accel. Conf.*, IEEE Cat. No. 89, CH2669-0, 292 (1989).
56. F. M. Bacon, *Rev. Sci. Instrum.* **49**, 427 (1978).
57. A. Septier, in *Focusing of Charged Particles* (A. Septier, ed.), Vol. 2, p. 123. Academic Press, New York, 1967.
58. A. Benninghoven, F. G. Rüdenauer, and H. W. Wehner, *Secondary Mass Spectrometry*, Chapter 4, p. 468. Wiley, New York, 1987.

59. R. A. Demirkhanov, Yu. V. Kursanov, and V. M. Blagoveschenskii, *Prib. Tekh. Eksp.* **1,** 30 (1964).
60. J. E. Osher and G. W. Hamilton, *Proc. Symp. Ion Sources Form. Ion Beams, 1971,* Rep. BNL-50310, p. 157 (1971).
61. O. B. Morgan, Jr., *Proc. Symp. Ion Sources Form. Ion Beams, 1971,* Rep. BNL-50310, p. 129 (1971).
62. W. L. Stirling, R. C. Davis, T. C. Jernigan, O. B. Morgan, T. J. Orzechowski, G. Schilling, and L. D. Stewart, *Proc. Int. Conf. Ion Sources, 2nd,* Vienna, *1972,* p. 278 (1973).
63. R. W. Bikes, Jr. and J. B. O'Hagan, *Rev. Sci. Instrum.* **49,** 435 (1978); **53,** 585 (1982).
64. H. Horiike, M. Akiba, Y. Arakawa, S. Matsuda, and J. Sakuraba, *Rev. Sci. Instrum.* **52,** 567 (1981).
65. B. H. Wolf, *Nucl. Instrum. Methods* **B9,** 13 (1976).
66. P. G. Weber, *Rev. Sci. Instrum.* **54,** 1506 (1983).
67. M. R. Shubaly, R. G. Maggs, and A. E. Wooden, *IEEE Trans. Nucl. Sci.* **NS-32**(5), 1751 (1985).
68. J. P. Ruffell, D. H. Douglas-Hamilton, R. E. Kaim, and K. Izumi, *Nucl. Instrum. Methods* **B21,** 229 (1987).
69. T. Taylor, J. S. C. Wills, E. C. Douglas, and Tran Ngoc, *Rev. Sci. Instrum.* **61,** 454 (1990).
70. G. Sidenius, *Nucl. Instrum. Methods* **151,** 349 (1978).
71. F. M. Penning, *Physica (Amsterdam)* **4,** 71 (1937).
72. M. L. Mallory and E. D. Hudson, *IEEE Trans. Nucl. Sci.* **NS-22**(3), 1669 (1975).
73. J. R. J. Bennett, *Proc. Int. Cyclotron Conf.,* Oxford, England, *5th, 1969,* p. 199 (1972).
74. P. I. Vasiliev, N. I. Venikov, D. V. Zevjakin, A. A. Ogloblin, N. N. Khaldin, B. I. Koroshavin, V. I. Chuev, and N. I. Chumakov, *Nucl. Instrum. Methods* **71,** 201 (1969).
75. A. S. Pasyuk, E. D. Vorob'er, R. I. Ivannikov, V. E. Zuzketsov, V. B. Kutner, and Y. P. Tret'yakov, *At. Energ.* **28,** 75 (1970).
76. T. Itahashi and S. Mine, *Nucl. Instrum. Methods Phys.* **A300,** 1 (1991).
77. B. H. Wolf, J. Bossler, H. Emig, K. D. Leible, M. Müller, P. M. Rüch, and P. Spädtke, *Rev. Sci. Instrum.* **61,** 406 (1990).
78. H. Baumann and K. Bethge, *Nucl. Instrum. Methods* **122,** 517 (1974).
79. K. O. Nielsen, *Nucl. Instrum. Methods* **1,** 289 (1957).
80. M. Ma, J. E. Mynard, and K. G. Stephens, *Nucl. Instrum. Methods* **A227,** 279 (1989).
81. J. P. O'Connor, N. Tokoro, J. Smith, and M. Sieradzi, *Nucl. Instrum. Methods* **B37/38,** 478 (1989).
82. J. H. Freeman, *Nucl. Instrum. Methods* **22,** 306 (1963).
83. D. Aitken, *Springer Ser. Electrophys.* **10,** 23 (1982).
84. P. H. Rose, *Rev. Sci. Instrum.* **61,** 342 (1990).
85. H. W. Savage, ed., *Separation of Isotopes in Calutron Units,* United States Atomic Energy Commission, Nucl. Energy Ser. 7 (1951).
86. E. Yabe, *Rev. Sci. Instrum.* **58,** 1 (1987).
87. R. Bernas and O. Nier, *Rev. Sci. Instrum.* **19,** 89 (1948).
88. I. Chavet and R. Bernas, *Nucl. Instrum. Methods* **51,** 77 (1967).
89. R. Pratap, K. B. Lal, and V. P. Salvi, *Rev. Sci. Instrum.* **61,** 481 (1990).
90. J. L. Tuck, *Picket Fence,* USAEC Rep. WASH-184, p. 77. U.S. At. Energy Comm., Washington, DC, 1954.

REFERENCES

91. R. Keller, P. Spädtke, and F. Nöhmayer, *Proc. Int. Ion Eng. Congr.*, Kyoto, *1983*, p. 25 (1983).
92. B. Torp, B. R. Nielsen, D. M. Ruck, H. Emig, P. Spädtke, and B. H. Wolf, *Rev. Sci. Instrum.* **61**, 595 (1990).
93. R. Jancel and T. Kahan, *Electrodynamics of Plasmas*. Wiley, New York, 1966.
94. P. M. Morozov, *Dokl. Akad. Nauk SSSR* **102**, 61 (1955).
95. J. Moreau and R. Vienet, *Rapp. CEA-AC-Fr. Commis. Energ. At.* **CEA-AC-4972** (1957).
96. L. K. Goodwin, *Rev. Sci. Instrum.* **24**, 635 (1953).
97. M. E. Abdelaziz, S. G. Zakhovy, and A. M. Addelghaffer, *Rev. Sci. Instrum.* **61**, 460 (1990).
98. D. Blanc and A. Degeilh, *J. Phys. Radium* **22**, 230 (1961).
99. S. N. Misra and S. K. Gupta, *Nucl. Instrum. Methods* **122**, 303 (1974).
100. I. H. Wilson, *Radiat. Eff.* **18**, 95 (1973).
101. E. Regenstreif, *Le Synchrotron a Protons du Cern*, Chapter 5. Eur. Org. Nucl. Res., Geneva, 1959.
102. F. Norbeck and R. C. York, *Nucl. Instrum. Methods* **118**, 327 (1974).
103. Y. Saito, Y. Mitsuoka, and S. Suganomata, *Rev. Sci. Instrum.* **55**, 1760 (1984).
104. K. Leung, D. A. Bachman, R. R. Herz, and D. S. McDonald, *Nucl. Instrum. Methods* **B74**, 291 (1993).
105. K. Izumi, *Nucl. Instrum. Methods* **B21**, 124 (1987).
106. M. Moisan and Z. Zakrzewski, *J. Phys. D* **24**, 1025 (1991).
107. R. Geller, B. Jacquot, and C. Jacquot, *Plasma Phys. Controlled Nucl. Fusion Res., Proc. Int. Conf., 4th*, Madison, WI, *1971*, CN-28, p. G6 (1971).
108. R. Geller, *IEEE Trans. Nucl. Sci.* **NS-23**, 904 (1976).
109. J. Parker, ed., *Proc. Int. Workshop ECR Ion Sources and their Applications, 8th*, MSU-CP 47, p. 151. NSCL Michigan State University, East Lansing, 1987.
110. F. W. Meyer and M. I. Kirkpatrick, eds., *Proc. Int. Workshop ECR Ion Sources, 10th*, Knoxville, TN, *1990*, CONF-9011136 (1991).
111. A. G. Drentje, ed., *Proc. Int. Workshop ECR Ion Sources, 11th*, Grønigen, The Netherlands, *1993*, KVI-Report 996 (1993).
112. C. M. Lyneis and T. A. Antaya, *Rev. Sci. Instrum.* **61**, 221 (1990).
113. R. A. Phaneuf, this manuscript.
114. Y. Jongen and C. Lyneis, in *The Physics and Technology of Ion Sources* (I. G. Brown, ed.). p. 207. Wiley, New York, 1989.
115. R. Geller, *IEEE Cat. No.* **CH2669-0**, 1088 (1989).
116. G. D. Alton and D. N. Smithe, *Rev. Sci. Instrum.* **65**, 775 (1993).
117. T. Taylor, *Rev. Sci. Instrum.* **63**, 2507 (1992).
118. H. Amemiya, K. Schimizu, S. Kato, and Y. Sakamoto, *Jpn. J. Appl. Phys.* **27**, 2927 (1988).
119. O. A. Popov, *J. Vac. Sci. Technol., A* **7**, 894 (1989).
120. P. Decrock, P. Van Duppen, F. Baeten, C. Dom, and Y. Jongen, *Rev. Sci. Instrum.* **61**, 279 (1990).
121. P. McNeeley, G. Roy, J. Soukup, J. M. D'Auria, L. Buchmann, M. McDonald, P. W. Schmorr, H. Sprenger, and J. Vincent, *Rev. Sci. Instrum.* **61**, 273 (1990).
122. V. Kobecky, J. Musil, and F. Zacek, *Phys. Lett. A* **50A**, 309 (1974).
123. S. Pesic, *Nucl. Instrum. Methods* **198**, 593 (1982).
124. R. Nishimoto, S. Hagashii, M. Tanaka, A. Komori, and Y. Kawai, *Proc. Conf. Plasma Process. Stud., 7th*, Jan. 25, 26, Tokyo, p. 17 (1990).

125. N. Sakudo, K. Tokiguchi, H. Koike, and I. Kanomata, *Rev. Sci. Instrum.* **48**, 762 (1977).
126. N. Sakudo, K. Tokiguchi, H. Koike, and I. Kanomata, *Rev. Sci. Instrum.* **49**, 940 (1978).
127. Y. Torii, M. Shimada, and I. Watanabe, *Nucl. Instrum. Methods* **B21**, 178 (1987).
128. Y. Torii, M. Shimada, and I. Watanabe, *Rev. Sci. Instrum.* **61**, 253 (1990).
129. J. Hipple, C. Hayden, G. Dionne, Y. Torii, M. Shimada, and I. Watanabe, *Rev. Sci. Instrum.* **61**, 294 (1990).
130. Y. Matsubara, H. Tahara, M. Takahashi, S. Nogawa, and J. Ishikawa, *Rev. Sci. Instrum.* **63**, 2595 (1992).
131. T. Taylor and J. S. C. Wills, *Nucl. Instrum. Methods Phys.* **A309**, 37 (1992).
132. S. R. Walther, K. N. Leung, and W. B. Kunkel, *Rev. Sci. Instrum.* **57**, 1531 (1986).
133. J. M. Lafferty, ed., *Vacuum Arcs: Theory and Applications*. Wiley, New York, 1980.
134. A. J. Dempster, *Rev. Sci. Instrum.* **7**, 46 (1936).
135. V. S. Venkatasubramanian and H. E. Duckworth, *Can. J. Phys.* **41**, 234 (1963).
136. R. G. Wilson and D. M. Jamba, *J. Appl. Phys.* **38**, 1976 (1967).
137. K. V. Saludze and A. A. Plyutto, *Sov. Phys.—Tech. Phys. (Engl. Transl.)* **10**, 1006 (1966).
138. R. H. Fowler and L. W. Nordheim, *Proc. R. Soc. London, Ser. A* **119**, 173 (1928).
139. H. J. Zwally, D. W. Koopman, and T. D. Wilkerson, *Rev. Sci. Instrum.* **40**, 1492 (1969).
140. C. A. Anderson, *Int. J. Mass Spectrom. Ion Phys.* **3**, 413 (1970).
141. D. A. Anderson and J. R. Hinthorne, *Anal. Chem.* **45**, 1421 (1973).
142. I. G. Brown, J. E. Galvin, and R. A. MacGill, *Appl. Phys. Lett.* **47**, 338 (1985).
143. I. G. Brown, J. E. Galvin, B. F. Gavin, and R. A. MacGill, *Rev. Sci. Instrum.* **57**, 1069 (1986).
144. H. Shiraishi and I. G. Brown, *Rev. Sci. Instrum.* **61**, 3775 (1990).
145. A. A. Vasilyev, *Rev. Sci. Instrum.* **63**, 2434 (1992).
146. L. W. Nordheim, *Proc. R. Soc. London, Ser. A* **121**, 628 (1928).
147. R. H. Good and E. W. Mueller, *Handb. Phys.* **211**, 176 (1956).
148. R. Gomer, *Field Emission and Field Ionization*. Harvard Univ. Press., Cambridge, MA, 1961.
149. M. J. Southon, Ph.D. Thesis, Cambridge University, Cambridge, UK (1963).
150. H. A. M. van Eekelen, *Surf. Sci.* **21**, 21 (1970).
151. C. D. Hendricks, in *Electron Beam Science Technology* (R. Bakish, ed.), p. 915. Wiley, New York, 1964.
152. R. Levi-Setti, *Nucl. Instrum. Methods* **168**, 1391 (1980).
153. S. Sato, T. Kato, and N. Igata, in *Ion Sources and Ion Assisted Technology* (T. Takagi, ed.), p. 91 (1987).
154. G. L. Allan, J. Zhu, and G. J. F. Legge, in *Proc. Nuclear Techniques of Analysis, 4th*, (R. J. McDonald, ed.), p. 49. Lucas Heights, NSW Australia, 1985.
155. G. L. Allan, G. J. F. Legge, and J. Zhu, *Nucl. Instrum. Methods* **B34**, 122 (1988).
156. R. A. D. MacKenzie and G. D. W. Smith, *Nanotechnology* **1**, 163 (1990).
157. P. D. Prewett and G. L. R. Mair, *Focused Ion Beams from Liquid Metal Ion Sources*. Wiley, Chichester, England, 1991.

158. J. Orloff, *Rev. Sci. Instrum.* **64**, 1105 (1993).
159. G. E. Taylor, *Proc. R. Soc. London, Ser. A* **280**, 383 (1964).
160. P. Marriot and J. C. Riviere, *Emission Mechanisms in the Liquid Metal Ion Source: A Review*. Rep. No. AERE-R11294. Harwell Laboratory, Oxfordshire, 1984.
161. R. G. Forbes, *J. Phys. D* **15**, 1301 (1982).
162. R. Gomer, *Appl. Phys.* **19**, 365 (1979).
163. P. D. Prewett, G. L. R. Mair, and S. P. Thompson, *J. Phys. D* **15**, 1339 (1982).
164. R. L. Seliger, J. W. Ward, and R. L. Kubena, *Appl. Phys. Lett.* **34**, 310 (1979).
165. T. Kashiwagi, Y. Gotoh, H. Tsuji, J. Ishikawa, and T. Takagi, in *Ion Sources and Ion Assisted Technology* (T. Takagi, ed.), p. 87 (1987).
166. G. D. Alton and P. M. Read, *Nucl. Instrum. Methods* **B54**, 7 (1991).
167. P. Marriott, *Angular and Mass Resolved Energy Distribution Measurements with a Gallium Liquid-Metal Ion Source*, Rep. No. AERE-R-12674. Harwell Laboratory, Oxfordshire, 1987.
168. L. W. Swanson, G. A. Schwind, A. E. Bell, and J. C. Brady, *J. Vac. Sci. Technol.* **16**, 1864 (1979).
169. H. Boersch, *Z. Phys.* **139**, 115 (1954).
170. K. H. Loeffler, *Z. Angew. Phys.* **27**, 145 (1969).
171. W. Knauer, *Optik* **59**, 335 (1981).
172. L. W. Swanson, G. A. Schwind, and A. E. Bell, *J. Appl. Phys.* **51**, 3453 (1980).
173. G. L. R. Mair, D. C. Grindod, M. S. Mousa, and R. V. Latham, *J. Phys. D* **16**, L209 (1983).
174. G. D. Alton and P. M. Read, *J. Appl. Phys.* **66**, 1018 (1989).
175. G. D. Alton and P. M. Read, *J. Phys. D* **22**, 1029 (1989).
176. A. E. Bell, G. A. Schwind, and L. W. Swanson, *J. Appl. Phys.* **53**, 4602 (1982).
177. P. Wagner and T. M. Hall, *J. Vac. Sci. Technol.* **16**, 1871 (1979).
178. P. D. Prewett and D. K. Jeffries, *J. Phys. D* **13**, 1747 (1980).
179. R. Clampitt, *Nucl. Instrum. Methods* **189**, 111 (1981).
180. R. Clampitt, *J. Vac. Sci. Technol.* **12**, 1208 (1975).
181. J. Ishikawa and T. Takagi, *J. Appl. Phys.* **56**, 3050 (1984).
182. J. Ishikawa and T. Takagi, *Vacuum* **36**, 825 (1986).
183. J. Ishikawa, H. Tsuji, and T. Takagi, *Rev. Sci. Instrum.* **61**, 592 (1990).
184. Y. Gotoh and J. Ishikawa, *Rev. Sci. Instrum.* **63**, 2438 (1992).
185. P. M. Read, J. T. Maskrey, and G. D. Alton, *Rev. Sci. Instrum.* **61**, 502 (1990).
186. A. Wagner, *Appl. Phys. Lett.* **40**, 440 (1982).
187. V. Wang, J. W. Ward, and R. L. Selinger, *J. Vac. Sci. Technol.* **19**, 1158 (1981).
188. R. L. Kubena, C. L. Anderson, R. L. Selinger, R. A. Jullens, and E. H. Stevens, *J. Vac. Sci. Technol.* **19**, 916 (1981).
189. E. Miyauchi, H. Hashimoto, and T. Utsumi, *Jpn. J. Appl. Phys.* **22**, L225 (1983).
190. E. Miyauchi, H. Arimoto, H. Hashimoto, T. Furuya, and T. Utsumi, *Jpn. J. Appl. Phys.* **22**, L287 (1983).
191. K. Gamo, T. Ukegawa, Y. Inomoto, Y. Ochiai, and S. Namba, *J. Vac. Sci. Technol.* **19**, 1182 (1981).
192. A. E. Bell, G. A. Schwind, and L. W. Swanson, *J. Appl. Phys.* **53**, 4602 (1982).
193. R. Clampitt, *Nucl. Instrum. Methods* **189**, 111 (1981).
194. T. Okutani, M. Fukuda, T. Noda, H. Tamura, H. Watanabe, and C. Shepard, *Symp. Electron, Ion, Photon Beam Technol., 17th*, Los Angeles, CA, 1983 (1983).
195. R. Clampitt and D. K. Jeffries, *Nucl. Instrum. Methods* **149**, 739 (1978).

196. K. Gamo, T. Ukegawa, Y. Inomoto, K. K. Ka, and S. Namba, *Jpn. J. Appl. Phys.* **19**, L595 (1980).
197. K. Gamo, Y. Inomoto, Y. Ochiai, and S. Namba, *Proc. Conf. Electron Ion Beam Sci. Technol., 10th*, Vol. 83-2 (1982).
198. N. D. Bhaskar, C. M. Klimcak, and R. P. Fraukolz, *Rev. Sci. Instrum.* **63**, 366 (1990).
199. A. Wagner and T. M. Hall, *J. Vac. Sci. Technol.* **16**, 1871 (1979).
200. L. W. Swanson, G. A. Schwind, A. E. Bell, and J. E. Brody, *J. Vac. Sci. Technol.* **16**, 1864 (1979).
201. J. Van de Walle and P. Sudrand, in *Proceedings of the 29th International Field Emission Symposium (IFES)* (H. O. Andren and Norden, eds.), p. 341. Almqvist & Wiksell, Stockholm, 1982.
202. M. F. Harrison, *U.K. At. Energy Res. Establ., Rep.* **AERE-GP/R 2505** (1958).
203. H. L. Daley, J. Perel, and R. H. Vernon, *Rev. Sci. Instrum.* **37**, 473 (1966).
204. V. T. Raiko, M. S. Ioffe, and V. S. Zolatarev, *Pribl. Tekh. Eksp.* **1**, 29 (1961).
205. G. Kuskevics and B. Thompson, *Rev. Sci. Instrum.* **37**, 710 (1966).
206. D. M. Jamba, *Rev. Sci. Instrum.* **40**, 1072 (1969).
207. G. S. Kino, "A Study of Systems for Space Propulsion" *NASA Contract.* **NAS-34100** 1964–65, (1965).
208. R. G. Wilson, *Proc. Int. Conf. Appl. Ion Beams*, Grenoble, p. 105 (1967).
209. R. Haug, *Rev. Sci. Instrum.* **41**, 670 (1970).
210. C. F. Cuderman, *Rev. Sci. Instrum.* **49**, 1475 (1969).
211. G. R. Brewer, *Ion Propulsion*. Gordon & Breach, New York, 1970.
212. H. L. Daley and J. Perel, *Rev. Sci. Instrum.* **42**, 1324 (1971).
213. G. D. Alton, *Rev. Sci. Instrum.* **59**, 1039 (1988).
214. G. D. Alton, M. T. Johnson, and G. D. Mills, *Nucl. Instrum. Methods* **A328**, 154 (1993).
215. P. Dyer and R. G. H. Robertson, *Nucl. Instrum. Methods* **189**, 351 (1981).
216. H. Liebl and B. Senftinger, *Rev. Sci. Instrum.* **59**, 2174 (1988).
217. C. H. Kunsman, *Science* **62**, 269 (1925).
218. E. J. Jones, *Phys. Rev.* **44**, 707 (1933).
219. J. P. Blewett and E. J. Jones, *Phys. Rev.* **50**, 464 (1936).
220. R. H. Dawton, in *Electromagnetically Enriched Isotopes* (M. L. Smith, ed.), p. 37. Butterworth, London, 1956.
221. G. Couchet, *Ann. Phys. (Leipzig)* [6] **9**, 931 (1954).
222. A. Septier and H. Leal, *Nucl. Instrum. Methods* **29**, 257 (1964).
223. D. W. Hughes, R. K. Feeney, and D. N. Hill, *Rev. Sci. Instrum.* **51**, 1471 (1980).
224. D. R. Peele, D. L. Adler, B. R. Litt, and B. H. Cooper, *Rev. Sci. Instrum.* **60**, 730 (1989).
225. G. D. Alton, unpublished.
226. A. E. Souziz, W. E. Carr, S. I. Kim, and M. Seidl, *Rev. Sci. Instrum.* **61**, 788 (1990).
227. M. N. Saha, *Philos. Mag.* [6] **40**, 472 (1920).
228. G. J. Beyer, E. Herrmann, A. Peotrowski, V. I. Raiko, and H. Tyroff, *Nucl. Instrum. Methods* **96**, 347 (1971).
229. P. G. Johnson, A. Bolson, and C. M. Henderson, *Nucl. Instrum. Methods* **106**, 83 (1973).
230. R. Kirchner, *Nucl. Instrum. Methods Phys.* **A292**, 203 (1990).
231. A. Peotrowski, R. L. Gill, and D. McDonald, *Nucl. Instrum. Methods* **224**, 1 (1984).

4. ADVANCED SOURCES OF HIGHLY CHARGED IONS

Ronald A. Phaneuf

Department of Physics, University of Nevada, Reno

4.1 Introduction

Most of the matter in the universe exists in high-temperature plasma environments that contain highly ionized ions. Collisional and radiative processes involving highly ionized (and therefore highly charged) ions play important roles in astrophysical environments and in laboratory plasmas for fusion-energy and X-ray laser research. A "map of the universe" [1], indicating the relationship between electron density and temperature for different kinds of plasmas, both naturally occurring and man-made, is presented in Figure 1. To produce highly ionized ions in the laboratory requires a significant investment of energy to remove the bound atomic electrons. For example, a minimum energy of 1030 eV is required to strip one carbon atom bare of its six electrons to form a C^{6+} ion. This sum of the binding energies of the removed electrons is referred to as the neutralization energy. This would be the energy required if each of the removed electrons carried away zero kinetic energy—an unlikely scenario in any plasma! It therefore represents a lower limit for the energy that must be supplied to the ion. For an argon atom, with 18 electrons, the neutralization energy is 14,398 eV.

It is evident that special measures must be taken to produce highly ionized ions for research. One method that has been employed for decades is to accelerate ions of modest charge to energies in the mega-electron volt range or higher and then to pass them through a thin foil or gas target to collisionally strip them of their electrons [2]. Another method involves directing such an energetic ion beam through a gas target in which slow, highly charged recoil ions are produced by multiple ionization in energetic collisions with these fast "hammer" ions [3]. The slow recoil ions of the gas target species are accelerated and formed into a low-energy beam for use in experiments that do not require high ion intensities. The production of *intense*, highly charged ion beams at *low* kinetic energies requires a special ion source. One advantage of producing highly charged ions directly in the ion source is the energy multiplication by the charge q that results when they are subsequently accelerated to form a beam. This

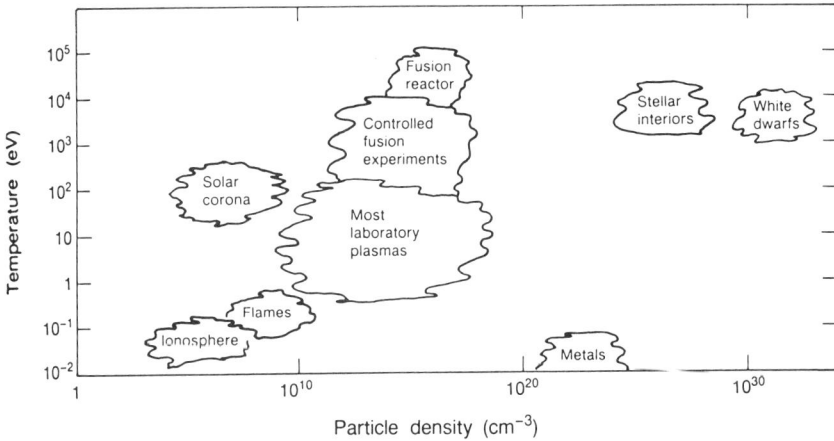

FIG. 1. "Map" of temperature versus density characteristics of some astrophysical and laboratory plasmas. [From Brown, I. G., *The Physics and Technology of Ion Sources*, Copyright © (1989 John Wiley & Sons, Inc.) Reprinted with permission from John Wiley & Sons, Inc.]

multiplication extends the energy range of ion accelerators used primarily for nuclear physics research and has been, in fact, a major motivation and the main source of funding for multicharged ion source development. These same multicharged ion sources may also be used for producing low-energy beams. For a more comprehensive technical review of the rapidly evolving field of ion source technology, the reader is referred to a compendium [4] and the *Proceedings of the Fourth International Conference on Ion Sources* [5]. An overview of the impact of advanced ion sources on the study of collisional interactions of highly charged ions at low energies is provided in Phaneuf [6].

Following a brief introduction of some basic physics considerations, this chapter will overview the operating principles and performances of several types of advanced highly charged ion sources that are currently employed in experimental physics research.

4.1.1 Basic Physics Considerations

The dominant ionization mechanism in most multicharged ion sources is stepwise impact ionization in successive collisions of atoms and ions with energetic electrons. Multiple ionization in a single collision with an electron must also be accounted for in accurate models of charge-state abundances in hot plasmas [7]. For comprehensive treatments of the physics of electron-impact ionization of ions, the reader is referred to two reviews [8,9]. Figure 2 compares experimental cross sections for electron-impact single ionization of Fe^{q+} ions for $0 \le q \le 15$. The ionization thresholds are seen to increase and the ionization cross sections to decrease dramatically with increasing ionic charge q.

In order to reach high ionization stages, the ions must undergo a sufficient number of collisions with energetic electrons during the time τ that they are confined in the ion source plasma. At each collision, the electron energy must be larger than the ionization potential of the ion for ionization to occur. Optimally, the electron energy should be near that at which cross-section maximum occurs. If the electron density is high enough that the mean time between collisions is short compared with ion excited-state lifetimes, ionization may also occur from excited levels, thereby increasing the ionization efficiency. Processes which compete with electron-impact ionization to reduce the ion charge state are electron–ion recombination and electron capture in collisions with neutral atoms. Important parameters for multicharged ion source operation are therefore the electron density n_e, the mean electron energy E_e or temperature T_e (if the electron velocity distribution is Maxwellian), the neutral atom density n_0, and the ion confinement time τ. Neglecting these competing processes and multiple

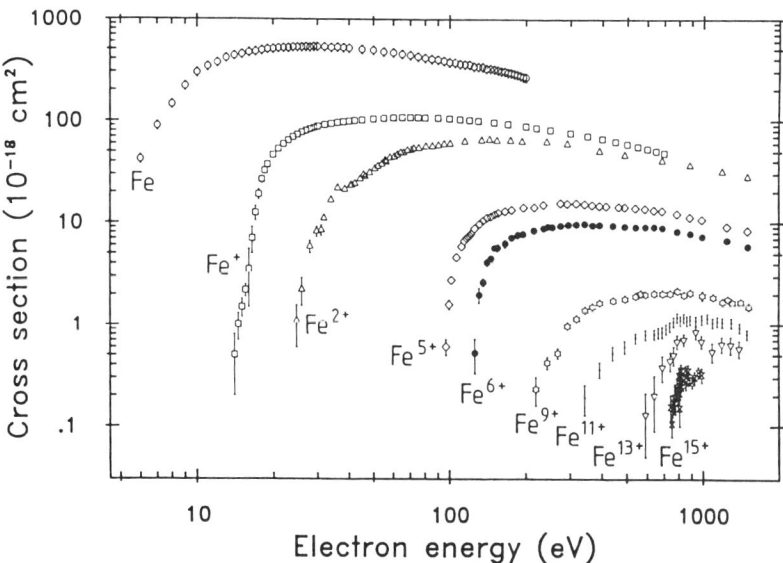

FIG. 2. Experimental cross-sections for electron-impact single ionization of Fe^{q+} ions for ionic charge q, ranging from 0 to 15. The figure is taken from Müller [9].

ionization, the required confinement time [10] to reach ion charge state q, $\tau_{0 \to q}$, may be approximated as

$$\tau_{0 \to q} = \frac{1}{n_e v_e} \sum_{i=0}^{q} \left(\frac{1}{\sigma_{i-1 \to i}} \right), \quad (4.1)$$

where v_e is the mean electron velocity and $\sigma_{i-1 \to i}$ is the cross-section for electron-impact ionization of the particular element from charge state $i-1$ to charge state i (at electron velocity v_e). A fundamental objective in the design of multicharged ion sources is the achievement of a sufficiently energetic electron population with the highest possible product of electron density n_e and ion confinement time τ. This is analogous to the well-known Lawson criterion for achieving energy breakeven in thermonuclear fusion reactors. Of course, recombination and electron capture place additional constraints on n_e and n_0 and must also be taken into account in multicharged ion source modeling, design, and operation.

4.2 The Electron-Beam Ion Source

Although the electron-beam ion source (EBIS) is conceptually the simplest of the advanced sources of highly charged ions, it is probably the most challenging from a technical standpoint. The EBIS holds the current record for the highest charge state ion beam accelerated from an ion source (hydrogen-like Xe^{53+}). One advantage of the EBIS over other multicharged ion sources is that it is excited by a monoenergetic electron beam of variable energy, which permits the determination of relative cross sections for electron–ion collision processes occurring within the ion source itself. This has been accomplished either by detecting photon emissions or by measuring the extracted ion current as a function of the electron-beam energy. A special configuration of this type of device designed for such collision measurements, the electron-beam ion trap (EBIT), is discussed briefly in Section 4.2.3. Comprehensive reviews of EBIS technology and sources of additional references have been presented elsewhere [10,11].

4.2.1 EBIS Design and Operating Principles

In an EBIS, multicharged ions are created by and radially trapped in the space charge well of an intense, high-energy, magnetically confined electron beam. A cylindrical electrostatic drift-tube structure traps the ions axially and is pulsed periodically to expel ions for acceleration and beam formation. The mechanical arrangement of an EBIS and typical

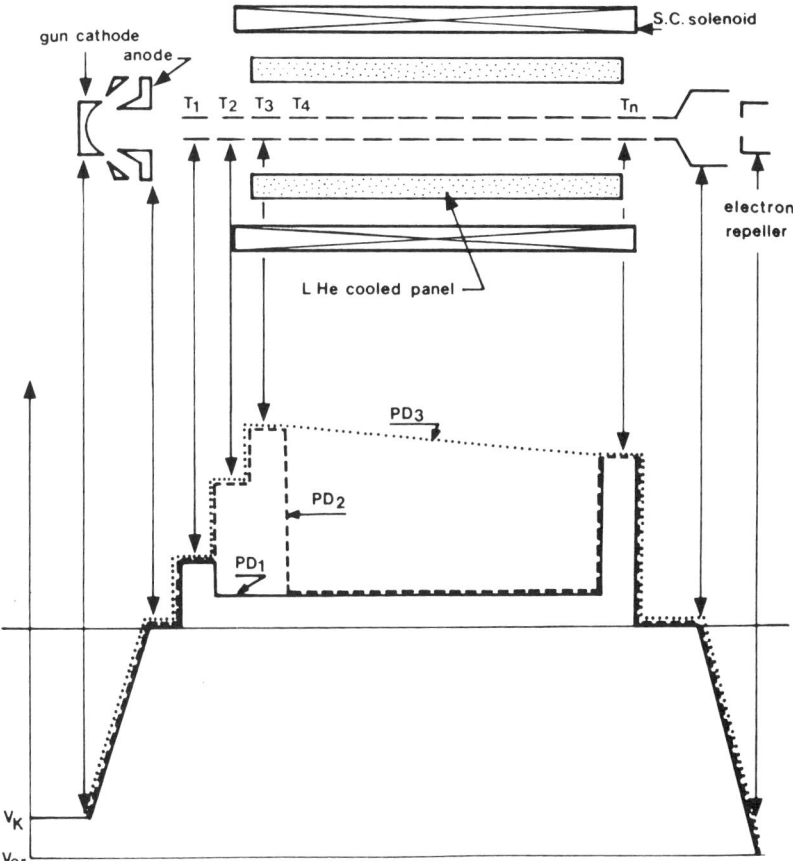

FIG. 3. Schematic showing the main features of an electron-beam ion source (EBIS). Drift-tube electrode potential distributions: PD1, filling trap with ions; PD2, trapping and "cooking" ions; PD3, expelling ions for beam formation. V_K is the cathode potential and V_{er} is the electron-repeller potential. The figure is adapted from Arianer and Geller [10].

electrode potential distributions are shown in Figure 3. The acronym CRYEBIS is frequently applied to current-generation EBIS devices, since a superconducting solenoid is most often used to generate axial magnetic field strengths as high as 10 T. Cryogenic pumping also facilitates the extreme ultra-high-vacuum conditions that are required to achieve the highest charge states. So-called "warm EBIS" sources using conventional solenoids have also been successfully constructed and operated. The high-power dissipation in the solenoid limits the magnetic field on axis in such ion sources to about 1 T, which in turn limits the achievable electron beam current density and ion source performance.

In a typical EBIS, an electron beam with an intensity of a few A/cm^2 is produced by a high-quality electron gun located outside, but on the axis of, the solenoidal magnetic field. This electron beam, whose energy may range from 1 to 50 keV, is accelerated into the solenoid, where it is compressed to a current density of 100–1000 A/cm^2 by the magnetic field. Upon leaving the magnetic field at the opposite end of the solenoid (~1 m long), it expands and is decelerated prior to impact on a cooled electron collector. Extreme care must be taken to align the electron gun with the axis of the magnetic field, which must have a high degree of uniformity. Otherwise, significant spiraling of the electron beam will be induced, causing a fraction of the beam electrons to impact the electrode structure. This is detrimental and usually fatal to effective EBIS operation due to localized heating of the cryostructure and resultant vacuum degradation. To produce highly charged ions and prevent their loss by electron-capture collisions, the background pressure in an EBIS must be held to less than 10^{-9} Torr with an electron beam present. The typical base pressure in cryogenic EBIS sources is in the 10^{-12}-Torr range. Ions of the desired element are produced by admitting a small quantity of gas or by injection of a low-energy feeder ion beam through a hole in the center of the electron-gun cathode. The latter method must be used to produce ions of solid metallic elements. The choice of the electron-beam energy and ion-interaction or "cooking" time by sequencing the potentials on the ion trap electrodes determines the relative ion charge-state abundances in an EBIS. A typical sequence first provides for ion injection, then formation of an axial trap to confine and ionize the ions, and finally ejection into the ion-beam extraction region after the desired charge state is reached. The corresponding electrostatic potential distributions on the drift-tube electrodes are shown in Figure 3.

4.2.2 EBIS Performance

Modeling the charge-state performance of an EBIS is more straightforward than that for other plasma sources because the electron beam is essentially monoenergetic (energy spreads of 20–50 eV are typical), and, as noted above, the ion confinement time can be controlled to a significant degree by the sequencing of potentials on the trap electrodes. This confinement/interaction time may typically be varied from milliseconds to tens of seconds, depending on the desired ion charge state. As noted above, extreme ultra-high-vacuum conditions are required to reduce the loss of ions in high charge states due to electron-capture collisions. Ignoring electron capture and electron–ion recombination, the fractional abundances of the individual ion charge states may be roughly estimated by solving a simple set of first-order coupled differential equations for the individual charge-state populations, N_i,

$$\frac{dN_i}{d(J_e t)} = -N_i(J_e t)\sigma_{i \to i+1} + N_{i-1}(J_e t)\sigma_{i-1 \to i}, \quad (4.2)$$

where J_e is the electron current density, t is the interaction time, and $\sigma_{i \to i+1}$ and $\sigma_{i-1 \to i}$ are the cross-sections for single ionization out of and into charge state i, respectively. Figure 4 presents the results of such a model calculation of the charge-state evolution of Ar ions in an EBIS with a 10-keV ion beam [10]. A more complete model would include electron–ion recombination, multiple ionization processes, and electron

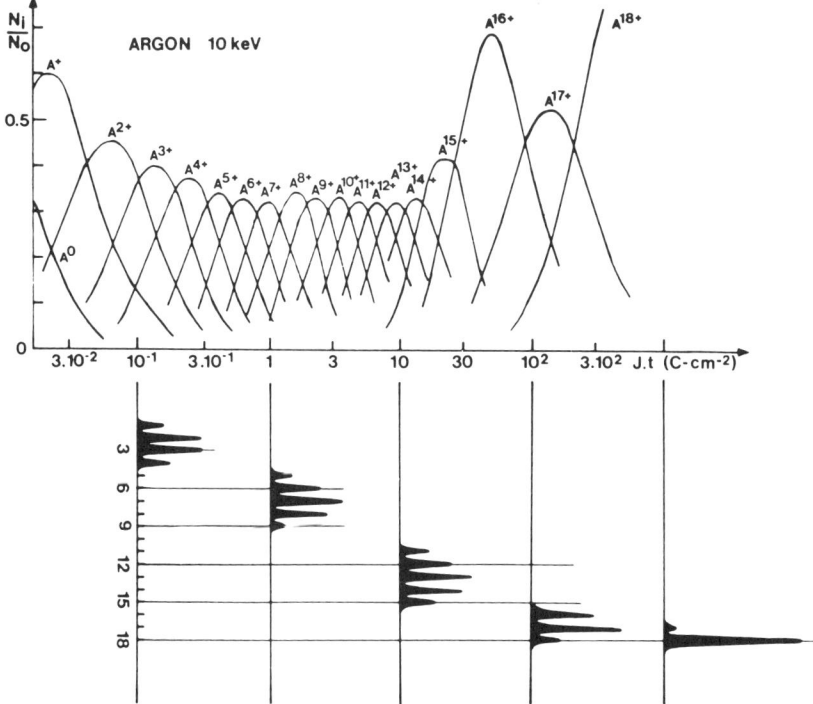

FIG. 4. Calculated evolution of Ar^{q+} ion charge-state populations in an EBIS for a 10-keV electron beam, based on solutions to coupled Equations 2. The fractional abundances are plotted versus product of electron current density and ion heating/confinement time. The figure is taken from Arianer and Geller [10].

TABLE I. Characteristics of Some Currently Operating EBIS and EBIT Devices[a]

Source	Location	E_e maximum (keV)	Ions produced	Application (remarks)
KRION-2	JINR, Dubna, Russia	52	Ar^{18+}, Kr^{36+}, Xe^{53+}	Atomic physics
DIONÉ	CEN, Saclay,	10	Ar^{18+}, Kr^{30+}	Synchrotron injector
CRYSIS	MSI, Stockholm	19	Ar^{18+}, Xe^{52+}	Heavy-ion storage ring injector, atomic physics
Kryo-EBIS	IAP, Frankfurt	6	Ar^{18+}, Kr^{26+}, Xe^{36+}	Atomic physics, testing (DC operation)
NICE-1	IPP, Nagoya, Japan	2	I^{42+}	Atomic physics (DC operation, L-N_2 cooled)
Mini-EBIS	Tokyo Metropolitan	2	Ne^{9+}, Ar^{16+}, I^{30+}	Atomic physics (DC operation, L-N_2 cooled)
KSU-CRYEBIS	KSU, Manhattan, KS	10	Ar^{18+}, Kr^{34+}, Xe^{46+}	Atomic physics, RFQ injector
CEBIS-II	Cornell, Ithaca, NY	5	Ar^{16+}, Xe^{28+}	Atomic physics
EBIT-II	LLNL, Livermore, CA	30	U^{70+}, Th^{80+}	Atomic physics (trapped ions or extracted beam)
Super-EBIT	LLNL, Livermore, CA	160	U^{92+}	Atomic physics

[a] Adapted from Stöckli [13].

capture from neutral species that may also be present. The charge-state distribution after a fixed interaction time is rather narrow, with only a few ionization stages having appreciable population. Depending on the ion charge state, extracted ion beam intensities from an EBIS can reach 10^9 ions per pulse, with the average intensity depending on the required interaction time (which in turn limits the pulse repetition rate).

4.2.3 The Electron-Beam Ion Trap

A variation of the EBIS, called the electron-beam ion trap (EBIT), has been developed and successfully applied to the quantitative study of electron collisions with very highly charged ions [12]. The EBIT is discussed elsewhere in this volume and is described only briefly here for completeness. The EBIT is similar in design and operation to the EBIS, but is made extremely compact and has a higher electron-beam energy (up to 100 keV or more). Electron–ion collision processes occurring within EBIT are studied via high-resolution X-ray spectroscopy. In normal EBIT operation, the ions are not extracted after a predetermined interaction time, as in an EBIS. Rather, the ions are held in the trap by the electrode potentials, and the electron-beam energy is alternately switched between two energies, the first to create and sustain the population of ions to be trapped and the second to study collisions of the beam electrons with those trapped ions. The second electron energy may be stepped to determine the energy dependence of the cross-section for the interaction of interest. The required interaction times are determined by the same considerations discussed above for the EBIS.

The compactness of EBIT permits the simultaneous attainment of extremely high electron-beam energies, current densities, and vacuum conditions, which are requirements for producing the highest ionization stages. Electron–ion recombination processes involving Th^{80+} have been studied in EBIT. Highly charged ion beams have also been successfully extracted by operating the device as a conventional EBIS. A comparison of the parameters and performance of EBIS and EBIT devices is presented in Table I [13].

4.3 The Electron–Cyclotron Resonance Ion Source

The electron–cyclotron resonance ion source (ECRIS) may be most simply described as a "magnetic bottle" in which electrons are confined and heated by a microwave-frequency field via the cyclotron resonance. Of the multiply charged ion sources, the ECRIS has probably made the largest impact on accelerator facilities and on the study of fundamental interactions of highly charged ions. While the ECRIS plasma is much more difficult to characterize, the technology is more forgiving than that of the EBIS, and approximately two dozen are successfully operating throughout the world at the current time. More comprehensive discussions of ECRIS technology may be found elsewhere [10,14]. ECRIS development has closely paralleled that of the EBIS, and these two major competing devices may in fact be regarded as complementary technologies, with the EBIS providing the highest ion charge states and the ECRIS the highest ion beam intensities.

4.3.1 ECRIS Design and Operating Principles

As in the EBIS, atoms in an ECRIS are stepwise ionized in multiple collisions with energetic electrons, which in this case are heated by cyclotron resonance acceleration by the microwave field. The desirable plasma parameters are the same: high electron densities and temperatures and long ion residence times in the source. Electron–ion recombination and electron-capture collisions with neutral species must also be minimized in order to achieve the highest ion charge states.

The main features of a typical two-stage ECRIS [15] are shown in Figure 5. Electron confinement in the main stage is achieved in a "minimum-B" configuration, in which an axial magnetic mirror field is created by suitably separated pair of solenoidal coils. Mirror ratios range typically between 1.2 and 2. Electron motion in the radial direction is confined by a multipole magnetic field created by a cylindrical array of permanent magnets. In the case shown, a hexapolar field is produced by 12 bars of $SmCo_5$ magnets, with the direction of magnetization of each successive bar rotated by 90°. Apart from their strong magnetization, an important characteristic of nonferromagnetic permanent magnet materials such as $SmCo_5$ and NdFeB is their near-unit magnetic permeability, which permits a direct superposition of the axial and radial magnetic fields and a resulting simplification of the magnetic structure. Typical axial and radial magnetic field configurations [16] in an ECRIS are shown in Figures 6 and 7, respectively. A vacuum vessel typically 5–10 cm in diameter and 20–30 cm long is placed within this magnetic structure and forms the main-stage plasma discharge chamber. A gas containing the element to be ionized is admitted either into this chamber or into a first or preionization stage. In the example shown, this first stage is differentially pumped from the main stage and operates at a pressure near 10^{-4} Torr, while the main stage operates in the range $1-3 \times 10^{-6}$ Torr.

In operation, the relative strengths of the radial and axial magnetic fields are arranged such that a closed ECR surface is contained within the

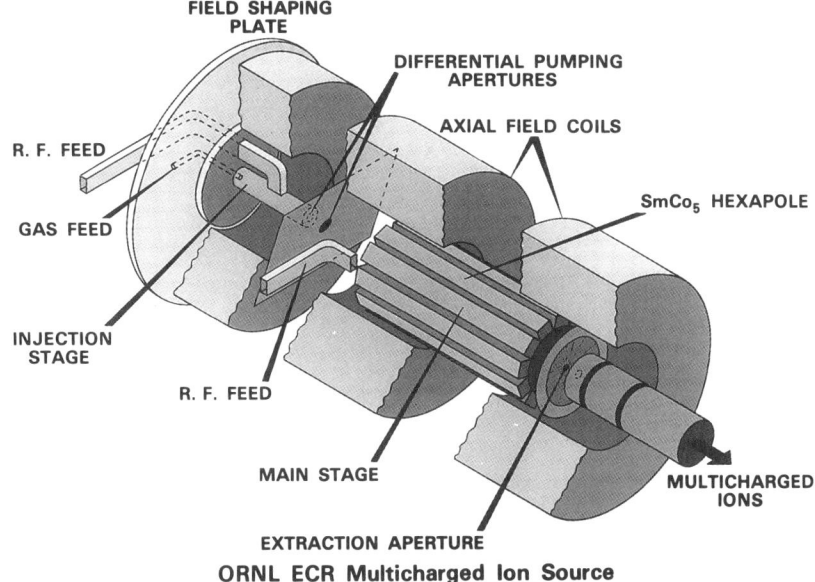

FIG. 5. Cutaway drawing illustrating the main features of a two-stage electron–cyclotron-resonance ion source (ECRIS). The figure is taken from Meyer and Hale [15].

main stage plasma chamber, which also serves as a multimode microwave cavity. The magnetic field and microwave frequency are related by the familiar equation for cyclotron resonance:

$$f = \frac{eB}{2\pi m_e} = 2.796 \cdot 10^{10} \cdot B \quad \text{(tesla)}. \qquad (4.3)$$

For a typical microwave frequency, f, of 10 GHz, the resonant field is 0.358 T, which is readily achieved by a combination of permanent magnets and water-cooled coils. ECRIS have been successfully operated at frequencies ranging from 2.45 to 18 GHz. In the latter case, superconducting solenoids were required to achieve the required field strength for resonance. This adds complexity to the design and operation of the source, but results in significant energy savings, since the coils typically dissipate 50–100 kW of electrical power, for which cooling must be also provided.

Electrons in the plasma chamber are stochastically heated by the microwave field, being accelerated each time they cross the ECR magnetic surface. The electron velocity distribution in such a plasma is non-Maxwellian, with a mean kinetic energy in the 1-keV range and a high-energy tail

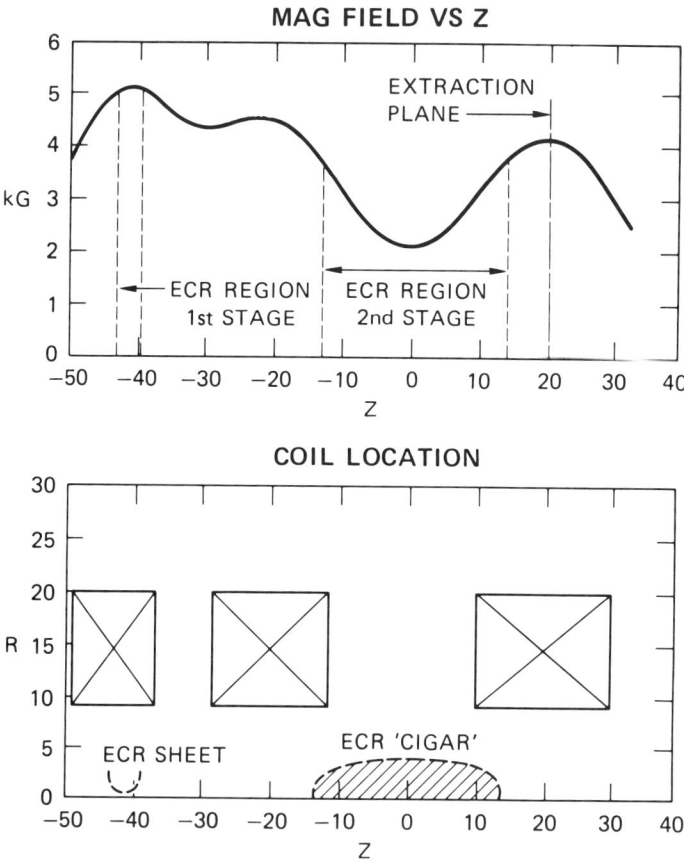

FIG. 6. Typical axial magnetic field configuration of an ECRIS, showing the coil placement, magnetic mirror, and ECRIS region for a microwave frequency of 10.6 GHz. The figure is adapted from Meyer [16].

that can reach several hundred kilo-electron volts. Ions drift at thermal energies through the plasma chamber, with residence times on the order of a millisecond. Due to their much larger mass, ions are essentially unconfined by the magnetic fields and do not absorb energy from the microwave field. The principal mechanism for ion translational heating is momentum transfer in collisions with hot electrons, which is very inefficient due to the large mass imbalance. The ions become multiply ionized in successive collisions with the hot electrons, but do not attain large kinetic energies in the source.

To form an ion beam, the plasma chamber is maintained at a positive

FIG. 7. Calculated radial profile of hexapole magnetic field lines for the ORNL-ECRIS in Figure 5. The positions and directions of magnetization of the $SmCo_5$ permanent magnet bars are indicated. The figure is adapted from Meyer [16].

potential of 1-25 kV, depending on the desired beam energy, and ions which drift downstream into the extraction region are accelerated by a conventional Pierce-type extraction electrode held at or near ground potential. Because there is no efficient mechanism for heating ions in the plasma chamber, thermal effects contribute very little to the energy spread of the extracted ion beam. The ion energy spread is typically 5 eV per ion charge and dominated by the plasma space potential gradients in the extraction region.

The gas pressure in the source is an important parameter which is adjusted to vary the mean free path of the electrons in the discharge, thereby optimizing the mean electron kinetic energy and the frequency of electron–ion interaction. Considering the relative magnitudes of the collisional reaction rates, the neutral density in the discharge must be maintained roughly two orders of magnitude lower than the electron density in order to keep the rate of ion production by electron impact equal to that of loss by electron-capture collisions. By control of the gas pressure and microwave power, the ion abundance in a given charge state may be optimized. Since the electron energy distribution is much broader in the

ECRIS than in the EBIS, a correspondingly broader charge-state distribution results. The microwave power deposited in the plasma may typically range from 10 to 1000 W, depending on the desired ion species and charge state. Beams of metallic ions are produced by the use of thermal ovens to vaporize elements such as Ca, Ba, or Pb or by inserting thin wires or foils of refractory metals such as Mo, Ta, or W into the edge of the plasma chamber, where they are vaporized by energetic electron bombardment. Readily-vaporized metal halides are also used in thermal ovens to produce metallic ions.

4.3.2 ECRIS Performance

While the ECRIS cannot compete with the EBIS or EBIT in terms of the highest charge states that can be produced, it operates in a continuous rather than a pulsed mode and is therefore capable of delivering average ion beam currents that are many orders of magnitude larger. This is a critical issue for experiments involving interacting beams. ECRIS devices have produced beams of fully stripped ions to Ar^{18+}. For intermediate-Z elements such as Xe and heavier metallic elements such as Pb or Ta, microampere-range beams in charge states as high as $+30$ have been produced. Some typical ion beam currents [17] measured from the CAPRICE ECRIS, developed in Grenoble [18], are presented in Figure 8.

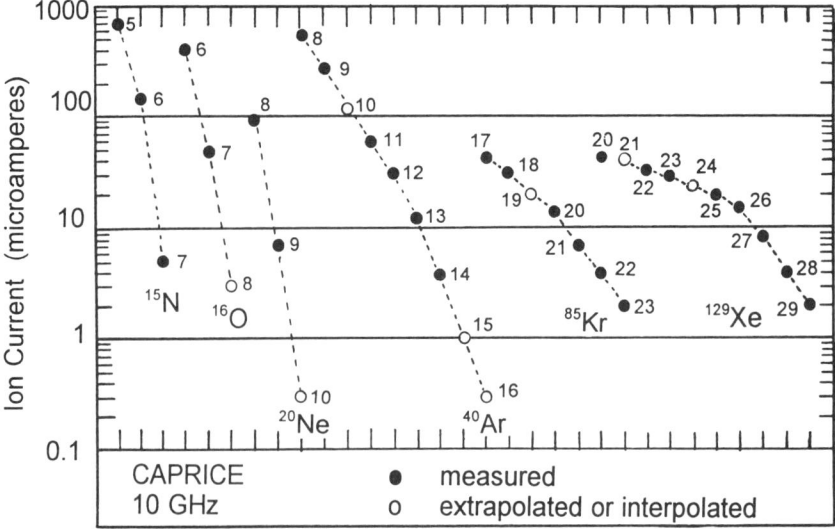

FIG. 8. Ion beam intensities measured in microamperes for the CAPRICE-$2\omega_{ce}$ ion source operating at 10 GHz. The figure is adapted from Hitz et al. [17].

The relative intensities of different ion charge states depends strongly on the ion source operating conditions, particularly the working gas pressure and microwave power. The modeling of such ECRIS discharges is a formidable task, and most of the "conventional wisdom" is derived from experimentation with different design and operating parameters. Gas mixing has been shown to improve charge-state performance and stability. The mixing gases of choice are either He for lighter ions or O_2 for heavier ions and are chosen to be lighter than the working gas. The role of the mixing gas is not yet fully understood.

Since the plasma chamber is a multimode microwave cavity, the number of possible discharge modes is enormous, each depending on the relative axial and radial magnetic field strengths, gas pressures, and microwave power (the main "tuning knobs"). Stability of the discharge is critical, and a useful figure of merit for stability is the ratio of the maximum magnetic field in the plasma chamber to the resonant field, B_{MAX}/B_{ECR}, which should be maximized for most stable operation. A first or preionization stage can also be effective in stabilizing and optimizing the main-stage plasma, and numerous configurations and approaches have been tried, ranging from overdense plasma production to hot filament and secondary electron emitting surfaces. These methods have achieved varying degrees of success and must be regarded as more art than science at the present time.

The microwave frequency is also an important parameter, with higher frequency generally correlating with better high-charge-state performance. The RF field cannot propagate in a plasma when the microwave frequency is higher than the plasma frequency ω_p, given by

$$\omega_p = \sqrt{\frac{4\pi n_e e^2}{m_e}}, \qquad (4.4)$$

where n_e is the electron density and e and m_e are the electron charge and mass, respectively. From Equation (4.4), the maximum plasma electron density n_{max} (in cm^{-3}) that can be sustained in the discharge at a given microwave frequency, f (in hertz), is therefore

$$n_{max} = 1.26 \cdot 10^{-8} f^2. \qquad (4.5)$$

For a typical microwave frequency of 10 GHz, this corresponds to an electron density of 1.3×10^{12} cm^{-3}. Since the maximum plasma electron density scales as the square of the RF frequency, higher frequencies should correspond to plasma conditions which favor multistep ionization. This general trend has been verified in a number of ECRIS devices, although direct comparisons are complicated by possible changes in plasma

TABLE II. Characteristics of Some Currently Operating ECRIS Devices

Source	Location	Frequency (GHz)	Ions produced	Application (remarks)
CAPRICE	CEN, Grenoble, France	10	Ar^{18+}, Ta^{30+}	Atomic physics (LI2A)
CAPRICE-$2\omega_{ce}$	CEN, Grenoble, France	10, 14	Ar^{18+}, Kr^{31+}, U^{52+}	Source development, atomic physics
	ORNL, Oak Ridge, TN	10.6	Ar^{18+}	Atomic physics
	GSI, Darmstadt, Germany	14.5		Linac/ion-storage-ring injector
MINIMAFIOS	CEN, Grenoble, France	18	Ar^{18+}	Source development, atomic physics
ECR4	GANIL, Caen, France	14.5	Xe^{29+}, Ta^{31+}	Cyclotron injector
ECRIS 2	KVI, Groningen, The Netherlands	10	Ar^{16+}	Atomic physics
ISIS	KFA, Jülich, Germany	14.4	Ar^{16+}, Kr^{27+}	Cyclotron injector (superconducting)
ECRIS	University of Giessen, Germany	10		Atomic physics
OCTOPUS	Louvain-la-Neuve, Belgium	8.5	Kr^{20+}, Xe^{23+}	Cyclotron injector, atomic physics
	JAERI, Takasaki, Japan	6.4		Cyclotron injector
AECR	LBL, Berkeley, CA	14		Cyclotron injector, atomic physics
PIIECR	ANL, Argonne, IL	10.25	Cs^{30+}, U^{32+}	Linac injector, atomic physics (on 300-kV platform)

stability due to a decrease in B_{MAX}/B_{ECR} as the frequency (and hence B_{ECR}) is increased. Table II presents a partial list of ECRIS devices currently in operation throughout the world and some of their characteristics.

4.4 Other Techniques

Numerous techniques and technologies have been utilized to produce highly charged ions, although none has been as widely developed and applied as the EBIS and ECRIS. Ion-accelerator-based methods were mentioned in Section 4.1. Noteworthy among the additional techniques are hot plasma generation by the focusing of intense laser pulses onto solid targets and the multiple photoionization of atoms by X rays from synchrotron radiation sources. These methods will be briefly described below.

4.4.1 Laser-Produced-Plasma Ion Source

When laser radiation is focused onto a solid target in vacuum so that the resulting power density exceeds approximately 10^8 W/cm^2, a plasma is produced on the surface which expands outwardly from the target. As the focused laser power density is increased beyond this threshold value, the temperature of the expanding plasma increases, and multiply charged ions are created. Currently, power densities in this range are achievable only using pulsed lasers, although inexpensive tabletop versions of the latter are sufficient. For a more detailed discussion of the application of pulsed-laser-produced plasmas as sources of highly charged ions, the reader is referred to a review [19].

The formation, heating, and expansion of such a transient plasma are not yet fully understood, and only a general description of the process is possible here. The laser radiation can penetrate the surface only to a depth at which the electron density in the solid is such that the plasma frequency becomes equal to the photon frequency (see Equations (4.4) and (4.5)). At this point the laser radiation locally heats the electrons in the solid via inverse bremmstrahlung. The resulting hot electrons produce intense ionization of nearby atoms in the solid, forming a localized plasma. This causes the target material to ablate violently and a dense plasma plume to expand outward from the surface. If the laser pulse is of sufficient duration, it can once again penetrate the plasma when the electron density in the expanding plume has dropped below the cutoff value. The laser frequency and power density are both important parameters in determining the characteristics of the expanding plasma, in which ions may routinely achieve kinetic energies of tens to hundreds of kilo-electron volts.

Using a commercial 10-J pulsed CO_2 laser (λ = 10.6 μm) focused to a power density of 3×10^{10} W/cm^2 on solid targets, beams of C^{6+} and O^{6+} ions and multicharged metal ion beams of Al^{10+} and Fe^{16+} have been analyzed from the plasma plume with sufficient intensity for collision cross-section measurements with gas targets [20]. The experimental setup that was used is shown in Figure 9. The pulsed nature of the laser-produced plasma lends itself ideally to time-of-flight charge and energy analysis. In these experiments, the maximum analyzed ion-beam intensities were on the order of 10^3 ions per pulse at a pulse repetition rate of 1 Hz.

Multicharged ion beam intensities have been increased by many orders of magnitude by the application of axial magnetic fields and ion extraction/acceleration methods [21]. To date, the application of laser-produced-

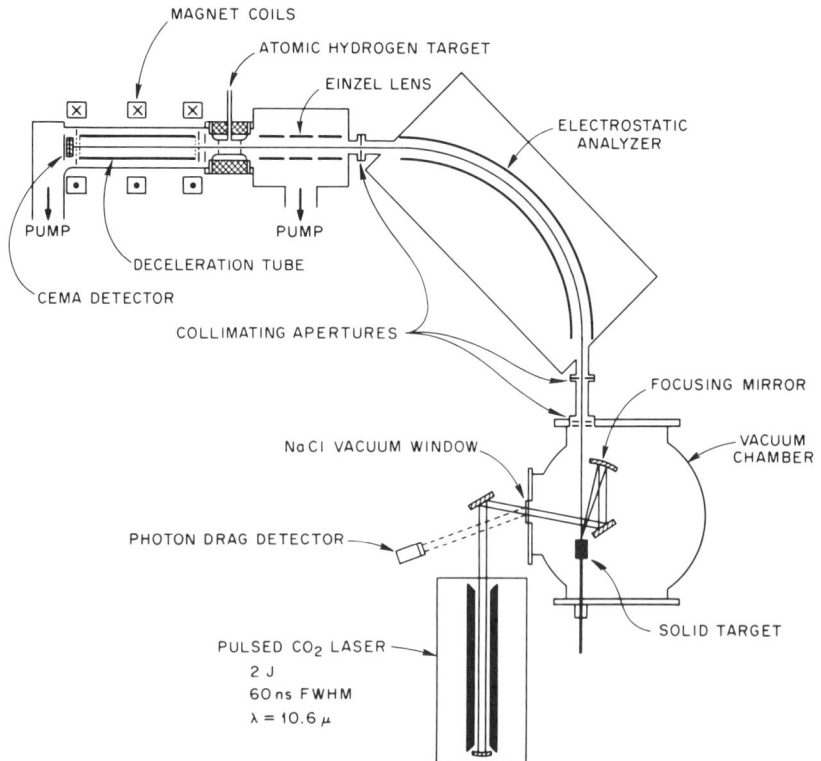

FIG. 9. Application of laser-produced plasma ion source and time-of-flight analysis to the measurement of electron-capture cross sections for multicharged ions in gases. The figure is taken from Phaneuf et al. [20].

plasma ion sources to produce highly charged ion beams for research still remains rather limited [22].

4.4.2 Photoionization Source

The development of intense synchrotron radiation sources for research makes possible the production of multiply charged ions by X-ray photoionization of atoms. In this case a tightly bound (e.g., K-shell) atomic electron is photoionized, and a multiply charged ion results from the ensuing Auger decay cascade as the core-excited ion relaxes. Charge-state reduction by subsequent electron-capture collisions places limits on the neutral gas density that can be used. At currently achievable X-ray intensities, single-collision conditions prevail, so the degree of ionization that is achievable by this method is limited by the energy of the incident photons. At the National Synchotron Light Source at Brookhaven National Laboratory, Ar^{q+} ions in ionization stages up to +6 have been produced using broadband synchotron X-ray radiation and subsequently stored in a Penning ion trap [23]. Approximately 10^4 ions were stored in the trap with this technique. Higher ion charge states should soon become possible as new synchrotron radiation sources having higher photon energies and intensities come on line (e.g., the Advanced Light Source at Lawrence Berkeley National Laboratory and the Advanced Photon Source at Argonne National Laboratory).

References

1. I. G. Brown, in *The Physics and Technology of Ion Sources* (I. G. Brown, ed.), pp. 1–22. Wiley, New York, 1989.
2. H. D. Betz, *Rev. Mod. Phys.* **44,** 465 (1972).
3. C. L. Cocke, *Phys. Rev. A* **20,** 749 (1979).
4. I. G. Brown, ed., *The Physics and Technology of Ion Sources*. Wiley, New York, 1989.
5. B. H. Wolf, ed., *Proceedings of the 4th International Conference on Ion Sources*, Bensheim, Germany, *1991, Rev. Sci. Instrum.* **63** (4, Part II) (1992).
6. R. A. Phaneuf, *AIP Conf. Proc.* **257,** 3–14 (1992).
7. A. Müller, *Phys. Lett. A* **113A,** 415 (1986).
8. G. H. Dunn, in *Electron Impact Ionization* (T. D. Mark and G. H. Dunn, eds.), pp. 277–319. Springer-Verlag, New York and Vienna, 1985.
9. A. Müller, in *Physics of Ion Impact Phenomena*, pp. 13–90. Springer-Verlag, Berlin, 1991.
10. J. Arianer and R. Geller, *Annu. Rev. Nucl. Part. Sci.* **31,** 19 (1981).
11. E. D. Donets, in *The Physics and Technology of Ion Sources* (I. G. Brown, ed.), pp. 245–279. Wiley, New York, 1989.
12. M. A. Levine, R. E. Marrs, J. N. Bardsley, P. Beiersdorfer, C. L. Bennett, M. H. Chen, T. Cowan, D. Dietrich, J. R. Henderson, D. A. Knapp, A.

Osterheld, B. M. Penetrante, M. B. Schneider, and J. H. Scofield, *Nucl. Instrum. Methods Phys. Res. Sect. B* **43,** 431 (1989).
13. M. Stöckli, *Z. Phys. D, Suppl.* **21,** S111, 111–115 (1991).
14. Y. Jongen and C. M. Lyneis, in *The Physics and Technology of Ion Sources* (I. G. Brown, ed.), pp. 207–238. Wiley, New York, 1989.
15. F. W. Meyer and J. W. Hale, *Proc. IEEE Part. Accel. Conf.*, pp. 319–321 (1987).
16. F. W. Meyer, *Nucl. Instrum. Methods Phys. Res., Sect. B* **9,** 532 (1987).
17. D. Hitz, G. Melin, M. Pontonnier, and T. K. Nguyen, *Proc. ECR Workshop, 11th,* Gröningen, the Netherlands, KVI Report 996, p. 91 (1993).
18. B. Jacquot and M. Pontonnier, *Nucl. Instrum. Methods Phys. Res., Sect. A* **287,** 341 (1990); **295,** 5 (1990).
19. R. H. Hughes and R. J. Anderson, in *The Physics and Technology of Ion Sources,* (I. G. Brown, ed.), pp. 299–311. Wiley, New York, 1989.
20. R. A. Phaneuf, I. Alvarez, F. W. Meyer, and D. H. Crandall, *Phys. Rev. A* **26,** 1892 (1982); R. A. Phaneuf, *ibid.* **28,** 1310 (1983); R. A. Phaneuf, M. Kimura, H. Sato and R. E. Olson, *ibid.* **31,** 2914 (1985).
21. O. B. Anan'in, Yu. A. Bykovskii, V. P. Gusev, Yu. P. Kozyrev, I. V. Kolesov, A. S. Pasyuk and V. D. Peklenkov, *Sov. Phys.—Tech. Phys. (Engl. Transl.)* **27,** 903 (1982).
22. T. R. Sherwood, *Rev. Sci. Instrum.* **63,** 2789 (1992).
23. S. D. Kravis, D. A. Church, B. M. Johnson, M. Meron, K. W. Jones, J. C. Levin, I. A. Sellin, Y. Azuma, N. Berrah-Mansour, H. G. Berry, and M. Druetta, *Phys. Rev. A* **45,** 6379 (1992).

5. ELECTRON AND ION OPTICS

George C. King

Department of Physics and Astronomy, University of Manchester, England

5.1 Introduction

In many practical situations it is necessary to transport and control beams of charged particles. This is often achieved most effectively by the use of lenses. The lenses have a focusing action that maximizes the transmission efficiency of the transport system and also allow the energy of the particles in the beam to be varied. Lenses may use electrostatic or magnetic fields or indeed a combination of both. This chapter deals exclusively with the use of electrostatic lenses (see also, for example, Heddle [1], Hawkes and Kasper [2], and Szilagyi [3]). These are suitable for many applications, are easy to construct and operate, and avoid stray magnetic fields. The energies of the particles considered here will generally be less than a few kilo-electron volts, and so relativistic effects will be neglected.

The differential equations that describe the deflection of charged particles by purely electrostatic fields do not include the charge or mass of the particles. So, with the appropriate polarity of the applied voltages an electrostatic lens system may equally well be used for charged particles of any charge or mass, e.g., electrons, positrons, and negatively or positively charged ions. The more massive ions will travel much more slowly than electrons, but they will follow the same trajectories.

5.1.1 Analogy with Optical Lenses

An electrostatic lens consists of two or more electrodes held at different potentials. When designing electrostatic lens systems it is often useful to recall the analogy that exists between these lenses and the more familiar optical lenses as used, for example, in an optical telescope. This analogy is clearly illustrated in Figure 1 which shows a cross-section through an electrostatic lens consisting of two cylindrical electrodes held at potentials of $V_1 = +100$ V and $V_2 = +10$ V with respect to a reference potential that corresponds to the zero of particle kinetic energy. Close to the gap between the two cylinders, the electric field varies strongly which results

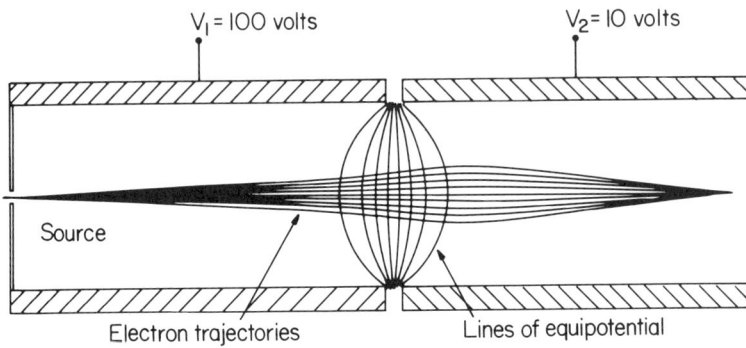

FIG. 1. An electrostatic lens consisting of two cylindrical electrodes held at potentials of 100 and 10 V, respectively. The resultant lines of equipotential are shown together with some typical electron trajectories.

in the sharply changing contours of the equipotential lines. Any charged particle crossing this region will be refracted, and a focusing action can be achieved. This focusing action is demonstrated in Figure 1 for 100-eV electrons emanating from a point source. (For the case of positively charged particles V_1 and V_2 would be -100 V and -10 V with respect to the reference voltage rail.) The analogy between this focusing action and an optical lens is clear, and the use of electron lenses allows the collection of electrons over a large solid angle. In the case of the electrostatic lens, however, focusing can be associated with a change in the kinetic energy of the particles, and it is possible to change the ratio of initial and final energies over several decades or so in a single lens. This is an advantage, for example, when inputting a beam to an energy analyzer whose resolution depends markedly on the particle energy.

As well as having similarities to optical lenses, electrostatic lenses have important differences. One difference concerns their respective aberrations. In general, electrostatic lenses have much larger aberrations than their optical counterparts. In terms of lens design, this means that the aperture, or f number, of an electrostatic lens is usually limited to keep these aberrations to a tolerable level. Fortunately the problem of aberrations is ameliorated by the fact that in most applications electrostatic lenses are used to transport beams of charged particles rather than to produce images. Here an increase in image spot size by say 10–20% because of lens aberrations is not so serious. Another important difference between optical and electrostatic lenses is that the latter deal with particles that repel each other due to their coulomb repulsion. This may limit the beam current that can be transmitted or modify the beam profile and energy distribution.

5.2 Collimation and Definition of a Charged Particle Beam

5.2.1 Windows and Pupils (Beam and Pencil Angles)

In any application it is necessary to define the spatial and angular extents of the charged particle beam. This can be accomplished by the use of just two physical defining apertures, whose roles are illustrated in Figure 2. The first aperture, A_1, defines the radial size of the beam and is called the *window* of the system. From each point within this window, particles are emitted which are assumed here to be emitted isotropically. The second aperture, A_2, determines the angular extent of the beam of particles that are transmitted and is called the *pupil* of the system. Each point within the window gives rise to a *pencil of rays*, whose pencil half-angle θ_p is given by $\theta_p \cong r_p/L$, where r_p is the radius of the pupil aperture and L is the distance between the window and the pupil. Since r_p is usually much smaller than L, the pencils of rays from all points within the window have approximately the same pencil angle. The angle that the central ray of a pencil makes with the optical axis is called the beam angle, θ_B, of the pencil. For the case of the pencil shown in Figure 2, the beam angle is given by $\theta_B \cong r_w/L$, where r_w is the radius of the window aperture. The beam angle will vary for points across the window, but the maximum value is often quoted. The sizes of the physical apertures will obviously

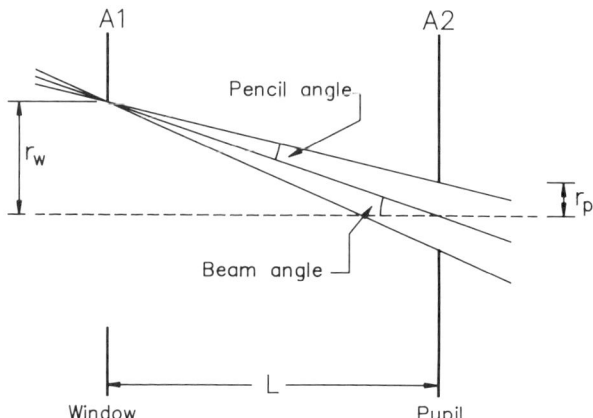

FIG. 2. The radial and angular definition of a beam by a window and pupil.

depend upon the particular situation, but typical diameters are about 1 mm, leading to pencil and beam angles of ~0.1 rad. It is important to note that two physical apertures will completely define a beam of charged particles. The use of more than two physical apertures to limit the beam should be avoided since this can lead to a nonuniform particle density over the beam cross-section which is called *vignetting*.

5.2.2 The Law of Helmholtz–Lagrange

An electrostatic lens will produce images of the physical apertures that define the beam, and lens design may be considered in terms of these windows and pupils and their images. This is illustrated in Figure 3, in which the lens produces an image of the window: such images are often referred to as *virtual* windows. The radius of the image, r_2, is related to the radius of the window, r_1, by the linear magnification of the lens, $M = r_2/r_1$. In passing through the lens, from potential V_1 to V_2, there has also been a change in particle energy and a change in pencil angle (θ) and beam angle. The law of Helmholtz–Lagrange relates the quantities r, θ, and V, and may be written as

$$r_1 \theta_1 V_1^{1/2} = r_2 \theta_2 V_2^{1/2}; \quad (5.1)$$

i.e., the product $r\theta V^{1/2}$ is a conserved quantity.

It is important to note that the angles in this relationship relate to the pencil angles and *not* to the beam angles. Once the pencil angle is determined at some point in the optical system by the physical apertures, at other points it is determined by the Helmholtz–Lagrange relationship. On the other hand the beam angles can be controlled by suitable positioning of the defining apertures. This is also illustrated in Figure 3, in which the pupil has been positioned at the focal plane of the lens. This places the image of the pupil at infinity and so produces a zero beam angle at the image. Equation (5.1) shows that the pencil angle will increase as the beam energy is reduced, and this may have important consequences for any optical components, e.g., energy analyzers that may follow a stage of retardation.

5.2.3 Definition and Measurement of Lens Voltages (Beam Energies)

It is essential that the applied lens voltages V_1, V_2, etc. are measured with respect to the correct reference which is frequently not ground. The correct reference is the zero of particle kinetic energy, i.e., the reference is chosen such that a particle will have a kinetic energy of qV when it is in a region of potential V. It is extremely important to distinguish between this reference and the ground potential.

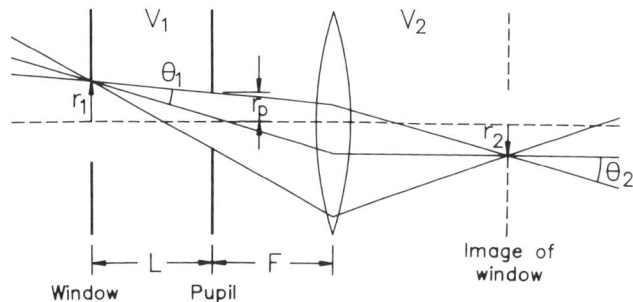

FIG. 3. An illustration of the Helmholtz–Lagrange law and the positioning of a pupil to produce a zero beam angle at the image plane.

5.3 Electrostatic Lenses

5.3.1 The Thick Lens Representation

In elementary optics, the refraction of a ray by a lens is assumed to occur abruptly at a single plane that lies at the center of the lens. In the case of an electrostatic lens (see, for example, Figure 1) the refraction occurs over an extended distance, and the lens is described as a thick lens. This description is illustrated in Figure 4. Here the details of the

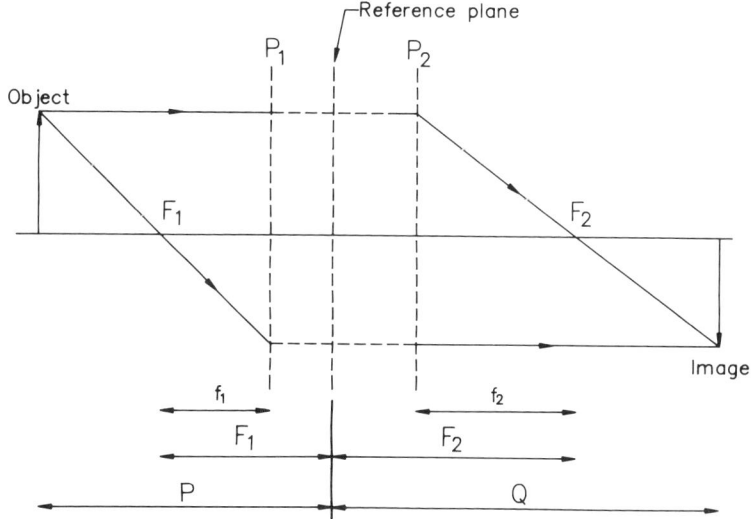

FIG. 4. The thick lens description of an electrostatic lens.

particle trajectories within the region of lens actions are not included. Instead only the asymptotic trajectories are considered, and the lens is represented by two principal planes, P_1 and P_2, each of which has a corresponding focal length, f_1 and f_2 with focal points F_1 and F_2, respectively. The positions of the principal planes, the focal points F_1 and F_2, and also the object distance (P) and image distance (Q) are measured with respect to a reference plane that is usually chosen to be the mechanical symmetry plane of the lens. For example, the reference plane of the two cylinder lens, shown in Figure 1, would lie at the center of the gap between the two cylinders. The distances F_1 and F_2 are called the mid-focal lengths.

The asymptotic trajectories of the particle can be determined as follows.

(i) A particle entering the lens parallel to the optical axis follows a straight line trajectory to principal plane P_2, where the trajectory is refracted such that it leaves the lens through focal point F_2.
(ii) A particle passing through focal point F_1 follows a straight line trajectory to principal plane P_1 and is then refracted such that it leaves the lens parallel to the optical axis.
(iii) Trajectories, parallel at the entrance side of the lens, cross each other at the same point in the focal plane, F_2. This allows an arbitrary trajectory to be traced.

Useful relationships that can be derived from the thick lens geometry shown in Figure 4 include

$$(P - F_1)(Q - F_2) = f_1 f_2 \tag{5.2}$$

$$M = \frac{-f_1}{(P - F_1)} = -\frac{(Q - F_2)}{f_2}, \tag{5.3}$$

where M is the linear magnification (r_2/r_1). The linear magnification of a real image is negative, but it is usual to speak of magnification as though it were positive and ignore the sign.

An example of the thick lens representation of a real lens is shown in Figure 5, in which the vertical scale has been multiplied by a factor of 3. The particular lens is a three-cylinder lens with applied voltages V_1, V_2, and V_3 of 1.0, 8.6, and 5.0 V, respectively, for which $f_1 = 1.061D$, $f_2 = 2.372D$, $F_1 = 2.197D$, and $F_2 = 0.865D$, where D is the diameter of the lens. All lengths associated with the lens, $P_1 Q_1 f_1 f_2$, etc., scale with D and hence are given in units of D. In practice D should be large compared with the sizes of the window and pupil, say by about a factor of 5 or more, and typical values lie in the range 3–50 mm. Note that in Figure 5 the two principal planes, P_1 and P_2, are crossed and lie on the low-voltage

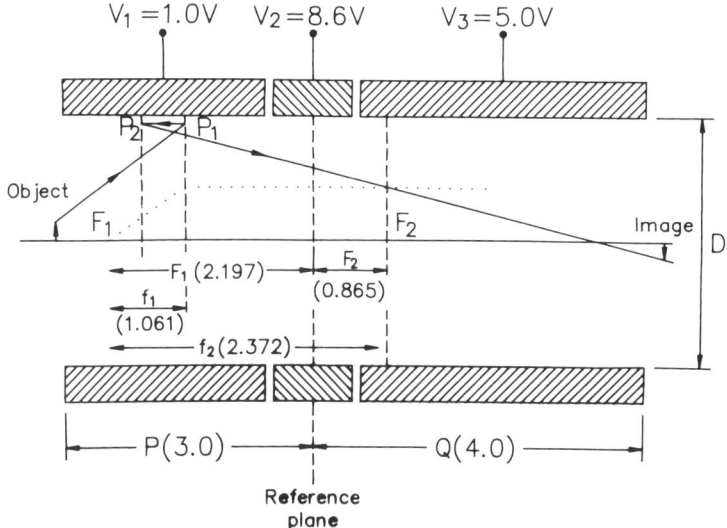

FIG. 5. An example of the thick lens representation of a real lens and the tracing of an arbitrary ray. The vertical scale has been magnified by a factor of 3, and all distances are given in units of D, the internal diameter of the lens.

side of the lens. This is a general characteristic of electrostatic lenses. Figure 5 also shows the tracing of an arbitrary ray. Note that an object in one principal plane is imaged onto the other principal plane with unit magnification. However, there is a change in the angle of the ray.

5.3.2 Lens Data

Values of f_1, f_2, F_1, and F_2 in tabular or graphical form for various electrode configurations and applied voltages are given in a number of sources. The most comprehensive is the book by Harting and Read [4] which, while no longer in print, is held in many libraries. This book includes data for the most common lens geometries including double- and triple-element lenses. The data are given for accelerating lenses only, but these can be easily converted to the case of decelerating lenses. Data have also been published for cylindrical lenses comprising two [5], three [6,7], four [8] and five [9,10] elements and for aperture lenses formed by two [11] or three [12,13] apertures.

5.3.3 Lens Geometries

The two most common types of electrostatic lens are the cylindrical lens and the circular aperture lens, their cylindrical symmetry being well suited to cylindrical beams of charged particles. The focal properties of a lens depend on the number of electrodes it contains, its shape, and the voltages applied. It may be noted, however, that cylinder lenses and aperture lenses of similar diameter have similar focal properties, although cylinder lenses tend to be slightly stronger and to have lower aberrations. In practice, therefore, the choice between the use of cylinder or the use of aperture lenses is usually determined by the mechanical aspects of the lens design, e.g., ease of manufacture and alignment of electrodes. The choice of the number of electrodes in a lens depends upon the particular application as discussed below. In general the more electrodes a lens has, the greater is the degree of control of its focusing and imaging properties. Cylinder lenses will be described in some detail here, but exactly the same principles apply to aperture lenses.

5.3.4 Two-Cylinder Lenses

Such lenses consist of two cylinders separated by a gap, g, which is typically $0.1D$. The length of each cylinder should be large compared with its diameter, so that the axial potential can reach its asympototic value. In practice this means that each length should be greater than about $1.5D$. The focal parameters of the lens, $f_1 f_2 F_1$ and F_2, depend on the ratio of voltages, V_2/V_1, applied to the two electrodes, and values of these parameters for various values of V_2/V_1 are given by Harting and Read [4]. For example, with V_2/V_1 equal to 8.0, $f_1 = 1.00D$, $f_2 = 2.84D$, $F_1 = 1.87D$, and $F_2 = 1.47D$. If the object distance, P, of the lens is $3.0D$, then Equations (5.2) and (5.3) can be used to deduce that the image distance, Q, will be $3.98D$ and the linear magnification of the lens, M, will be 0.88. If D were 5 mm, a typical value, then P and Q would be 15 and 19.9 mm, respectively.

Lens data can be equivalently and more conveniently displayed in the form of PQ curves. In elementary lens optics the object and image distances are related by a rectangular hyperbola, i.e., $(1/P) + (1/Q) = (1/f)$, where f is the focal length. Similarly PQ curves for a two-cylinder electrostatic lens, for example, are represented by rectangular hyperbolas, each one corresponding to a particular value of V_2/V_1. Such a family of PQ curves for a double-cylinder lens, deduced from the data of Harting and Read [4], is shown in Figure 6. These curves also give directly the linear magnification, M. For example, with $V_2/V_1 = 8$ and $P = 3.0D$, the PQ curve gives $Q \cong 4.0D$ and $M \cong 0.9$. The disadvantage of the two-cylinder lens is that, for a fixed object position, the position of the image will change if V_2/V_1 is varied. This problem is overcome in the three-cylinder lens.

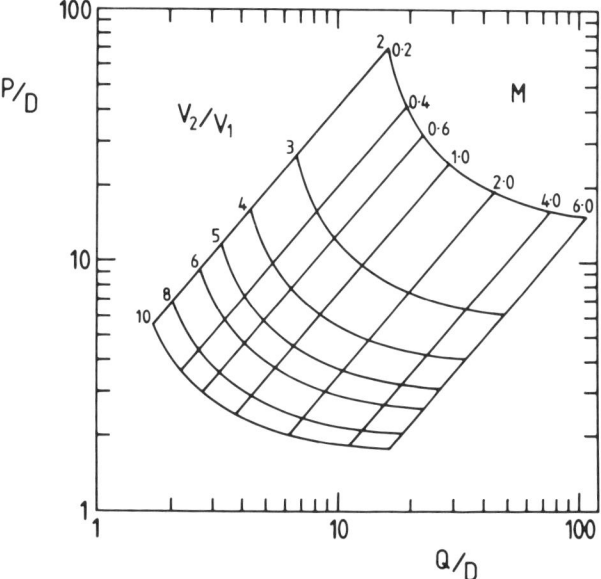

FIG. 6. PQ curves for a double-cylinder lens.

5.3.5 Three-Cylinder Lenses

The three-cylinder lens is shown schematically in the inset of Figure 7. The length, A, of the central cylinder is typically 0.5 or $1.0D$, with the larger value providing the greater range of overall voltage ratio. The focal properties of the three-cylinder lens now depend on two voltage ratios, namely, V_3/V_1 and V_2/V_1, and this gives the lens a very useful property; i.e., the overall voltage ratio V_3/V_1 can be varied while maintaining constant values of object and image distance by suitable adjustment of the focusing voltage V_2/V_1. Because of this property they are often referred to as zoom lenses. It is a very important property since in practical situations the objects and images, e.g., particle sources, detectors, and energy analzyers, are usually fixed in position. The value of V_3/V_1 can be greater or less than unity or equal to unity, when the lens is called an einzel lens. Even though there is no overall voltage change in an einzel lens there is still focusing, and this lens finds many useful applications.

The focal properties of three electrode lenses, as functions of V_3/V_1 and V_2/V_1, are presented in tabular form by Harting and Read [4] and in a more convenient graphical form of a zoom lens curve for particular values of P and Q. An example of such a zoom lens curve is shown in

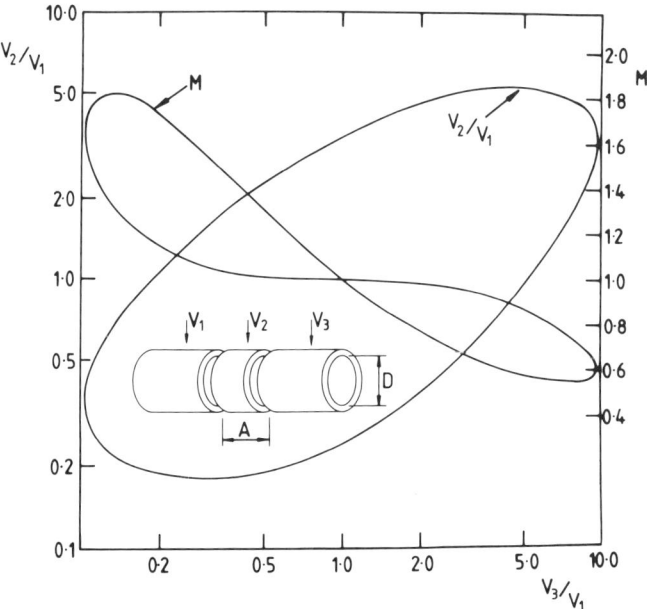

FIG. 7. An example of a zoom lens curve for a three-cylinder lens with $P = 5D$ and $Q = 5D$.

Figure 7 for a triple-cylinder lens with A/D equal to 1.0 and P and Q both equal to $5D$. The zoom lens curve gives the values of V_2/V_1 that should be applied for a given value of V_3/V_1. Note that for a given value of V_3/V_1 there are two values of V_2/V_1. Although both values of V_2/V_1 would provide the necessary focusing action, the numerically higher value usually results in the lower lens aberration. This is because the higher value of V_2 constrains the particles to move closer to the optical axis within the central element when aberrations will be smaller.

Also shown on the zoom lens curve of Figure 7 is the variation in linear magnification, M, of the lens with V_3/V_1. M depends on the values of P and Q but also on the focal properties which change with V_3/V_1 and V_2/V_1. The magnification is also double valued, each value corresponding to the higher or lower value of V_2/V_1. The flatter part of the magnification curve corresponds to the higher range of V_2/V_1, and here the magnification can be held constant to within about $\pm 10\%$ of its average value over the entire range of V_3/V_1. This is another reason for choosing the higher value of V_2/V_1. Any variation in M needs to be taken into account when the variation in the transmission efficiency of the lens with V_3/V_1 is being considered. Values of M close to unity, over the range 0.5 to 2.0, are usually chosen in order to minimize lens aberrations.

The range of the overall voltage ratio V_3/V_1 depends on the strengh of the lens. In light optics the strength of a lens is given by the reciprocal of its focal length $(1/f) = (1/P) + (1/Q)$. Similarly, for electrostatic lenses, the value of $(1/P) + (1/Q)$ gives a measure of the strength of the lens. For values of P and Q of $5D$ (Figure 7) the maximum value of V_3/V_1 is about 10 and the minimum value is about 1/10, i.e., the reciprocal of the maximum value. The stronger the lens is, the greater is the maximum value of V_3/V_1. There is a limit to this, however. If the lens becomes too strong, ray paths start to cross the axis within the region of the lens where the electric field is strongly varying. This should be avoided, especially if the object or image is defined by physical pieces of metal that could distort the electric field. If on the other hand P and Q become too large, the lens becomes weak, leading to a low maximum value of V_3/V_1, and also the lens aberrations increase. An important part of lens design is the choice of the optimum values of P and Q which will include the above considerations. For a given lens application, a particular design criterion can be adopted (e.g., minimum aberration of the image) which can be represented by a figure of merit. The lens data compilation of Harting and Read [4] provides such figures of merit for some important lens applications. Minimum values of P or Q are about $2D$ with typical values in the range 2 to $10D$.

5.3.6 Four-Cylinder Lenses

Four-cylinder lenses provide yet one more degree of freedom with the addition of the extra electrode. For the overall voltage ratio V_4/V_1, there are now two focusing potentials, V_2/V_1 and V_3/V_1. In practice this means that while varying the overall voltage ratio V_4/V_1 it is possible to keep the linear (or angular) magnification of the lens constant as well as maintaining constant values of P and Q. This is particularly valuable when it is important to have a lens with a transmission function as uniform as possible. Four-cylinder lens data for several useful combinations of P, Q, and M are available [8].

5.3.7 Lens Aberrations

The aberrations of an electrostatic lens are analogous to those of optical lenses, e.g., spherical aberration, coma, and barrel distortion. Electrostatic lenses, however, are usually used to transport rather than image charged particle beams. Consequently most of these aberrations can be neglected and usually only spherical aberration is considered and mini-

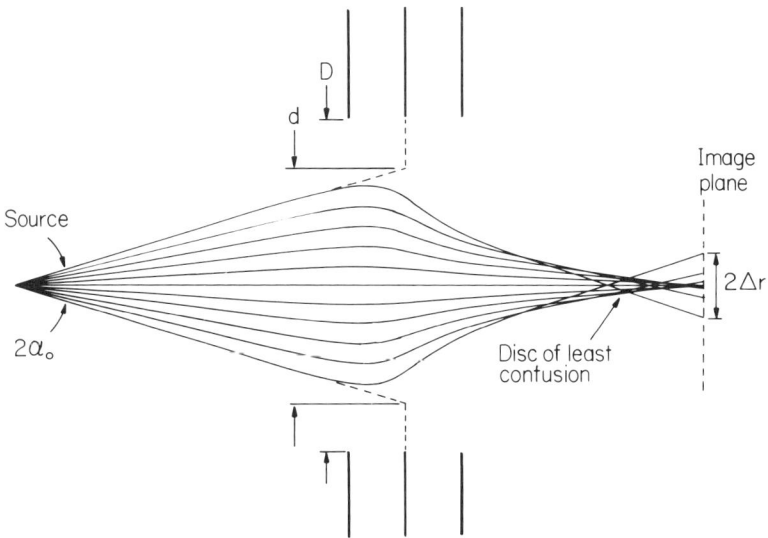

FIG. 8. Schematic diagram of the effects of spherical aberration in an electrostatic lens.

mized. The effects of spherical aberration are illustrated in Figure 8 for the case of a three-aperture lens. Paraxial rays from the object come to a focus at the image plane, whereas nonparaxial rays cross the optical axis before the image plane. This leads to an aberrated image in which the aberration, defined as Δr (see Figure 8), is given by $\Delta r = MC_s\alpha_0^3$, where M is the linear magnification, α_0 is the maximum half-angle of the rays from the object, and C_s is the third-order spherical aberration coefficient. Values of C_s are tabulated alongside the focal parameters of the lens by, for example, Harting and Read [4]. This expression for Δr may be used when the filling factor, η, of the lens is less than about 50%. η is defined as the maximum diameter of the beam in the lens, d, divided by the lens diameter, D, as deduced from the asymptotic trajectories (see Figure 8). Clearly, spherical aberration increases dramatically with increasing α_0 which corresponds to the total angle of the trajectory, i.e., pencil angle plus beam angle, and which should therefore be restricted in size.

As can be seen in Figure 8 there is a position in front of the image plane at which the beam has a minimum diameter, and this is called the *disc of least confusion*. This disc has a diameter that is one quarter that of the aberrated spot at the image plane. This situation can be used to advantage by placing any aperture through which the beam has to pass at the position of the disc of least confusion rather than at the image plane. Alternatively when operating the lens in practice, the mid-focusing potential (V_2) may be adjusted empirically so as to make the lens less strong and place the disc of least confusion at the anticipated image plane.

5.4 Designing Electrostatic Lens Systems

5.4.1 Useful Design Rules

Some of the design rules that have arisen so far are collected together here.

- Two physical apertures should be used to define the beam (the window and the pupil) and no more. Additional limiting apertures can lead to undesirable vignetting.
- Location of these physical apertures in regions at potentials of less than about 10 V should be avoided because the deleterious effects of patch fields, etc., are more damaging at low particle energies. In particular, use should be made of virtual apertures, corresponding to images of physical apertures placed at regions of higher potential.
- Physical apertures should be placed sufficiently far from a lens that the electric field of the lens is not appreciably disturbed. In practice this means that any physical aperture should be placed more than about one or two lens diameters from the gap at which the lens action is occurring.
- At critical points in the optical system, e.g., at the entrance aperture of an energy analyzer, it is important to minimize the angular divergence of the beam by making the beam angle zero at that point.
- Filling factors of electrostatic lenses should be less than about 50% to reduce the effects of spherical aberration.
- It is usually better to use a three- rather than a two-element lens since the former generally has lower aberrations and allows a greater control of focusing properties.
- When using triple-element lenses, which have two possible values of focus potential (V_2/V_1), it is usually better to use the higher of the two values in order to minimize lens aberrations.

5.4.2 Combining Electron Lenses

A single two- or three-element lens can typically operate with overall voltage ratios of up to ~25 (or, correspondingly, down to ~$\frac{1}{25}$). For higher voltage ratios it is necessary to combine two or more lenses. A schematic diagram of a two-lens combination is shown in Figure 9a, in which the

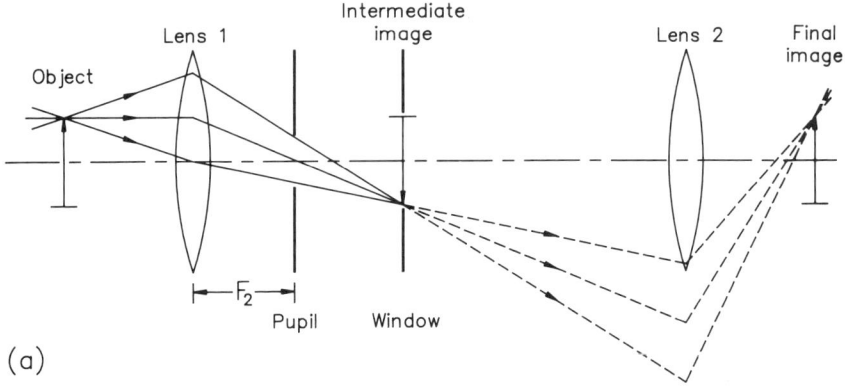

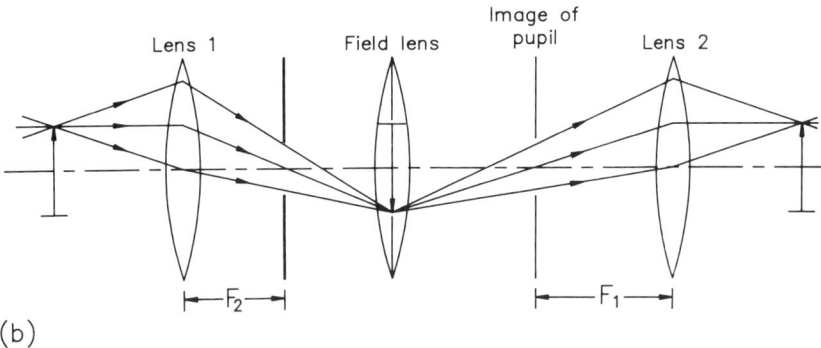

FIG. 9. (a) A two-lens combination with an intermediate image—note the large angles of the rays in the second lens. (b) The two-lens combination with the addition of a field lens.

voltage on the last electrode of the first lens must be the same as the voltage on the first electrode of the second lens. The first lens produces an intermediate image that is then imaged by the second lens to produce the final image. The two physical defining apertures are placed between the lenses. The window produces virtual apertures at the object and final image planes which is particularly useful when it is not possible or advisable to place physical apertures at these two positions. The pupil defines the pencil angle in the region between the two lenses. In other regions the pencil angle is determined by the Helmholtz–Lagrange relationship. By placing the pupil at the focal plane of the first lens in Figure 9a the beam angle at the object has been made equal to zero. The beam angle at the final image, however, is nonzero as shown in exaggerated form in Figure 9a. This leads to a large angular spread of the beam and an increase in the filling factor of the second lens with increased aberration. This problem can be solved using a technique borrowed from light optics: the addition of a *field lens*. This field lens is placed at the position of the intermediate image as shown in Figure 9b or as close as possible to it if a physical aperture is placed there. The field lens does not change the size of the coincident, intermediate image, but images the physical pupil onto the focal plane of the second lens, producing in turn a zero beam angle at the final image.

An alternative arrangement of two lenses that produces a zero beam angle at both the object and the final image, but which does not require a field lens, is shown schematically in Figure 10. The two lenses are separated by the two physical apertures A_1 and A_2. A_2 is placed at the image plane of lens 1 while A_1 is placed at its focal plane. This produces a zero beam angle at the object, and the size of the object is determined by the size of the image of A_2. Lens 2 is placed so that the physical aperture A_1 is its object, and its focal length is chosen so that A_2 falls at its focal plane. Since A_2 is now at the focal plane of lens 2 the exit pupil is at infinity which results in a zero beam angle at the final image. The size of this image is determined by the size of the image of A_1. In effect there has been an interchange of window and pupil: the window of lens

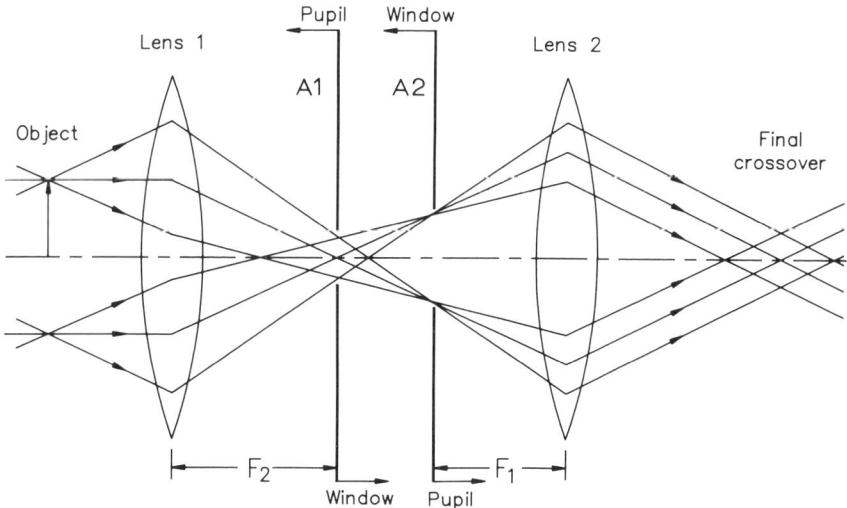

FIG. 10. A two-lens combination with an interchange of windows and pupils.

1 has become the pupil of lens 2 and vice versa. In this arrangement the spatial information in the object has been lost but this is usually not significant in the transport of charged particles.

5.4.3 Electrode Materials and Alignment

Any surface that is exposed to the particle beam must be electrically conducting to avoid any buildup of charge which would lead to spurious potentials. It has been found that some conducting surfaces are better than others for use in electrostatic lenses. These include the surfaces of molybdenum, high-conductivity oxygen-free copper, gold, and colloidal graphite. In particular thin molybdenum sheet (~0.075 mm in thickness) is suitable for lens and defining apertures. These materials are also compatible with ultra-high-vacuum requirements. Aluminum has advantages as a construction material in that it is cheap, light, nonmagnetic, vacuum compatible, and easy to machine. However, it has not been found to be a good electrostatic lens material, and so when using this material it is advisable to coat its surface with colloidal graphite. It is also important to align the lens electrodes accurately. Cylinders may be aligned using three equispaced insulating (e.g., ruby) balls. Aperture lenses may be positioned on three accurately ground ceramic rods that carry insulating spacers.

5.5 Computer Simulation Programs

5.5.1 The Use of Computer Simulation Programs

Computer simulation programs have revolutionized the design of electrostatic lens systems. They solve Laplace's equation for a given set of electrodes and applied voltages to obtain the potential distribution, and then they numerically integrate and display the charged particle trajectories. These programs offer a number of useful features and advantages.

- They enable the trajectories of the particles to be visualized as they pass through the system which is especially important when more than a single lens is involved.
- They show directly the formation of an image and the associated lens aberrations which can be difficult or tedious to evaluate otherwise.
- They allow nonstandard electrode shapes to be incorporated into a design: the available, tabulated lens data are necessarily limited to a relatively few standard electrode configurations, e.g., two- and three-electrode lenses.
- They allow the effects of additional electrodes, such as defining apertures

and deflector plates, on the particle trajectories to be observed: such additional electrodes may also have a focusing action.
- They allow a particular design to be tested and optimized interactively *before* any mechanical part has been manufactured. Perhaps this is the most useful strategy for using a computer simulation program, i.e., to design the system using well-defined design principles and then to optimize this design using the program.

There are a number of simulation programs currently available. Some are two dimensional and exploit the symmetry of a given configuration, e.g., cylindrical or planar, while others are fully three dimensional. Usually the programs allow the voltages on the electrodes to be easily and quickly adjusted. They may have other features such as the addition of magnetic fields and the inclusion of space-charge effects. These programs will be exemplified here by SIMION [14] and CPO-3D [15]. These are highly interactive and versatile programs, and only their basic modes of operation will be described. They run on PC-type machines and require a mouse, a math coprocessor, a VGA display, and, in the case of SIMION, drivers for the display and hard copy.

5.5.2 SIMION

This is a widely used program that is two dimensional and uses the finite difference (FD) method to calculate the electrostatic potential distribution. In this method the space enclosed by the electrodes is not treated as continuous but as a lattice of discrete points. The potentials of all the points are evaluated using a numerical approximation to Laplace's equation, making use of the known potentials at the boundary points. Interpolation can then be used to find the potential of any intermediate point. SIMION, version 4.0, for example, has a mesh size of up to 16,000 points. A mouse is used to specify the electrode configuration, which may have cylindrical or planar symmetry, and assign the applied voltages. The program solves Laplace's equation for the system and displays the computed equipotentials. The user then specifies the initial parameters of the charged particle, e.g., mass, charge, position, energy, and angle, and the program integrates the subsequent trajectory. This trajectory can be displayed and a hard copy taken, or values of the particles parameters at a specific position, e.g., at the image plane, can be obtained. The voltages on the electrodes and also their shapes can be quickly and easily adjusted. The SIMION program has a useful zoom facility which enables a particular area, for example, that containing the image, to be enlarged for closer inspection. The computer simulations of the two-cylinder lens shown in Figure 1 and the spherical aberration in Figure 8 were obtained using the SIMION program. Further details of SIMION and updated versions are available from D Dahl [14].

5.5.3 CPO-3D (Charged Particle Optics Program, Three-Dimensional)

This is a fully three-dimensional program that uses the boundary charge (BC) method to solve Laplace's equation. In this method the electrodes are replaced by charges on their surfaces. The program subdivides the electrodes into flat rectangles or triangles called segments and computes the charge on each segment. Flat segments are used, for example, for cylinders, while triangular segments are used for spheres and cones. Having produced this arrangement of segments with known charges, the program can then calculate the electrostatic potentials and fields *anywhere* in space. The BC method has some advantages over the FD method. It generally requires less memory for a given accuracy, it can handle curved surfaces more easily, and it avoids problems that can arise with unbound systems, i.e., those which are not surrounded by an outer electrode, e.g., an isolated sphere.

Other important features of the CPO-3D program include its three dimensionality and its inclusion of space-charge effects. The three dimensionality allows nonsymmetric electrodes to be used and also more than one symmetry to be present in a simulation, e.g., the cylindrical symmetry of a lens and the spherical symmetry of a hemispherical analyzer. The inclusion of space charge is especially important and allows the program to be applied, for example, to electron guns which includes cathode sur-

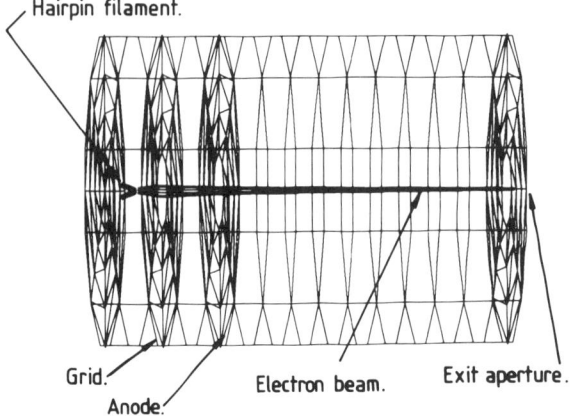

FIG. 11. The use of the CPO-3D program to simulate a triode electron gun.

faces that are notoriously difficult to model. An example of the use of the CPO-3D program is shown in Figure 11. The figure corresponds to a hairpin cathode of tip radius 0.05 mm placed behind a circular aperture (grid) of diameter 0.5 mm which is in turn placed 1.00 mm behind an anode, also of diameter 0.5 mm. The cathode is held at 0 V, the anode is held at 100 V, and the grid voltage has been adjusted *by the program* to give the smallest focus at an exit aperture placed 4 mm in front of the anode. Figure 11 includes some typical electron trajectories. The slight asymmetries in the trajectories are caused by the current in the hairpin (1 A in this example) and the potential drop across it. The figure also shows the way in which the program has divided the circular apertures into triangular segments and the cylinders into rectangular segments. Further details of the CPO-3D program can be obtained from R. B. Consultants, Ltd. [15].

References

1. D. W. O. Heddle, *Electrostatic Lens Systems*, Adam Hilger, Bristol, 1991.
2. P. W. Hawkes and E. Kasper, *Principles of Electron Optics*, Vols. 1 and 2. Academic Press, London, 1989.
3. M. Szilagyi, *Electron and Ion Optics*. Plenum, New York, 1988.
4. E. Harting and F. H. Read, *Electrostatic Lenses*. Elsevier, Amsterdam, 1976.
5. F. H. Read, A. Adams, and J. R. Soto Montiel, *J. Phys. E* **4**, 625 (1971).
6. A. Adams and F. H. Read, *J. Phys. E* **5**, 150 (1972).
7. A. Adams and F. H. Read, *J. Phys. E* **5**, 156 (1972).
8. G. Martinez and M. Sancho, *J. Phys. E* **16**, 625 (1983).
9. D. W. O. Heddle, *J. Phys. E* **4**, 981 (1971).
10. D. W. O. Heddle and N. Papadovassilakis, *J. Phys. E* **17**, 599 (1984).
11. F. H. Read, *J. Phys. E* **2**, 165 (1969).
12. F. H. Read, *J. Phys. E* **2**, 679 (1969).
13. F. H. Read, *J. Phys. E* **3**, 127 (1970).
14. D. A. Dahl, Idaho National Engineering Laboratory, EG & G Idaho Inc., Idaho Falls, 1988.
15. R. B. Consultants Ltd., c/o Integrated Sensors Ltd., PO Box 88, Sackville St., Manchester M60 1QD.

6. ELECTRON ENERGY ANALYZERS

J. L. Erskine

Department of Physics, University of Texas, Austin

6.1 Introduction

Electron energy analyzers are principal components in a broad range of important spectroscopic tools used in scientific research and analytical applications. To obtain optimum performance from an analyzer, it is usually necessary to understand not only its basic operating principles but also some of the more subtle details that ultimately limit performance. This chapter provides a practical guide to understanding, selecting, constructing, and using electron energy analyzers. Essential aspects of their design and operation are described. Useful formulas relating to the more common deflection-type analyzers, and an extensive bibliography of relevant literature, are included.

6.2 Electron Energy Analyzing Systems

Various options are available for measuring the energy and angular distribution of electrons emitted from some source, and careful consideration of the source characteristics and the experimental requirements can lead to improved experimental capabilities by optimizing the choice of electron energy analyzer used. This section discusses several of the more common types of electron energy analyzers and outlines some of the criteria that should be considered in selecting an electron analyzer for a specific application.

6.2.1 Time-of-Flight Devices

Time-of-flight techniques are generally limited to ion detection due to the extremely high velocity of electrons even at low kinetic energies. However, in cases in which electrons are produced from the source region in pulses, a time-of-flight technique might be appropriate for energy analysis. Suitable sources for time-of-flight-based electron energy analyzers include pulsed lasers, Kerr-cell-switched CW lasers, and synchrotron radiation from an electron storage ring. In storage ring light sources, electrons are stored in "bunches" which produce light emission in narrow pulses. Figure 1 illustrates the basic features of a time-of-flight electron energy analyzer [1]. Electrons produced by a pulsed source are allowed to drift at constant velocities in a field-free region (drift tube) of fixed length. The electrons are detected by a microchannel plate (MCP), and their arrival times are recorded by a suitable waveform digitizer. The angular resolution is determined by geometrical considerations such as source region geometry and aperture and detector sizes, but the energy resolution and dynamic range are governed primarily by the detection system response. In practice, geometrical factors typically do not affect the energy resolution, and the source pulse width will be short compared with the overall detector δ-function response time. Therefore, the primary concerns in electron time-of-flight measurements involve possible space-charge effects in the source region (which are more legitimately considered as part of the emission process rather than a factor in the instrument response function) and the response function of the detection system. Both temporal response and dynamic range including possible nonlinearities are important.

Space-charge effects are associated with mutual couloumb repulsion, which, in turn, are associated with high electron densities. Manifestations of space-charge effects are observed in electron guns (especially at the cathode), in high-current electron monochromators [2], and in pulse-laser-excited electron emission from solids [1,3]. A discussion of space-charge effects is beyond the scope of the present chapter; however, the references indicated provide entry into the literature.

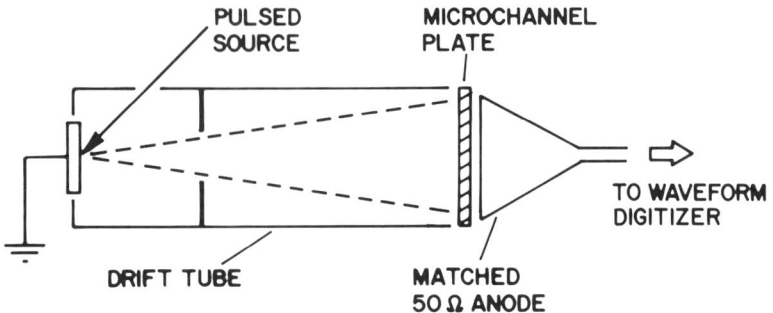

FIG. 1. Schematic diagram showing essential features of a time-of-flight electron energy analyzer. Electrons are excited by a pulsed source. They travel at constant velocity in a drift tube, are amplified and detected by a microchannel plate (MCP) system, and are digitized by suitable electronics.

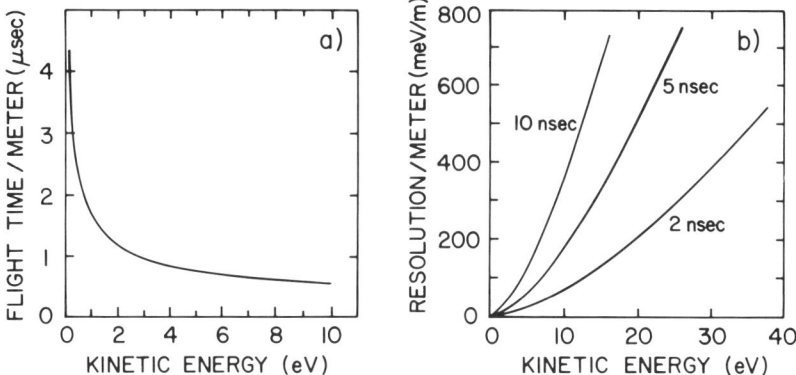

FIG. 2. Graphs that characterize time-of-flight spectrometers. (a) Relationship between flight time per meter and electron kinetic energy; (b) energy resolution per meter of drift tube length for time-of-flight instrument response functions of 2, 5, and 10 nsec.

Figure 2 illustrates the relationships among flight time through the drift tube, electron kinetic energy, energy resolution, and the detection system response time. The major limiting factor in time-of-flight electron energy analysis is the detection system response. Transient waveform digitizers [4] are available with step response times of ≈1 nsec. These devices and fast digital oscilloscopes require amplifications of $\sim 10^7$. Microchannel plate systems [5] can provide gains of 10^6–10^7 and pulse widths of 1 nsec; 300 psec can be achieved in a single MCP amplifier at lower gain. The 1-nsec response time limit imposed by MCP and transient waveform digitizer technology limits time-of-flight electron energy measurements to applications involving low energies and low to moderate resolution. The choice of drift tube length introduces some flexibility, but for practical reasons, such as magnetic shielding requirements, drift tube lengths greater than a few meters are not practical.

A time-of-flight spectrum recorded by the waveform digitizer consists of a convolution of the actual signal, the MCP response, the anode and cable response, and the digitizer response. The most difficult technical problem that must be solved is proper design and construction of the anode/transmission line, including an appropriate anode bias relative to the MCP detector. In order to minimize ringing due to multiple reflection, the anode assembly must present a matched load (usually 50 Ω) to the waveform digitizer, including vacuum feedthroughs, cables, and the anode.

Construction details, time-domain reflectometer tests, and δ-function response measurements, using a 100-fsec laser source, of a 0.5-m time-of-flight instrument that achieves a δ-function response of 2 nsec FWHM, have been described [1]. Nonlinear response and dynamic range limits associated with MCP response to electron pulses are also considered. While mechanisms responsible for limiting the dynamic range and linear response of MCPs are not fully understood [6], manifestations of these effects are straightforward to identify and characterize if suitable attenuators are installed in the drift tube.

6.2.2 Transmission and Retarding Grid Devices

Retarding field devices [7,8] are the simplest type of electron energy analyzers and also serve as models for understanding the effects of retardation in lens columns. Figure 3 illustrates a common retarding grid analyzer configuration. The source is located at the center of two concentric spherical grids and a spherical collector. The source and inner grid are grounded to produce a field-free region. The second grid, biased with a negative potential, $-V_R$, serves as the retarding grid, and the collector is biased positively with respect to the ground.

The retarding grid functions as a high-pass filter. This is illustrated by the inset in Figure 3 which shows the transmission as a function of the kinetic energy of electrons emerging from the source. Ideally if the electrons are emitted from a point source at the center of the grid system,

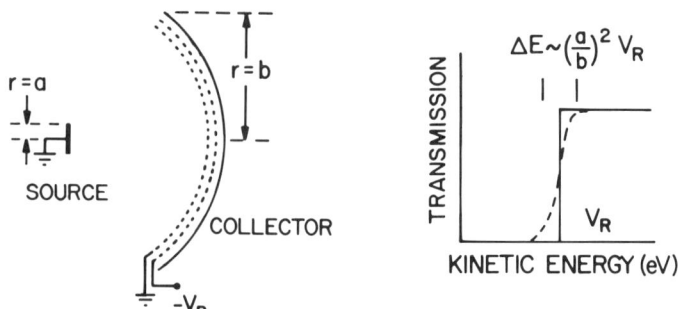

FIG. 3. (left) Retarding grid analyzer. Electrons emitted from the source travel in the field-free region between the source and the first grid. A negative voltage, $-V_R$, applied to the second grid creates a retarding electric field that decelerates the electrons. The collector is biased positive with respect to the source to collect electrons that pass through the grid. (right) The transmission function of the grid system. A finite source size introduces the rounding of ΔE in the ideal step response.

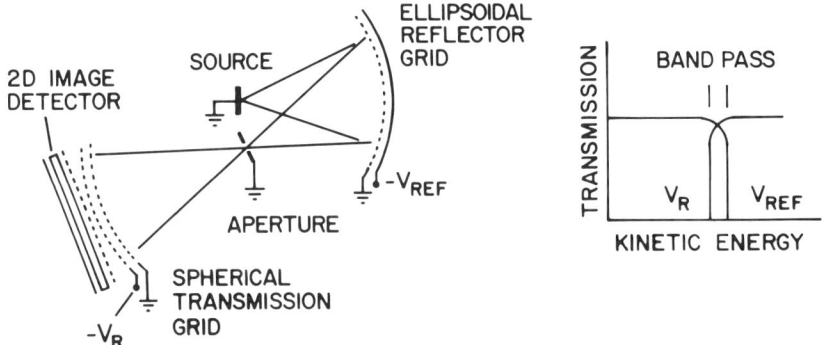

FIG. 4. Schematic diagram of a display-type grid energy analyzer system and associated band pass transmission function.

the transmission function will be a step function. In practice, sources have finite size which results in momentum components perpendicular to the radial field giving rise to a finite energy width, ΔE, in the transmission function. Other factors such as stray magnetic fields, improper alignment, and imperfections in grid shape can also affect trajectories and limit performance.

Figure 4 illustrates a tandem reflection/transmission analyzer that allows bandpass energy filtering. The first grid system functions as a low-pass filter. Electrons having kinetic energy higher than eV_{REF}, where $-V_{REF}$ is the bias applied to the retarding grid, can overcome the retarding field and are collected. Lower energy electrons are reflected and focused through a grounded aperture. The remainder of the system is identical to the high-pass retarding grid analyzer of Figure 3. If a suitable two-dimensional (imaging) detector is placed behind the low-pass analyzer, the angular and energy distribution of emitted electrons can be obtained simultaneously. Energy analyzer systems [8] based on the configuration shown in Figure 4 have been developed that achieve 100-meV resolution and substantial angular acceptance.

6.2.3 Electrostatic Deflection Devices

Electrostatic deflection devices represent the most popular approach to electron energy analysis [9]. Essentially, all electron spectrometers utilize one of the four basic electrostatic deflection analyzer configurations illustrated in Figure 5. These devices are versatile (not limited to low energies as is the time-of-flight technique), are relatively compact, are compatible with ultrahigh vacuum (which presents some problems for

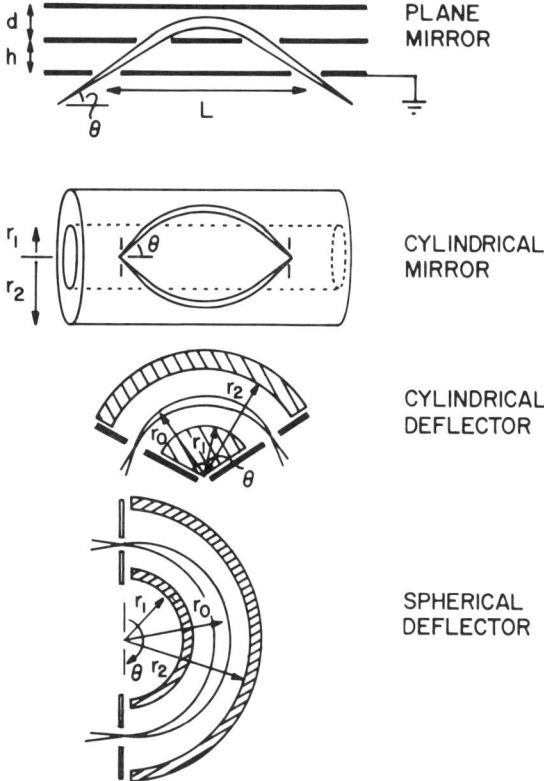

FIG. 5. Four of the most commonly used electrostatic electron energy analyzers. Geometrical parameters shown for each analyzer are used in the text.

magnetic devices that require insulated wire), and offer a variety of attractive features. For these reasons, only electrostatic deflection type analyzers are considered. Magnetic deflection or hybrid (Wein filter) analyzers are not discussed.

Electrostatic deflection and mirror analyzers function by establishing different trajectories for input electrons with different kinetic energies. In mirror-type analyzers, electrons enter the energy dispersing region through a surface corresponding to a natural equipotential. In deflection-type analyzers, electrons enter through slits in a plane perpendicular to the equipotential surfaces. This introduces technical problems involving fringing fields which are discussed later. The plane mirror analyzer [10] provides a simple model for illustrating how field dispersive analyzers function. Electrons enter a field-free region between two grounded plates. Apertures define the angle and the angular spread of trajectories corresponding to electrons that are admitted to the second region, in which a uniform electric field is established by application of a negative potential to the upper plate. Electrons experience constant acceleration and follow parabolic trajectories in the uniform field. Electrons having a specific energy, E_p, the analzyer pass energy, arrive at the exit slit a distance, L, from the entrance slit and escape the analyzer to be detected. Electrons having $E > E_p$ or $E < E_p$ fail to escape the analyzer exit slit.

Geometrical parameters (deflection angles, angles at which electrons enter the analyzer, radii, slit shapes and widths, etc.) determine the energy resolution, the response function, and focusing properties of the various analyzers. The plane mirror [10] and cylindrical [11] deflector analyzers have fields with translational symmetry and therefore offer only one-dimensional focusing. The cylindrical mirror [12] and spherical deflection [13] analyzers offer point-to-point (stigmatic) focusing which can be useful in certain applications. The axial symmetry of the cylindrical mirror analyzer and the convenience of being able to locate an on-axis electron gun inside the inner cylinder have made this configuration very popular for Auger electron spectroscopy including scanning applications.

In many practical applications analyzer design allows for preretardation of the electron energies by suitable grids or lenses. Later, it will be shown that the performance of an analyzer may be significantly improved if electrons are decelerated and energy is analyzed at relatively low kinetic energies. Preretardation permits two distinct operating modes. In one, the voltage that defines the analyzer pass energy is swept to measure the electron kinetic energy distribution, and the relative energy resolution defined by $\Delta E/E_p$ is constant; in the second, the analyzer pass energy, E_p, is held fixed, and the electrons are accelerated or decelerated to E_p to measure the kinetic energy distribution. This second mode maintains a constant absolute energy resolution, ΔE.

6.3 Trajectories and Focusing in Dispersive Deflection Analyzers

Optimum design of deflection-type electron energy analyzers results from applying certain constraints to the analyzer configurations and electron trajectories. This section describes how proper analyzer design leads to optimum performance and summarizes useful formulas for the four analyzer configurations in Figure 5.

6.3.1 Focusing Conditions

The plane mirror analyzer offers the simplest mathematical treatment of a deflection-type analyzer and will be used as an example. It is straightforward to show using the equations of motion for an electron of charge e in a uniform field that the distance L between the two grounded slits (entrance slit and exit slit) is given in terms of the electron kinetic energy E and the angle θ, at which the electron enters the analyzer by

$$L = 2h \cot \theta + 2Ed/eV \sin 2\theta, \qquad (6.1)$$

where V is the potential difference between the plates and the geometrical parameters are as defined in Figure 5. Solving for the energy E yields the expression for the analyzer pass energy E_p,

$$E_p = \frac{e}{d}\left[\frac{L - 2h \cot \theta}{2 \sin 2\theta}\right] V. \qquad (6.2)$$

All of the deflection-type analyzers shown in Figure 5 are characterized by an equation similar to Equation (6.2): $E_p = FV$, where F is a geometrical parameter appropriate to the analyzer configuration. Table I summarizes pass energy formulas for the analyzer configurations in Figure 5.

In addition to the pass energy formulas, it is important to have expressions that characterize analyzer energy resolution and transmission properties. These formulas are also obtained from the trajectory equations. The sensitivity of an electron energy analyzer is proportional to the number of electrons with the correct kinetic energy, here defined as the pass energy E_p, that can pass through the entrance slit and be detected after leaving the exit slit. It is therefore desirable to choose analyzer configurations and analyzer parameters that maximize the angular acceptance at the entrance slit while maintaining conditions prescribed by Equations (6.1) and (6.2) that govern the energy filtering. Such constraints,

TABLE I. Pass Energy Formulas for the Analyzer Configurations in Figure 5

Plane mirror, $\theta = 30°$	$E_p = \left[\dfrac{L - 2h\sqrt{3}}{2d\sqrt{3}}\right] V$
Cylindrical mirror, $\theta = 42.3°$	$E_p = \left[\dfrac{1}{0.763 \ln(r_2/r_1)}\right] V$
Cylindrical deflection, $\theta = 127°$	$E_p = \left[\dfrac{1}{2 \ln (r_2/r_1)}\right] V$
Spherical deflection, $\theta = 180°$	$E_p = \left[\dfrac{r_2}{r_1} - \dfrac{r_1}{r_2}\right]^{-1} V$

TRAJECTORIES AND FOCUSING IN DISPERSIVE DEFLECTION ANALYZERS   217

$$dL/d\theta = 0 \quad \text{(first-order focus condition)} \tag{6.3}$$

$$d^2L/d\theta^2 = 0 \quad \text{(second-order focus condition)}, \tag{6.4}$$

lead to specific geometrical parameters that optimize the analyzer acceptance angle and therefore maximize the instrument sensitivity.

For the plane mirror analyzer, the focus conditions in Equations (6.3) and (6.4) lead to two homogeneous linear equations having a solution given by

$$\det \begin{vmatrix} -\csc^2\theta & \cos 2\theta \\ \csc^2\theta \, \text{ctn}\theta & -\sin 2\theta \end{vmatrix} = 0. \tag{6.5}$$

This leads to $\tan^2\theta = \frac{1}{3}$; i.e., the optimum angle θ for the plane mirror analyzer is 30°. Corresponding analysis for the other analyzer configurations lead to optimum angles for those devices, and these are listed in Table I.

6.3.2 Energy Resolution

Trajectory analysis leads to formulas that define the energy resolution of electrostatic deflection and mirror analyzers [14,15]. The resolving power depends on the entrance and exit slit widths, w_a and w_b, a linear scale factor that specifies the physical size of the analyzer (mean radius, or length) and the angular acceptance of the analyzer. It is sometimes useful to distinguish between the base energy resolution ΔE_B, which results from considering trajectories associated with uniform illumination of the entrance aperture, and the more conventional full-width-at-half-maximum resolution ΔE. For the deflection analyzers in Figure 5 which exhibit essentially linear energy dispersion and at least first-order focusing, generally $\Delta E = \Delta E_B/2$. The base energy resolution ΔE_B for all four analyzers can be expressed generally as [9]

$$\frac{\Delta E_B}{E_p} = Aw + B\alpha^n + C\beta^n, \tag{6.6}$$

where E_p is the pass energy, w is the slit width, and α and β characterize the entrance beam divergence (semi-) angles in the deflection plane (α) and perpendicular to it (β). Equation (6.6) is based on the assumption that $w_a = w_b$; in cases in which $w_a \neq w_b$, w should be replaced by $w = (w_a + w_b)/2$. The parameters A, B, C, and n depend on the specific analyzer configuration and are included in Table II. Table II illustrates two important features of the various analyzer configurations: plane mirror and cylindrical mirror analyzers yield second-order focusing (first terms

TABLE II. Numerical Values of Parameters in Equation (6.6) for the More Popular Electrostatic Deflection Electron Energy Analyzers

	A	B	C	n
Plane mirror, $\theta = 30°$	$3/L$	9.2	1	3
Cylindrical mirror, $\theta = 42.3°$	$2.2/L$	5.55	0	3
Cylindrical deflection, $\theta = 127°$	$2/r_0$	4/3	1	2
Spherical deflection, $\theta = 180°$	$1/r_0$	1	0	2

in the angular contributions appear for $n = 3$), and cylindrical mirror and spherical deflector analyzers exhibit out-of-plane focusing.

6.3.3 Figure of Merit and Luminosity

While Equation (6.6) and Table II characterize analyzer energy resolution in terms of slit width, analyzer size, and angular acceptance, they do not directly specify the optimum choice of these parameters. For a fixed slit width, w, the intensity of electron flux transmitted through the analyzer will increase with the solid angle acceptance at the entrance slit (α and β), but the resolution will begin to diminish when the angle-dependent term contributions in Equation (6.6) approach or exceed the slit-width-dependent term. A strategy is required to optimize the transmitted flux while maintaining the desired resolution.

The parameter that most directly characterizes the figure of merit for an electron energy analyzer is the luminosity. This is defined as the product of the entrance slit area, the entrance solid angle, and the transmission of the analyzer. In most practical situations, the effects of scattering and attenuation from grids in the analyzer optics can be neglected, as well as other factors that can affect transmission (magnetic fields, fringe fields, etc.). The transmission can then be assumed to be unity, and the figure of merit becomes the product of the entrance area and the solid angle. This product is also known as the *étendue* E' of the analyzer. Several authors [15–17] have compared the *étendue* of various analyzer configurations, including those shown in Figure 5, and have examined optimization strategies that govern the choice of how large the angular terms in Equation (6.6) should be in relation to the slit-width term. The results depend on specific incident flux characteristics (the angular and spatial dependence of the intensity at the entrance slit), which in turn depend both on the source characteristics and on the lens behavior, which is discussed in Section 6.5. However, a few general conclusions can be stated for model source distributions based on reasonable assumptions.

Polaschegg [16] has shown that, for uniform slit illumination conditions, the ratio $\alpha^2 r_0/w$ can be optimized for maximum *étendue* by fixing the resolution $\Delta E_B/E_0$ and solving for an extremum of the *étendue* $E' = 2\alpha \cdot 2\beta \cdot lw$. Using the image equation for a 180° spherical deflection analyzer, the resolving power can be shown to be $R_B = E_0/\Delta E_B = 2r_0/(w_a + w_b + 2\alpha_m^2 r_0)$, where w_a and w_b are the entrance and exit slit widths and α_m represents the maximum allowed value of α. The *étendue* becomes, assuming $w_a = w_b$, $E' = \text{const}\,(1 - \alpha_m^2 R_B)\alpha_m r_0/R_B$, and the requirement $\partial E'/\partial \alpha_m = 0$ yields

$$w_a = 2\alpha_m^2 r_0. \tag{6.7}$$

If the exit slit is replaced by a photographic plate or a high-resolution position-sensitive detector, a better assumption is that $w_b \ll w_a$, and a corresponding analysis leads to

$$w_a = 4\alpha_m^2 r_0. \tag{6.8}$$

In other words, when $w_a = w_b$, one seeks input conditions that yield equal contributions to $\Delta E_B/E_0$ from the slit width and α_m^2 terms, but for multichannel detectors, one should choose a smaller (by a factor of 2) value of α_m^2.

Numerical evaluation is required in order to optimize the ratio of $w_a:w_b:\alpha_m^2 r_0$ for maximum luminosity. In this case, the convolution of entrance slit intensity distribution with the exit slit transmission function yields $I(E)$, the intensity of monoenergetic electrons being analyzed. For uniform entrance slit intensity distributions ($dI/d\alpha = \text{const}$, $dI/dw_a = \text{const}$) the optimization procedure leads to nearly the same result as that obtained by optimizing the *étendue* for $w_a = w_b$, $\alpha_m^2 r_0 : w_a : w_b = 0.4:1:1$.

Heddle [15] has examined the relative figure of merit for several dispersive spectrometers. Here again, specific details of source sizes, source distributions, etc. enter into the analysis, but a few general conclusions can be obtained based on reasonable assumptions. If the *étendues* appropriately normalized to account for analyzer physical size (E'/r_0^2, where r_0 is the mean radius of the analyzer) are compared for various spherical deflection and cylindrical mirror configurations, it is apparent that cylindrical mirror analyzers offer the greatest *étendue*. This, coupled with favorable geometry for incorporating a coaxial electron gun, accounts for choosing a CMA for high-resolution-scanning electron beam spectroscopic applications such as scanning AES. Spherical deflection analyzers with suitable lens systems offer especially favorable properties for experiments requiring angular resolution, and their stigmatic focusing properties are important in achieving high performance in both energy resolution and increased sensitivity based on multichannel detection.

Considerable improvements in the total accepted phase space from the source (beam area × collected solid angle) can be achieved by using retardation before analysis. For example, the luminosity of a spectrometer may be increased by the factor $(E/E_0)^2$ if the slit dimensions and acceptance angles are increased to maintain the same resolution, ΔE [17]. This strategy compensates for low source brightness, and transmitted flux is increased by (E/E_0). In general, if the *étendue* of a given spectrometer is $(\Delta E/E_0)^n$, preretardation improves it by $(E/E_0)^{n-1}$ [15]. Additional discussions of preretardation are presented in Section 6.5 in relation to input lens systems.

6.4 Fringing Fields, Terminations, and Guard Rings

Analytical treatments of deflection-type analyzers neglect the perturbing effects of slits and slit plates on the electric fields responsible for focusing and energy dispersion within the analyzer. In mirror-type analyzers, the electrons are injected through a screen-covered aperture that forms a natural equipotential, and fringing field effects are usually negligible. Figure 6 illustrates fringing effects in a practical deflection-type analyzer, in which electrons are admitted into the energy dispersing region through an aperture which is parallel to the desired analyzer electric field. The slit plate not only distorts the fields near the slit, but also shortens the effective path length through which electrons are deflected. The departure of actual electric fields and potentials inside the analyzer from the ideal can affect the ultimate resolving power of the instrument. Two approaches have been used to minimize undesirable effects of fringing fields. One approach is to adjust geometrical parameters such as the total deflection

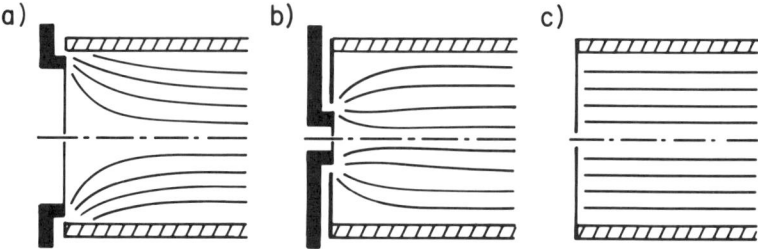

FIG. 6. Equipotentials associated with three slit terminations for deflection-type energy analzyers: (a) conventional equipotential slit, (b) improved design [20] incorporating guard electrodes connected to inner and outer electrodes, and (c) ideal termination based on continuous variation of guard ring potential.

angle and slit-to-analyzer entrance plane distance to compensate for or minimize defocusing effects resulting from fringing fields. Alternatively, suitable guard electrodes and field terminations can be employed.

The most widely used method for correcting errors in electron trajectories resulting from fringing field effects is based on work by Herzog [18], who studied field distributions in parallel plate capacitors, and others [19,20] who have investigated practical deflection-type analyzers. The advent of high-speed digital computers has permitted highly accurate numerical simulation of electron trajectories through realistic cylindrical [21] and spherical [22] deflection analyzer models. These calculations yield optimum values of the parameters that characterize the analyzer deflection angle and the gap as a function or $\Delta r/r_0$ to minimize the effects of fringing fields.

An alternate approach to dealing with fringing fields is to add suitable guard electrodes or field terminations to force the electric fields to maintain the desired spatial distribution near the slits. This strategy preserves mechanical designs based on calculated optimum geometry. Jost [20] has treated a simple approach to field termination which utilizes equipotential plates extending from the outer and inner electrodes of a deflection analyzer. This approach can significantly reduce the effects of fringing fields without changing the deflection angles.

A more accurate field termination can be achieved using individually biased guard rings or suitable resistive films that yield voltage gradients along the slit plane that accurately reproduce the desired electric field dependence. One approach which is relatively simple, yet very effective, is to coat an insulating surface with carbon and clamp the surface to the inner and outer spherical or cylindrical electrodes. While this does not yield the correct potential distribution, it is close enough to yield a very effective terminating field.

6.5 Input Lenses and Operating Modes

A detailed discussion of electron optics is presented elsewhere in this volume, and this section is limited to addressing basic properties of lens systems specifically related to electron energy analyzers.

6.5.1 Energy Analyzer Operating Modes

Figure 7 presents a simplified schematic diagram of a practical electron energy analyzer. Electrons emitted from a source region are focused by a lens system onto the analyzer entrance slit. The lens system defines the maximum source area and source solid angle viewed by the energy ana-

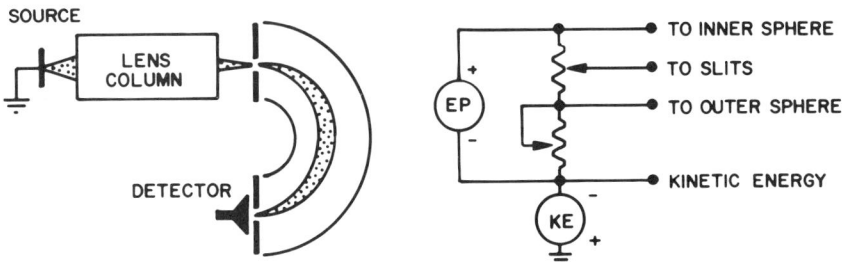

FIG. 7. Schematic diagram of a practical electron energy analyzer system. The resistors are adjusted to yield voltages specified in Table III. When the output of power supply KE is zero, the slit voltage is equal to the pass energy in electron volts.

lzyer and defines the beam angle and cross-section at the entrance slit, and also it accelerates or decelerates electrons of selected initial kinetic energy to the analyzer pass energy E_p. Specific analyzer parameters (slit widths, mean radius, and pass energy) and other factors including field terminations, power supply stability, stray electric and magnetic fields, and the uniformity of analyzer component surfaces (contact potential) all affect the analyzer resolving power. The lens system determines primarily the analyzer angular resolution and principal transmission characteristics. In some cases in which the entrance slit is set to a large width or in which no slit is used and the beam focus at the entrance plane determines a virtual slit width, the lens system can also affect energy resolution through the $w/2r_0$ term in Equation (6.6). A poor lens design or an improperly functioning lens system can cause a very substantial and unnecessary loss of electron analyzer performance.

The bias arrangement for a detection-type analyzer is illustrated schematically in Figure 7. Two independent voltage sources determine the analyzer pass energy and the initial kinetic energy of the transmitted electrons. When the voltage source labeled KE in Figure 7 is set equal to zero, the voltage source EP and the resistor divider establish the correct slit and electrode potentials given by Table III, permitting electrons having energy E_p to pass through the analyzer.

Two options, called analyzer modes, are available for sweeping the analyzer over the desired range of kinetic energies. The constant pass energy mode is achieved by keeping E_p constant and sweeping the KE power supply over the desired range of kinetic energy. If the resistive divider is properly set up, the slit voltage measured to the ground with KE set equal to zero will be equal to the pass energy in electron volts.

Table III. Deflection-Type Analyzer Operating Voltages

Analyzer type	Constraint	Operating voltages
Cylindrical deflection [2]	$\theta = 127°$	$V_0 = V\left(1 + 2\ln\dfrac{r_2}{r_0}\right)$
		$V_i = V\left(1 + 2\ln\dfrac{r_1}{r_0}\right)$
Spherical deflector [13]	$\theta = 180°$	$V_0 = V\left(\dfrac{2r_0}{r_2} - 1\right)$
		$V_i = V\left(\dfrac{2r_0}{r_1} - 1\right)$

Electrons having zero kinetic energy will be accelerated to the analyzer pass energy and will be detected. Electrons having higher kinetic energy are detected by changing the retarding voltage KE to decelerate electrons from KE to E_p. The energy resolution for this mode is constant (independent of retardation voltage) and is given by Equation (6.6). The constant retardation ratio operating mode is achieved by setting the KE power supply to zero and ramping E_p. In this operating mode, the ratio of slit voltage to pass energy is constant, and both the analyzer pass energy and its energy resolution increase proportionally with the measured kinetic energy.

6.5.2 Input Optics

Figure 8 displays three input lens configurations. In configuration A, a grounded grid near the analyzer entrance slit forms a parallel plate lens (with the slit) and establishes a field-free region between the sample and the grid. This arrangement is not of practical interest because the angle at which electrons enter the analyzer is determined by the source position, and, without a source angle defining aperture, there is not a good criterion for judging mechanical alignment. Also, the solid angle and surface area viewed by the analyzer are not defined. Figure 8B represents a more useful lens system and is employed in angle-resolved photoemission [23–26], an application in which an accurately known angular resolution of ~1° is required. The grounded apertures A_1 and A_2 define the source area and maximum accepted solid angle from which electrons are collected. A particularly versatile lens configuration is shown in Figure 8C and offers an electronically variable collection angle [25,27], linear magnification, and a means of limiting the angle at which electrons enter the slit.

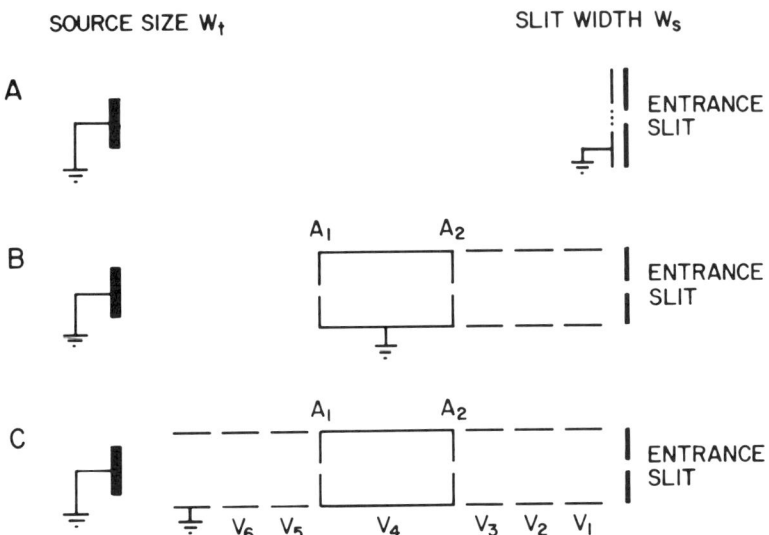

FIG. 8. Input lens configurations. In each case, the slit voltage defines the retarding potential. Lens column B provides a fixed well-defined angular resolution; lens column C offers variable angular acceptance.

6.5.3 Lens System and Analyzer Constraints

Two important constraints on electron trajectories must be considered in designing an input lens system and establishing lens operating modes. One constraint is the maximum angle permitted at the entrance slit, which is given by Equation (6.7) for the specific case of a spherical deflection analyzer. This condition is not a fundamental constraint, but is imposed to optimize analyzer performance as discussed in Section 6.3.3. The second constraint is fundamental and governs the collection efficiency that can be achieved with a specific lens system. This constraint is expressed by the Helmholtz–Lagrange relation which states that the product of beam cross-section, beam angle, and beam energy is conserved. A two-dimensional form applicable to discussing the lens in Figure 8C is

$$\alpha_t w_t \sqrt{E_t} = \alpha_a w_a \sqrt{E_a} = \alpha_s w_s \sqrt{E_s}, \qquad (6.9)$$

where t, a, and s designate the target (source), aperture, and slit, respectively, α_t is the analyzer angular acceptance (angular resolution), w_t characterizes the source region viewed by the analyzer or the source size (whichever is smaller), and E_t is the kinetic energy of the electrons to be detected. In the specific example of the lens system in Figure 8C, α_a and w_a are fixed by the internal apertures, and E_a is the intermediate electron kinetic energy in the field-free region between the apertures. The last set of parameters correspond to the analyzer entrance slit. The parameter $E_s = E_p$ is the pass energy, w_s is the slit width, and α_s is generally chosen to be given by Equation (6.7). Once the analyzer resolution is chosen by selecting w_s and E_p, the total available collection "phase space" at a given kinetic energy from the source is constrained by Equation (6.9).

6.5.4 Practical Limitations

Equation (6.9) establishes the fundamental constraints on electron trajectories associated with any lens system. However, the actual lens configuration (number of lens elements and their relative spacings) determines the practical range over which the parameters can be effectively controlled. In general, lenses with more elements offer greater flexibility and a greater range of operation than simpler lenses. For example, the lens system in Figure 8C offers a lens close to the source region that can be used to vary the collection angle at the sample [25,27] subject to the constraint

$$\alpha_t = \alpha_a \left(\frac{w_a}{w_t}\right)\left(\frac{E_a}{E_t}\right)^{1/2}. \quad (6.10)$$

A minimum requirement for an electron energy analyzer lens system is that it provide energy retardation capability over a desired range while maintaining the ability to image the source at the entrance slit at a fixed magnification. A three-element lens offers two variable voltage ratios and can be used as a zoom lens to focus a source of electrons so as to preserve the image distance while varying the ratio of initial to final kinetic energy. A three-element zoom lens operated in this manner cannot, in general, maintain a constant linear or angular magnification and can maintain its zoom feature (constant image distance) only over a limited range of retardation ratios.

Multielement lenses having more than three elements (and more than two voltage ratios) generally permit the zoom feature to be extended over a greater retardation ratio range [27–30]. In addition, the added degree of freedom of a four (or more)-element complex lens permits an additional parameter (either the linear magnification or the angular magnification) to be held constant while changing the retardation ratio. It is clear that a lens having at least four elements is a good starting point for a flexible electron analyzer design.

6.5.5 Evaluation of Lens Column Operating Modes

Commercial electron energy analyzer systems usually include a control unit that provides all voltages required to operate the analyzer. The low cost of high-performance personal computers, the availability of precision programable voltage sources, and the new generation of laboratory-oriented software (Lab View), however, now offer an attractive option to turnkey analyzer control units.

Optimum lens column operating voltages can be determined empirically in analyzers having simple lens configurations and angle limiting apertures [26]. The counting rate as a function of lens voltages for a fixed kinetic energy and given pass energy can be used as a criterion for finding the combination of lens voltages that yields the highest transmission. Equation (6.9) and the constraint on entrance slit angle (Equation (6.7)) can be used to define the range of kinetic energies over which a lens operating mode can yield a prescribed energy resolution. Electron trajectory analysis of the empirically determined mode can be used to ensure that the required angular conditions are met at the entrance slit and that the electron trajectories in the lens column are consistent with a high-transmission mode.

6.6 Multichannel Energy Detection and Imaging

Minor modifications to conventional electron energy analyzers can yield significant improvements in performance by permitting image detection or multichannel energy detection [31,32]. There are other strategies such as dispersion compensation [33] that can improve analyzer detection sensitivity, but these techniques will not be discussed here. If the exit slit of a hemispherical analyzer is replaced by a suitable field termination and a spatially sensitive detector having an active width $\geq \Delta r = r_2 - r_1$, then, in principal, all electrons that do not collide with the outer or inner hemisphere can be detected as shown in Figure 9. A second possibility is to keep the slit and place a spatially sensitive detector a suitable distance away from the slit to sense an image of the source region.

Hemispherical and cylindrical mirror analyzers are especially suitable for imaging and multichannel detection due to their stigmatic focusing characteristics. The imaging properties of hemispherical analyzers in relation to multichannel energy detection have been studied in detail [31]. Energy dispersion across the analyzer exit plane is slightly nonlinear, and the image size and shape also depend on various parameters including the exit plane location and entrance slit angles. However, in practical cases, these factors are small, and it is possible to achieve essentially uniform multichannel energy detection over nearly the entire exit plane.

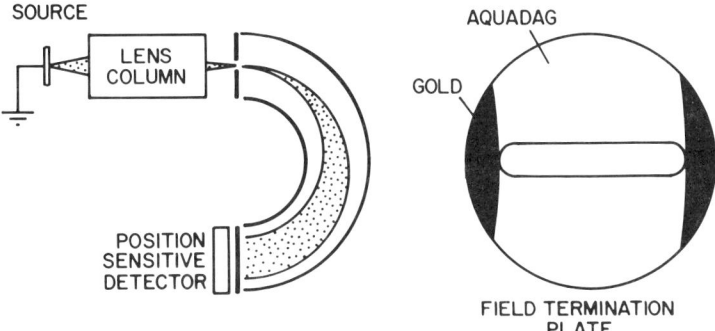

FIG. 9. (left) Schematic representation of a multichannel detection electron analyzer. (right) A field termination technique that has yielded 2 meV energy resolution. Shaded regions are gold plated to contact the inner and outer spheres. The unshaded region is coated with carbon using Aquadag or Aerodag. A high-transmission grid at the slit potential is placed in the aperture to terminate fields from the channel plate bias.

Critical factors that limit the quality of multichannel detection and the fraction of exit plane area over which multichannel detection can be carried out are the field termination at the exit plane and the beam angles at the entrance slit. The lens column must also provide good focusing characteristics for all electrons within the energy detection window (low chromatic aberrations). Channel plates are used in conjunction with anode arrays [32] or resistive anodes [31] to achieve multichannel energy detection. To obtain maximum benefits from multichannel detection, it is desirable to choose a large value for $\Delta r = r_2 - r_1$. However, fringing field problems become more pronounced as $\Delta r/r_0$ increases, and careful attention to field terminations is required at both the entrance slit and the exit plane. Suitable field terminations [31] have permitted energy resolution below 3 meV to be achieved in a high-resolution electron energy loss spectrometer.

6.7 Construction Hints and Practical Details

There are a few practical considerations affecting electron energy analyzer performance that are extremely important and are summarized here for completeness [34]. Stray magnetic fields can dramatically degrade the performance of a spectrometer. The transmission can be reduced, especially in analyzers having high angular resolution, by magnetic fields between the source and the entrance slit, and resolution as well as transmission can be affected by magnetic fields in the energy dispersive region of the analyzer. Several authors have considered magnetic fields and shielding in relation to energy analyzers [35]. As a practical guide for estimating the maximum tolerable magnetic field, the following expression is useful:

$$H_m = 6.74 E^{1/2} \, d/(L)^2, \qquad (6.11)$$

where H_m is the field (in gauss), E is the electron kinetic energy (in electron volts), d is the allowed beam deviation (in centimeters), and l is the path length (in centimeters).

Adequate electrostatic shielding is also extremely important. Charging of insulators or applied voltages on wires can create strong perturbing fields. Care must be exercised not only at the sample, at which mounting insulators, biased wires, etc. must be shielded, but also in the lens column, near the slits, and in the analyzer itself.

In cases in which extremely high energy resolution is required, inhomogeneities in the electric field that defines the analyzer pass energy can reduce the ultimate resolving power [36]. Molybdenum appears to be the most desirable metal for surface potential uniformity; however, all spectrometers that achieve extremely high energy resolution (a few millielectron volts) are traditionally coated with carbon. Alcohol suspended graphite (Aerodag) is very convenient for this purpose. Equally important in achieving high resolution is the stability of the power supplies that provide lens and analyzer voltages. The slit and pass energy voltages (refer to Figure 7) are the most critical. Power supplies having microvolt resolution and stability and very low noise are available.

Finally, it is important to realize that careless or inaccurate mechanical assembly or alignment can reduce spectrometer performance. Small inaccuracies in lens alignment can move a beam focus position and cause severe intensity loss. Distorted grids or a deformed deflection electrode can result in significant loss in ultimate analyzer performance.

Acknowledgments

This work was supported by NSF DMR 9303091 and the R. A. Welch Foundation.

References

1. D. C. Anacker and J. L. Erskine, *Rev. Sci. Instrum.* **62,** 1246 (1991).
2. H. Ibach, *High Resolution Electron Energy Loss Spectrometers*. Springer-Verlag, Berlin, 1991.

3. K. Giesen, F. Hage, F. J. Himpsel, H. J. Riess, and W. Steinmann, *Phys. Rev. B* **33**, 5241 (1986); G. Farkas and C. Toth, *Phys. Rev. A* **41**, 4123 (1990); D. M. Riffe, X. Y. Wang, M. C. Downer, D. L. Fisher, T. Tajima, and J. L. Erskine, *J. Opt. Soc. Am.* **10**, 1424 (1993).
4. Tektronix offers several wide bandwidth (200 MHz) Transient Waveform Digitizers that are suitable for time-of-flight measurements (Tektronix 7912 AD).
5. G. Beck, *Rev. Sci. Instrum.* **47**, 849 (1976); J. P. Boutot, J. D. Delmotte, J. A. Miehe, and B. Sipp, *ibid.* **48**, 1405 (1977).
6. J. L. Wiza, *Nucl. Instrum. Methods Phys. Res.* **162**, 587 (1979); G. W. Fraser, *ibid.* **221**, 115 (1984).
7. J. A. Simpson, *Rev. Sci. Instrum.* **32**, 1283 (1961); T. H. DiStefano and D. T. Pierce, *ibid.* **41**, 180 (1970); C. C. Chang, *Surf. Sci.* **25**, 53 (1971); N. J. Taylor, in *Techniques of Metals Research* (R. F. Bunshah, ed.), Vol. 7, pp. 117–159. Wiley (Interscience), New York, 1972; N. J. Taylor, *Rev. Sci. Instrum.* **40**, 792 (1969).
8. D. E. Eastman, J. J. Donelan, H. C. Hein, and F. J. Himpsel, *Nucl. Instrum. Methods* **172**, 327 (1980); also see A. Clarke, G. Jennings, and R. F. Willis, *Rev. Sci. Instrum.* **58**, 1439 (1987), for other wave-vector imaging analyzers.
9. A comprehensive review of electron energy analzyers for surface analysis has been prepared by D. Roy and J. D. Carette, in *Topics in Current Physics* (H. Ibach, ed.). Springer-Verlag, Berlin, 1977.
10. N. V. Smith, P. K. Larsen, and M. M. Traum, *Rev. Sci. Instrum.* **48**, 454 (1977); T. S. Green and G. A. Proca, *ibid.* **41**, 1409 (1970).
11. A. L. Hughes and V. Rojansky, *Phys. Rev.* **34**, 284 (1929); A. L. Hughes and J. H. McMillen, *ibid.*, p. 291.
12. P. W. Palmberg, *J. Vac. Sci. Technol.* **12**, 379 (1975); V. V. Zashkvara, M. I. Korsunski, and O. S. Kosmachev, *Sov. Phys.—Tech. Phys. (Engl. Transl.)* **11**, 96 (1966); H. Sar-el, *Rev. Sci. Instrum.* **38**, 1210 (1967); **39**, 533 (1968); J. H. Risley, *ibid.* **43**, 95 (1972); E. Blauth, *Z. Phys.* **147**, 228 (1957); W. Mehlhorn, *ibid.* **160**, 247 (1960).
13. E. M. Purcell, *Phys. Rev.* **54**, 818 (1938); C. E. Kuyatt and J. A. Simpson, *Rev. Sci. Instrum.* **22**, 952 (1951); J. A. Simpson, *ibid.* **35**, 1698 (1964).
14. Equations governing trajectories, focussing, energy resolution, pass energies, and analyzer voltages are given in the above references for the plane mirror analyzer (PMA) [10]; the cylindrical deflection analyzer (CDA) [11]; the cylindrical mirror analyzer (CMA) [12]; the spherical deflection analyzer (SDA) [13], and in various reviews; i.e., CDA [2].
15. D. W. O. Heddle, *J. Phys. E* **4**, 589 (1971).
16. H. D. Polaschegg, *Appl. Phys.* **9**, 223 (1976).
17. J. C. Helmer and N. H. Weichert, *Appl. Phys. Lett.* **13**, 266 (1968).
18. R. Herzog, *Z. Phys.* **97**, 596 (1935).
19. H. Wollnick and H. Ewald, *Nucl. Instrum. Methods* **36**, 93 (1965).
20. K. Jost, *J. Phys. E* **12**, 1001 (1979).
21. D. Roy and J.-D. Carette, *Appl. Phys. Lett.* **16**, 413 (1970); C. Oshima, R. Franchy, and H. Ibach, *Rev. Sci. Instrum.* **54**, 1042 (1983); C. Oshima, R. Souda, M. Aono, and Y. Ishizawa, *ibid.* **56**, 227 (1985).
22. M. J. Sablik, J. P. Winningham, and C. Gurgiolo, *Rev. Sci. Instrum.* **56**, 1320 (1985); S. Nishigaki and S. Kanai, *ibid.* **57**, 225 (1986).

23. H. A. Stevens, A. W. Donoho, A. M. Turner, and J. L. Erskine, *J. Electron Spectrosc. Relat. Phenom.* **32**, 327 (1983).
24. E. W. Plummer, *Nucl. Instrum. Methods* **177**, 179 (1980).
25. S. D. Kevan, *Rev. Sci. Instrum.* **54**, 1441 (1983).
26. G. K. Ovrebo and J. L. Erskine, *J. Electron Spectrosc. Relat. Phenom.* **24**, 189 (1981).
27. A. Sellidj and J. L. Erskine, *Rev. Sci. Instrum.* **61**, 49 (1990).
28. G. Martinez and M. Sancho, *J. Phys. E* **16**, 625 (1983); G. Martinez, M. Sancho, and F. H. Read, *ibid.*, p. 632.
29. M. V. Kurepa, M. D. Tasic, and J. M. Kurepa, *J. Phys. E* **7**, 940 (1974).
30. D. W. O. Heddle and N. Papadovassilakis, *J. Phys. E* **17**, 599 (1984).
31. F. Hadjarab and J. L. Erskine, *J. Electron Spectrosc. Relat. Phenom.* **36**, 227 (1985).
32. L. J. Richter and W. Ho, *Rev. Sci. Instrum.* **57**, 1469 (1986).
33. S. D. Kevan and L. Dubois, *Rev. Sci. Instrum.* **55**, 1604 (1984).
34. J. H. Moore, C. C. Davis, and M. A. Coplan, *Building Scientific Apparatus.* Addison-Wesley, Reading, MA, 1989.
35. C. J. Powell, *Methods Exp. Phys.* **7B**, 275–305 (1968); M. E. Rudd, in *Low Energy Electron Spectrometry* (K. D. Sevier, ed.), pp. 32–34. Wiley (Interscience), New York, 1972; W. G. Wadley, *Rev. Sci. Instrum.* **27**, 910 (1956).
36. G. J. Schulz, *Rev. Mod. Phys.* **45**, 378 (1973).

7. ELECTRON POLARIMETRY

Timothy J. Gay[1]

Physics Department, University of Missouri, Rolla

7.1 Introduction

The use of electron spin as an experimental variable in studies of atomic physics provides detailed information about a variety of phenomena which is unavailable if spin is ignored. For example, analysis of the dynamics of electron exchange or tests of the importance of spin–orbit forces on continuum electrons in electron–atom collisions rely crucially on knowledge of the spins of the participating electrons. Such measurements require a source of polarized electrons, an electron polarimeter, or both. Spin experiments are being reported with increasing frequency. This is due in large part to advances over the last two decades in polarized electron technology, with regard to both sources and polarimeters. In particular, the advent of GaAs polarized electron sources and compact, efficient Mott polarimeters has dramatically lessened the difficulty associated with such experiments. These advances have led to major developments in condensed matter, nuclear, and particle physics as well as atomic physics. The physics of polarized electrons and their applications have been reviewed thoroughly in an excellent book by Kessler [1].

An ensemble of electrons is said to be spin polarized if, relative to an arbitrary axis of quantization, i, the number of electrons with spin up, $N\uparrow$, differs from that with spin down, $N\downarrow$. The degree of polarization relative to i is defined to be

$$P_i \equiv \frac{N\uparrow - N\downarrow}{N\uparrow + N\downarrow}. \tag{7.1}$$

More generally, the polarization vector is given by $\vec{P} = tr[\rho\vec{\sigma}]$, where ρ is the electron ensemble's density matrix and $\vec{\sigma}$ is the Pauli spin operator. Electron polarimetry is the measurement of P_i or $\vec{P}$.

Electron polarimeters can be separated roughly into two classes: those for the analysis of high-energy electrons (>1 MeV) based on electron–

---

[1] Current address: Department of Physics and Astronomy, Behlen Laboratory, University of Nebraska, Lincoln.

electron (Møller) or electron–photon scattering and those designed for analysis of relatively low-energy electrons (<1 MeV), based on Mott scattering or inelastic exchange excitation of atoms. High-energy devices are used exclusively in nuclear and particle physics experiments, whereas low-energy polarimeters are used in atomic, condensed matter, and some nuclear physics measurements. It is the latter type that we consider here.

We take the phrase "Mott scattering" to refer generally to collisions between electrons and single atoms or groups of atoms in which spin–orbit forces act on the continuum electrons, causing a left–right scattering asymmetry in the plane perpendicular to $\vec{P}$, an effect first predicted by Mott in 1929 [2]. For such an asymmetry to be appreciable, the target atoms must have high Z (>50). The scattering asymmetry A, which may be measured using the arrangement shown in Figure 1a, is defined as

$$A = \frac{N_L - N_R}{N_L + N_R}, \qquad (7.2)$$

where $N_L(N_R)$ is the intensity measured at the "left" ("right") detector. The component of electron polarization perpendicular to the scattering plane defined by the detectors is given in turn by

$$P = A/S_{\text{eff}}(E, \theta), \qquad (7.3)$$

where S_{eff}, or the "effective Sherman function," is the polarimeter's analyzing power. It depends both on the polar angle θ at which the detectors are placed and the incident electron energy E.

The dynamical cause of the Mott asymmetry is best understood by considering the semiclassical scattering potential experienced by an electron in its rest frame when scattering from a bare nucleus: $V = [(k/r) + (k'/r^3) M_l M_s]$. Here M_l and M_s are the orbital and spin angular-momentum projections of the continuum electron perpendicular to the scattering plane, k and k' are constants, and r is the electron–nucleus separation. For electrons with a given spin polarization, the second (spin–orbit) term of V will be positive or negative relative to the first (coulomb) term, depending on the sign of M_l. This depends in turn on whether the impact parameter of the incident electron is to the left or right of the target nucleus in the scattering plane (Figure 1b). For heavy nuclei, the spin–orbit and coulomb terms can be comparable, and the normal coulomb potential can be sufficiently modified to result in a significant left–right asymmetry.

Two reviews of Mott polarimetry have been published [3,4], and the topic is discussed in some detail in a number of other, more general reviews [5–10]. Journal articles of note, either for their overviews of the field or for their particularly detailed, careful discussions, are referenced below [11–18].

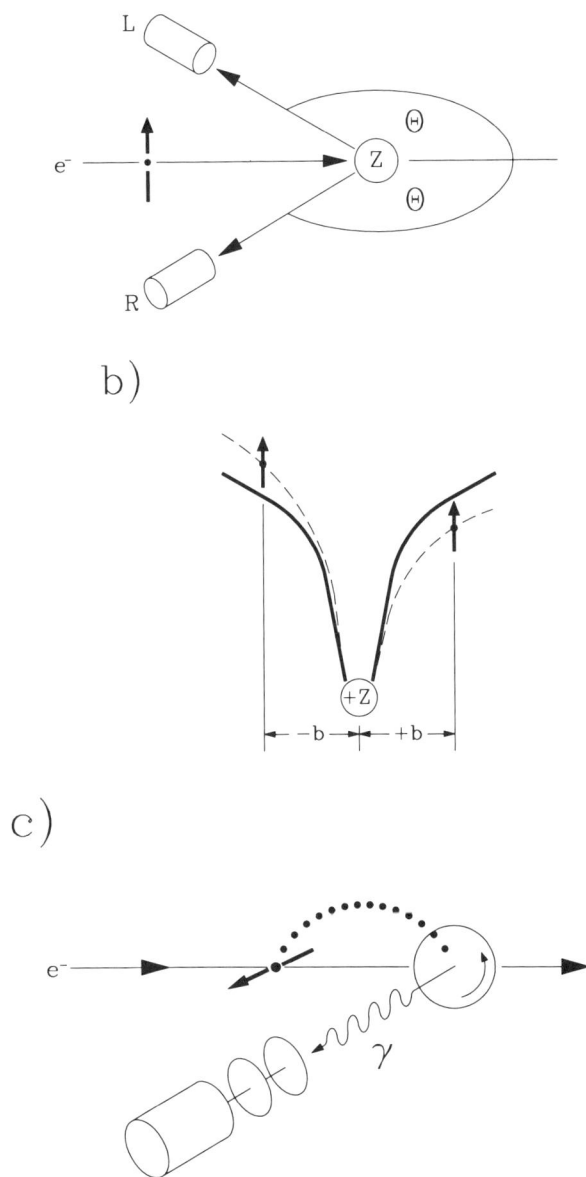

FIG. 1. (a) Mott scattering geometry. (b) Electron scattering from a bare nucleus. The coulomb potential is indicated by the boldface lines; impact parameter-dependent spin–orbit perturbations are indicated by the dashed lines. (c) Geometry for optical electron polarimetry.

Optical electron polarimetry, as yet a nascent technology, involves exchange excitation of atoms by the electrons whose polarization is to be measured. In essence, the spin angular momentum of the incident electrons is converted to the oriented orbital angular momentum of the atomic excited states after the incident electrons have "taken up residence" in the atom. This conversion of spin to orbital orientation is accomplished by the spin–orbit forces active in the atom, as opposed to those acting on a scattering continuum state. The oriented atomic state will emit light with circular polariation along $\vec{P}$ when it decays (Figure 1c). In this case,

$$P = \eta_2 \Lambda, \qquad (7.4)$$

where η_2 is the relative Stokes parameter associated with the circular polarization of the light and Λ, the analyzing power, is a constant dependent on the quantum numbers of the excited state. The Stokes parameter η_2 is analogous to A (Equation (7.2)) and is given by

$$\eta_2 = \frac{I^+ - I^-}{I^+ + I^-}, \qquad (7.5)$$

where I^+ (I^-) is the detected intensity of light emitted along $\vec{P}$ with right (left)-handed circular polarization. Only a few reports of optical electron polarimetry in the literature exist at this time [19–23].

The choice of which polarimeter to use in a given experiment can depend on a number of often nonorthogonal factors, including cost, available space, required accuracy, incident electron flux and energy, and vacuum environment. In this chapter, we present a comprehensive review of the variety of electron polarimeters which are applicable to atomic physics experiments and consider their respective advantages and disadvantages in given situations. We also discuss techniques for the calibration of these devices and consider potential sources of error in polarimetric measurements.

7.2 Electron Polarimeters

7.2.1 The "Standard" Mott Polarimeter

The type of electron polarimeter which has been used most often in atomic physics is that of the "standard" configuration (Figure 2) [3,16,24–35]. The electrons to be analyzed are accelerated to an energy in the range 100–150 keV and then strike a thin gold (high-Z) target whose normal is parallel to the electron beam axis. Those electrons which are scattered to a polar angle of 120° in the plane perpendicular to $\vec{P}$ can strike the left and right detectors, usually of the surface-barrier type. Since the

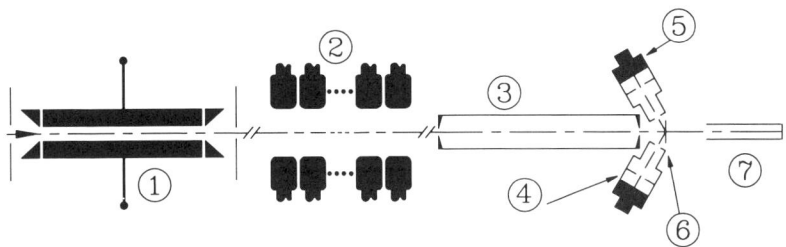

FIG. 2. The "standard" configuration for Mott polarimetry with a (1) Wien filter spin rotator. Other elements include (2) accelerator column; (3) collimator; (4) scattered-electron defining apertures; (5) surface-barrier detectors; (6) Au film target; and (7) Faraday cup.

target chamber is generally at the high-voltage end of the accelerator column, the detector signals must be transformed to ground potential.

Measuring A yields only the component of $\vec{P}$ perpendicular to the scattering plane. To measure all three components of $\vec{P}$, the vector must be rotated with some combination of electric and magnetic fields. Often Wien filters, 90° electrostatic benders, axial magnetic fields, or a combination of these are used to accomplish spin rotation prior to acceleration, where the required fields are lower. All of the components of $\vec{P}$ can be measured if the electron spins can be rotated to any direction [34]. Alternately, additional detectors can be used to determine two components of $\vec{P}$ simultaneously [25,33].

The measurement of P requires that the polarimeter's analyzing power, S_{eff}, be known. With standard polarimeters, S_{eff} is determined by measuring A for a series of target foil thicknesses, t, and using a calculated value of S, the Mott asymmetry corresponding to elastic scattering from single target atoms by electrons with $P = 1$. By extrapolating A to $t = 0$, one attains single-atom scattering conditions [35]. Calculations of S have been made for a variety of targets, electron energies, and scattering angles [36]. While detection of scattered electrons with energy resolution typical of surface barrier detectors (~10 keV) obviously does not discriminate against inelastic scattering (one assumption upon which calculations of S are based), systematic errors caused by this are small, especially at incident energies above 100 keV.

The evolution of the standard configuration is best understood by considering Figure 3. Analyzing powers with gold targets become appreciable only for angles >90° and have broad maxima in their angular dependencies between about 120° and 160°. Thus placement of the electron detectors at 120° accomplishes two things: the signal rate is maximized and the

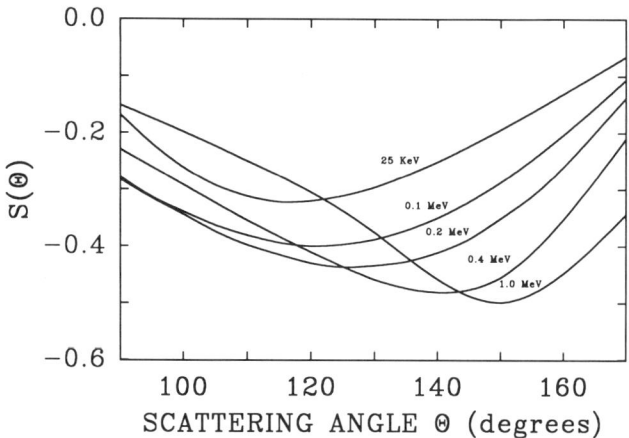

FIG. 3. Calculated values of S for Au with different values of θ and E [36].

variation of S_{eff} (or A) over the solid angles subtended by the detectors is minimized. The energy range of 100–150 keV represents a compromise between several considerations. While A increases monotonically with energy at 120° until E exceeds 1 MeV, the differential scattering cross-section decreases monotonically with energy. Moreover, multiple and plural scattering effects in the target, which reduce the accuracy with which S_{eff} can be determined, and sensitivity to target surface contamination, become increasingly bothersome as E is lowered below 100 keV.

7.2.2 The Concentric-Electrode Mott Configuration

In 1979, following a suggestion of Farago, a group at Rice University reported the design of a relatively compact Mott polarimeter which employed concentric cylindrical electrodes (Figure 4) ([14]; see also References [15,35,37]). In this configuration, a transversely polarized electron beam passes through an outer cylinder, nominally at ground potential, and is accelerated and focused by the radial field between the outer cylinder and an inner cylinder at positive high voltage between 20 and 120 kV. The beam enters a hole in the inner cylinder, where it scatters from a target foil. Electrons backscattered at 120° emerge from the inner cylinder, are decelerated by the radial field, and enter the retarding field analyzer/detectors. Those electrons that have lost too much energy in the target cannot surmount the intercylinder potential well or the potential barrier set up by the retarding fields and are thus not detected by the channel

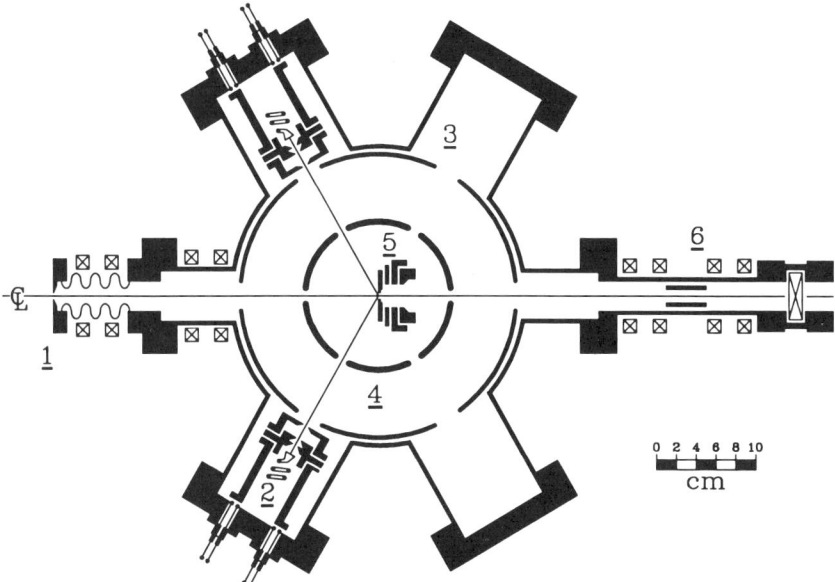

FIG. 4. Concentric-cylindrical-electrode Mott polarimeter [35]: (1) entrance aperture and input optics; (2) retarding-field analyzer/electron detector assembly; (3) grounded outer cylinder; (4) high-voltage inner cylinder; (5) movable target holder; and (6) exit electron optics.

electron multipliers (CEMs). Most of the incident electrons are scattered to angles <5° in the target and exit the polarimeter along the beam center-line.

In addition to its compactness, this design offers several advantages over standard polarimeters. Signal handling is much easier because the electrons are detected near ground potential, and the device can be used in an "in-line" orientation. Perhaps the most important advantage, though, is the good energy resolution with which the electrons are analyzed. Resolutions better than 3 eV have been reported, which essentially ensures that only elastic events are registered [35]. A foil-thickness extrapolation in conjunction with complete rejection of inelastic scattering ensures experimentally the conditions assumed in calculations of S.

To improve the efficiency and compactness of the concentric-electrode design, the Rice group has developed low-voltage devices based on spherical [23,38] and conical [39–41] (see Figure 5) geometries. These use typical inner-electrode potentials of 40 and 20 kV respectively, and are much smaller than cylindrical polarimeters; the spherical "mini-Mott" analyzer

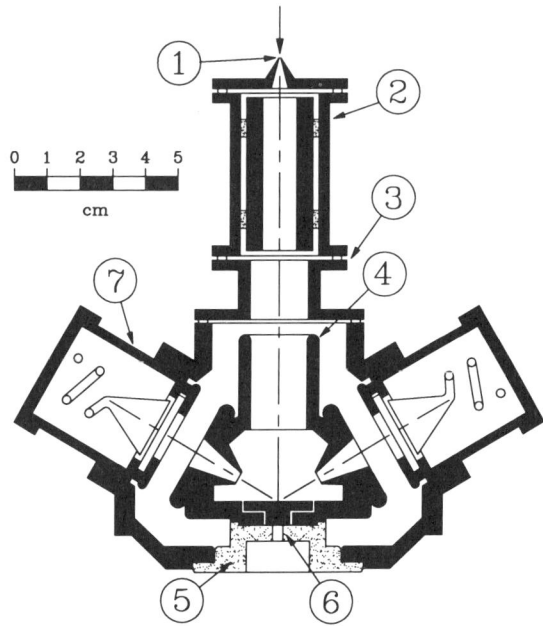

FIG. 5. Conical-electrode Mott polarimeter [39]: (1) input aperture; (2 and 3) input lenses; (4) high-voltage target assembly; (5) Macor insulator; (6) target mount; and (7) CEM housing.

has a characteristic dimension of ~15 cm while that of the conical "micro-Mott" detector is 10 cm. The CEMs in these designs subtend much larger solid angles about the beam–target interaction point than do those in the cylindrical configuration. This, in conjunction with the low operating voltages (and hence higher scattering cross sections), results in significantly higher detected electron signal for a given incident electron current.

The geometry of these smaller polarimeters permits the use of four CEMs so that two components of $\vec{P}$ can be measured simultaneously. The spherical design provides second-order focusing of the beam on the target, as opposed to the single-plane focusing of the cylindrical design. Such focusing reduces the potential for systematic error due to instrumental asymmetries (see Section 7.4.2). Focusing is provided by electrostatic lenses in the conical and standard configurations. A variant of the mini-Mott analyzer which uses cylindrical electrodes to mimic spherical fields has been developed [21]. This geometry also makes it easier to employ target-switching motion feedthroughs for foil-thickness extrapolations; the mini- and micro-Mott designs developed at Rice use single bulk-gold targets. An interesting hybrid between concentric-electrode and standard configurations, in which concentric ground and high-potential spherical electrodes provide focusing of the incident beam, while the detectors are placed in the hollow high-voltage terminal, has been reported [27].

The retarding-field analyzers used in concentric-electrode polarimeters permit the measurement of A vs ΔE, the maximum energy loss of electrons in the target. By extrapolating A to $\Delta E = 0$, one obtains the Mott asymmetry corresponding to purely elastic (albeit possibly multiple) scattering. This has been proposed as an alternative to the more difficult foil thickness extrapolation procedure. While ΔE measurements can, at high E, replace t extrapolations, care must be exercised in their application (see Section 7.4.1 and Reference [35]).

7.2.3 Atomic-Target Mott Polarimeters

While high-Z targets are required for A to be appreciable, relativistic incident electron energies are not required. Large asymmetries can also be observed with low (<1 keV)-energy electrons in elastic scattering from, e.g., single Hg atoms [7]. Polarimeters based on this principle have been in use since 1961 [42–44]. Such a device, designed for optimum efficiency and compactness, is shown in Figure 6 [17]. Transversely polarized elec-

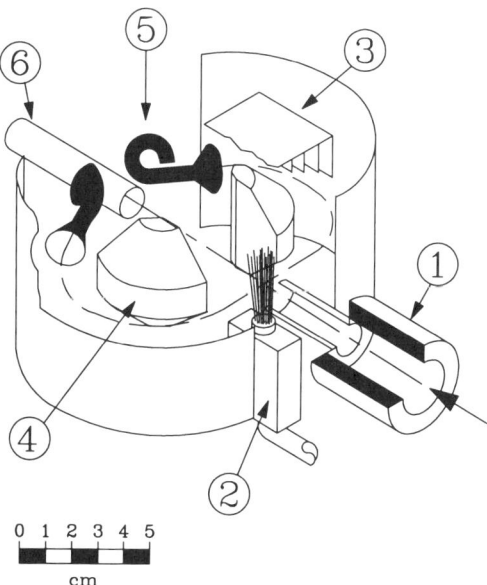

FIG. 6. Hg-vapor Mott polarimeter [17]: (1) input optics; (2) Hg vapor source; (3) L-N_2-cooled Hg trap; (4) electrostatic analyzers; (5) CEMs; and (6) Faraday cup.

trons impinge on an effusive Hg beam emerging from a heated reservoir. The Hg vapor is collected above the interaction region in a liquid-nitrogen-cooled trap. Electrons which are scattered into a fairly large solid angle centered about 90° in the plane perpendicular to $\vec{P}$ are bent in pseudospherical electrostatic analyzers and detected by CEMs. The analyzers serve three functions in this design: they shield the CEMs from UV photons generated in the interaction region, increase the detection solid angle, and provide some discrimination against inelastic scattering.

Because vapor-target densities are much lower than those associated with solid gold films, low E is a requirement for single-atom polarimeters in order to achieve reasonable efficiency. This is generally not a problem for atomic physics measurements. A notable variation in Hg vapor polarimeters is that of Gehenn [45], which has the advantage of being extremely compact and rotatable about the incident electron beam axis, so that all transverse components $\vec{P}$ can be measured with two detectors. It uses a diffuse evaporative Hg background target.

7.2.4 Low-Energy Variants with Solid Targets

A number of other polarimeters which are based on scattering asymmetries arising from an averaged nonzero spin–orbit coupling between the continuum electrons and the bulk targets have been developed. In the "diffuse scattering" Mott polarimeter (Figure 7a), the input beam (~150 eV) is focused and steered onto a vapor-deposited polycrystalline Au film target that is renewed periodically so that the device's analyzing power does not change [18,46,47]. Electrons scattered at all azimuthal angles and to polar angles between 90° and ~150° enter a collection region where they are guided by electrostatic potentials into a retarding field region, above which is placed a channelplate electron detector. The retarding field is used to discriminate against some fraction of the inelastically scattered electrons in order to maximize S_{eff}. The channelplate uses a four-segment anode to give information about both components of the transverse electron polarization.

Polarimeters based on polarized low-energy electron diffraction (PLEED) use a single-crystal target (most often W) from which the incident beam diffracts [30,48]. The left–right asymmetry of the Bragg peaks is caused by spin–orbit coupling between the crystal lattice and the incident electrons. Target surface conditions are critical, and the incident beam must have good energy and angular definition for the diffraction peaks to be resolved. While PLEED polarimeter input conditions are thus more stringent than those of diffuse-scattering polarimeters, their typical analyzing power is significantly better: ~0.3 vs 0.1.

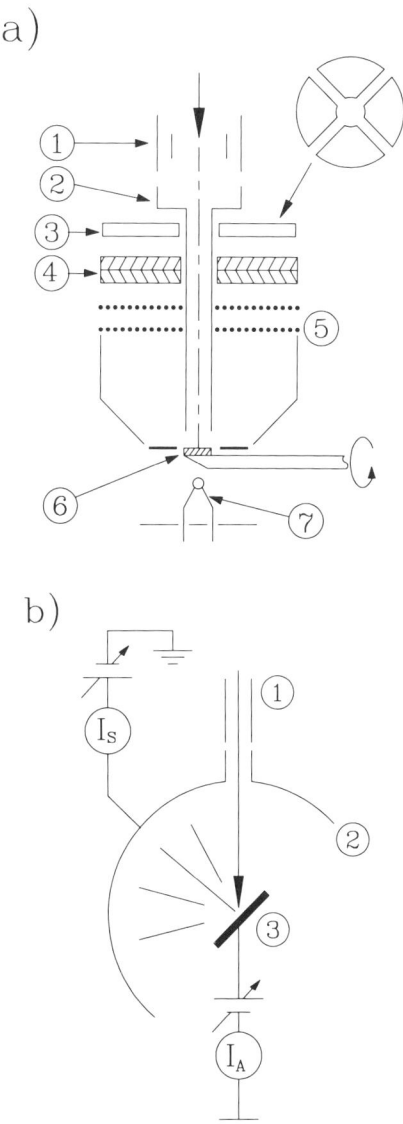

FIG. 7. (a) "Diffuse-scattering" low-energy Mott polarimeter [46]: (1) input optics; (2) drift tube; (3) segmented anode (see detail); (4) chevron channel plates; (5) scattered-electron extraction, guiding, and energy-discrimination electrodes; (6) target; and (7) Au evaporator. (b) "Secondary-electron" or "absorbed-current" polarimeter [52]: (1) input optics; (2) scattered/emitted current (I_s) collector; and (3) target, absorbing current I_A.

"Secondary-electron" polarimeters (Figure 7b) are based on the principle that the net current absorbed by a metallic target on which a polarized electron beam impinges is spin sensitive [48–52]. By operating at an energy (typically 160 eV for polycrystalline gold) for which the secondary electron emission from the target equals the incident current with unpolarized electrons, i.e., for which the absorbed current is zero, the absorbed current with polarized electrons will depend dramatically on P. This effect has been observed with ferromagnetic targets and with nonmagnetic crystalline and polycrystalline targets. Absorbed current polarimeters can be used only in an analog mode, but are extremely sensitive to small changes in P.

7.2.5 Optical Polarimeters

Several experiments have proven the principle of optical electron polarimetry, but no apparatuses built specifically to analyze electron polarization as an ancillary component of an experiment have been reported. Initial studies used Zn and Hg targets, but later developments involved He [20–23]. In this scheme, the ground-state target atoms are excited to the 3^3P_J multiplet, and the circular polarization η_2 of the subsequent $3^3P \to 2^3S$ florescence is measured. The electron polarization is then given by

$$P = \left(\frac{1}{2} - \frac{\eta_3}{6}\right)\eta_2 \equiv \Lambda\eta_2, \quad (7.6)$$

where η_3 is the linear polarization fraction of the light with respect to the axes parallel and perpendicular to the electron-beam axis. The incident electron energy is restricted to a range between 23.0 (threshold) and 23.6 eV to prevent cascading of higher-lying excited states into the 3^3P levels, which would invalidate Equation (7.6). In this energy range, $|\eta_3| < 0.15$, so Λ is about 0.5. Since η_3 depends only weakly on E and can be measured, an *in situ* measurement of Λ is possible.

Figure 8 shows the prototype we have used to test He for polarimetric measurements [22]. The electron beam crosses an effusive He target which is dumped into a large diffusion pump. The axis of the optical polarimeter is parallel to $\vec{P}$. Light emitted along $\vec{P}$ is collected and before being detected by a photomultiplier tube passes through various polarizing elements and an interference filter. Because the measurements must be made near the 3^3P excitation threshold, count rates for a given electron current are relatively low (typically 40 Hz/μA). This problem could be ameliorated by using a differentially pumped gas-cell target, as long as pressures were kept low to eliminate radiation trapping.

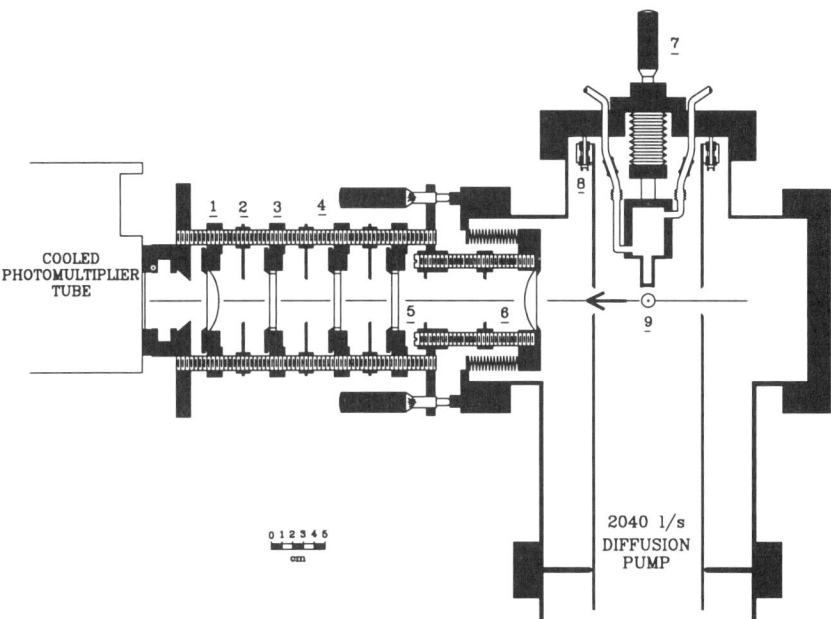

FIG. 8. He optical polarimeter [22]. Photon polarimeter elements are the (1) refocusing lens; (2) collimators; (3) interference filter; (4) linear polarizer; (5) retarder; and (6) collection lens. Also indicated are the (7) effusive gas target; (8) electrostatic target shield; and (9) electron beam (emerging from the plane of the diagram). $\vec{P}$ is indicated by an arrow.

7.3 Calibration Methods

An absolute measurement of P requires that S_{eff} or Λ be known absolutely, i.e., that the polarimeter be calibrated. There are essentially four methods for accomplishing this.

7.3.1 Calculation

If the analyzing power results from single collisions, it can be calculated directly from basic atomic scattering theory. This approach is directly applicable only to optical polarimeters and to single-atom Mott analyzers (for which S_{eff} equals S), in which the incident and scattered electron trajectories are known with confidence. Thus Hg-vapor analyzers which accept scattered electrons over a large solid angle, energy range, or both cannot be calibrated in this manner. Errors associated with calculated values of S are typical of low-energy electron–atom dynamical scattering calculations: somewhat better than 20% but worse than 2% [53]. The values of Λ, on the other hand, are determined kinematically by angular-momentum algebra and have essentially no uncertainty. There is a weak dependence of Λ on η_3, but η_3 can in turn be measured to high precision [22].

7.3.2 Asymmetry Extrapolations and Calculated S

In Mott polarimeters which allow t or ΔE extrapolations and in which the scattered electron trajectories are well characterized, S_{eff} is given to a good approximation (see Section 7.4.1) by S/A_0, where A_0 is the value of A extrapolated to t or $\Delta E = 0$. Thus the effects of multiple and plural scattering in the solid target are experimentally eliminated, and one deals in effect with single scattering. Calculations of S at the higher energies of the standard and concentric-electrode configurations are reliable to better than 5%, but the extrapolation procedure introduces additional error [35].

7.3.3 Double-Scattering Measurements

In a double-scattering measurement, unpolarized electrons scatter to a well-defined angle, θ, from a first target, thus developing a polarization $\vec{P}(\theta)$ perpendicular to the scattering plane. A is determined in a subsequent, experimentally identical scattering into angles $\pm \theta$. Mott first suggested this experiment and showed that $A = S_{\text{eff}}^2$ [2]. This method can be used to measure the analyzing power of a Mott polarimeter directly, without resort to assumption or theoretical calculation, if a first scattering of unpolarized electrons which mimics the Mott detector's experimental geometry can be arranged. This has been done (with considerable difficulty) by several investigators to calibrate standard Mott polarimeters [12,34,42,43,54,55]. The extremely careful measurements of Gellrich et al. [54] represent the highest accuracy [0.3%] measurements of S_{eff} to date. A variant of the double-scattering method, involving specular diffraction of the initially unpolarized electrons from a single crystal of W, has been reported [56].

7.3.4 Use of Electrons with Known Polarization

The measurement of A with incident electrons of known $\vec{P}$ yields S_{eff}. In calibrations of this type, $\vec{P}$ must be measured in a preliminary experiment or produced in such a fashion that its value can be safely assumed. This has been accomplished in a variety of ways. β electrons from allowed Gamow–Teller or Fermi nuclear decays have a longitudinal polarization equal to their relativistic velocity, $\beta = v/c$, and have been used to measure

S [57]. Optical pumping of metastable He [58] and Cs [59] has been used to produce electrons with $P \approx 1$, which can then be liberated by chemi- or photoionization. In the case of He, a check on P was made by measuring the atomic polarization with a Stern–Gerlach polarimeter. Finally, optical polarimeters, which measure $\vec{P}$ to good absolute accuracy (~1%), can be used to calibrate other devices [21]. These techniques are particularly useful for the calibration of low-energy polarimeters such as the LEED, diffuse, and high-efficiency Hg vapor designs, which cannot be calibrated using other methods.

7.4 Systematic Errors

Errors in the measurements of P can arise in a number of ways. In Mott measurements, errors in the calculated value of S (if used) must be combined with uncertainties due to extrapolation procedures, instrumental asymmetries, and background signals. Optical electron polarimetry involves uncertainties related primarily to the measurement of Stokes parameters.

7.4.1 Mott Scattering: Extrapolation Errors

Because of multiple and plural scattering in solid targets, S_{eff} is less than the true atomic elastic-scattering analyzing power S. As ΔE or t increases, the discrepancy becomes larger, as shown in Figure 9. An attempt to determine S_{eff} absolutely by extrapolating A to $\Delta E = 0$ will result in error if S_{eff} for elastic scattering depends on t (curve A, Figure 9). Such errors are small (i.e., curve A is fairly flat) for $E > 100$ keV. Thickness extrapolations can lead to error if the upper bound on the range of thicknesses used is not sufficiently thin to guarantee linearity in the extrapolation form. Care must also be taken to understand the energy resolution characteristics of the RFAs used for ΔE extrapolations. Generally, errors due to extrapolation procedures are small (<5%), and decrease as E increases. These issues have been considered in detail by several authors [11,13,15,16,35].

7.4.2 Mott Scattering: Instrumental Asymmetries

Instrumental asymmetries (IAs) can be of three types: geometric, electronic, or temporal [1,3,54]. Of these, geometric IAs are most common and treacherous. Angular misalignments and/or spatial displacements of the beam on the target cause the detectors to be at different polar scattering angles and to subtend different solid angles about the beam–target inter-

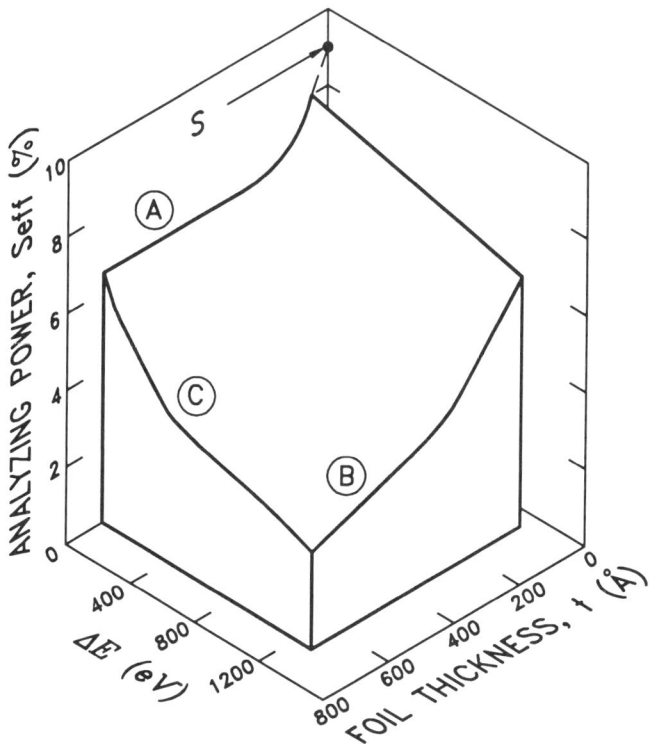

FIG. 9. Schematic representation of the effects of target thickness and electron energy loss on S_{eff} for 20-keV scattering from Au [35]. Curve B is characteristic of foil-thickness extrapolations with poor energy resolution; curve C represents a ΔE extrapolation using a bulk target (see text).

ception point. Errors in S_{eff} of ~2% per degree of angular deviation and 6% per millimeter of spatial displacement are typical. Geometric IAs could also include effects due to poorly characterized electron trajectories (due to either poor knowledge of the experimental geometry or the actions of, for example, spurious magnetic fields) that might result in misestimation of scattering angles in calculating S or S_{eff}. Monte Carlo simulation can be useful in understanding some of these effects.

Experimentally, IAs can be reduced or eliminated by the use of forward-angle monitor counters, low-Z targets for which the Mott asymmetry is small, or both; precise reversal of $\vec{P}$; the use of ZnS screens or electrodes to characterize beam positions; rotation of the entire apparatus about the incident beam axis, or a combination thereof [11,13].

Electronic asymmetries alter A in essentially the same way that geometric ones do, and they can be dealt with in a similar fashion. Temporal variation of P can also lead to errors in A and is most effectively reduced by frequent (and perhaps random) spin reversal.

7.4.3 Mott Scattering: Backgrounds

Background is defined as any detected signal not resulting from direct target scattering of the incident beam into the detector [1,3,11–13,16,32,55]. It can represent a significant (~20% in some cases) fraction of the total signal and is usually caused by scattering from the chamber walls or the target backing films. Three common background reduction techniques are: (1) good collimation of the incident beam and the detectors (the latter measure reduces polarimeter efficiency), (b) careful design of the electron-beam trap if it is housed in the polarimeter, and (c) chamber construction or wall coating with low-Z materials to minimize backscattering. In this context, an important advantage of concentric-cylinder polarimeters is their low background: the primary beam exits the chamber, eliminating the main source of backscattered electrons, and any remaining ones usually lose sufficient energy in wall collisions to be rejected by the RFAs.

7.4.4 Optical Polarimetry

The analyzing power of optical polarimeters can be determined very accurately, so uncertainty in P comes primarily from the photon polarimetry. With extreme care, Stokes parameters can be measured to 0.1%, but more typical uncertainties are 2–3%, due to uncertainties in the optical constants of the polarizing elements and collection lens. Procedures for the measurement of these constants are outlined in a number of references; manufacturers' claims regarding the optical properties of their wares should be regarded with caution [60]. A careful analysis of the ensemble of photon paths through the apparatus must be undertaken in the most precise work. Finally, the independence of the light polarization on target pressure must be established to eliminate the possibility of radiation trapping.

7.5 Comparison of Polarimeters

Table I compares the polarimeters discussed above. While there are no hard-and-fast rules for selecting a polarimeter for a given application, the following guidelines are generally valid. Probably the most crucial

TABLE I. Electron Polarimeters for Atomic Physics

Type	Operating energy (keV)	Required vacuum (Torr)	Size (m^3)	δ (eV)	Σ ($mm^2 \cdot sr \cdot eV$)	I/I_0	Analyzing power	Efficiency, ε	Comments	References
Mott "standard"	90–150	10^{-6}	2–10	5000	10^3	10^{-3}–10^{-2}	0.3–0.4	10^{-4}	Good efficiency without requiring external calibration; very large	3, 16, 24–35
Cylindrical concentric electrode	20–120	10^{-6}	1–2	1000	10^4	10^{-7}–10^{-6}	0.2–0.4	10^{-7}	Best for high-accuracy absolute measurements; low background; low efficiency	14, 15, 35, 37
Spherical concentric electrode	10–50	10^{-6}	10^{-1}	1000	10^4	10^{-4}–10^{-3}	0.2–0.3	10^{-5}	Compact; double focusing reduces instrumental asymmetries	21, 23, 38
Conical concentric electrode	10–35	10^{-6}	10^{-2}	500	10^3	10^{-3}	0.1–0.3	10^{-5}	Very compact	39–41
Hg beam	0.01–1	10^{-6}	10^{-2}	5	10	10^{-4}	0.1–0.4	10^{-5}	Low-efficiency versions do not require calibration; not easily UHV compatible	17, 42–44
Diffuse	0.1	10^{-9}	10^{-2}	40	10^2	10^{-2}	0.1	10^{-4}	Optimizes compactness and efficiency; requires UHV and calibration	18, 46, 47
PLEED	0.1	10^{-10}	10^{-2}	2	10	10^{-3}	0.3	10^{-4}	UHV and calibration required	30, 48
Absorbed current	0.1	10^{-9}	10^{-3}	10	1	—	—	10^{-4}	UHV and calibration required; very compact; very sensitive to changes in polarization	48–52
Optical	0.01–0.02	10^{-6}	1	0.2	10	10^{-8}	0.5–0.7	10^{-9}	Potential for best accuracy; inefficient; must be operated at low energy	20–23

requirement for a polarimeter is that it be efficient enough to measure P with reasonably small uncertainty in reasonable time. (The definition of "reasonable" being, of course, experiment dependent.) The amount of time required to measure P to a given statistical precision is inversely proportional to the "efficiency" or "figure of merit," $\varepsilon \equiv [a^2 I/I_0]^{1/2}$, where a is the device's analyzing power (either S_{eff} or Λ), I_0 is the incident beam current, and I is the detected signal rate [1]. This parameter is maximized by standard and diffuse Mott polarimeters. The efficiencies of concentric-electrode devices can be improved if Au is replaced by higher-Z targets [41,61,62]. Two related parameters are the maximum energy width of the beam, δ, that can actually be analyzed, and the electron-optical acceptance, Σ. Even if ε is large, the polarimeter is not useful if most of the beam to be analyzed is rejected by its input optics (Σ) or if the input energy range is too broad (δ) to allow a well-defined a. Sometimes the individual quantities (I/I_0) and a may be more important than their combination in ε. Experiments with extremely small values of I_0 may depend crucially on a high value of I/I_0 (the "sensitivity") alone. Beams with small P need maximal a so that systematic errors due to, for example, IAs are minimized.

Accuracy alone may be the most important requirement of an experiment. One must then pick a technique which combines good efficiency, to reduce statistical uncertainty, with good knowledge of a. For accuracy requirements of 1–2% (which is usually more than adequate in atomic physics), the cylindrical Mott analyzer is the best choice. More stringent requirements dictate the use of a double-scattering calibration or optical polarimetry.

Space, safety, cost requirements, or a combination thereof often rule out the use of a standard Mott polarimeter. If, because of beam intensity problems or transport difficulties, it is advisable to attach the analyzer directly to the experimental vacuum chamber or even place it inside an existing chamber, then use of, for example, a PLEED device may be mandatory. Vacuum requirements also play a role here. It would not be possible to use a PLEED polarimeter in most crossed-beam target chambers. Similarly, a Hg analyzer is difficult to interface directly with a UHV system. This must be kept in mind if a GaAs polarized electron source is to be used.

High-energy Mott polarimeters minimize a number of potential systematic errors, can be internally calibrated, and, in the case of the standard design, have good efficiency. Their major drawback is their size and expense. Low-energy polarimeters are small and efficient, but must ultimately be calibrated (with the exception of the optical and some Hg polarimeters). The solid target Mott devices operating below 1 keV require

UHV, are sensitive to target surface conditions, and require good control of the input beam characteristics.

A good general polarimeter which represents a nice compromise among all of the above factors is a UHV-compatible concentric-cylinder Mott analyzer run at 100 keV. With the use of low-angle monitor counters and one Al target in conjunction with a series of Au ones, it can measure P with an uncertainty associated almost entirely with the calculation of S. Its only drawback is a relatively low figure of merit.

Acknowledgment

The author acknowledges the support of the National Science Foundation.

References

1. J. Kessler, *Polarized Electrons*, 2nd ed. Springer-Verlag, Berlin, 1985.
2. N. F. Mott, *Proc. R. Soc. London, Ser. A* **124**, 425 (1929).
3. T. J. Gay and F. B. Dunning, *Rev. Sci. Intrum.* **63**, 1635 (1992).
4. D. T. Pierce, R. J. Celotta, M. H. Kelley, and J. Unguris, *Nucl. Instrum. Methods Phys. Res., Sect. A* **266**, 550 (1988).
5. P. S. Farago, *Adv. Electron. Electron Phys.* **21**, 1 (1965).
6. P. S. Farago, *Rep. Prog. Phys.* **34**, 1055 (1971).
7. J. Kessler, *Rev. Mod. Phys.* **41**, 3 (1969).
8. H. Frauenfelder and R. M. Steffan, in *Alpha, Beta, and Gamma Ray Spectroscopy* (K. Siegbahn, ed.). pp. 1431–1452. North-Holland Publ., Amsterdam, 1965.
9. H. Frauenfelder and A. Rossi, *Methods Exp. Phys.* **5B** 214–274 (1963).
10. H. F. Schopper, *Weak Interactions and Nuclear Beta Decay*. North-Holland Publ., Amsterdam, 1966.
11. A. R. Brosi, A. I. Galonsky, B. H. Ketelle, and H. B. Willard, *Nucl. Phys.* **33**, 353 (1962), and references therein.
12. J. van Klinken, *Nucl. Phys.* **75**, 161 (1966), and references therein.
13. D. M. Lazarus and J. S. Greenberg, *Phys. Rev. D* **2**, 45 (1970), and references therein.
14. L. A. Hodge, T. J. Moravec, F. B. Dunning, and G. K. Walters, *Rev. Sci. Instrum.* **50**, 5 (1979).
15. D. M. Campbell, C. Hermann, G. Lampel, and R. Owen, *J. Phys. E* **18**, 663 (1985).
16. G. D. Fletcher, T. J. Gay, and M. S. Lubell, *Phys. Rev. A* **34**, 911 (1986).
17. K. Jost, F. Kaussen, and J. Kessler, *J. Phys. E* **14**, 735 (1981).
18. See, e.g., J. Unguris, D. T. Pierce, and R. J. Celotta, *Rev. Sci. Instrum.* **57**, 1314 (1986).
19. M. Eminyan and G. Lampel, *Phys. Rev. Lett.* **45**, 1171 (1980).
20. J. Goeke, J. Kessler, and G. F. Hanne, *Phys. Rev. Lett.* **59**, 1413 (1987).
21. M. Uhrig, A. Beck, J. Goeke, F. Eschen, M. Sohn, G. F. Hanne, K. Jost, and J. Kessler, *Rev. Sci. Instrum.* **60**, 872 (1989).

22. J. E. Furst, W. M. K. P. Wijayaratna, D. H. Madison, and T. J. Gay, *Phys. Rev. A* **47**, 3775 (1993); T. J. Gay, J. E. Furst, H. Geesmann, M. A. Khakoo, D. H. Madison, W. M. K. P. Wijayaratna, and K. Bartschat, in *Correlations and Polarization in Electronic and Atomic Collisions and (e,2e) Reactions* (P. J. O. Teubner and E. Weigold, eds.), p. 265. IOP, Bristol, 1992.
23. I. Humphrey, C. Ranganathaiah, J. L. Robbins, J. F. Williams, R. A. Anderson, and W. C. Macklin, *Meas. Sci. Technol.* **3**, 884 (1992).
24. J. Kessler, J. Lorenz, H. Rempp, and W. Bühring, *Z. Phys.* **246**, 348 (1971).
25. R. Möllenkamp, W. Wübker, O. Berger, K. Jost, and J. Kessler, *J. Phys. B* **17**, 1107 (1984).
26. B. Reihl, M. Erbudak, and D. M. Campbell, *Phys. Rev. B* **19**, 6358 (1979).
27. M. Landolt, R. Allenspach, D. Mauri, *J. Appl. Phys.* **57**, 3626 (1985); M. Landolt, private communication, (1993).
28. R. Raue, H. Hopster, and E. Kisker, *Rev. Sci. Instrum.* **55**, 383 (1984).
29. M. J. Alguard, J. E. Clendenin, R. D. Ehrlich, V. W. Hughes, J. S. Ladish, M. S. Lubell, K. P. Schüler, G. Baum, W. Raith, R. H. Miller, and W. Lysenko, *Nucl. Instrum. Methods* **163**, 29 (1979).
30. M. R. O'Neill, M. Kalisvaart, F. B. Dunning, and G. K. Walters, *Phys. Rev. Lett.* **34**, 1167 (1975).
31. T. Nakanishi, K. Dohmae, S. Fukui, Y. Hayashi, I. Hirose, N. Horikawa, T. Ikoma, Y. Kamiya, M. Kurashina, and S. Okumi, *Jpn. J. Appl. Phys.* **25**, 766 (1986).
32. N. Ludwig, A. Bauch, P. Naß, E. Reichert, and W. Welker, *Z. Phys. D* **4**, 177 (1986); P. Naß, Unpublished Diplomarbeit Thesis, Universität Mainz (1982).
33. E. Kisker, G. Baum, A. H. Mahan, W. Raith, and B. Reihl, *Phys. Rev. B* **18**, 2256 (1978).
34. K. Bartschat, G. F. Hanne, and A. Wolcke, *Z. Phys. A* **304**, 89 (1982).
35. T. J. Gay, M. A. Khakoo, J. A. Brand, J. E. Furst, W. V. Meyer, W. M. K. P. Wijayaratna, and F. B. Dunning, *Rev. Sci. Instrum.* **63**, 114 (1992).
36. See, e.g., A. W. Ross and M. Fink, *Phys. Rev. A* **38**, 6055 (1988); J. W. Motz, H. Olsen, and H. W. Koch, *Rev. Mod. Phys.* **36**, 881 (1964); G. Holzwarth and H. J. Meister, *Nucl. Phys.* **59**, 56 (1964).
37. J. J. McClelland, M. R. Scheinfein, and D. T. Pierce, *Rev. Sci. Instrum.* **60**, 683 (1989).
38. L. G. Gray, M. W. Hart, F. B. Dunning, and G. K. Walters, *Rev. Sci. Instrum.* **55**, 88 (1984).
39. F. B. Dunning, L. G. Gray, J. M. Ratliff, F.-C. Tang, X. Zhang, and G. K. Walters, *Rev. Sci. Instrum.* **58**, 1706 (1987).
40. F.-C. Tang, X. Zhang, F. B. Dunning, and G. K. Walters, *Rev. Sci. Instrum.* **59**, 504 (1988).
41. D. P. Pappas and H. Hopster, *Rev. Sci. Instrum.* **60**, 3068 (1989).
42. H. Deichsel, *Z. Phys.* **164**, 156 (1961).
43. H. Deichsel and E. Reichert, *Z. Phys.* **185**, 169 (1965).
44. M. Düweke, N. Kirchner, E. Reichert, and S. Schön, *J. Phys. B* **9**, 1915 (1976).
45. W. Gehenn, R. Haug, M. Wilmers, and H. Deichsel, *Z. Angew. Phys.* **28**, 142 (1969).
46. M. R. Scheinfein, D. T. Pierce, J. Unguris, J. J. McClelland, R. J. Celotta, and M. H. Kelley, *Rev. Sci. Instrum.* **60**, 1 (1989).
47. J. Woods, M. Tobise, and R. C. O'Handley, *Rev. Sci. Instrum.* **60**, 688 (1989).

48. See, e.g., R. Feder, ed., *Polarized Electrons in Surface Physics*. World Scientific, Singapore, 1985.
49. H. C. Siegmann, D. T. Pierce, and R. J. Celotta, *Phys. Rev. Lett.* **46**, 452 (1981).
50. M. Erbudak and N. Müller, *Appl. Phys. Lett.* **38**, 575 (1981).
51. M. Erbudak and G. Ravano, *J. Appl. Phys.* **52**, 5032 (1981).
52. D. T. Pierce, S. M. Girvin, J. Unguris, and R. J. Celotta, *Rev. Sci. Instrum.* **52**, 1437 (1981), and references therein.
53. D. H. Madison, private communication (1993).
54. A. Gellrich, K. Jost, and J. Kessler, *Rev. Sci. Instrum.* **61**, 3399 (1990); A. Gellrich and J. Kessler, *Phys. Rev. A* **43**, 204 (1991).
55. P. E. Spivak, L. A. Mikaelyan, I. E. Kutikov, V. F. Apalin, I. I. Lukashevich, and G. V. Smirnov, *Sov. Phys.—JETP (Engl. Transl.)* **14**, 759 (1962), and references therein.
56. H. Hopster and D. L. Abraham, *Rev. Sci. Instrum.* **59**, 49 (1988).
57. V. Eckardt, A. Ladage, and U. V. Moellendorff, *Phys. Lett.* **13**, 53 (1964), and references therein.
58. D. M. Oro, W. H. Butler, F.-C. Tang, G. K. Walters, and F. B. Dunning, *Rev. Sci. Instrum.* **62**, 667 (1991).
59. E. H. A. Granneman, M. Klewer, and M. J. Van der Wiel, *J. Phys. B* **9**, 2819 (1976).
60. See, e.g., H. G. Berry, G. Gabrielse, and A. E. Livingston, *Appl. Opt.* **16**, 3200 (1977), and references therein.
61. R. Loth, *Z. Phys.* **203**, 66 (1967).
62. W. Eckstein, *Z. Phys.* **203**, 59 (1967).

8. POSITION-SENSITIVE PARTICLE DETECTION WITH MICROCHANNEL-PLATE ELECTRON MULTIPLIERS

Ken Smith

Departments of Physics and Space Physics and Astronomy and the Rice Quantum Institute, Rice University, Houston, Texas

The development of microchannel plate (MCP) electron multipliers in the 1960s spawned a new generation of detectors for charged and neutral particles. These are position-sensitive or "imaging" devices that indicate the positions of particle impact on the detector surface. There are several configurations of MCP-based position-sensitive particle detectors (PSDs), which differ in the specific means for determination of the position information. In most of these, position resolutions on the order of 50 μm are achieved, and some of the detectors also permit measurement of particle arrival times with subnanosecond resolution.

8.1 Microchannel Plates

Initially developed as an image intensifier, the microchannel plate typically consists of a disk of lead glass, 10 to 100 mm in diameter, having a thickness between 0.5 and 2 mm and is fabricated by a multistep process. A bundle of glass tubes containing a solid cores is drawn and fused, forming a single fiber. These fibers are stacked in an hexagonal assembly which is fused and drawn further. These hexagonal multifiber bundles are then assembled and fused together, forming a larger billet. The billet is fused and sliced into wafers at a specific bias angle of a few degrees with respect to the tube axes. The wafers are chemically etched to remove the solid core material, leaving a disk that is approximately 60% transparent with holes having diameters of a few micrometers. Through further processing, the interior walls of the holes are made conductive and coated with a thin SiO_2 layer having a high secondary electron emission coefficient [1]. As indicated in Figure 1a, the resulting glass disk is perforated by a regular array of tubular channels, typically between 5 and 20 μm in diameter. The channel orientation is 5° to 10° from the normal to the disk. A metal coating is deposited on the front and back surfaces of the glass wafer, and the electrical resistance between these surfaces is on the order

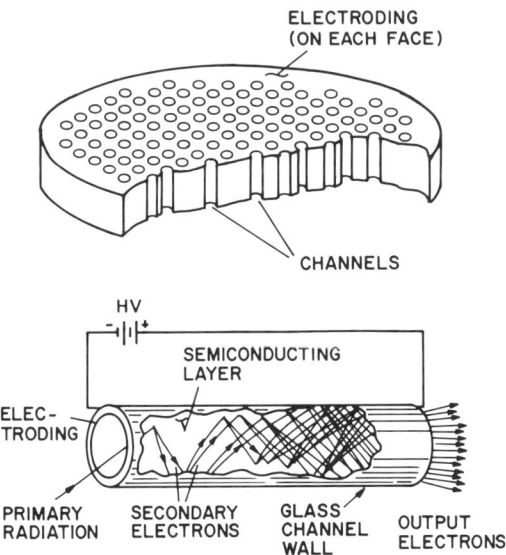

FIG. 1. (a) Schematic view of a single microchannel plate. (b) Scenario for electron multiplication within a single microchannel. It functions much as a miniature "channeltron" electron multiplier.

of 10^7–10^8 Ω. A potential difference of about 1000 V applied between the front and the back electrodes produces an electric field in the channels. As shown in Figure 1b, if a primary particle or photon enters one of the channels at the negatively biased end, strikes the channel wall, and ejects a secondary electron, that electron will be accelerated down the channel and will collide with the wall, emitting additional secondary electrons, which are themselves accelerated, leading to an avalanche of electron current in the channel. Under normal operating conditions, emission of a single electron at the negatively biased microchannel entrance will result in an output of about 1000 electrons at the positively biased end. In image intensifiers or in particle detectors in which the position information is read with a camera, the electrons exiting the MCP are accelerated to a phosphor screen which is then viewed or photographed. If the readout mechanism for position information is electronic, however, an additional MCP(s) is usually placed in tandem with the input one to provide greater charge output.

When plates are stacked with their surfaces parallel, the channels in adjacent plates are oriented for a maximum angle between them. This channel orientation is intended to minimize the possibility of ions formed in the channel in one plate from traveling into the channels in the adjacent plate, thereby causing spurious output counts. This "ion feedback" effect can alternatively be mitigated by fabricating the MCP with curved channels, although this approach may lead to slight distortions in the spatial position information ultimately provided by the detector [2]. The literature identifies stacks by the relative orientation of the channels in the plates, calling two-plate stacks a "chevron" configuration and three-plate stacks a "z" configuration. The characteristics of stacks of two and three MCPs have been investigated by many groups [3–9]. These studies show that the output pulse height amplitudes depend on the number of plates used, the bias voltages applied across the plates, the spacing between the plates, the interplate electric field, and the count rate. A typical pulse height distribution for a z-configuration detector is shown in Figure 2a [10]. Most of the pulse amplitudes lie within a range of about ±25% of the mean pulse amplitude. This relative uniformity in pulse height is due to a dynamic suppression of the gain of an individual channel at large electron flux due to the combined effects of space charge in the channel and depletion of the available charge on the channel walls. This "saturation" effect simplifies work with position-sensitive detection because it limits the necessary dynamic range of the electronics required. For most position-sensing applications, a narrow pulse-height spectrum is desirable. The mean pulse-height amplitude for a chevron-configuration plate stack clearly depends greatly on the potential difference applied to the detector. Typical gains for single and chevron-stacked pairs of plates are shown in Figure 2b for the normal range of voltage applied across each plate. Clearly, the output charge pulse amplitude can vary by orders of magnitude under different bias conditions.

If detection efficiencies are determined by pulse-counting methods that record a count when the output pulse exceeds a threshold value, microchannel plate stacks are excellent detectors of charged particles, energetic neutral particles, and photons. For singly charged ions the detection efficiency increases with ion impact energy, reaching a plateau between 3 and 20 keV. The detection efficiency at the "plateau" is found to be equal to the ratio of the area of the channel entrances to the total detector area, indicating that, if an ion with adequate energy enters a channel, then it can be detected with essentially unit efficiency [6,11]. For ions, the detection efficiency remains high out to impact energies on the mega-electron volt range, in which efficiencies above 75% have been observed for 5.4-MeV α particles [12]. The detection efficiency for incident electrons, however, drops by about a factor of 2 as their energy is increased from 100 to 10,000 eV [13].

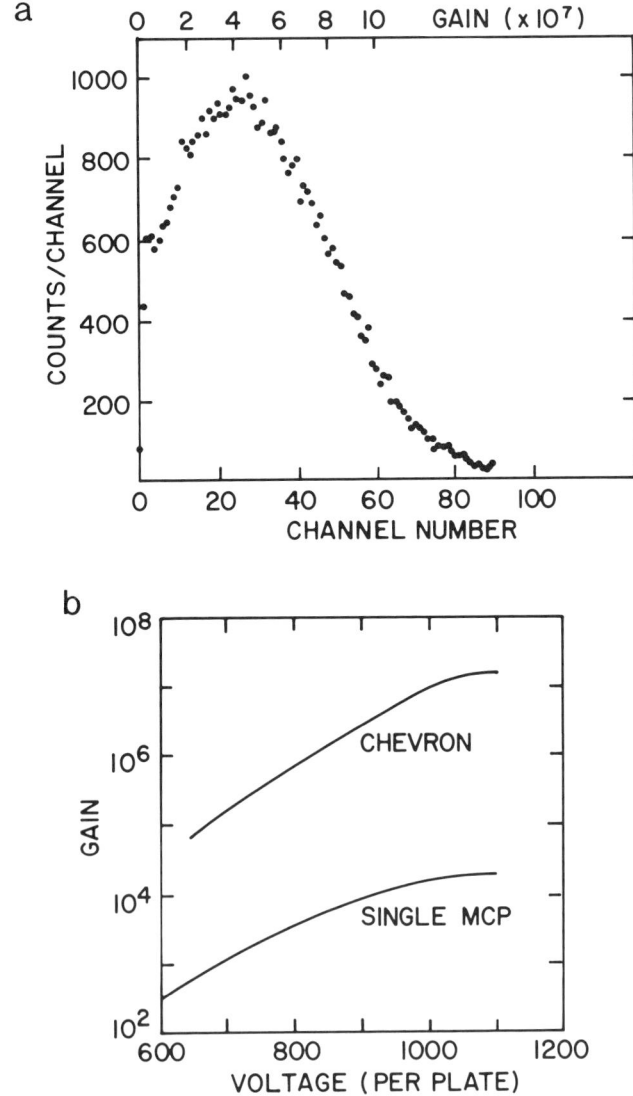

FIG. 2. MCP output characteristics. (a) Typical charge pulse-height spectrum observed for a chevron-stacked pair of microchannel plates. Average pulse heights under normal operating conditions are in the vicinity of 1 pC. (b) Gain (electrons out per input event) of single and stacked MCPs as a function of the bias voltage applied across the single plate and across each plate in the pair.

The output pulse amplitude from a specific channel in the microchannel plate depends both on the total count rate integrated over the detector surface and on the count rate in the local area of the channel. In a pulse counting arrangement, small output pulses can fall below the detection electronics threshold, resulting in an apparent decrease in efficiency at higher count rates. In microchannel plates or any other electron multiplier, the output pulse amplitude will decrease if the output current becomes comparable to the current available to charge the dynode structure. In microchannel plates, the dynode consists of the inner walls of the channels, which themselves have very high electrical resistances. When an individual channel is excited, significant charge is removed from the output end of the channel, and some time is required for it to recharge. If the channel is excited again before it can fully recharge, the resulting output pulse will be smaller. Some of these effects were modeled by Eberhardt [14,15] and by Giudicotti et al. [16], indicating that the dynamic characteristics of an individual channel's charging are complex and depend on the states of the adjacent channels. The output pulse height distribution depends significantly on the count rate as indicated in Figure 3, which is taken from the work of Fraser et al., who performed detailed measurement and analysis of the effect of count rate on gain in microchannel plate detectors [17,18]. In our atomic collisions work using chevron-configuration plates [19] we routinely measure beams of 1000 particles/sec that impinge on

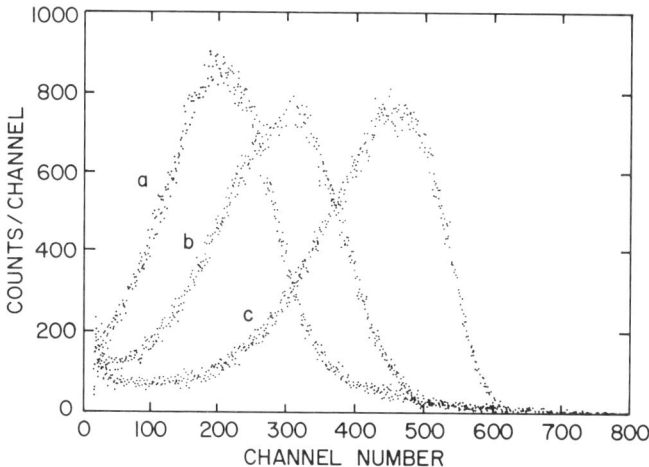

FIG. 3. Dependence of the output pulse-height spectrum on the count rate: (a) 3.1 counts/channel/sec; (b) 1.3×10^{-2} counts/channel/sec; and (c) 3.7×10^{-5} counts/channel/sec.

areas containing about 300 channels. Under these circumstances, there is a decrease in output pulse height of less than 30%, indicating a channel recharge time on the order of a few tenths of a second. This result is consistent with measurements by Siegmund et al. [3], who report a "gain droop" of about 25% at an output count rate of 4 counts/channel/sec, and with similar results of Pearson et al. [20]. One important virtue of microchannel plates is that they can respond quantitatively to short bursts of a few hundred counts at rates exceeding 10^{10}/sec^{-1} [21]. This capability is obtained only when each detected particle strikes a different channel.

The uniformity of MCP detector efficiency at all points on the detector surface is superior to that of most particle multipliers. Spatial variations in gain of up to 50% are, however, not uncommon. Operated in the pulse counting mode, spatial variations of detection efficiency can be maintained below 2% of the total detection efficiency [6]. As with any detector based on secondary electron emission, the nature and condition of the first surface impacted by the detected particle can influence the detector efficiency. (Because, obviously, if no electron is emitted by the primary particle impact, detection cannot occur, and the secondary electron emission coefficient characteristics of the dynode surface depend upon its condition.) Degradation of the detection efficiency can occur after exposure of the MCP to relatively intense particle fluxes which damage the secondary electron emitting surfaces [22]. Work with 1- to 5-keV ions indicates that surface damage increases with increasing ion energy and ion mass. Ion fluxes required to damage the surface in a few hours are in the neighborhood of 10^6–10^7 cm^{-1} sec^{-1}. Therefore, in critical applications in which the input surface of the detector is exposed to large ion fluxes or to potential contamination (e.g., pump oil vapors), periodic scanning of the surface or illumination with a uniform particle flux is advisable to check detector uniformity. In detectors using two or more MCPs the uniformity of detector efficiency can be adversely affected if the MCP surfaces are not parallel. MCPs have been observed to warp after long exposure to air. Under these conditions the active portion of the MCP expands more than the solid glass in which it is mounted, and the resulting stresses produce a bowl-shaped distortion of the plate. The warpage is presumably due to absorption of water by the MCP material and can in fact be sufficient to cause spontaneous fracturing of the plate. (It is therefore recommended that plates be stored in vacuum or in an inert atmosphere.) If warped plates are used in a stacked arrangement, they should be oriented so that their distortions are in the same direction.

The performance of microchannel plates as timing detectors is truly outstanding. The small dimensions and large electric field within the channels ensure rapid electron transit times. Typical time delays between channel excitation and channel output are in the vicinity of 0.18 nsec, and the time dispersion of the output electron pulse is about 0.06 nsec. These values have been obtained by both calculation [23] and direct measurement [24] for microchannel plates that have 12-μm-diameter channels having a length-to-diameter ratio of about 40. Transit time spreads approaching 0.020 nsec have been reported for MCPs with 6-μm pores [25]. Timing resolution in the range of 50 psec for photon detection [26] and resolutions of 70–100 psec for ions have been reported [12]. These results were obtained by observation of the MCP output on anodes specifically designed for fast response. Alternate methods [6,27] in which the timing signal is extracted from the MCP electrode opposite the anode permit subnanosecond timing resolution in conjunction with position-sensitive detection. For timing kilo-electron volt energy ions, one should note that the ion may travel less than 100 μm in a nanosecond. Therefore the flatness of the detector surface and the time required for the ion to penetrate a multiplier channel before it strikes the wall may present inherent timing uncertainties that exceed 1 nsec. Additionally, the ion penetrates the channel prior to striking the channel wall; uncertainty in the penetration distance on the order of 100 μm makes kilo-electron volt energy ion detection times uncertain by about a nanosecond.

In applications such as collision experiments, Raman scattering, and lidar, gating of the detector is desirable. Rapid gating can be achieved by depositing electrodes on an electrodeless MCP in the form of a strip transmission line and allowing the MCP operating voltage to be modulated by pulses passing through the transmission line [28]. Gating on a 0.1-nsec time scale has been achieved in this way [24]. The MCP gain observed under this type of gating is estimated to be about half that observed under "normal" DC conditions. Clearly, because the gate pulse duration is comparable to the electron transit time in the channels, the dynamics of channel operation are different. More conventional gating in a few nanoseconds can be accomplished by standard grid-modulation techniques for charged input particles [23], by modulation of the potential applied to the MCP face [29], or by modulation of the field at the interface between two stacked plates.

8.2 Position-Sensitive Readout Systems

The remarkable characteristic of MCPs is that they essentially represent a two-dimensional array of literally millions of individual electron multipliers, providing the opportunity to use the detector as an imaging device. The only inherent limitations in resolution are derived from the spacing between channels and irregularities in the MCP fabrication [30]. An avoidable limitation is obtained if the electric field in front of the detector is oriented to return secondary electrons emitted from the interchannel surface back to the front of the detector. Unless this field is quite strong (several kV/cm) the secondary electrons have enough energy to leave the surface and return at a position substantially different from that at which they were originally emitted. Detection of such straying electrons will compromise the detector's resolution, and in most critical imaging applications, the electric field at the detector surface is simply oriented to keep these electrons from returning to the surface.

There are several different operational systems for obtaining information on the positions of particle impact on the MCP. These systems have different capabilities, some of which are outlined below. As a class of instruments, these detectors are usually referred to as microchannel-plate-based position-sensitive detectors (PSDs).

8.2.1 Optical Systems

A straightforward system for position-sensitive detection is one in which the MCP output electrons impact a phosphor that is viewed with a CCD camera [29,31,32]. CCD cameras of either one- or two-dimensional format provide information on the pattern of MCP output electrons impacting the phosphor. The spot size on the phosphor is determined by the number of MCP channels excited and the electron trajectories between the MCP output face and the phosphor. In a typical proximity-focused arrangement, the separation between these surfaces may be a millimeter and the potential difference applied can be 3 to 4 kV. Under these conditions, the spot size from a single channel's output is expected to be on the order of 50–100 μm [33]. The detector's position resolution for detection of single events could, in principle, be increased to the MCP channel spacing by use of a high-resolution video camera and image-analysis software that locates the centroids of the spots observed on the phosphor.

The CCD camera can give excellent position resolution, but information on the arrival times of individual particles is limited by the framing rate of the camera. The optical systems can, however, permit measurements of higher fluxes of particles than electronic readout systems do and can accurately report position information from multiple particle impacts that are essentially simultaneous [21].

Quantitative analysis of optical position-sensitive detection systems is fundamentally similar to that applied to MCP-based image intensifiers and must consider the behavior of the MCP, the phosphor, and the CCD camera [34]. In applications in which very stable position-measuring accuracy is required, fiber-optic devices have been used [22,32] to couple the phosphor output to the CCD camera chip itself. This technique eliminates any aberrations introduced by the camera lens and also more efficiently transmits the phosphor light output to the optical sensor. Forand *et al.* [22] have demonstrated that systems in which a phosphor that is fiber-optically coupled to an optical sensor can be quite linear and can provide quantitative results at input fluxes up to several microamperes per square millimeter.

The most straightforward system for electronic interpretation of the MCP output is a multianode array placed behind the MCP stack [35,36]. In most applications, each anode is connected to an individual amplifier and to processing electronics. This arrangement has the advantage of dimensional stability and can achieve time resolution in the subnanosecond range [25,37].

Individual anodes can be fabricated with standard integrated circuit processing techniques that permit an anode assembly affixed directly to (but insulated from) the MCP output surface. This mounting eliminates any compromise of resolution resulting from spreading of the electron cloud emerging from the MCP output. Hybrid assemblies of a linear array of 128 145- × 2000-μm anodes, amplifiers, and counting electronics have been fabricated, providing a position resolution of 160 μm and a pulse pair time resolution less than 250 nsec [38]. Implementations that combine greater position resolution with large detector area are, however, not practical because of the number of individual amplifier circuits required.

8.2.2 Interpolative Anode Arrays

One approach to achieving higher spatial resolution with a limited number of amplifiers is to use a multiple anode structure connected to a current divider network as shown in Figure 4a. The anode spacing is arranged so that the MCP output impacts several anodes, and the output position coordinate is taken as the difference output, $Q_A - Q_B$. As long as the MCP output is detected by several anodes, the detector resolution (i.e., the uncertainty in the measured particle impact position) can be considerably smaller than the anode spacing because the current divider operation acts to interpolate position information, establishing the approximate location of the MCP output's centroid. The accuracy with which the centroid location is established depends primarily on the number of anodes impacted, the amount of charge intercepted, and the noise figure of the divider/amplifier combination.

An early implementation of the interpolation technique was developed by R. W. Wijandts van Resandt *et al.* [39] and applied by de Bruijn and

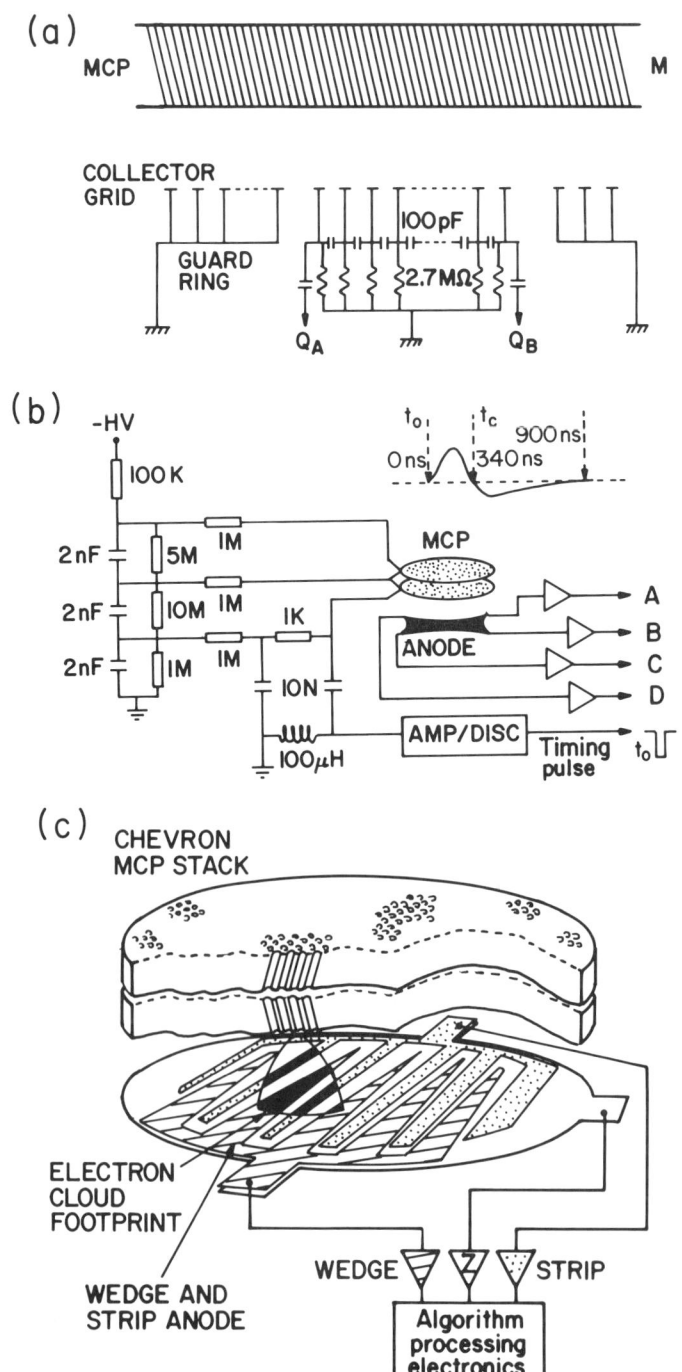

FIG. 4. Common anode configurations: (a) discrete interpolative anode; (b) resistive anode; and (c) wedge-and-strip anode.

Los [40] to microchannel plate detectors. The work of de Bruijn and Los elegantly demonstrates how the interpolation method permits electrode design that speaks to geometry of the images under study, providing yet another dimension of detector optimization.

A more elaborate version of the interpolation method is exemplified by an imaging X-ray detector described by Kellogg et al. [41,42]. This detector, developed for satellite-based X-ray astronomy, achieves a 20-μm resolution (limited primarily by the MCP pore spacing) in a 26- $\times$ 26-mm image space and can accommodate count rates of 10,000/sec. Two independent anode arrays, each consisting of 30 100-μm-diameter wires spaced on 200-μm centers were used to measure coordinates of the MCP output. One array is orthogonally mounted above the other, and their relative potentials are adjusted so that about half the MCP output charge is collected by each array. Individual wires within each array are connected by 10-kΩ resistors, and every eighth wire is also connected to the input of an AC-coupled charge-sensitive amplifier. The detected position information is provided by digital and analog circuits that, respectively, identify the pair of amplifiers between which the MCP output falls and interpolate the output position between that pair.

8.2.3 Resistive Anode

The logical limit of interpolative anode arrays is one in which the anode is in fact continuous. If the anode takes the form of a two-dimensional sheet of material having a uniform resistivity, ρ (ohms per square), the coordinates of the charge cloud impinging on the anode can be determined from measurement of the amounts of charge arriving at three or more points around the periphery of the anode. The anode's geometry determines the specific algorithm required for position determination, but Hasebe et al. have shown that arbitrarily shaped anodes can provide position data [43]. By far, the most commonly used resistive anode is one that provides a linear output of coordinate information. Figure 4b shows the usual detector configuration, and Figure 5a details the anode which consists of a sheet of uniform resistivity ρ, bounded by arc-shaped resistors having a resistance, R_L (ohms per unit length). Gear [44] showed that if the radius of curvature a and the anode's resistance parameters are related by

$$R_L = \frac{\rho}{a}, \qquad (8.1)$$

then current flows in the anode as if it were flowing in an infinite resistive sheet. Consequently the ratio of charge collected at two adjacent corners of the electrode to the total charge collected on all four corners is directly

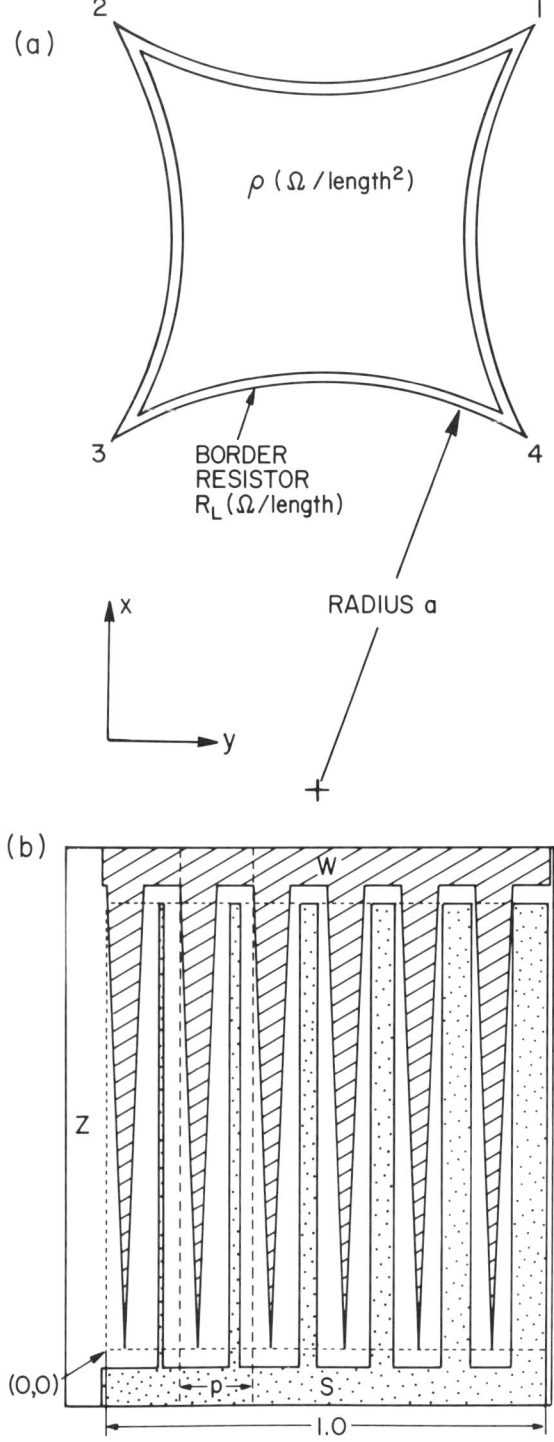

proportional to the displacement of the centroid of the charge cloud in the direction perpendicular to the line connecting the adjacent corners. Specifically for the designation of corners shown in Figure 5a, the x and y coordinates are given by charge ratios

$$x = \frac{(Q_1 + Q_4) - (Q_2 + Q_3)}{Q_1 + Q_2 + Q_3 + Q_4}$$
$$y = \frac{(Q_1 + Q_2) - (Q_3 + Q_4)}{Q_1 + Q_2 + Q_3 + Q_4},$$

(8.2)

where the origin ($x = 0$, $y = 0$) is the center of the anode. Gear's theorem applies strictly to a static solution of current flow in the sheet, and Equation (8.2) holds only for the sums of charge collected at the electrode contacts, but does not hold at any specific time *during* the charge-collection process. (While detectors that determine position from the arrival times of charge at the electrode contacts have been fabricated [45], no effective two-dimensional position-sensing detector using delay line anodes has been reproted, although its feasibility and promise has been discussed [46].) Most PSD implementations rely on electronics that measures the amplitudes of charge signals arriving at the contacts and determines coordinates by Equation (8.2). This ratiometric scheme requires only that the signals from the electrodes be proportional to the charges arriving at the contacts, alleviating the requirement for absolute measurements of charges or currents.

The accuracy of position determination in resistive-encoded anodes depends on the uniformity of the surface resistivity ρ and linear resistance R_L in the anode structure. Other limiting factors include electronic noise and inaccuracy in analog circuitry that may be used for evaluating the position coordinates along the lines of Equation (8.2). The anode's ρ and R_L can be evaluated by testing the anode for its response by placing a current source at a series of known locations on the sensitive surface and determining absolute errors in the coordinates of the source given by Equation (8.2). This can be done only on a case-by-case basis for individual anodes and is tedious, at best. Nonuniformities in ρ and R_L can be expected to lead only to smooth variations in the linearity of the detector's position coordinate output. Deviations from linearity are most easily established by actually operating the detector, placing a grid of accurately known spacing either on or immediately adjacent to the detector's input surface,

---

FIG. 5. Detail of anode geometries: (a) gear-format resistive anode (see text); (b) wedge-and-strip anode (see text).

illuminating the detector uniformity with the particles to be detected, and examining the shadow pattern made by the grid. This procedure also serves to establish a precise calibration of the detector system's position output. Further, observing the sharpness of the grid shadow can give relatively good information on the position resolution of the detector and on the mean uncertainty associated with an individual position datum [47].

Most resistive-encoded-anode detector systems achieve (or at least claim) spatial resolutions in the vicinity of 50–100 μm. The resolution is limited primarily by electrical noise in the anode itself and in the electronics that determines position. This electronics must be relatively sensitive, as the typical amount of charge generated by a chevron-stacked microchannel plate pair in response to a single input particle is typically between 10^{-13} and 10^{-12} C. Interpretation of the position output to 0.2% (which would be 50 μm on a 25-mm-diameter detector) requires a charge measurement uncertainty in the range of 1.5×10^{-16} to 10^{-15} C (1000–10,000 electrons) in each analog channel. This is feasible, but not trivial. Employing sensitive amplifiers requires careful attention to mounting and shielding all portions of the anode circuit. The position-measurement accuracy can be systematically compromised if the anode is capacitively coupled to electrodes whose potentials change during the amplifiers' charge-integration period. For instance, charge-integration periods are typically 1 μsec. If the potential on an electrode coupled to the anode by 0.001 pF is changing at a rate of 100 V/μsec, the anode will receive a spurious charge of 10^{-13} C during the charge-integration period. This charge will appear as a common signal on all four anode terminals and will add to the denominators, but not the numerators in Equation (8.2), moving the apparent position indication along a line between the actual event position and the coordinate system origin. The direction of this motion toward or away from the origin depends on whether the spurious charge is, respectively, the same as or opposite the signal electron charge. The spurious position offset observed depends on the relative magnitude of the charges involved. If the intended signal pulse has an amplitude of only 10^{-13} C, then the spurious signal will have the effect of shifting the position indication to a point halfway between the actual event position and the detector center. Unfortunately, many applications for PSDs involve pulsing electrodes near the detectors at or around the same time as particles are detected; and pickup distortion of position information is a particularly insidious problem if gating pulses are applied to the MCP stack electrodes.

The thermal noise of the anode is primarily Johnson noise and cancels out in the denominators of Equation (8.2), but adds in the numerators. If one simply considers the noise in the x and y coordinates independently, the fluctuation in the numerators of Equation (8.2) (in coulombs) is given by

$$Q_{\text{noise}} = \sqrt{\frac{4kT\tau}{r}}, \qquad (8.3)$$

where r is the resistance of the anode as measured between two adjacent terminals and the other two (in ohms), τ is the shaping time constant of the charge amplifiers (in seconds), T is the absolute temperature (in kelvins), and k is Boltzmann's constant. The most effective energy transfer between the anode and the preamplifier is obtained when the anode resistance r is equal to τ/C, where C is the capacitance between the anode and the ground. Then the anode noise will be about

$$Q_{\text{noise}} = \sqrt{2kTC}. \qquad (8.4)$$

For a typical case, of a time constant of 1 μsec and an anode resistance of 1 MΩ, the noise contribution is on the order of 2500 electrons rms [48].

The thermal noise will introduce a randomly oriented offset between the computed position and the actual particle impact coordinates. The magnitude of this offset is proportional to the ratio of the noise charge to the actual signal charge. In a commercial system [49] the distribution of position indications for a 20-μm-diameter beam held in a fixed position on a PSD as a function of signal pulse height was measured. For pulse heights of the mean height and greater, the position indication was consistent within ± 50 μm; the position indication consistency deteriorated at smaller pulse heights, ultimately reaching ± 1000 μm for the minimum height pulse at which position information was recorded. Thus, the uncertainty of an individual position datum depends significantly on the magnitude of the charge that generated it. In most instances, the electronics used for position evaluation will have a limited dynamic range. It is often difficult to set the gain of the detector so that all pulses fall within the dynamic range and so that all pulses produce a position indication of acceptable uncertainty.

In addition to the Johnson noise of the anode itself, noise associated with the amplifier circuitry and electrical interference that is uncorrelated with the particle detection time will also provide additional uncertainty to the position measurement. The amplifier noise has the same effect as the Johnson noise described above, but the action of electrical interference is (usually) to cause an uncertainty that increases with the distance of the event from the coordinate origin at the anode center. This effect occurs again because the electrical interference is most likely to appear as a common signal on all the amplifier inputs and will add to the denominators in Equation (8.2), but will cancel in the numerators. This spurious signal therefore does not introduce a random uncertainty in the indicated position, rather the uncertainty lies along a line that passes through the event coordinates and those of the origin.

8.2.4 Wedge and Strip Anodes

Figure 4c shows another popular type of anode arrangement, and Figure 5b shows the most common configuration of the anode. Called a wedge-and-strip anode, this device consists of three electrodes arranged as shown. Numerous variations of geometries are possible [50,51] to produce a two-dimensional position. Anodes are configured as "wedges," which receive a share of the circular electron cloud's charge that is proportional to the charge cloud's position in the direction parallel to the wedge axis, and as "strips" of width increasing with the dimension perpendicular to the wedge axis, so that the fraction of charge received by the strips is proportional to the charge cloud's position in that direction. If the electron signal impinges on an area significantly larger in diameter than the characteristic repeat dimension of the array, then the location of its centroid is given by

$$x = \frac{2Q_S}{Q_W + Q_S + Q_Z}$$

$$y = \frac{2Q_W}{Q_W + Q_S + Q_Z},$$

(8.5)

where the coordinates are referenced to a point at the lower left of the anode. The advantages of this anode configuration include the absence of a resistive element with its associated noise, relatively straightforward fabrication techniques, high electronic speed, and low cost.

The resolution of these devices is somewhat greater than that of the resistive-encoded anode because of the absence of Johnson noise from the anode. Ignoring the spatial resolution quantitization due to the discrete nature of the channel array in the input MCP, the spatial resolution of a wedge-and-strip anode is determined by the combined effects of the electronic noise in the amplifiers (which is exacerbated by the fact that these anodes have relatively large electrode capacitances). The FWHM resolution in the electronic noise is given [52] by

$$R_x(\text{electronic}) = \frac{4.71L}{f_{max} - f_{min}} \frac{1}{Q} \sqrt{N_s^2(1 - 2f_i + f_i^2) + f_i^2(N_w^2 + N_z^2)},$$

where L is the length of the anode in the direction perpendicular to the wedge axis; Q is the total charge gain of the MCP stack; f_i is the fractional area of the S electrode at the coordinates of particle impact; N_s, N_w, and

N_z are the noise; and f_{max} and f_{min} are the maximum and minimum values of f_i at the ends of the anode. Another source of uncertainty in coordinate determination for this anode is the so-called "partition noise." In a wedge-and-strip anode, the MCP output is divided unequally among different electrodes. This partitioning of the signal is subject to shot noise, which is the statistical variation of the number of counts that may fall within each partition (i.e., on each section of the anode). The FWHM resolution component due to this uncertainty is given [52] by

$$R_x(\text{partition}) = \frac{2.355L}{f_{max} - f_{min}} \sqrt{\frac{1}{Q} f_i(1 - f_i)}.$$

In general the device resolution is given by the sum of these quantities as indicated in Figure 6. The ultimate resolution limitation, that of the MCP pore separation, has been observed using wedge-and-strip anodes [53].

Other sources of position measurement uncertainty for wedge-and-strip detectors include interelectrode cross-talk due to capacitive coupling and

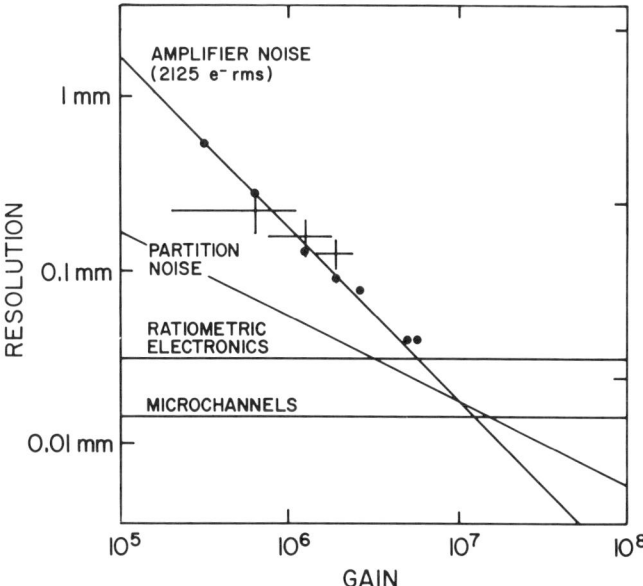

FIG. 6. Typical position-coordinate measurement uncertainties due to noise for a wedge-and-strip anode as a function of the MCP stack's gain [50]. This figure does not take account of the systematic effects (mentioned in the text) which manifest themselves as image distortions. Data points are taken with artificial pulses that are capacitively introduced to a detector anode that is 25 mm square with a pattern period (p) of 1.5 mm.

effective preamplifier input capacitances, which can result in a diagonal elongation of the image. An additional difficulty arises from the fact that the electron cloud size may be comparable to that of the anode period p (in Figure 5b). If the charge cloud is too small, the image will have a distortion that is periodic with spatial period p.

8.3 Conclusion

This brief survey of microchannel-plate-based position-sensitive detectors is by no means exhaustive and has not dealt, for instance, with anode schemes that incorporate direct digital encoding of the image [54]. It has, however, outlined the mainstream applications of these detectors which have truly opened new vistas in atomic physics, physical chemistry, and astronomy. Some of the results reported here were obtained under support of grants from the National Science Foundation, Atmospheric Sciences Section, and NASA.

References

1. B. N. Laprade, *Proc. SPIE–Int. Soc. Opt. Eng.* **1072**, 102 (1989).
2. D. M. Hassler, G. J. Rottman, and G. M. Lawrence, *Appl. Opt.* **30**, 3575 (1991).
3. G. H. W. Siegmund, K. Coburn, and R. F. Malina, *IEEE Trans. Nucl. Sci.* **NS-32**, 443 (1985).
4. M. L. Edgar, R. Kesse, J. S. Lapington, and D. M. Walton, *Rev. Sci. Instrum.* **60**, 3673 (1989).
5. D. F. Rogers and R. F. Malina, *Rev. Sci. Instrum.* **53**, 1438 (1982).
6. R. S. Gao, P. S. Gibner, J. H. Newman, K. A. Smith, and R. F. Stebbings, *Rev. Sci. Instrum.* **55**, 1756 (1984).
7. M. Hellsing, L. Karlsson, H. O. Andren, and H. Norden, *J. Phys. E* **18**, 920 (1985).
8. A. Muller, N. Djuric, G. H. Dunn, and D. S. Belic, *Rev. Sci. Instrum.* **57**, 349 (1986).
9. K. Tobita, H. Takeuchi, H. Kimura, Y. Kusama, and M. Nemoto, *Jpn. J. Appl. Phys.* **26**, 509 (1987).
10. J. L. Wiza, *Nucl. Instrum. Methods* **162**, 587 (1979).
11. T. Sakurai and T. Hashizume, *Rev. Sci. Instrum.* **57**, 236 (1986).
12. R. H. Kraus, Jr., D. J. Vieira, H. Wollnik, and J. M. Wouters, *Nucl. Instrum. Methods Phys. Res., Sect. A* **264**, 327 (1988).
13. J. O. McGarity, A. Huber, J. Pantazis, M. R. Oberhardt, D. A. Hardy, and W. E. Slutter, *Rev. Sci. Instrum.* **63**, 1973 (1992).
14. E. H. Eberhardt, *IEEE Trans. Nucl. Sci.* **NS-28**, 712 (1981).
15. E. H. Eberhardt, *Appl. Opt.* **18**, 4818 (1979).
16. L. Guidicotti, M. Bassan, R. Pasqualotto, and A. Sardella, *Rev. Sci. Instrum.* **65**, 247 (1994).

17. G. W. Fraser, M. T. Pain, J. E. Lees, and J. F. Pearson, *Nucl. Instrum. Methods Phys. Res., Sect. A* **306**, 247 (1991).
18. G. W. Fraser, M. T. Pain, and J. E. Lees, *Nucl. Instrum. Methods Phys. Res., Sect. A* **327**, 328 (1993).
19. R. S. Gao, L. K. Johnson, K. A. Smith, and R. F. Stebbings, *Phys. Rev. A* **40**, 4914 (1989).
20. J. F. Pearson, J. E. Lees, and G. W. Fraser, *IEEE Trans. Nucl. Sci.* **35**, 520 (1988).
21. S. Ichimura, K. Goto, K. Kokubun, H. Shimizu, and H. Hashizume, *Jpn. J. Appl. Phys.* **29**, 1209, (1990); *Rev. Sci. Instrum.* **61**, 1192 (1990).
22. J. L. Forand, C. Timmer, E. Wahlin, B. D. DePaola, G. H. Dunn, D. R. Swenson, and K. Rinn, *Rev. Sci. Instrum.* **61**, 3372 (1990).
23. M. Ito, H. Kume, and K. Oba, *IEEE Trans. Nucl. Sci.* **NS-31**, 408 (1984).
24. B. K. F. Young, R. E. Stewart, J. G. Woodworth, and J. Bailey, *Rev. Sci. Instrum.* **57**, 2729 (1986).
25. H. Kume, K. Koyama, K. Nakatsugawa, S. Suzuki, and D. Fatlowitz, *Appl. Opt.* **27**, 1170 (1988).
26. D. Bebelaar, *Rev. Sci. Instrum.* **57**, 1116 (1986).
27. W. G. McMullan, S. Charbonneau, and M. L. W. Thewalt, *Rev. Sci. Instrum.* **58**, 1626 (1987).
28. R. Luppi, F. Pecorella, and I. Cerioini, *Rev. Sci. Instrum.* **55**, 2034 (1984).
29. P. H. Kobrin, G. A. Schick, J. P. Baxter, and N. Winograd, *Rev. Sci. Instrum.* **57**, 1354 (1986).
30. As noted, the fabrication process entails sintering of many hexagonal boules into a larger boule. There is an inherent irregularity at the boundaries of the smaller boules, that can result in a so-called "chicken-wire" pattern in images resulting from lower gain in the immediate vicinity of those boundaries.
31. D. Murphy and K. Mausersberger, *Int. J. Mass Spectom. Ion Processes* **76**, 85 (1987).
32. M. Parkinson, *Appl. Opt.* **28**, 2087 (1989).
33. N. Koshida and M. Hosobuchi, *Rev. Sci. Instrum.* **56**, 1329 (1985).
34. U. Ellenberger, A. Glinz, and J. E. Balmer, *Meas. Sci. Technol.* **4**, 1430 (1993).
35. J. G. Timothy and R. L. Bybee, *Proc. SPIE–Int. Soc. Opt. Eng.* **265**, 93 (1981).
36. J. O. McGarrity, A. Huber, J. Pantazis, M. R. Oberhardt, D. A. Hardy, and W. E. Slutter, *Rev. Sci. Instrum.* **63**, 1973 (1992).
37. H. Brocknaus and A. Glasmachers, *IEEE Trans. Nucl. Sci.* **NS-39**, 707 (1992).
38. J. V. Hatfield, J. Comer, T. A. York, and P. J. Hicks, *Rev. Sci. Instrum.* **63**, 792 (1992).
39. R. W. Wijandts van Resandt, H. C. den Harink, and J. Los, *J. Phys. E* **9**, 503 (1976).
40. D. P. de Bruijn and J. Los, *Rev. Sci. Instrum.* **53**, 1020 (1982).
41. E. Kellogg, P. Henry, S. Murray, L. Van Speybroeck, and P. Bjorkholm, *Rev. Sci. Instrum.* **47**, 282 (1976).
42. E. Kellogg, S. Murray, U. Briel, and D. Bardas, *Rev. Sci. Instrum.* **58**, 550 (1977).
43. N. Hasebe, K. Kiso, E. Kaneda, and T. Doke, *Jpn. J. Appl. Phys.* **32**, 2162 (1993).
44. C. W. Gear, *Proceedings for the Skytop Conference on Computer Systems in Experimental Nuclear Phys.*, USAEC Conf-670301, p. 552 (1969).

45. H. Keller, G. Klingelhofer, and E. Kankeleit, *Nucl. Instrum. Methods Phys. Res., Sect. A* **258,** 221 (1987).
46. R. Raffanti and M. Lampton, *Rev. Sci. Instrum.* **64,** 1506 (1993), and references therein.
47. see G. W. Fraser, M. A. Barstow, and J. F. Pearson, *Nucl. Instrum. Methods Phys. Res., Sect. A* **273,** 667 (1988).
48. M. Lampton and F. Paresce, *Rev. Sci. Instrum.* **45,** 1098 (1974).
49. Quantar, Inc., Santa Cruz, CA.
50. see C. Martin, P. Jelinsky, M. Lampton, R. F. Malina, and H. O. Anger, *Rev. Sci. Instrum.* **52,** 1067 (1981).
51. H. O. Anger, U. S. Pat. 3,209,201 (1965).
52. O. H. W. Siegmund, S. Clothier, J. Thornton, J. Lemen, R. Harper, I. M. Mason, and J. L. Culhane, *IEEE Trans. Nucl. Sci.* **NS-30,** 503 (1983).
53. J. S. Lapington, R. Kessel, and D. M. Walton, *Nucl. Instrum. Methods Phys. Res., Sect. A* **273,** 663 (1988).
54. W. E. McClintock, C. A. Barth, R. E. Steele, G. M. Lawrence, and J. G. Timothy, *Appl. Opt.* **21,** 3071 (1982).

9. SWARM TECHNIQUES

David Smith and Patrik Španěl

Department of Biomedical Engineering and Medical Physics, University of Keele,
Stroke-on-Trent ST4 7QB, United Kingdom

9.1 Introduction

We are concerned in this chapter with the so-called "swarm" techniques used to study the reactions between ions and neutral atoms and molecules; positive ions and electrons; electrons and molecules; and positive ions and negative ions at low interaction energies. By low energies we mean center-of-mass collision energies, E_{cm}, of less than 1 eV and indeed mostly less than 0.1 eV, i.e., at thermal energies, and kinetic temperatures of the reactants lower than 1000 K and even down to temperatures of ~10 K. These swarm techniques are the most valuable and productive methods for obtaining kinetic data at thermal energies for the charged particle reactions, data which are necessary to properly understand the reactions occurring in ionized gases such as the terrestrial atmosphere, interstellar gas clouds, and laboratory discharge plasmas.

There are two distinct methods by which studies of such reactions can be conducted at low temperatures: beam methods [1], which yield cross-sections from which rate coefficients can be obtained indirectly, and swarm methods [2], which yield rate coefficients directly. A third method involves the use of various types of ion trap [3,4], which are discussed in detail elsewhere in this volume. The essential difference between beam and swarm experiments is that in the former only a single bimolecular collision occurs between the reactants at some well-defined E_{cm}, whereas in the latter multiple collisions of an ensemble of charged particles occur either with some inert buffer gas, thus establishing a Maxwellian speed distribution among the charged particles with a characteristic temperature equal to the buffer gas temperature, T_g, or in the case of electrostatic drift tubes at some higher "temperature" or mean energy depending on the electric field strength and the buffer gas pressure (see Section 9.4). Then the "charged particle gas" at a particular temperature, T_r, reacts with a reactant gas at the buffer gas temperature, T_g, a spread of interaction energies (E_{cm}) characterized by these temperatures being involved.

To clarify the meaning of a swarm of charged particles, it is valuable to first describe the beam method. Beam techniques involve the generation of well-collimated beams of the reacting species—ions, electrons, or neutrals—which cross at some well-defined angle under very low ambient pressure conditions [1]. In the intersection region the species may react, and from the reduction of intensity of one (or both) of the beams or from the intensity of the product species, the cross-section for the reaction is determined as a function of E_{cm}. By varying the intersection angle and determining the distribution of the products of the reactions as a function of the scattering angle (differential scattering methods), a great deal of information about the dynamics and mechanisms of the reactions as well as reaction cross-sections, σ, can be obtained [1,5]. A disadvantage of conventional beam methods is that the experiments cannot be performed accurately at very low E_{cm} (that is, at E_{cm} equivalent to room temperature and below) to give rate coefficients, k, of acceptable accuracy. (Note that k can be calculated from σ using the relation

$$k = \langle \sigma v \rangle = \int_0^\infty f(v)\sigma(v)v\,dv, \qquad (9.1)$$

where v is the relative velocity of the colliding particles and $f(v)$ is the Maxwell–Boltzmann speed distribution function [5]). Low-energy collisions can be studied using merged-beam techniques which have been developed to study binary ion–ion and ion–electron neutralization reactions. The reactive species are formed in beams which are merged to overlap collinearly [6,7]. Thus, by adjusting the laboratory velocities of the two beams to be similar, the E_{cm} can in principle be made very small. Merged beam methods have had some success, although uncertainty still surrounds the data obtained at very low E_{cm}, principally because of the extreme difficulty of ensuring that the beams are precisely collinear [6]. Questions arise concerning the internal energy states of the reactant ions, a very important point considering that the reactivity of positive ions with neutrals, electrons, and negative ions can be very dependent on internal energy [8]. It is for these reasons that kinetic data obtained from swarm experiments are strongly favored for modeling the ion chemistry of thermal and near-thermal plasmas.

The "ideal" swarm experiment usually involves the creation of an ensemble of charged particles of number density n_1 in an inert buffer gas of number density n_2 such that $n_1 \ll n_2$. Multiple collisions between the charged particles and the buffer gas ensure randomization (Maxwellianization) of the charged particles velocities and the relaxation of the charged particle mean energies (which may be high initially) to those appropriate to the buffer gas temperature, T_g. Then the introduction of reactant species

INTRODUCTION

whose number density n_3, can be controlled but which again is much less than n_2, initiates the reaction. (In some experiments reactions with the buffer gas are studied). If the rate of change of n_1 as a function of n_3 is determined, the rate coefficient, k, for the reaction can be derived using

$$\frac{dn_1}{dt} = -kn_1n_3. \tag{9.2}$$

The k so determined is appropriate to the reaction at the defined temperature T_g. Although these swarm experiments are often termed "multicollision" experiments (in contrast to the single-collision beam experiments) this does not imply that a particular charged particle necessarily undergoes many collisions with the reactant species, although this can occur if n_3 is high. Rather, many collisions occur between the ensemble of charged particles and the reactant particles, and the experiment observes the net effect of these many collisions which occur for a spread of E_{cm} governed by T_g. Such ideal swarm experiments include the stationary afterglow (SA), which is used to study ion-neutral and ion-electron reactions, the flowing afterglow (FA), and the selected ion flow tube (SIFT) experiments, which are used to determine the rate coefficients and productions for ion-neutral reactions, and the flowing afterglow/Langmuir probe (FALP) experiment, which is used to study electron–ion recombination, electron attachment, and ion–ion recombination. These techniques are discussed in detail later.

Some swarm experiments are "nonideal" in the sense that the charged particle velocity distributions may not be closely Maxwellian and that their corresponding mean energy $\bar{\varepsilon}$ can be much greater than that equivalent to T_g. These experiments are very valuable since they provide kinetic data at "equivalent temperatures" much greater than those accessible by the conventional heating of reaction vessels. Among these experiments are the ion drift tubes, notably the selected ion flow drift tube (SIFDT) which is used to study ion-neutral reactions at E_{cm} up to about 1 eV, and the electron drift tube which is used to study electron attachment reactions at suprathermal E_{cm}. These techniques also are described later in this chapter.

Swarm methods have been used to study a wide range of gas phase reaction processes at and near thermal energies. These methods can be grouped under just four headings: (1) afterglow plasmas, e.g., the SA, FA, and FALP techniques; (2) the SIFT technique; (3) static drift tubes (SDT) and flow drift tubes (FDT), including the SIFDT; and (4) expanding (supersonic) jet experiments, especially the *Cinetiqué de reaction en ecoulement supersonic uniformé* (CRESU) [reaction kinetics in uniform supersonic flow] technique. Each has its important place in the study of ionic and electronic reactions; the afterglow methods are the most versatile in that they are used to study ion-neutral, ion–electron, electron-neutral, and ion–ion reactions. The SIFT technique can be used to study an amazing variety of ion-neutral reactions; the drift tube methods can be exploited to study ion-neutral and electron-neutral reactions at suprathermal energies; and the supersonic jet experiments can be used to study reactions at very low temperatures. We now discuss these individual techniques.

9.2 Afterglow Plasma Techniques

Afterglow plasma techniques fall into two distinct categories, the pulsed ionization techniques, specifically the SA and similar methods, and the continuous ionization flow techniques, notably the FA methods, of which the FALP method is especially versatile.

9.2.1 The Stationary Afterglow

The SA plasma technique was the first to be used to study thermal energy reactions (see the review in Reference [9]), and typical apparatuses are shown in Figure 1. The principle of operation is simple. Ionization is created in a pure gas or a gas mixture by a short-duration gas discharge (DC, RF, or microwave discharges are used). Pulse lengths are usually within the range of microseconds to milliseconds depending on the size of the discharge vessel and the pressure of the gas. Following the cessation of the discharge, the electron temperature, T_e, which is initially high, quickly relaxes in collisions with the gas atoms and molecules to the gas temperature, T_g [10]. Thus an afterglow plasma is formed in which $T_e = T_+ = T_g$, and the electron number density, n_e, is equal to the number density of the singly charged positive ions, n_+, as long as no negative ions are formed in the discharge or in the afterglow. If negative ions are present (number density, n_-), then $n_+ = n_e + n_-$. Thus reactions between the composite charged and the neutral particles can be studied under thermalized conditions using suitable diagnostic techniques. The positive ions that reach the walls of the vessel (by ambipolar diffusion [11]) are sampled via a pinhole orifice, behind which is a differentially pumped mass spectrometer, and thus the loss and production rates of the various ion species, e.g., dn_+/dt, can be determined. When the dimensions of the vessel are relatively large (Figure 1), n_e and dn_e/dt in the afterglow plasma can be determined using Langmuir probes [12]. Afterglow plasmas are also created in small microwave cavities (dimensions of a few centimeters; see Figure 1) when n_e and dn_e/dt are determined by monitoring the temporal change in the resonant frequency of the cavity.

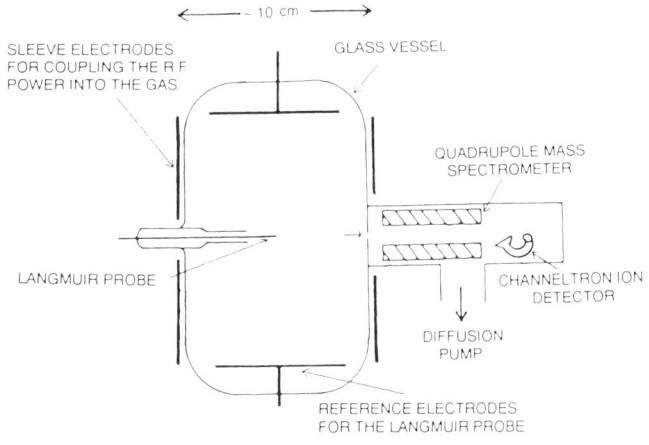

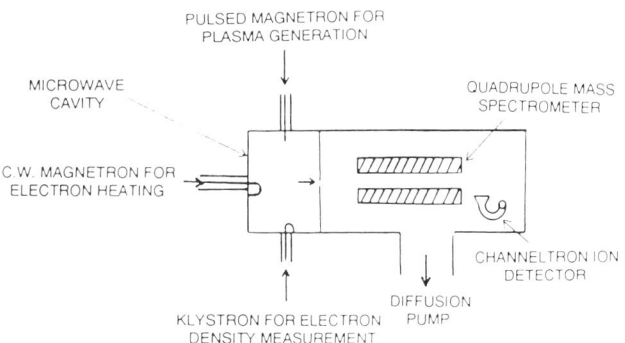

FIG. 1. Schematic representations of two types of stationary afterglow (SA) apparatus. Ionization is created in the larger glass vessel by pulses of radiofrequency power coupled into the gas via the external electrodes, and the Langmuir probe is used to determine the electron and ion number densities and the electron temperature in the afterglow plasma. In the smaller microwave cavity, the ionization is created by pulsed magnetron power, and the electron number density is measured by monitoring the shift of the resonant frequency of the cavity using the variable-frequency klystron. A CW source of microwave power is used to heat the electrons in the afterglow. In both apparatuses, the wall currents of ions are sampled through an orifice and mass analyzed and detected by a differentially pumped mass spectrometer/ion detection system.

As an example of the use of the SA to study ionic reactions, consider the following typical case. Helium buffer gas at a pressure of a few Torr with a small admixture of oxygen (typical pressure of a few mTorr) is ionized, producing essentially only He^+ primary ions (and, of course, electrons). These ions are then lost from the afterglow plasma by ambipolar diffusion with electrons to the walls of the containing vessel and via the reaction

$$He^+ + O_2 \rightarrow O^+ + O + He. \quad (9.3)$$

Thus the appropriate continuity equation for the He^+ ions is

$$\frac{d[He^+]}{dt} = D_a \nabla^2 [He^+] - k[He^+][O_2], \quad (9.4)$$

where the square brackets signify the number densities of the enclosed species, D_a is the ambipolar diffusion coefficient for He^+ ions [11], and k is the rate coefficient for Reaction (9.3). In the late-time afterglow (i.e., for fundamental mode diffusion), provided that $[He^+] \ll [O_2]$, the solution of Equation (9.4) yields the simple exponential form

$$[He^+]_t = [He^+]_0 \exp - \left(\frac{D_a}{\Lambda^2} + k[O_2]\right) t. \quad (9.5)$$

Since $[He^+]_t$ is directly proportional to the diffusive current of He^+ ions to the walls, which is detected by the mass spectrometer (see Figure 1), the overall decay constant $(D_a/\Lambda^2 + k[O_2])$ can be determined. Actually, the pressure of the buffer gas (helium in this example) is usually high enough that the diffusive contribution D_a/Λ^2 is very small (Λ is the characteristic diffusion length of the plasma vessel, dependent only on its dimensions [11]), and so k can be directly determined. However, when D_a/Λ^2 is significant, the experiment must be repeated for several $[O_2]$, and then both D_a and k can be determined. By this procedure, the D_a for many ions diffusing in various buffer gases (many values are listed in Reference [11]) and the rate coefficients for many ion–molecule reactions over the temperature range from 200 to 600 K [13] have been determined. The various SA techniques have been described in the textbooks [9,11] together with original references to the work.

It is important to note that to study negative ion reactions using the SA, free electrons must be absent from the plasma, and then the diffusion of the negative ions to the plasma boundary (the walls) is not seriously inhibited by space-charge (ambipolar) electric fields [14]. Such has been accomplished using the high-pressure mass spectrometer (HPMS) technique [15], in which ionization is created in a relatively high-pressure gas mixture (several Torr and above) by a pulsed high-energy electron beam (typically 2 keV), and both the positive ions and the negative ions diffusing in the afterglow to the walls of the chamber (typical dimensions of a few centimeters) are sampled, as usual, by a differentially pumped mass spectrometer. This approach has permitted the study of the kinetics of a large number of reactions and also, of special note, the equilibria in many reactions. From the latter studies, much critical thermochemical data have been obtained (see the excellent review in Reference [15]).

Continuing with the example of the He/O_2 afterglow in which O^+ ions are produced via Reaction (9.3), the following secondary reaction is rapidly promoted if $[O_2]$ is sufficiently high:

$$O^+ + O_2 \rightarrow O_2^+ + O. \tag{9.6}$$

O_2^+ becomes the only ionic species in the afterglow. If n_e is high, these ions react with electrons in the afterglow via the dissociative recombination reaction

$$O_2^+ + e \rightarrow O + O. \tag{9.7}$$

The continuity equation for $[O_2^+] = n_e$ in the quasineutral plasma) is

$$\frac{dn_e}{dt} = D_a \nabla^2 n_e - \alpha_e n_e^2, \tag{9.8}$$

where α_e is the dissociative recombination coefficient for O_2^+ ions. Again, if the buffer gas pressure is sufficiently high, the diffusion term can be neglected and Equation (9.8) has the simple solution

$$\frac{1}{(n_e)_t} - \frac{1}{(n_e)_0} = \alpha_e t, \tag{9.9}$$

and so by determining n_e as a function of time, t, α_e can be obtained. Much of the early thermal energy α_e data were obtained using the SA/microwave cavity technique [16], in which a gas or gas mixture is ionized in a microwave cavity (see Figure 1) by a pulse of microwave energy, and the time variation of n_e in the afterglow plasma is monitored using a microwave probing signal to determine the change in the resonant frequency of the cavity as n_e decreases. The ions in the afterglow plasma are monitored using a pinhole orifice in the wall and a differentially pumped mass spectrometer. The temperature of the cavity (and hence of the afterglow plasma) can be varied over a limited range. These experiments established the magnitudes of α_e as a few $\times 10^{-7} cm^3 sec^{-1}$ for the majority of simple diatomic and triatomic ions and a few $\times 10^{-6} cm^3 sec^{-1}$ for polyatomic ions [16,17]. Further, heating the electrons in the microwave cavity using microwave energy has allowed α_e for some reactions to be determined up to values of $T_e \sim 5000$ K [17,18]. In some early SA experiments [19,20], n_e was determined using Langmuir probes. However, the marrying of the Langmuir probe technique to the FA (forming the FALP) has brought much greater scientific returns (see below).

An SA/microwave cavity technique has also been used to study electron attachment reactions. Known as the pulsed radiolysis technique [21,22], it utilizes a pulsed high-energy electron beam to generate X rays at a thin tungsten foil. The X rays then partially ionize a gas mixture in a microwave cavity, and, as before, the resonant frequency of the cavity is followed to determine the electron number density in the afterglow; hence, the attachment coefficient is deduced. The temperature of the gas can be varied over the range 77–373 K, and the electrons can be heated by microwave energy. Therefore, the separate influences of the electron energy and the attaching molecule temperature on the attachment rates can be investigated, if the electron energy can be determined (see Reference [22] for details). Measurements of the rate coefficients for electron attachment to several molecular gases within the mean electron energy range 0.03–1.0 eV have been reported [22]. However, this experiment does not include mass analysis of the product ions of the attachment reactions.

A technique related to the SA technique but which does not qualify as a plasma experiment because of the very low electron number densities involved (typically a few cm^{-3}) is the Cavalleri technique [23], developed to obtain accurate data on electron attachment reactions at thermal energies [24,25]. In this method, ionization is created in a gas mixture (usually N_2 with a small admixture of attaching gas) by a burst of X rays, and the rate of loss of the electrons is determined by observing the light intensity resulting from a time-delayed RF discharge through the N_2. The light intensity is proportional to the number of unattached electrons which is dependent on the time delay. Hence by correlating the number of electrons with the time delay the attachment coefficient can be deduced. The variation with temperature of the attachment coefficient for a few molecular gases has been accurately determined using this method. Again, mass analysis of the product ions has not been carried out in these experiments, although this would be very difficult in view of the very low level of ionization in the gas.

9.2.2 The Flowing Afterglow

The *continuous* ionization flow techniques differ from the *pulsed* ionization techniques in that ionization is continuously established upstream in a flowing gas so that the time variation of ionization density in the stationary systems is transformed into a spatial variation in flowing systems via the flow velocity of the ionized gas. This approach has been very successful and has many advantages.

Very successful is the FA technique [26]. A typical FA apparatus is shown in Figure 2. Ionization is continuously created in the upstream

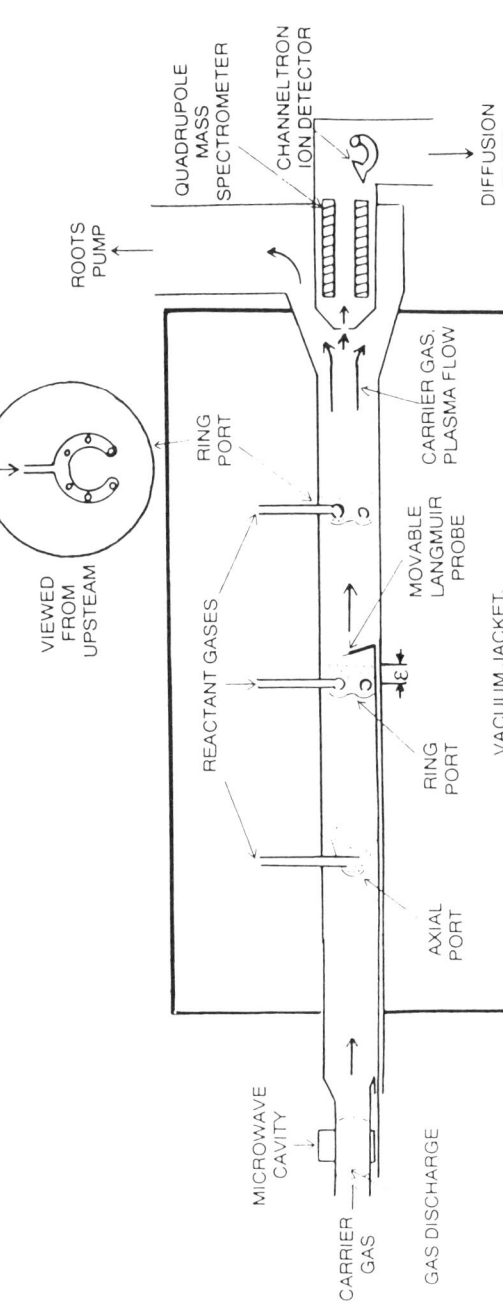

FIG. 2. Schematic representation of a flowing afterglow (FA) apparatus and, with the inclusion of the Langmuir probe diagnostic, a flowing afterglow/Langmuir probe (FALP) apparatus. Ionization is created upstream in the flowing carrier gas, and an afterglow plasma is created along a flow tube with a typical length of 100 cm and a diameter of 8 cm. A microwave discharge is the ionization source of choice for FALP experiments, but hot cathode electron emitters and DC discharges are used for FA experiments. Reactant gases are introduced into the plasma via the entry ports shown; the ring ports are most suitable for FALP studies because of their smaller end correction (ε). The mass spectrometer system monitors the positive and negative ions in the downstream afterglow plasma, and the movable Langmuir probe is used to measure the absolute electron and ion number densities and the electron temperature at any position along the axis of the flow tube. The Roots pump moves the carrir gas (and plasma) along the flow tube at a velocity of typically 10^4 cm s^{-1}. The vacuum jacket thermally insulates the whole flow tube for high- and low-temperature operation.

region of a flowing gas usually via a DC or microwave electrical discharge. The momentum of this carrier gas (usually an inert gas, commonly helium or argon at a pressure of ~1 Torr) ensures that ionization (positive ions and electrons) is distributed along the length of the flow tube (which is usually ~1 m long), thus creating an afterglow plasma. Because the afterglow plasma is remote from the electrical discharge, the charged particles are thermalized in the carrier gas (as in the late SA). At the downstream end of the flow tube is located a differentially pumped mass spectrometer/ ion detection system which samples the ions from the afterglow plasma through a pinhole. (The pinhole is very small, ~0.3 mm in diameter, to minimize carrier gas flow into the mass spectrometer.) The mass spectrometer system therefore detects a time-invariant current of ions which, for example, in pure helium carrier gas at 300 K is a current of He^+ ions (and some He_2^+ ions). To study the reaction of He^+ ions with O_2 (Reaction (9.3)), O_2 is added in measured quantities to the thermalized afterglow plasma via an entry port located at some position along the flow tube, and the decrease in the He^+ current at the mass spectrometer detector is monitored as a function of the O_2 flow rate. The entry port can be designed in several different ways: as a wall inlet, an axial tube inlet, or a ring inlet (see Figure 2) with small apertures pointing upstream which ensures faster mixing of the reactant gas with the carrier gas. The type of inlet used determines the extent of the mixing region for the reactant gas (the "end correction," ε, to the reaction length, z; see below). It can be shown [26] that the He^+ current at zero O_2 flow, i_0, and the He^+ current at a finite O_2 flow, i, are related by the expression

$$i = i_0 \exp - k[O_2]z/v, \qquad (9.10)$$

where $[O_2]$ is the O_2 number density in the flow tube (determined from the O_2 flow rate), z (the reaction length) is the distance of the O_2 entry port from the mass spectrometer sampling orifice adjusted by the end correction, v is the plasma flow velocity (determined from the carrier gas flow rate and the pressure [27]), and k is the rate coefficient for the reaction. Thus a plot of ln i against $[O_2]$ provides a value for k. The experiment can also be performed by fixing $[O_2]$ and by varying z. The advantages of the FA method over the SA method are that the reactant gases are not exposed to the electrical discharge (thus avoiding problems such as the excitation and dissociation of the reactant gases), and that sequential addition of gases into the afterglow at various positions along the flow tube (see Figure 2) allows the production of reactant ions by ion–molecule reactions which cannot be made by electron impact ionization. This permits the study of the reactions of a wider variety of ions with stable molecules [13] and even with reactive radicals [28].

Flowing afterglows have been exploited to study a very large number and variety of positive ion and negative ion reactions at thermal energies (in some cases over the wide temperature range from 80 to 900 K [29]). The ion-neutral chemistry of the terrestrial ionosphere and stratosphere has been elucidated largely as a result of FA studies (coupled with *in situ* determinations of ionic concentrations using rocket- and satellite-borne mass spectrometers) [30]. Extensive compilations of the rate coefficients and product ions obtained from FA (and SIFT; see below) experiments are available [13].

9.2.3 The Flowing Afterglow/Langmuir Probe Apparatus

The FA was initially used to study only ion-neutral reactions, the only diagnostic tool necessary being the downstream mass spectrometer. Subsequently, the FALP technique was developed [31], in which a Langmuir probe is included in the FA (see Figure 2) to measure n_e, n_+, and T_e along the length of the afterglow plasma column. The development of the Langmuir probe technique for use in flowing afterglow plasmas represented a major step forward in the study of ionic reactions at thermal energies and has permitted the study of positive ion–electron recombination [32,33], electron attachment [34,35], and positive ion–negative ion recombination [7,36] (in addition to the conventional study of ion-neutral reactions) in one apparatus. Hence the FALP is arguably the most versatile apparatus yet devised for the study of ionic reactions at thermal energies. Further, the operation of the FALP is very straightforward. To determine, for example, the recombination coefficient, α_e, for Reaction (9.7), the chemical versatility of the FA is used to create an afterglow plasma containing only O_2^+ ions and electrons. This is readily achieved using helium carrier gas and by exploiting the ion–molecule Reactions (9.3) and (9.6). If n_e (and hence the O_2^+ number density) is sufficiently large and if the helium pressure is also large enough to inhibit ambipolar diffusive loss of ionization, then the variation of n_e along the plasma column (i.e., along z) is given by Equation (9.8) but with t replaced by $(z_0 - z)/v$:

$$\frac{1}{n_{ez}} - \frac{1}{n_{e0}} = \frac{\alpha_e(z_0 - z)}{v}. \tag{9.11}$$

Here n_{e0} and n_{ez} are the electron number densities at positions z_0 (a reference position) and z, respectively, and v is the plasma flow velocity (which is readily measured using the Langmuir probe [27]). Thus α_e can be determined by measuring the z gradient of n_e. It has been possible using this approach to determine α_e for many molecular ions, including O_2^+, NO^+, H_3O^+, and $NH_4^+(NH_3)_2$ (some over the wide temperature range from 90 to 550 K [32,33]) and also for many ions which are considered to be important in intersteller chemistry [37]. Details of the operation of the FALP, the physics of the Langmuir probe [12], and many experimental results are given in some key research papers [32,33,37].

It is also desirable to know the neutral products of dissociative recombination reactions. This is a much more challenging experimental problem, but some such products have been identified by applying vacuum ultraviolet (VUV) and laser-induced fluorescence (LIF) spectroscopy to recombining FALP plasmas. Such spectroscopic experiments are possible since n_e (and n_+) in these plasmas are relatively high (typically 10^{10} cm^{-3}) and hence so are the number densities of the neutral products of recombination reactions. Thus the H atoms (by VUV) and OH radicals (by LIF) produced in the recombination of several ionic species including H_3O^+, HCO_2^+, and N_2OH^+ have been quantitatively studied [38], as have the H atoms produced in the recombination of several protonated ion species [39]. These are the first measurements of the neutral products of interstellar ions and as such represent another success of the FALP technique.

In all the FALP experiments discussed above, the carrier gas used was helium which ensured that in the afterglow $T_e = T_+ = T_g$. However, recent work [40] has shown that the use of argon carrier gas allows elevated electron temperatures (i.e., $T_e > T_+$ and T_g) up to 3000 K to be established in the afterglow. This is possible because of the deep Ramsauer minimum in the momentum transfer cross-section for electron–argon atom collisions [41]). This advance has proceeded in parallel with the further development of the Langmuir probe technique to rapidly determine electron energy distribution functions and electron temperatures in these afterglows [42]. With this development, it is now possible to study recombination (and attachment reactions; see below) over a wide range of T_g (80–600 K) and T_e (from T_g to 5000 K). This has allowed studies of α_e for the important ionospheric ions O_2^+ and NO^+ over the T_g and T_e ranges that are appropriate to the ionospheric plasma [43].

The FALP can also be used to determine electron attachment rate coefficients, β. If an electron attaching gas is added to the afterglow plasma, then the electrons in the plasma are converted to negative ions as, for example, in the dissociative attachment reaction

$$CCl_4 + e \rightarrow Cl^- + CCl_3. \tag{9.12}$$

The Langmuir probe is used to determine the dependence of n_e on z. It can be shown [34] that, if both diffusion and recombination losses are negligible (i.e., for high carrier gas pressures and small ionization densities), then

$$n_{ez} = n_{e0} \exp - \beta[\text{CCl}_4] \frac{z_0 - z}{v}, \qquad (9.13)$$

where the symbols have the same meaning as those in Equation (9.11). In this way, β for many attachment reactions, including those for SF_6 and several "freon" gases, and their variation with T_g have been determined using the FALP technique [34]. The experimental details and many of the results of these attachment studies are described in a recent review [35].

Following the development of the FALP to study reaction processes at elevated T_e, some electron attachment studies have been extended to elevated T_e (as well as over a range of T_g). Remarkable results are being obtained in these studies, the β for some reactions increasing rapidly with T_g and decreasing with increasing T_e [40].

If sufficient electron attaching gas is added to the afterglow plasma, all the electrons can be converted to negative ions, thus creating a flowing positive ion–negative ion afterglow plasma. The FALP can then be used to determine ionic recombination coefficients, α_i [36]. Again this is achieved by determining the z gradient of the ion number densities using the Langmuir probe. The analysis of the data to obtain the α_i is exactly the same as that to determine α_e (Equation (9.11)). Using this technique, the most comprehensive study of ionic recombination at thermal energies has been carried out (including the determination of α_i for several stratospheric reactions [44]). Indeed, the data obtained are the most reliable data on this process at thermal energies and are essential to the calculation of deionization rates in the stratosphere and the troposphere [45].

9.3 The Selected Ion Flow Tube Technique

Notwithstanding the great success of the FA for the study of ion-neutral reactions (reviewed in Reference [46]), it has a serious limitation in that reactant ions have to be created by introducing an ion source gas either over the ion source or into the afterglow plasma. If the ions so formed react with their parent (source) gas, as is often the case, this can create serious problems in the interpretation of the mass spectrometric data. For example, if the reactions of CH^+ ions are to be studied, the CH^+ ions can be created in the plasma by adding one of several different hydrocarbons such as C_2H_2 and CH_4. However, more than one ion species are often formed and the CH^+ ions react rapidly with the parent gases, making it difficult to study their reactions with other molecules, and it is particularly difficult to determine the products of the reactions. A similar problem occurs in the study of negative ions. This problem has been overcome by

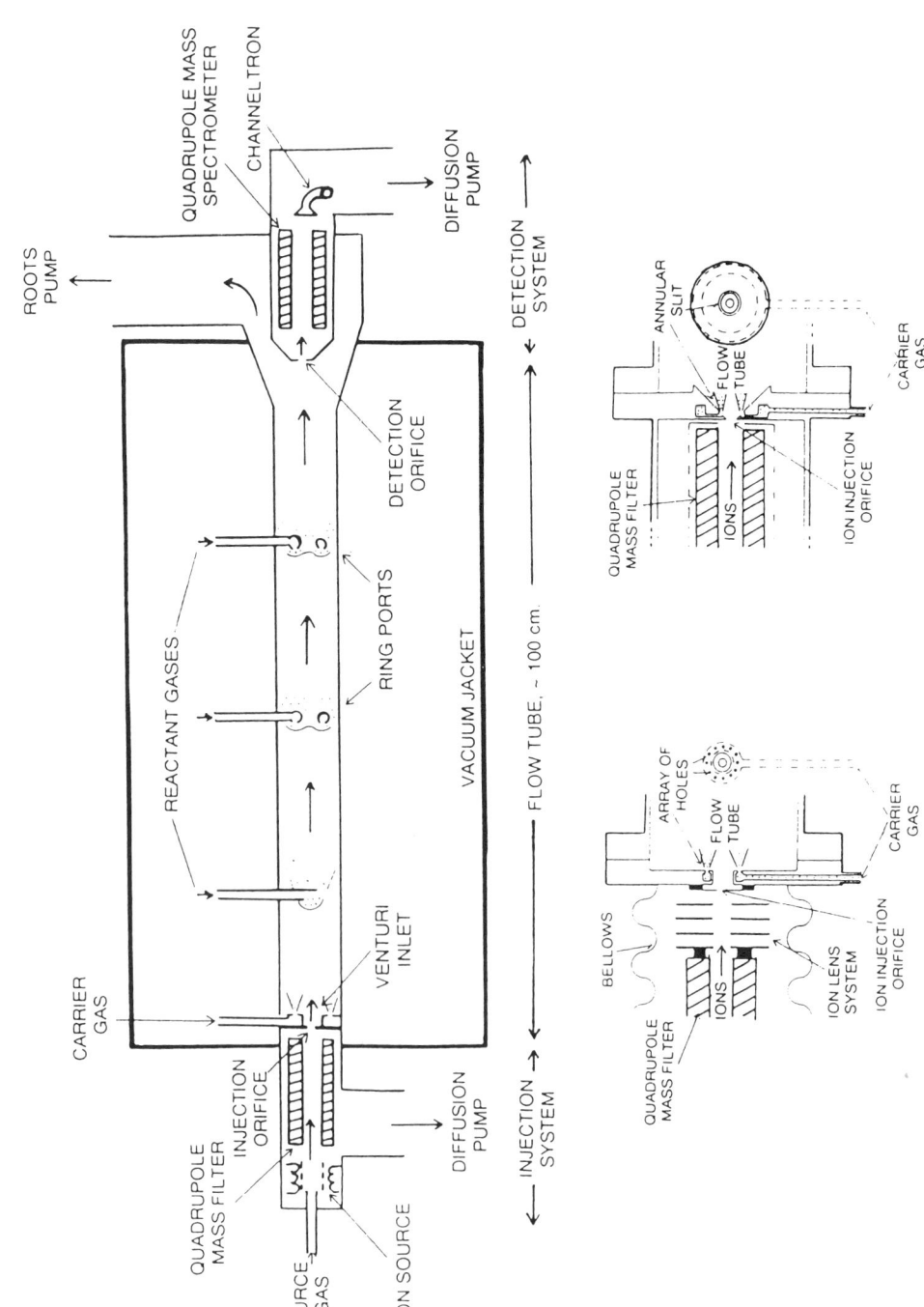

the development of the SIFT [47,48], illustrated schematically in Figure 3, in which the ions are created in an ion source which is external to the flow tube. The ions are then extracted from the ion source, selected according to their mass-to-charge ratio, using a quadrupole mass filter, and injected into the flowing carrier gas via a small orifice (typically ~1 mm in diameter). The carrier gas is inhibited from entering the quadrupole mass filter chamber (which would prevent it from operating) by injecting it into the flow tube through a venturi-type inlet at supersonic velocity in a direction away from the orifice. Two basic types of venturi inlet are in current use (see Figure 3). In the original one, the carrier gas is "injected" into the flow tube through several small apertures surrounding the ion injector orifice, whereas in a later design the gas is injected through an annular slit.

Continuing with the above hydrocarbon example, if methane is introduced into the ion source, C^+, CH^+, CH_2^+, CH_3^+, or CH_4^+ can be selectively injected into the flowing carrier gas and convected down the flow tube as a swarm of thermalized ions (the ion number densities are typically 10^2–10^4 cm^{-3}). The significant point is that the CH_4 source gas is not present in the helium carrier gas. The reactions of each of these ions can now be studied individually with any gas or vapor using an experimental approach identical to that used for the FA (Equation (9.10)). The ion products of the reactions can be determined unambiguously since confusion is avoided because the CH_4 parent gas is not present in the flow tube. Some SIFT apparatuses can be operated over the wide temperature range 80–600 K [48]. Constructional details for a SIFT are given in a major review [49].

Several important experimental points should be made concerning the operation of a SIFT. The reactions of practically any positive or negative ion species that can be extracted from an ion source at a sufficient current ($\sim 10^{-9}$ A is the practical lower limit) can be studied, but with the proviso that these ions are injected into the (helium) carrier gas at a sufficiently low energy that they do not fragment on collision with the helium atoms. This, of course, depends on the dissociation energy of the ion, but even

---

FIG. 3. Schematic representation of a selected ion flow tube (SIFT) apparatus, in which ions are created in the ion source, selected according to their mass-to-charge ratio by the mass filter, and injected into the carrier gas through an orifice which is typically 1 mm in diameter. Two forms of the Venturi-type inlets by which carrier gas is introduced into the flow tube are shown in the lower part of the diagram. Reactant gases are added to the flowing swarm of thermalized ions via the entry ports shown, and the loss rate of primary (injected) and product ions are monitored by the downstream detection system. The vacuum jacket facilitates high- and low-temperature operation.

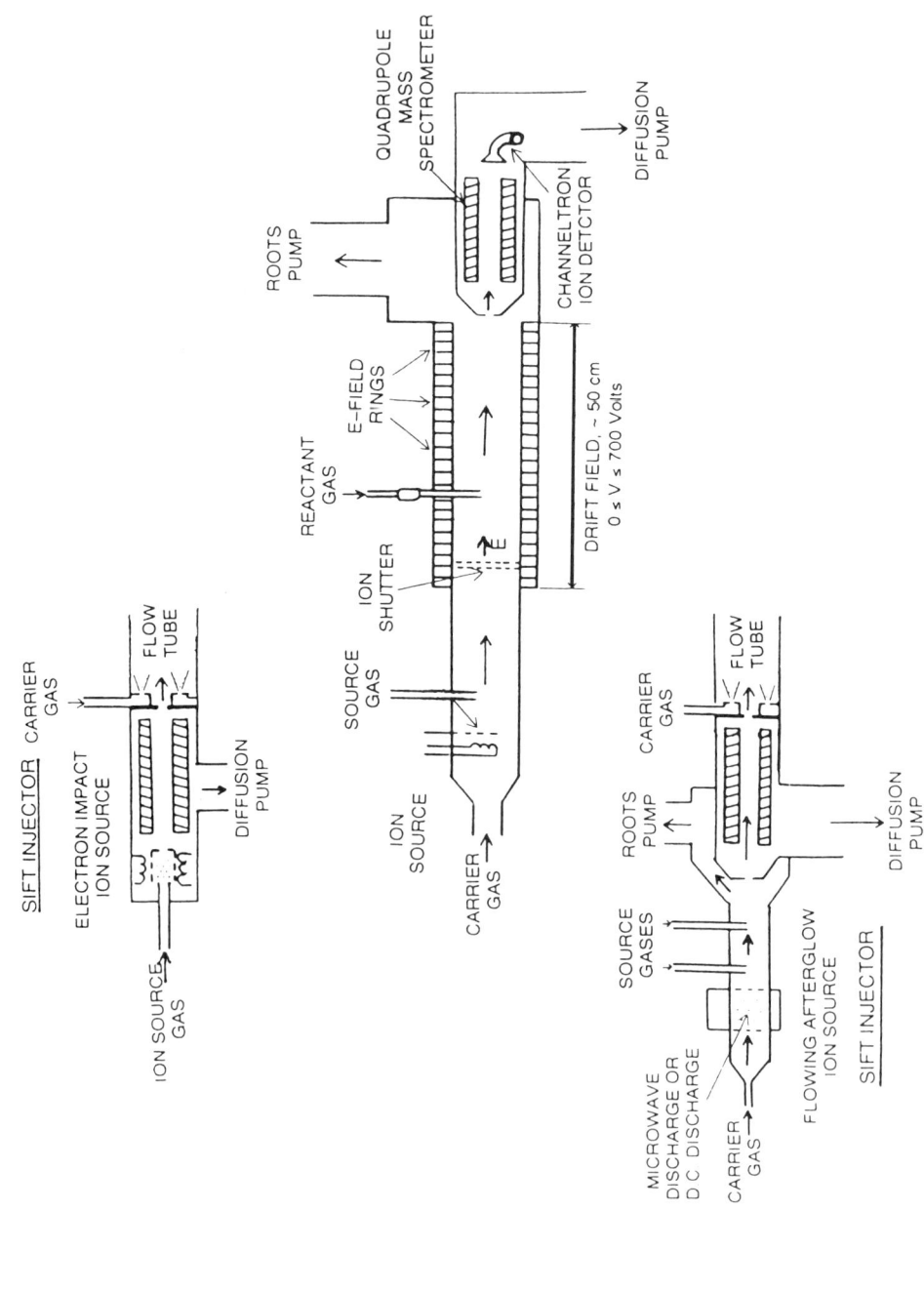

weakly bound species such as $H_3O^+(H_2O)_3$ ions have been successfully injected and their reactions studied [50]. Low-pressure and high-pressure electron impact sources are routinely used to prepare a wide variety of positive and negative ions, and the chemical versatility of the FA has been very successfully exploited as an ion source for the SIFT (see Figure 4) to prepare ions that cannot be created by electron impact, such as the cluster ions mentioned above [51,52].

Ions produced in the collision-dominated medium which is the FA plasma are almost invariably relaxed to their ground vibronic states, but this is not so likely for ions produced in low-pressure electron impact ion sources. Thus excited ions (electronically and vibrationally) are sometimes a significant component of the injected ions in SIFT experiments. Then the convected ion swarm will consist of ions of different reactivities. This is manifested by the nonlinearity of the decay curves from which the rate coefficients are derived (following Equation (9.10)). Because these excited ions are in contact with only the inert helium carrier gas, they are able to survive in the SIFT and so their reactions can be studied [53].

Using the SIFT, which is a simple and well-controlled swarm technique, the rate coefficients and product ions for a large number of positive and negative ion-neutral reactions have been studied. They include the bimolecular reactions of doubly charged ions, electronically and vibrationally excited ions, cluster ions, and termolecular reactions, some over a wide range of temperature. This has led to a greater understanding of the mechanisms, kinetics, and energetics of these reactions and to a clearer insight into the chemistry of planetary atmospheres, intersteller gas clouds, and laboratory plasmas such as gas lasers and surface etchant plasmas. Some useful reviews of the SIFT technique and the results of the numerous experimental studies have been published [48,49,53,54].

FIG. 4. Schematic representation of a flow drift tube (FDT) apparatus is shown as the center figure. Ions are created by a simple ion source immersed in the carrier gas and are convected (by the gas flow) to the uniform drift field which is established by suitable potentials on the E-field rings. The ion shutter (see text) facilitates the measurement of the ion drift velocity. Rate coefficients for the ion-neutral reaction are obtained by introducing reactant gas into the ion swarm and monitoring the ion loss and production rates downstream. Two forms of SIFT injector that can be coupled to the FDT to form a selected ion flow drift tube (SIFDT) apparatus are also illustrated. The first includes only a simple electron impact ion source, whereas the second includes a flowing afterglow ion source which extends the variety of ions that can be studied.

9.4 Drift Tube Techniques

The basic principles of the operation of drift tubes are simple. Electrons or ions are created or introduced into a buffer gas at pressures typically 0.1 to a few Torr. A homogeneous electric field, E, is established along the axis of the drift tube by a number of equally spaced metal rings held at appropriate potentials. Thus the electrons (or the ions) are constrained to move in the direction of E, and in doing so they undergo multiple collisions with the buffer gas particles. Therefore, rather than constantly being accelerated by E, the charged particles reach a steady mean drift velocity, v_d, which is dependent on the ratio of E/N, where N is the buffer gas number density. If E/N is not too great, the v_d of both electrons and ions are much smaller than their mean random velocities, v_i, although both v_d and v_i increase with E/N. It is a straightforward matter to measure v_d as a function of E/N for charged particles in a particular buffer gas and thus to derive the mobility, μ, of the charged particles, since $\mu = v_d/E$. The essential point is that the drifting electrons or ions acquire some velocity (energy) distribution, $f(v)$, and hence some mean energy, ε, which is related to E/N (and therefore to v_d and μ), and so reactions of the electrons or ions in these swarms can be studied as a function of E/N and therefore of ε. The SDT (contrast with the FDT) has been used for decades for such studies [9,11,55]. The SDT has also been used to study ion-neutral reactions [11], but here we refer only to its use for the study of electron attachment. We then discuss the use of FDT and SIFDT techniques for the study of ion-neutral reactions.

9.4.1 The Static Drift Tube

In the study of electron attachment reactions using the SDT method [56,57], a pulse of electrons is created at one end of a drift tube (photoionization, photoemission, gas discharges, and α particles have all been used) in an inert (nonattaching) buffer gas to which very small amounts of electron attaching gas (number density, n_a) can be added. Since $N \gg n_a$, the drifting electron swarm attains an equilibrium energy distribution, $f(\varepsilon)$, which is dependent on E/N, the nature of the buffer gas, and its temperature, T, but not on the nature of the attaching gas since it is in very low concentration. Ar and N_2, are commonly used as buffer gases since $f(\varepsilon)$ for electrons in these gases can be calculated quite accurately over a wide range of E/N [57]. The electron current pulse amplitude at the downfield electrode is measured first for pure buffer gas and then after the small fraction of attaching gas has been added. From the ratio of these amplitudes, the probability of attachment per centimeter traveled by the electrons in the E field direction (α') can readily be obtained as a function of E/N. The ratio α'/N_a corresponds to the attachment cross-section (σ_a) averaged over the electron distribution function $f(\varepsilon)$, and from it the rate coefficient for electron attachment, β, can be calculated. Thus β can be obtained as a function of E/N and hence of the mean electron energy ε (utilizing the known $f(\varepsilon)$) [57]. Using this technique, σ_a and β have been determined as a function of ε for many molecular gases over a wide range of ε from thermal up to about 4 eV, mostly at an attaching (and buffer) gas temperature equal to room temperature, but some important studies have been carried for attaching gas temperatures up to 700 K [58].

Attention has been focused on the differences in the measured β for some particular molecules as determined in truly thermalized FALP experiments and in SDT experiments [59]. It was noted in FALP experiment that for some halogenated methanes and ethanes, the β increased rapidly when the attaching gas temperature, T_g, and the electron temperature, T_e, were increased together, whereas in SDT experiments when T_g was held constant the β actually decreased with increasing ε (equivalent to increasing T_e). This forcibly demonstrates the importance of vibrational excitation of the molecules (by heating) in the dissociative attachment process. For further discussion of this, see References [35] and [59].

9.4.2 The Flow Drift Tube

The SDT has been used to study ion-neutral reactions [11,53]. However, the FDT (and SIFDT, described later) has been much more productive for ion-neutral reaction studies. A typical FDT apparatus is shown in Figure 4 [60]. The FDT is very similar in dimensions to a conventional FA apparatus but with the vital difference that the flow tube wall is constructed from nearly 100 metal rings separated by O rings (vacuum seals) and Mylar spacers (electrical insulators). Through the application of suitable potentials to the rings, a uniform electric field, E, can be established along the axis of the flow tube. The upstream half of the flow tube (usually field free) is the ion production region, in which a flowing afterglow plasma is created as a source of thermalized ions which are convected by the carrier gas flow downstream to the drift field region (electrons in the plasma are excluded from the reaction region by the E field). Thus a variety of positive and negative ions can be synthesized, and their reactions with a variety of neutrals can be studied, again as a function of E/N in a manner very similar to that used for the conventional FA and SIFT experiments described previously (i.e., following Equation (9.10)). To extract quantitative results the ion velocity (and ion residence time) in the drift region is required, and this is obtained by the following procedure.

An ion shutter is located between the upstream field-free region and the drift field region and consists of two closely spaced grids ~1 mm apart. An intense electric field is created between the grids and prevents the flow of ions into the drift field. This field can be pulsed off to allow the passage of a burst of ions into the drift section, and the arrival time of these ions at the downstream mass spectrometer detection system provides a value for the ion velocity. Since the carrier gas is flowing at some velocity, v_0, the measured ion velocity is the sum of v_0 and the drift velocity, v_d, of the ions through the carrier gas. (Note that v_0 is only a significant fraction of v_d at low E/N.) Using this approach, the v_d and hence the mobilities of many ionic species in He and Ar carrier gas have been determined [61].

By monitoring the decrease of the primary ion signal at the downstream detection system as a function of the reactant gas number density in the drift field region, the rate coefficient, k, for an ion-neutral reaction can readily be determined as a function of E/N. However, it is desirable to relate k to the center-of-mass energy between the reactant ion and the reactant molecule, E_r. This can be obtained by adopting the Wannier expression [62] which describes the mean kinetic energy, E_i, of an ion drifting through a buffer gas with drift velocity v_d as

$$E_i = \tfrac{1}{2}m_i v_d^2 + \tfrac{1}{2}M v_d^2 + \tfrac{3}{2}k_b T, \qquad (9.14)$$

where m_i and M are the masses of the ion and the buffer gas atom, respectively, and k_b is the Boltzman constant. E_r is simply given by

$$E_r = \tfrac{1}{2}[m_r m_i/(m_r + m_i)](v_i^2 + v_r^2), \qquad (9.15)$$

where m_r is the mass of the reactant neutral, v_r^2 its mean square velocity, i.e., $\tfrac{1}{2}m_r v_r^2 = \tfrac{3}{2}k_b T$, and v_i^2 is the ion mean square velocity, i.e., $\tfrac{1}{2}m_i v_i^2 = E_i$. Combining these equations, one obtains

$$E_r = [m_r/(m_r + m_i)](E_i - \tfrac{3}{2}k_b T) + \tfrac{3}{2}k_b T. \qquad (9.16)$$

From a measurement of v_d, E_i is obtained and hence E_r can be calculated. Initially it was seriously questioned if the Wannier expression (9.14) sufficiently accurately described E_i, but its value has been confirmed by theoretical work [63] coupled with experimental observations [64]. The FDT technique has provided rate coefficients for many ion–molecule reactions over the E_r range from thermal (0.05 eV) to ~3 eV and at a fixed buffer gas temperature of 300 K [53,60,65].

9.4.3 The Selected Ion Flow Drift Tube

The FDT has the disadvantage that, like the FA, more than one ionic species is often created in the upstream afterglow source, and these species reach the reaction zone and can confuse the determination of product distributions for ion-neutral reactions. This can be circumvented by using a SIFT-type ion injector such as that illustrated in Figure 4, the basis of the SIFDT technique.

Ions are created in a remote ion source, filtered to produce a beam of ions of a given mass-to-charge ratio, and injected via the venturi inlet into the carrier gas. In most of the SIFDT apparatuses in current use, the ions first enter a field-free region where they thermalize before entering the drift field region, where their reactions can be studied as a function of E/N in the usual way. Clearly, since an E field must be established along the flow tube axis, either the upstream ion injector or the downstream mass spectrometer detector must be at a high potential relative to the ground. In the original SIFDT [48,49,66] the downstream detection system was at the high potential, whereas in later apparatuses [67] the ion injector was held at the high potential. The former approach is more convenient if various ion sources, such as a FA source, are to be easily used. SIFDT apparatuses have been operated over a carrier gas temperature range from 80 to 600 K [49,66,68].

The versatility of the SIFDT apparatuses in current use is striking. They permit the study of a wide range of reactions of neutrals with positive ions and negative ions, including endothermic reactions [53,69], collisional (vibrational) excitation and deexcitation of molecular ions [70], collisional dissociation of molecular ions [71], and, in the case of variable temperature SIFDT apparatuses, the separate influence of increasing temperature of the reactants and increasing ion energy (E_r, see Section 9.4.2) on the rate of ion–molecule reactions [72,73]. Brief summaries of some of these studies together with some experimental details are given in some reviews [49,66,69]. Of course, by reducing the E field to zero, a SIFDT becomes a conventional SIFT apparatus in which reactions can be studied as a function of temperature under truly thermalized conditions.

9.5 Swarm Experiments at Very Low Temperatures: The CRESU Technique

There has developed a strong interest in the study of ion-neutral reactions at very low temperatures, stimulated largely by the discovery of many species of molecules in cold interstellar clouds and by the desire to understand how molecules can be formed in such harsh regions [54]. The variable-temperature SIFT can provide data down to 80 K for reactions which do not involve condensable reactants. This temperature is in the upper part of the range of dense interstellar cloud temperatures, and there is a need for kinetic data at even lower temperatures; these can be obtained by the supersonic expansion of gases.

The CRESU technique mentioned in the Introduction [74] exploits the fact that a gas can be cooled to very low temperatures if it is rapidly

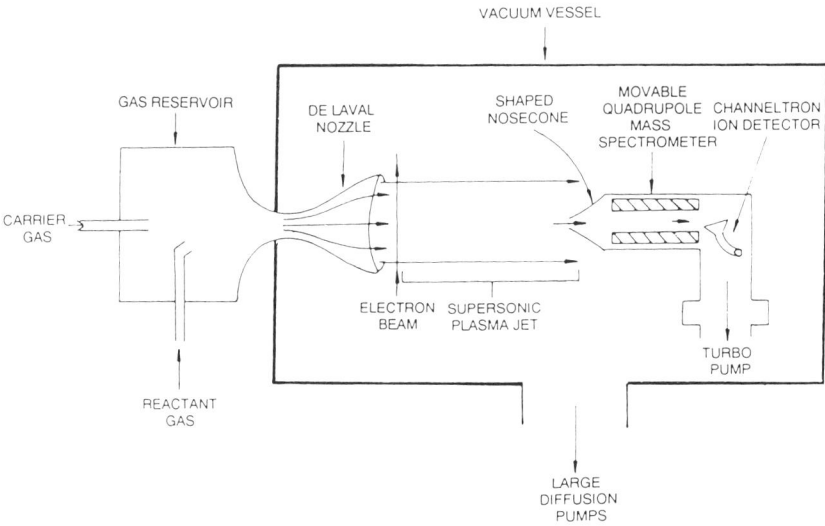

FIG. 5. A schematic representation of the CRESU apparatus (not to scale). Carrier gas is expanded through the de Laval nozzle and cools to a very low temperature. The cold gas is partially ionized by a high-energy electron beam, forming a fast-flowing plasma jet ~10 cm in diameter. Ion-neutral reactions are studied either by varying the concentration of reactant gas (introduced into the relatively warm carrier gas as shown) or by fixing the concentration of the reactant gas and moving the mass spectrometer detector along the axis of the plasma jet. The nosecone of the ion sampling system is specially shaped to avoid perturbations from supersonic shocks.

expanded. Gas (which is usually helium, argon, or nitrogen) from a reservoir (which can be precooled) is expanded via a specially shaped de Laval nozzle into a continuously pumped vacuum vessel (as shown in Figure 5). A cold supersonic jet of gas is created which is then partially ionized with a high-energy electron beam, thus generating a supersonic plasma. Downstream of the electron beam a cold flowing afterglow plasma exists in which reactions at very low temperatures can be studied. A mass spectrometer sampling orifice (in a specially shaped housing to minimize supersonic shock effects) can be moved along the axis of the flowing plasma to sample the ions. Reactant gases are introduced (via the relatively warm reservoir) into the carrier gas and hence into the cold plasma, and ion-neutral reaction rate coefficients are determined, as in the case of the FA and SIFT, by measuring the ion current to the mass spectrometer either as a function of the reactant gas concentration or by fixing the flow of the reactant gas and moving the mass spectrometer. Temperatures as low as 8 K have been achieved, and several reactions have been studied over the temperature range 8–160 K [75]. Reactions of condensable vapors (H_2O and NH_3) have even been studied down to 27 K [74]. This is possible because the vapor is introduced into the warm reservoir which is held above the condensation temperature of the vapor. A development of this technique is the inclusion of a SIFT-type ion injector which greatly increases the type of ions that can be introduced in the cold gas. Early results obtained using the technique are summarized elsewhere [76].

9.6 Concluding Remarks

Swarm methods have made major contributions to the study of ionic and electronic reactions in the thermal and near-thermal energy regimes. Indeed it is only from swarm media such as afterglow plasmas and from low-temperature swarm experiments such as the SIFT and the CRESU experiments that reliable rate coefficients for ion-neutral reactions applicable to the ion chemistry of low-temperature media can be obtained. A large amount of data on other reaction processes has been obtained using the very versatile and productive FALP apparatus. New swarm techniques are being conceived to study reactions at ultralow temperatures and at very high temperatures. A very recent application of swarm techiques, specifically the application of a modified form of the SIFT apparatus, is being used to detect and quantify trace gases in the atmosphere and on human breath [77,78]. The clinical value of this is great, as is described in a very recent review [79]. With the increasing use of such swarm apparatuses, an even more rapid growth in the understanding of reaction kinetics and the mechanistics of ion and electron reactions at low energies will surely occur.

Acknowledgments

A review of this exciting topic could not have been contemplated and would not have been required but for the inventiveness, experimental skills, and productivity of the many workers who have used and continue to use swarm methods to study gas phase reactions. We are grateful to all these many colleagues and friends for their implicit support in the preparation of this chapter.

References

1. J. M. Farrar, in *Techniques for the Study of Gas-Phase Ion-Molecule Reactions* (J. M. Farrar and W. H. Saunders, Jr., eds.), p. 325. Wiley, New York, 1988.

2. W. Lindinger, T. D. Märk, and F. Howorka, eds., *Swarms of Ions and Electrons in Gases*. Springer-Verlag, Wien, 1984.
3. P. R. Kemper and M. T. Bowers, in *Techniques for the Study of Gas-Phase Ion-Molecule Reactions* (J. M. Farrar and W. H. Saunders, Jr., eds.), p. 1. Wiley, New York, 1988.
4. D. Gerlich and S. Horning, *Chem. Rev.* **92**, 1509 (1992).
5. E. W. McDaniel, V. Cermak, A. Dalgarno, E. E. Ferguson, and L. Friedman, *Ion-Molecule Reactions*, p. 107. Wiley, New York, 1970.
6. J. T. Moseley, R. E. Olson, and J. R. Peterson, *Case Stud. At. Phys.* **5,** 1 (1975).
7. J. B. A. Mitchell and J. W. McGowan, in *Physics of Ion-Ion and Electron-Ion Collisions* (F. Brouillard and J. W. McGowan, eds.), p. 279, Plenum, New York, 1983.
8. C-Y. Ng and M. Baer, *State-Selected and State-to-State Ion-Molecule Reaction Dynamics*, Parts 1 and 2. Wiley, New York, 1992.
9. E. A. Mason and E. W. McDaniel, *Transport Properties of Ions in Gases.* Wiley, New York, 1988.
10. D. Smith, A. G. Dean, and N. G. Adams, *Z. Phys.* **253**, 191 (1972).
11. E. W. McDaniel and E. A. Mason, *The Mobility and Diffusion of Ions in Gases.* Wiley, New York, 1973.
12. J. D. Swift and M. J. R. Schwar, *Electrical Probes for Plasma Diagnostics.* Iliffe, London, 1970.
13. Y. Ikezoe, S. Matsuoka, M. Takebe, and A. A. Viggiano, *Gas Phase Ion-Molecule Reaction Rate Constants Through 1986.* Maruzen, Tokyo, 1987.
14. H. J. Oskam, *Philips Res. Rep.* **13**, 335 (1958).
15. P. Kebarle, in *Techniques for the Study of Gas-Phase Ion-Molecule Reactions* (J. M. Farrar and W. H. Saunders, Jr., eds.), p. 221. Wiley, New York, 1988.
16. J. N. Bardsley and M. A. Biondi, *Adv. At. Mol. Phys.* **6**, 1 (1970).
17. R. Johnsen, *Int. J. Mass Spectrom. Ion Processes* **81**, 67 (1987).
18. J. L. Dulaney, M. A. Biondi, and R. Johnsen, *Phys. Rev. A* **36**, 1342 (1987).
19. D. Smith, C. V. Goodall, and M. J. Copsey, *J. Phys. B* **1**, 660 (1968).
20. D. Smith and I. C. Plumb, *J. Phys. D* **6**, 196 (1973).
21. M. Toriumi and Y. Hatano, *J. Chem. Phys.* **82**, 254 (1985).
22. H. Shimamori, Y. Tatsumi, Y. Ogawa, and T. Sunagawa, *J. Chem. Phys.* **97**, 6335 (1992).
23. G. Cavalleri, *Phys. Rev.* **179**, 86 (1969).
24. R. W. Crompton and G. N. Haddad, *Aust. J. Phys.* **36**, 15 (1983).
25. Z. Lj. Petrovic and R. W. Crompton, *J. Phys. B* **20**, 5557 (1987).
26. E. E. Ferguson, F. C. Fehsenfeld, and A. L. Schmeltekopf, *Adv. At. Mol. Phys.* **5**, 1 (1969).
27. N. G. Adams, M. J. Church, and D. Smith, *J. Phys. D* **8**, 1409 (1975).
28. T. J. Bevilacqua, D. R. Hanson, and C. J. Howard, *J. Phys. Chem.* **97**, 3750 (1993).
29. W. Lindinger, F. C. Fehsenfeld, A. L. Schmeltekopf, and E. E. Ferguson, *J. Geophys. Res.* **78**, 4753 (1974).
30. E. E. Ferguson, F. C. Fehsenfeld, and L. Albritton, in *Gas Phase Ion Chemistry* (M. T. Bowers, ed.), Vol. 1, p. 45. Academic Press, New York, 1979.
31. D. Smith, N. G. Adams, A. G. Dean, and M. J. Church, *J. Phys. D* **8**, 141 (1975).
32. E. Alge, N. G. Adams, and D. Smith, *J. Phys. B* **16**, 1433 (1983).
33. N. G. Adams, D. Smith, and E. Alge, *J. Chem. Phys.* **81**, 1778 (1984).

34. D. Smith, N. G. Adams, and E. Alge, *J. Phys. B* **17**, 461 (1984).
35. D. Smith and P. Spanel, *Adv. At. Mol. Phys.* **32**, 307 (1994).
36. D. Smith, M. J. Church, and T. M. Miller, *J. Chem. Phys.* **68**, 1224 (1978).
37. N. G. Adams and D. Smith, *Chem. Phys. Lett.* **144**, 11 (1988).
38. C. R. Herd, N. G. Adams, and D. Smith, *Astrophys. J.* **349**, 388 (1990).
39. N. G. Adams, C. R. Herd, M. Geoghegan, D. Smith, A. Canosa, J. C. Gomet, B. R. Rowe, J. L. Queffelec, and M. Morlais, *J. Chem. Phys.* **94**, 4852 (1991).
40. P. Spanel and D. Smith, *Int. J. Mass Spectrom. Ion Processes* **129**, 193 (1993).
41. E. W. McDaniel, *Collision Phenomena in Ionized Gases.* Wiley, New York, 1964.
42. P. Spanel, *Int. J. Mass Spectrom. Ion Processes* (1995), in print.
43. P. Spanel, L. Dittrichova, and D. Smith, *Int. J. Mass Spectrom. Ion Processes* **129**, 183 (1993).
44. D. Smith and N. G. Adams, in *Physics of Ion-Ion and Electron-Ion Collisions* (F. Brouillard and J. W. McGowan, eds.), p. 501. Plenum, New York, 1983.
45. D. Smith and N. G. Adams, *Geophys. Res. Lett.* **9**, 1085 (1982).
46. E. E. Ferguson, *J. Am. Soc. Mass Spectrom.* **3**, 479 (1992).
47. N. G. Adams and D. Smith, *Int. J. Mass Spectrom. Ion Phys.* **21**, 349 (1976).
48. D. Smith and N. G. Adams, in *Gas Phase Ion Chemistry* (M. T. Bowers, ed.), Vol. 1, p. 1. Academic Press, New York, 1979.
49. D. Smith and N. G. Adams, *Adv. At. Mol. Phys.* **24**, 1 (1987).
50. D. Smith, N. G. Adams, and M. J. Henchman, *J. Chem. Phys.* **72**, 4951 (1980).
51. D. Smith and N. G. Adams, *J. Phys. D* **13**, 1267 (1980).
52. J. M. Van Doren, S. E. Barlow, C. H. DePuy, and V. M. Bierbaum, *Int. J. Mass Spectrom. Ion Processes* **81**, 85 (1987).
53. W. Lindinger and D. Smith, in *Reactions of Small Transient Species* (A. Fontijn and M. A. A. Clyne, eds.), p. 387. Academic Press, London, 1983; N. G. Adams and D. Smith, *ibid.*, p. 311.
54. D. Smith, *Chem. Rev.* **92**, 1473 (1992); D. K. Bohme, *ibid.*, p. 1487.
55. E. W. McDaniel, in *Swarms of Ions and Electrons in Gases* (W. Lindinger, T. D. Märk, and F. Howorka, eds.), p. 1. Springer-Verlag, Wien, 1984.
56. L. G. Christophorou, E. L. Chaney, and A. A. Christodoulides, *Chem. Phys. Lett.* **3**, 363 (1969).
57. L. G. Christophorou, D. L. McCorkle, and A. A. Christodoulides, in *Electron-Molecule Interactions and their Applications. Vol. 1.* (L. G. Christophorou, ed.) *Academic Press, New York, 1984.*
58. P. G. Datskos, L. G. Christophorou, and J. G. Carter, *Chem. Phys. Lett.* **195**, 329 (1992).
59. D. Smith, C. R. Herd, and N. G. Adams, *Int. J. Mass. Spectrom. Ion Processes* **93**, 15 (1989).
60. M. McFarland, D. L. Albritton, F. C. Fehsenfeld, E. E. Ferguson, and A. L. Schmeltekopf, *J. Chem. Phys.* **63**, 6610, 6620, 6623 (1973).
61. W. Lindinger and D. L. Albritton, *J. Chem. Phys.* **62**, 3517 (1975).
62. G. H. Wannier, *Bell Syst. Tech. J.* **32**, 170 (1953).
63. S. L. Lin and J. N. Bardsley, *J. Chem. Phys.* **66**, 435 (1977).
64. D. L. Albritton, I. Dotan, W. Lindinger, M. McFarland, J. Tellinghuisen, and F. C. Fehsenfeld, *J. Chem. Phys.* **66**, 410 (1977).
65. D. L. Albritton, in *Kinetics of Ion-Molecule reactions,* (P. Ausloos, ed.), p. 119. Plenum, New York, 1979.
66. N. G. Adams and D. Smith, in *Techniques for the Study of Gas-Phase Ion-*

Molecule Reactions (J. M. Farrar and W. H. Saunders, Jr., eds.), p. 165. Wiley, New York, 1988.
67. F. Howorka, I. Dotan, F. C. Fehsenfeld, and D. L. Albritton, *J. Chem. Phys.* **73,** 758 (1980).
68. A. A. Viggiano, R. A. Morris, and J. F. Paulson, *J. Chem. Phys.* **90,** 6811 (1989).
69. D. Smith and N. G. Adams, in *Rate Coefficients in Astrochemistry* (T. J. Millar and D. A. Williams, eds.), p. 153. Kluwer Academic Publishers, Dordrecht, The Netherlands, 1987; N. D. Twiddy, A. Mohebati, and M. Tichy, *Int. J. Mass Spectrom. Ion Processes* **74,** 251 (1986).
70. W. Federer, H. Ramler, H. Villinger, and W. Lindinger, *Phys. Rev. Lett.* **54,** 540 (1985); E. E. Ferguson, *Adv. At. Mol. Phys.* **25,** 61 (1988).
71. S. C. Smith, M. J. McEwan, K. Giles, D. Smith, and N. G. Adams, *Int. J. Mass Spectrom. Ion Processes* **96,** 77 (1990); J. Glosik, A. Jordan, V. Skalsky, and W. Lindinger, *ibid.* **129,** 109 (1993).
72. N. G. Adams, D. Smith, and E. E. Ferguson, *Int. J. Mass Spectrom. Ion Processes* **67,** 67 (1985).
73. A. A. Viggiano, R. A. Morris, F. Dale, J. F. Paulson, K. Giles, D. Smith, and T. Su, *J. Chem. Phys.* **93,** 1149 (1990).
74. J. B. Marquette, B. R. Rowe, G. Dupeyrat, G. Poissant, and C. Rebrion, *Chem. Phys. Lett.* **122,** 431 (1985).
75. B. R. Rowe, J. B. Marquette, G. Dupeyrat, and E. E. Ferguson, *Chem. Phys. Lett.* **113,** 403 (1985).
76. B. R. Rowe, in *Rate Coefficients in Astrochemistry* (T. J. Millar and D. A. Williams, eds.), p. 135. Kluwer Academic Publishers, Dordrecht, The Netherlands, 1987.
77. P. Španěl, M. Pavlik, and D. Smith, *Int. J. Mass Spectrom. Ion Processes* (1995), in press.
78. P. Španěl, and D. Smith, *J. Phys. Chem.* (1995), in press.
79. D. Smith and P. Španěl, *Int. Rev. Phys. Chem.* (1996), in press.

10. ACCELERATOR-BASED ATOMIC PHYSICS

C. R. Vane and S. Datz

Physics Division, Oak Ridge National Laboratory, Oak Ridge, Tennessee

10.1 Introduction

Our understanding of the structure of matter has progressed in large part through the use of beams of rapidly moving particles to probe (and sometimes to produce) the various states of matter being explored. High-energy-charged particles have been employed as fundamentally important tools in studies of atomic scale objects and interactions since the beginning of the modern era, when Geiger and Marsden measured large angle elastic scattering of α particles in thin gold foils to investigate the atomic structure of matter [1], verifying Rutherford's nuclear atom model [2]. Energetic atomic physics has come to include research involving energies from milli-electron volt to Tera-electron volt (TeV) regions and projectiles from electrons and protons to uranium ions and "bucky-balls." It encompasses traditional areas of atomic ionization and excitation and spectroscopies of resulting deexcitation pathways, as well as studies of entirely new phenomena, such as electromagnetic production of leptons in continuum and bound atomic states in relativistic heavy ion collisions. The central goal of this broad field of research is the development, advancing through interplay of theory and experiment, of a precise understanding of the dynamics of interactions among charged atomic bodies and of the complex atomic and molecular structures which are formed in these collisions. The field of energetic atomic collisions interacts well with many diverse branches of science, such as biology, astrophysics, chemistry, medicine, nuclear physics, plasma physics, and solid-state physics.

Research programs centered on studies of high-energy (>1 MeV/nucleon) atomic collisions are mainly accelerator based and rely heavily on flexible, routine operation of medium- to very-large-scale facilities to provide a wide variety of beams of particles with parameters suitably controlled to allow precise measurements. Historically, this field of research has developed primarily at facilities which were originally designated and funded for nuclear physics research. Initially the high-energy accelerator-based atomic physics (ABAP) community consisted almost entirely of parasitic users at these nuclear physics accelerators. In the 1960s, 1970s, and 1980s, accelerator-based atomic physics prospered, and the main use of several laboratories shifted from nuclear physics toward atomic physics and interdisciplinary applications. Virtually every type of particle accelerator has been successfully used for research in high-energy atomic physics, and a wide array of technologies has been developed to produce, transport, and monitor the high-energy beams needed. An excellent historical account of the development of various accelerators up to 1980 is given by Richard [3]. In the last decade a new class of facilities dedicated to atomic physics, based on heavy-ion storage rings, has been added. Except for storage rings, which are located primarily in Europe and Japan, laboratories pursuing high-energy accelerator-based atomic physics are now distributed worldwide.

In this chapter, we will discuss some of the particular advantages afforded using high-energy heavy projectiles in studies of atomic collisions. We will also discuss several kinds of accelerators used in ABAP, general principles of operation for each, and attributes which make each especially suited for particular areas of ABAP research. It is especially important to recognize these unique advantages. Low-energy ion sources such as electron beam ion traps (EBIT) [4] and electron beam ion sources (EBIS) [5] are now capable of producing high-charge states of virtually all atoms. These new sources may be viewed as competing technology for some measurements historically accomplished using beams accelerated to high energies. The availability of these relatively low-energy sources of highly charged ions has, in fact, opened new areas of research and has absorbed an important fraction of ABAP work at arguably significant cost savings, both in initial capital outlay and in expense of normal operation. However, there remain important advantages of high-energy beam techniques for atomic physics measurements, which make high-energy accelerators absolutely necessary for a growing list of research areas. We list below some of these particularly advantageous attributes along with examples of implementation found at various laboratories.

10.2 Advantages of Accelerator-Based Atomic Physics (ABAP)

For atomic physics, naturally, establishing the specific ionic state of the projectile, including the number of and levels occupied by the ion's complement of electrons, is usually vital to controlling and understanding the interactions being studied. In many cases, accelerators can be tuned to act as generators for uniquely defined atomic species, from fully stripped

Coulomb point charges to negative ions, molecular ions, and clusters of atoms. A number of different types of charged particles sources have been developed for injection of particular atomic projectiles into accelerators. Several of these sources are described in detail in other chapters of this volume. Subsequent acceleration, beam transport, and analysis stages can also often be adjusted to provide even greater selectability of projectile species. The rapidly growing field of accelerator mass spectroscopy (AMS), which is based primarily on the ability to separate and count individual specific atoms out of large numbers of background contaminates, clearly attests to the power of the accelerated beam technique for isotopic selection. AMS facilities now routinely achieve sensitivities for isotopic selection of 10^{-15}, i.e., one atom detected out of 10^{15} source background atoms [6].

High-energy beams of particles are used to deliver enough energy in a single collision to produce atomic and ionic states lying far from equilibrium. These may include core vacancy states, in which one or more of the relatively tightly bound target or projectile electrons is either excited or ionized. Decay of such highly excited systems is usually studied via measurements of electrons, photons, or both emitted in subsequent relaxation of the state. Research of this type gives us important information about the (fundamentally many-body) collision interactions themselves, the nature of transient excited states formed, and the relaxation processes by which the ions and atoms return to neutral or charged ground states. The excitation process may be resonant—as in dielectronic excitation [7,8] and resonant transfer and excitation (RTE) [9]–or otherwise manifest marked energy dependence—as in threshold studies of direct Coulomb excitation [9]. In such cases control of the exact energy of the colliding partners becomes an essential parameter of the investigation. Collisional formation of inner-shell vacancies in target or projectile atoms typically requires impact velocities near or exceeding the orbital speed of the bound electron being excited or removed. Figure 1 indicates ion energies giving velocity matching for K, L, and M electrons in neutral targets as a function of the target element (Z_T). Removal of electrons from projectile ions also requires similar energies, while electron capture cross-sections are typically several orders of magnitude lower, so that production of few-electron heavy ions by stripping in thin targets is achieved easily only at relatively high energies.

In collisions of fast, heavy, highly charged ions with stationary target atoms, the classical release distances for bound electrons may occur at relatively large impact parameters. At high projectile energies, interaction times are short, and inner-shell ionizing collisions with heavy targets result primarily in multiple ionization, with most of the collision energy going

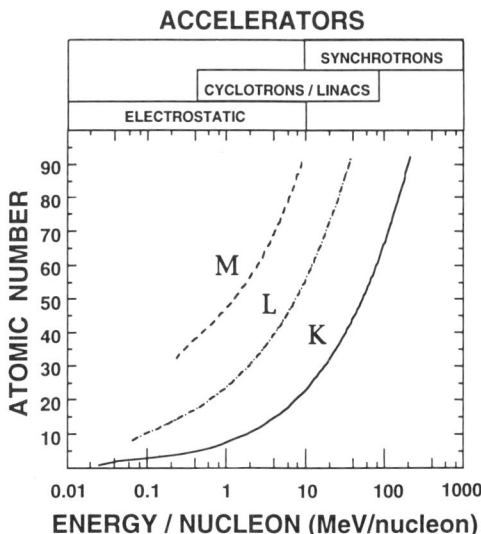

FIG. 1. Collision energies necessary to produce significant direct excitation and ionization of atomic K-, L-, and M-shell electrons. Approximate useful ranges of various accelerators are also indicated.

to ionization and to electron final kinetic energies. Relatively little energy is transferred to the recoiling target nucleus. The primary, high-energy "hammer" beams produce copious, very-low-energy, highly charged recoil ions, which are routinely used as primary beams for charge-changing and energy-loss measurements. Remarkably, some of the lowest energy studies with multiply charged ions have been made possible using these secondary recoil ions [10]. One advantage of these sources is the automatic coincidence timing signal available when rates are sufficiently low to tolerate single particle counting of the primary high-energy ions.

High projectile energies are also necessary to study phenomena occurring in bulk materials and to discriminate against signals arising from ion–surface interactions. Penetration of energetic ions into solids is a very broad and important field, impacting directly on basic research in a variety of subfields and on technological applications. Fundamental understanding of ion penetration in solids is especially necessary for technologies used in modification and analysis of materials, for example, in ion implantation, radiation damage studies, ion sputtering, and Rutherford backscattering analysis [11].

Measurements of high-energy ion–solid interactions under especially controlled conditions can permit studies of environments unavailable by other methods. For example, heavy ion channeling in perfect single crystals affords an opportunity for measurements of ions interacting with the extremely dense, quasi-free electron gas lying between rows of atoms, while avoiding close collisions with target atom cores. Preparation of specific few-electron ion-excited states through resonant coherent excitation (RCE) of the ions in passing down the crystalline axes or planes allows measurements of electron collisions with aligned short-lived states [12], virtually impossible by any other technique. This is achieved using "reverse kinematics" in which the target electrons become projectiles when viewed in the projectile frame. Any target with a relatively narrow electron Compton profile may be used in this technique. At ion velocities much higher than orbital velocities of the weakly bound target electrons, we can view the collision as a beam of electrons (with a relatively narrow velocity profile centered at the ion velocity) striking an ionic target. These "weakly bound electrons" can appear on atomic or molecular targets such as H_2 or He, or they may exist as conduction electrons in crystalline channels. For example, Hülskötter et al. [13] have inferred the electron–electron interaction components of ionization cross-sections for Li^{2+}, C^{5+}, O^{7+}, $Au^{52+.75+}$, $U86^+$, and U^{90+} at 0.75 to 405 MeV/nucleon, using a molecular hydrogen target. The corresponding projectile-frame equivalent electron energies vary from 0.4 to 222 keV. These measurements are carried out at the Stanford Van de Graaff and the Lawrence Berkeley Laboratory (LBL).

Significant experimental advantages may be obtained for laboratory-frame measurements made on atomic systems moving at high velocity. The mature field of beam-foil spectroscopy provides a number of excellent examples. Ions are passed through a thin-foil target, and photons or electrons emitted from the moving ions in subsequent relaxation of the excited states formed are detected in the stationary frame of the laboratory. Transit time in the thin foil is very short ($\sim 10^{-15}$ sec), and beam energy loss is small. The distance downstream of the exciting foil is directly convertible to time through the beam velocity, so that measurements of specific decay channels as a function of decay time after creation can be readily made for lifetimes as short as 10^{-12} sec. A thorough review of this subject through 1980 is given by Pegg [12].

Additional advantage can be achieved by using various kinematic effects which shift photon and electron energies and angles from emission-frame values. For example, autoionization electrons emitted from the moving ions (velocity, v_1) experience angular compression in the forward

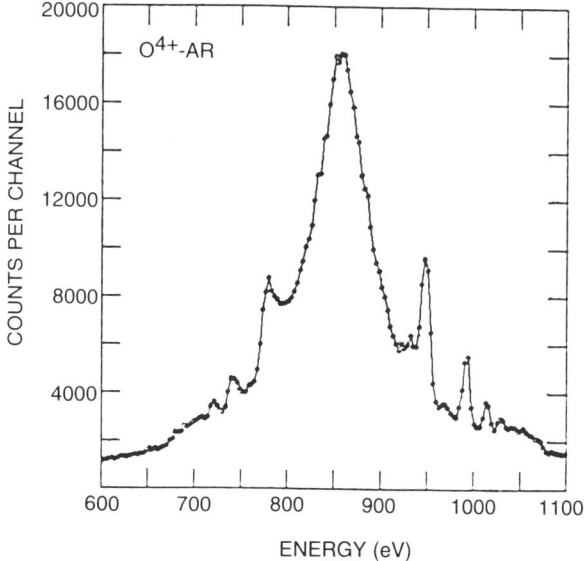

FIG. 2. Zero-degree electron spectrum from 25 MeV O^{4+} colliding with Ar.

direction, allowing enhanced detection efficiencies in the laboratory frame for properly designed electron analyzers. Simultaneously, electrons ejected in the projectile frame with velocity $\vec{v}_P$, have velocity $\vec{v}_L$ in the lab, such that

$$\vec{v}_L = \vec{v}_P + \vec{v}_I.$$

When $v_L \gg v_P$, low-energy electrons ejected in the projectile frame are boosted and energy separations among transitions are magnified. Higher energy electrons often are more efficiently analyzed and detected.

Figure 2 shows an electron spectrum taken at 0° (i.e., along the beam axis) for 25-MeV O^{4+} excited in collisions with an Ar target [14]. The central cusplike peak occurs at $v_L = v_I$ and arises from the processes of electron loss (from the projectile) to the continuum (ELC). The additional sharp peaks symmetrically spaced about the ELC cusp are due to autoionization of $1s^2 2pnl$ states formed in the collisions, with electrons emitted near 0° and 180° in the projectile frame. The electron energies in the projectile frame range from 2.1 to 6.3 meV. The solid-angle enhancement amounts to a factor of ~500 for this case.

Using the technique of 0° electron spectroscopy to study projectile Auger electrons has facilitated detailed studies of resonant transfer and excitation processes (RTEA). For example, in collisions of 5- to 25-MeV

Li-like O^{5+} ions with He at the ORNL tandem Van de Graaff, Auger electrons arising from dielectronic population of Be-like $1s2s2p^2$ ^3D and $1s2s2p^2$ ^1D states were analyzed to measure and separate RTEA and nonresonant transfer and excitation process (NTEA) contributions to the collisional excitation process [15].

In addition, many advantages may occur by carrying out low-energy (center-of-mass) collisions at high laboratory energies. When working with highly charged ions, the fact that electron capture cross-sections fall rapidly with increasing velocity allows one to maintain charge integrity for larger distances and times when working at high energies. A form of system isolation is obtained even in the presence of residual gas backgrounds which would make similar studies at low energies unfeasible. The center-of-mass energies are obtained by using merged beam techniques. In the case of electron ion collisions, this is accomplished by merging a relatively low-energy electron beam with a high-energy ion beam (e.g., the velocity of a 1-keV electron is equal to that of 1.8-MeV protons). The experiments are carried out either in a single pass through the electron "target" or by multiple passes as in a storage ring where milli-electron volt collision energies can be achieved. The fact that the ions are moving rapidly also means that higher laboratory energy electrons and hence higher space-charge-limited currents are available for the experiment.

At sufficiently high energies, relativistic effects become important and enormous transient Coulomb fields are generated in grazing but nonnuclear collisions between heavy partners. Considerable theoretical progress has been made in calculations describing these intense Coulomb collisions, especially with respect to identification and treatment of higher order effects and of perturbative and nonperturbative regions of the interaction [16–18]. Effectively, the highly charged ions become sources for intense transverse beams of virtual high-energy photons. Succeeding interactions of the photons with the Coulomb fields of the ions, essentially photon–photon collisions, result in copious production of pairs of leptons, electron–positron pairs being the overwhelmingly dominate mode. In the parlance of atomic excitation, the electromagnetic pulse generated in the collision drives transitions of electrons, initially bound in the Dirac negative-energy continuum, to states of positive total energy. Free pairs are formed when the final state is a continuum state, which is predicted to occur for the majority of pairs. However, for a few percent—depending on the ion charge—the electron occupies an ion bound state, and the electron (or other lepton) has been captured from the negative continuum. At energies above ~10 GeV/nucleon, this is expected to be the dominate electron capture cross-section. It is also the only capture process which increases with energy. Ions which have changed charge through electron capture are usually lost from the usable beam at the first subsequent bending magnet. For the future generations of ultrarelativistic heavy-ion colliders, such as the Relativistic Heavy-Ion Collider (RHIC) at Brookhaven National Laboratory and the Large Hadron Collider (LHC) at CERN, such capture processes are expected to present the limit for storage lifetimes of the heaviest ions [19].

The yields of free lepton pairs produced at intersection points in these machines are expected to be enormous and will present significant challenges to experiment and data acquisition design and operation. Brown [16] has estimated that for 100 GeV/nucleon Au + Au ions at RHIC electron–positron pairs will be generated at a rate of $\sim 10^7$ pairs/sec, assuming expected beam parameters. Fortunately, the pairs formed are expected to be confined mainly to emission in a forward angular cone of ≤20°. Transverse momenta are small and possibly separable from other signals of interest; for example, from much lower yield lepton pairs formed in nucleus–nucleus collisions carrying direct information on the formation and decay of a postulated quark–gluon plasma phase of matter [20]. Measurements of pair production at ~200 GeV/nucleon and of electron capture at 1 and 160 GeV/nucleon have been reported using beams at LBL [21] and at CERN [22] (~1 and 200 Gev/nucleon), respectively. The latter studies represent the highest energy atomic physics measurements performed to date using accelerated beams of heavy ions.

10.3 Types of Accelerators Used in Accelerator-Based Atomic Physics

Classified according to the method of acceleration, there are basically two kinds of high-energy particle accelerators used in ABAP. Electrostatic accelerators rely on static electric fields generated by various techniques to accelerated charged particles to a kinetic energy given by the product of the particle's charge and the electric potential through which it passes. Acceleration occurs in a single (or at most a few) step of potential, and the final energy attained is limited by the magnitude of the electric potential difference which can be maintained between the source of the accelerated particles and the target region, usually at ground potential. Current technology allows potentials up to 25–30 MV. Electrodynamic accelerators rely on motions of charged particles interacting with time varying electric and magnetic fields to accelerate the projectiles. Acceleration can occur many times, either repetitively, as in closed path machines, in which the particle's orbit is confined by magnetic fields to repeatedly pass through regions of acceleration, or consecutively, as in linear path machines, in which multiple regions of acceleration are coupled in tandem. Energy limits on dynamic accelerators are set primarily by economic constraints due to the increasing physical size required for these machines with increasing energy. Both types have advantages and disadvantages for specific areas of ABAP research, some of which we will cover below.

10.4 Electrostatic Accelerators

Electrostatic accelerators fall into three main types. Examples of electrostatic accelerators used for ABAP are listed in Table I along with the nominal operating energy limit. Cockroft–Walton (CW) [23] accelerators use cascade rectifier circuits to directly produce a potential drop across which ions are accelerated. The projectile source (typically a high-current positive-ion source) is normally located at the high-potential terminal. The advantages of using a Cockroft–Walton machine are clean, well-defined potentials for accleration, the resulting ease of stable operation, the possibility of continuous energy changes, and simple positive ion sources. The typical upper limit for terminal potential is ~1 MV. CW machines are used in many high-energy facilities as injection sources for other main accelerators which require high-velocity particles for efficient injection.

The second type of electrostatic accelerator, and the most prominent in high-energy ABAP, is the Van de Graaff accelerator [24]. In a Van de Graaff accelerator, high potential is developed on a conducting terminal by means of a mechanically driven charged belt made of insulating material. Charge is sprayed onto the belt at near-ground potential and carried to the high potential, where it is internally transferred to the terminal. Various means are used to regulate the exact potential on the terminal, but most rely on controlling the terminal discharge current rate. All modern high-voltage Van de Graaff machines rely on pressurized insulating gas surrounding the terminal to prevent electrical breakdown. A similar accelerator design manufactured by the National Electrostatics Corporation (NEC) uses a chain of insulated conducting pellets to transport the charge. More charge can be delivered, and therefore higher potentials than those in standard belt-charged Van de Graaff accelerators can be reached. Single-ended Van de Graaff machines accelerate positively or negatively charged particles from their terminals to ground, yielding projectiles with energy equal to the product of the charge and the terminal potential (q eMV). Tandem Van de Graaff accelerators rely on charge changing of the projectile at the terminal to substantially boost the final energy beyond the terminal potential. Typically, tandem accelerators operate in the following way. Singly charged negative ions are formed in a source at a negative

TABLE I. Examples of Accelerators Used in Atomic Physics Research Taken from a Random Sampling of Publications

Location	Type	Energy limit
Electrostatic accelerators		
Argonne National Laboratory, IL	Dynamitron	5 MV
Ruhr-Universität, Bochum, Germany	Dynamitron	4 MV
Royal Melbourne Institute of Technology, Australia	Tandem Van de Graaff	1 MV
University of Nevada, Reno	Van de Graaff	2 MV
University of Connecticut, Storrs	Van de Graaff	2 MV
University of North Texas, Denton	Tandem pelletron	3 MV
ATOMKI, Debrecen, Hungary	Van de Graaff	5 MV
University of Tokyo, Japan	Tandem pelletron	5 MV
University of Aarhus, Denmark	EN tandem Van de Graaff	6 MV
Kansas State University, Manhattan	EN tandem Van de Graaff	6 MV
Oak Ridge National Laboratory, TN	EN tandem Van de Graaff	6 MV
Stanford University, CA	FN tandem Van de Graaff	7 MV
ATLAS, Argonne National Laboratory, IL	MP tandem Van de Graaff	14 MV
TASCC, Chalk River National Laboratory, Canada	MP tandem Van de Graaff	14 MV
JAERI, Tokai, Japan	Tandem pelletron UD	18 MV
HHIRF, Oak Ridge National Laboratory, TN	Tandem pelletron UD	25 MV
Linear accelerators		
Super HILAC, Lawrence Berkeley Laboratory, CA	LINAC	8.5 MeV/nucleon
UNILAC, GSI, Darmstadt, Germany	LINAC	10 MeV/nucleon
ATLAS, Argonne National Laboratory, IL	LINAC	27 MeV/nucleon
Cyclotrons and synchrotrons		
Texas A&M University, College Station	Superconducting cyclotron	$k = 500$
TASCC, Chalk River National Laboratory, Canada	Superconducting cyclotron	$k = 500$
RIKEN Ring Cyclotron, Tokyo, Japan	Superconducting cyclotron	$k = 540$
GANIL, Caen, France	Multistage coupled cyclotrons	$2 \times k = 400$
SIS, GSI, Darmstadt, Germany	Heavy-ion synchrotron	14 GeV/nucleon
SPS, CERN, Geneva	Heavy-ion synchrotron	200 GeV/nucleon

potential of 50–300 kV measured with respect to ground. These ions are extracted from the source and accelerated to ground potential, where mass selection is performed using a bending magnet. The selected negative ions are accelerated as they move from ground potential to the terminal of the Van de Graaff. Within the terminal the ions lose some fraction of their electrons in a stripper gas or thin foil. A charge distribution of positive ions is thus formed at the terminal, having potential energies given by q eMV, where q is the charge of each ion after stripping, and all moving with kinetic energies given approximately by electron megavolts. Higher charges are typically produced for higher terminal potentials. The final energies after acceleration to ground are thus $\sim(q + 1)$ eMV, with correction necessary for the additional energy given by extraction of the negative ions from the source. Precise energy selection is usually accomplished using a momentum analyzing magnet located at the exit of the accelerator. Absolute beam energies can be selected in this way with a precision of 1 part in 10^4 [24]. A review of atomic physics experiments using electrostatic accelerators is given in Datz [25].

Electrostatic accelerators have been continually improved, from terminal voltages of 80 kV in 1929 [26] to present-day machines developing potentials in excess of 25 MV. The Holifield Heavy Ion Research Facility (HHIRF) 25-MV NEC accelerator at Oak Ridge National Laboratory, shown schematically in Figure 3, is the highest energy Van de Graaff or "pelletron" which has attained routine operation. Experiments there on radiative electron capture by fully stripped oxygen ions have been carried out at 25.5 MV [27]. This accelerator is in the process of being converted to a radioactive beams facility (the Holifield Radioactive Ion Beams Facility (HRIBF), with the first beams expected to become available in 1995.

Tandem Van de Graaff accelerators are particularly well suited for studies of moderately high-energy atomic collisions. Advantages include ease of energy variation and quality of beam emittance. Considerable development of negative ion source technology [28] has lead to the availability of virtually all elements as projectiles for tandem accelerators. Figure 4 shows an example of measurements [23] made at the HHIRF tandem accelerator which required rapid, nearly continuous changes in beam energy, along with severe controlled variations in beam intensity, while demanding constant beam collimation. An Mg^{11+} ($1s$) beam was accelerated, analyzed, and collimated to <1 mrad divergence and passed through $\sim$3000-Å-thick gold crystal aligned along the $\langle 111 \rangle$ axis. Transmitted Mg^{q+} ions exiting the crystal were magnetically analyzed and counted according to the final charge state. The fraction of bare Mg^{12+} ions formed by electron impact ionization in the crystal channels is shown in Figure 4a as a function of beam energy. The sharp enhancements in the Mg^{12+}

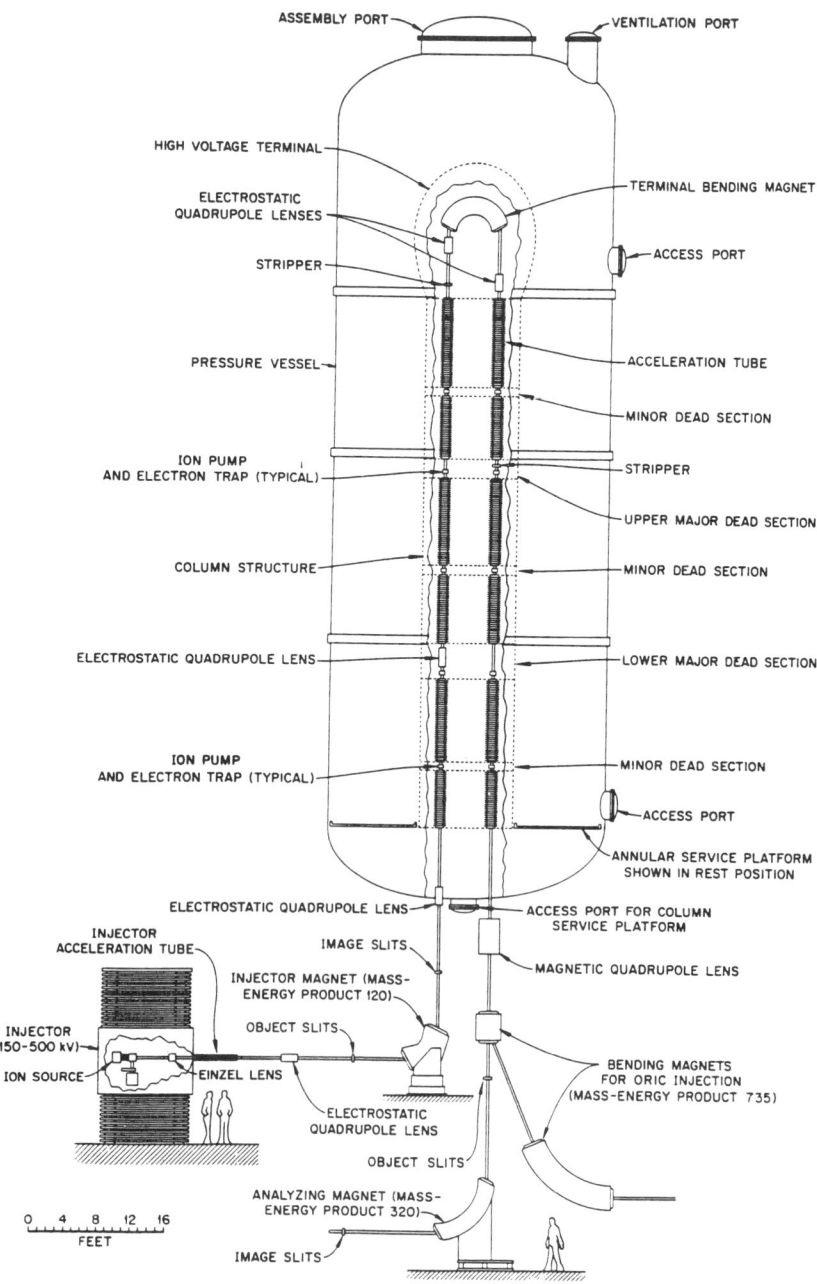

FIG. 3. Sectional view of the 25-MV Holifield Heavy-Ion Research Facility tandem electrostatic accelerator.

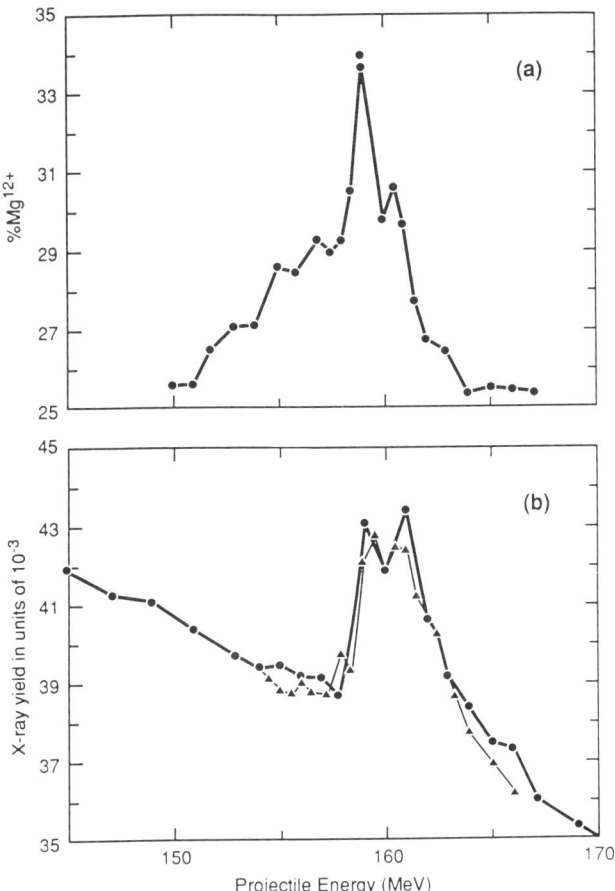

FIG. 4. Resonant coherent excitation of Mg^{11+} in $\langle 111 \rangle$ Au. (a) Fraction of Mg^{12+} in the beam emerging from the crystal as a function of ion beam energy. (b) Yield of Mg^{11+} Lyα X rays as a function of ion beam energy.

fraction observed near ~160 MeV occur because of RCE of Mg^{11+*} ($n = 2$) sublevels by the time-varying crystal fields experienced by the moving ions. Increased Mg^{12+} formation occurs because subsequent ionization of $n = 2$ excited states by the impact of crystal channel electrons on Mg^{11+*} ($n = 2$) is much more probable than that on unexcited Mg^{11+} (1s). These studies required measurements of more than 50 different energies with energy steps as small as 0.25 MeV over the region encompassing resonances. Repeated measurements over wide energy changes to show reproducibility were also necessary. During the same experiment, X rays emitted from the excited Mg^{11+} ($n = 2$) ions surviving transit through the crystal were measured (Figure 4b). To acquire these data, ion beam currents were switched using injection beam tuning, from projectile ion rates of a few thousand/sec used for the charge-state measurements to tens of picoamps ($>10^6$ ions/sec) for the X-ray measurements. These intensity switches were made without significant energy or beam steering variations.

A third type of electrostatic accelerator is the Dynamitron, built by Radiation Dynamics. The accelerating potential (1–4.5 MV) is developed from an RF generator on a high-voltage terminal containing a positive-ion source. A pressure chamber containing insulating gas also contains an acceleration tube and a string of diodes connected in series. The diode string is coupled to the ground and to the terminal. Electrodes connected to individual stages of the diode string are mounted to form a resonant circuit, such that an RF field ($\sim$100 kHz) coupled into the pressure vessel induces voltages across the individual electrodes which are summed in the rectifier circuit to produce the acceleration high voltage. The main advantages of Dynamitrons for accelerator-based atomic physics research are the ability to provide high-current ion beams with simple and continuous energy variability. Energies are limited, however, by a maximum potential of $\leq$5 MV.

10.5 Electrodynamic Accelerators

10.5.1 Cyclotrons and Synchrocyclotrons

Cyclotron accelerators reach high projectile energies by repeated relatively weak accelerations of charged particles between two electrodes which switch potential back and forth at radio frequencies. Particles are injected, stripped of electrons, and accelerated by the field in the gap between the electrodes. During periods in which the electric field is out of phase with the desired motion, the particles circulate inside hollow conducting D-shaped electrodes. The particles are held to circular or spiral trajectories by a confining magnetic field until the selected final energy is achieved. At this point, the extraction radius is reached, and either the field is shaped to cause a deviation in trajectory leading to extraction or a deflection field is ramped, directing the beam out of the cyclotron. Most modern cyclotrons obtain the extraction of ions by executing motion in a selected single turn after many orbits and achieve extremely high-energy selection for the extracted ions.

Considerable developments and improvements have been made in cyclotrons over the original fixed-energy device constructed by E. O. Lawrence in 1931 which produced 80-keV protons [29]. Cyclotrons now include particle- and energy-variable, superconducting magnet machines, which can reach fully relativistic energies for most ions. Conventional cyclotrons use constant RF frequency driver fields and magnetic confining fields which may be spatially tailored for specific applications, especially with respect to retaining focused beams of particles at extraction. As heavy ions are accelerated, relativistic variation of mass begins to limit the number of orbits which remain in phase with the RF and hence to limit the maximum energy at a fixed extraction radius. The maximum energies attainable with conventional cyclotrons lie at approximately a 1% increase in the relativistic mass of the accelerated particle which occurs at about 10 MeV/nucleon.

This limitation on upper energy has been overcome in most present-day cyclotrons by compensating for the relativistic mass increase by either time varying the RF frequency or spatially varying the strength of the magnet field. In synchrocyclotrons the magnetic field is held fixed while the accelerating RF frequency is adjusted in time to achieve continued synchronism for a pulse of projectiles, throughout acceleration as the mass of the particles increases. Theoretical energies attainable by synchrocyclotrons can be as high as 1 GeV/nucleon. The real limits are imposed by economics since the ultimate energy scales as the square of the magnet size or largest orbit radius.

Isochronous cyclotrons achieve high energies by tailoring azimuthal variation of the magnet field to high accuracy for a radially increasing field. The required azimuthal fields are modulated, usually by precise shaping of the pole tip faces or variable field shimming coils. Accelerated particles in isochronous cyclotrons trace out complicated trajectories with constant orbit period, hence the name. Modulated RF frequency is therefore unnecessary for maintenance of synchronism between particle motion and the accelerating fields.

The maximum energy for heavy ions of mass A in an isochronous cyclotron is given nonrelativistically by

$$E \text{ (MeV/nucleon)}_{max} = k(q/A)^2, \qquad (10.1)$$

where

$$k = (BR/0.144)^2. \qquad (10.2)$$

Here B is the magnetic field in teslas and R is the maximum radius in meters. The isochronous cyclotron ($k = 90$, ORIC) at Oak Ridge National Laboratory is an early example of a such an accelerator using room temperature magnet technology. Its design allows variable energy and variable particle mass operation, with energies up to 80 MeV for protons, but limited to ~1 MeV/nucleon for ions heavier than lead.

Advances in superconducting magnet technology have progressed sufficiently that nearly all high-energy heavy-ion cyclotrons now being developed use superconducting main magnet coils. This allows for an order of magnitude increase in the maximum energy for the same-size device. Superconducting cyclotrons are now operational, for example, at Michigan State University (k = 500 and 1200), Texas A&M University (k = 500), and Chalk River Laboratories (k = 520). Values of the coefficient k range from tens to about 1000. With spiral sector azimuthal field shaping, isochronous cyclotrons can attain heavy-ion energies of nearly 1 GeV/nucleon. Examples of cyclotrons used in accelerator-based atomic physics are given in Table I. The main advantages of cyclotrons arise because these accelerators can provide very high, very stable energy projectiles of uniquely defined mass. Cyclotrons inherently possess very high mass resolution, as the tuned RF field must remain synchronized with projectile motion over many cycles before extraction. Isotopic selection is therefore ensured. Substantial technical investment has been made in the development of frequency-stabilized RF power sources which have led to extremely stable beam energies. Cyclotron disadvantages include the relative difficulty of changing beam energy and high power consumption.

10.5.2 Radiofrequency Linear Accelerators

RF linear accelerators (LINACS) operate by generating consecutive accelerations of charged particles traveling along a linear path. This is accomplished using RF electric fields which oscillate in synchronism with movement of the particles through regions of high field restricted to gaps between metallic electrodes. Particles must be injected in pulsed bunches occurring in phase with the RF fields. In heavy-ion LINACS, the projectile velocity increases with passage through each gap, so that path lengths between gaps must also increase systematically to maintain the proper phase relationship. Discrete velocities are in design selected by the geometry and RF frequency chosen or in operation by the number of acceleration cavities (gaps) energized for a particular application. Figure 5 shows an example of a segment of the superconducting LINAC used at the ATLAS heavy-ion accelerator facility, Argonne National Laboratory. Other examples are listed in Table I.

The primary disadvantages of using LINACS in atomic physics research lie in the inherent difficulties of providing continuously variable beam energies and in the relatively poor energy resolution (~0.1%) of the beam normally associated with these accelerators. The main advantage is the

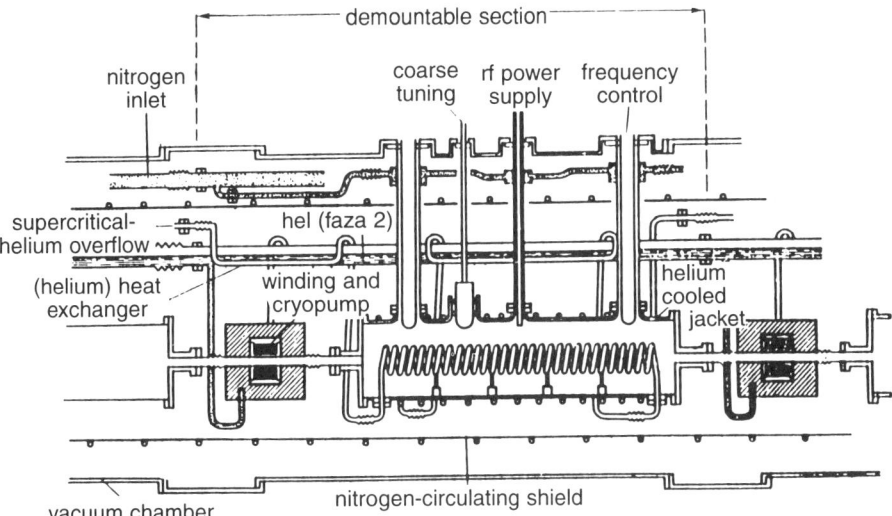

FIG. 5. Construction of a segment of a heavy-ion linear accelerator with superconducting elements (ATLAS, Argonne National Laboratory, France).

ability to accelerate intense beams of positive ions of essentially any charge and mass to high energy (~10 MeV/nucleon). LINACS can also be tuned for deceleration of heavy ions. Thus projectiles may be accelerated to energies sufficient to produce charge states of particular interest and then decelerated to substantially lower energies, at which collisional studies may be carried out or radiative or autoionizing relaxation signatures may be more precisely measured. LINACS are often used as energy boosters in conjunction with lower energy machines, such as Cockroft–Walton or Van de Graaff accelerators. In principle, LINACS could be constructed to achieve almost any beam energy. In practice, realistic energies per unit mass are limited to a few tens of MeV/nucleon. The UNILAC at GSI in Darmstadt, Germany, for example, has a useful energy range of up to ~20 MeV/nucleon for heavy ions.

10.5.3 Synchrotrons

In heavy-ion electrostatic accelerators, LINACS, and single-stage cyclotrons, the limiting energies are ~100 MeV/nucleon for light elements and ~10–20 MeV/nucleon for the heavier elements. The highest projectile energies (currently exceeding 100 GeV/nucleon) have been achieved by inducing repeated small accelerations of particles moving in stable orbits of constant radius. These constant radius circular accelerators are collectively known as synchrotrons. The accelerations are realized using short RF field sections as in RF LINACS, interposed in magnet ring structures producing the accelerator confining fields. The confining magnet fields must therefore be time dependent, increasing to offset the increasing rigidity of an accelerated beam pulse. For heavy particles the velocity changes during an acceleration cycle, so that the RF field frequency must also be modulated to retain in-phase conditions for continued efficient acceleration throughout a cycle.

Synchrotrons were originally developed in the 1940s for acceleration of protons in particle physics research. In the mid-1960s work was begun at Princeton and Berkeley to modify existing proton synchrotrons for heavy-ion use [30]. Energies of ~1 GeV/nucleon were obtained. At CERN, the Super Proton Synchrotron (SPS) has been modified, mainly through addition of injection accelerators designed for heavy-ion operation, to obtain beams as heavy as lead at energies up to 160 GeV/nucleon. Such heavy beams require substantially higher vacua as various loss mechanisms arising from Coulomb scattering interactions with residual background molecules scale typically as the square of the projectile change. At energies beyond ~1 GeV/nucleon, essentially all accelerated ions are fully stripped of electrons and become simple point charges for atomic physics' purposes. Thus, for heavy-ion energies in excess of ~1 GeV/nucleon, synchrotrons are the only available controlled sources. Higher energy heavy cosmic rays exist, but these obviously suffer from unpredictability and extremely low fluence. Examples of heavy-ion synchrotrons are given in Table I.

10.5.4 Storage Rings

The advent of ion storage rings has opened up many new venues in accelerator-based atomic physics. These facilities use the synchrotron principle to accelerate injected ions and then circulate them in a magnetically confined ring. They are specifically designed to provide the ultrahigh vacuum necessary for reasonably lengthy storage of highly charged ions. One special feature of these systems is that experiments can be carried out with the stored ion beam.

The nature of these facilities is best illustrated by the ring shown in Figure 6. The ions are formed either in a typical source or, in the case of multicharged ions, in a CRYEBIS or an ECR source. They are then preaccelerated, in this case, by radiofrequency quadrupole (RFQ) and then injected into the ring. There they can be accelerated up to the maximum energy E (MeV/nucleon) which can be contained by the maximum available magnetic field and the radius of curvature fixed by the lattice

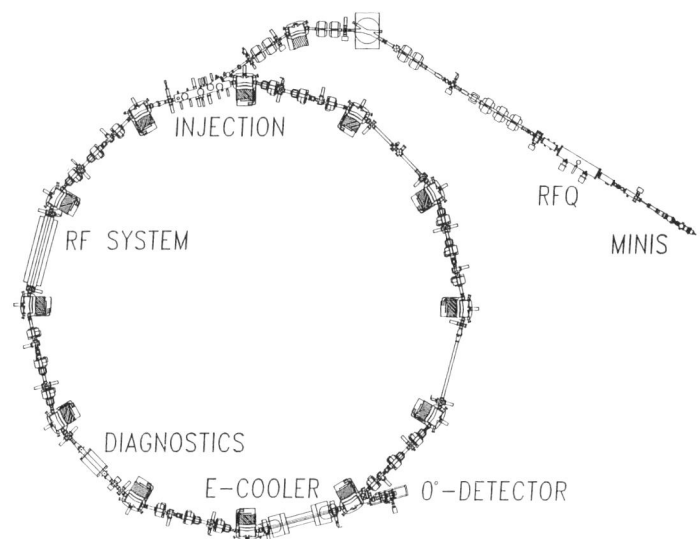

FIG. 6. Schematic diagram of the CRYRING heavy-ion storage ring of the Manne Siegbahn Laboratory in Stockholm. The MINIS is an ion source. An EBIS source is also available to provide highly charged ions.

design of the ring. Rings are generally classified in terms of their strength in tesla meters, BR [see Equations (10.1) and (10.2)]. Table II gives a list of storage rings which are currently used in atomic physics experiments. These facilities include the following unique features.

1. Storage time permits relaxation of metastable states of atomic ions and of vibrational states of molecular ions.
2. The ions can circulate at $\sim 10^6$ Hz so that even a small number of stored ions can yield large circulating currents; e.g., $\sim 10^{-9}$ ions circulating at 1 MHz is equal to ~ 170 μA.

TABLE II. Installations and Strengths of Ion Storage Ring Facilities

Institution	Location	Size (T/m)
Max-Planck Institut für Kernphysik	Heidelberg, Germany	1.6
University of Aarhus	Aarhus, Denmark	2.1
Manne Siegbahn Laboratory	Stockholm, Sweden	1.4
Institute for Nuclear Science	Tokyo, Japan	7
Gesellshaft für Schwerionenforschung	Darmstadt, Germany	10

3. The storage and circulation lead to advantages for ions which are difficult to produce in quantity and for dilute targets (e.g., electrons).

All the facilities mentioned above employ electron "cooling" to reduce the momentum spread in the injected beam. The cooling is accomplished by injecting an electron beam at the same velocity parallel to the stored ion beam in one leg of the ring. The electrons, being "cool" because of their acceleration, interact with the "hot" ions and carry off momentum by Coulomb scattering. Laser cooling of stored beams has also been carried out. By varying its laboratory energy, the electron beam can be used as an electron target for the study of electron–ion interaction, e.g., excitation, ionization, radiative recombination, dielectronic recombination, and dissociative recombination of molecular ions. These processes can also be moderated by collinear laser beams interacting in the merged electron–ion beam region as, for example, laser-induced radiative or dielectronic recombination. Relative energies on the order of <1 meV can be achieved with circulating beams of many MeV/nucleon [31]. A review of atomic physics work on storage rings has been given by Schuch [32].

10.5.5 Hybrid Accelerator Facilities

In order to reach the very high projectile energies (1 Gev/nucleon) needed to produce few-electron charge states of very heavy atoms, accelerator facilities have added secondary and tertiary stages of acceleration to the main accelerator. In some cases linear accelerator boosters have been added after primary acceleration by tandem Van de Graaffs, as at the Kansas State University J. R. Macdonald Laboratory. Cyclotrons have also been coupled with Van de Graaffs, as at the HHIRF at Oak Ridge National Laboratory and the Tandem Accelerator Super-Conducting Cyclotron Facility at Chalk River Laboratory. Typical operation in these multiaccelerator facilities proceeds from a primary source of ions and first-stage acceleration to stripping of a substantial portion of the projectile's electrons at the moderately high first-stage energies, followed by a second (or third) stage boost, giving the desired final high-energy ions. For relativistic and ultrarelativistic energies synchrotrons are usually used as the final stage of acceleration. For example, the SUPERHILAC–BEVALAC combination at Lawrence Berkeley Laboratory (decommissioned in 1993) coupled a LINAC yielding ions with 8.5 MeV/nucleon with the BEVATRON to provide ions as heavy as uranium at energies up to 1 GeV/nucleon. As noted the highest energy heavy ion accelerator facility currently operating is the SPS accelerator at CERN. A schematic of the CERN SPS and associated accelerators is shown in Figure 7. Positive ions from an ECR are accelerated in a LINAC to as

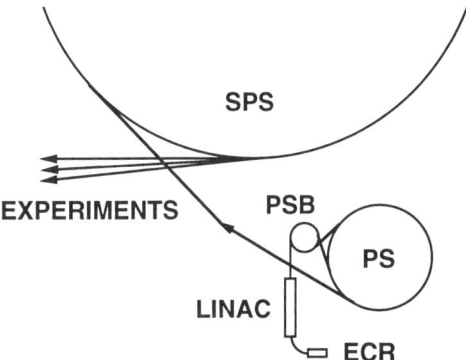

FIG. 7. Schematic layout (not to scale) of the CERN Super Proton Synchrotron (SPS) and associated injection accelerators. Heavy ions from an electron cyclotron resonance (ECR) source are boosted in energy by an ~50 MeV/nucleon linear accelerator (LINAC), an ~1 GeV/nucleon booster synchrotron (PSB), and an ~15 GeV/nucleon synchrotron (PS) for injection into the SPS, for which final energies of ~200 GeV/nucleon are obtained.

high as 50 MeV/nucleon, injected into the proton synchrotron (PS), and boosted to ~15 GeV/nucleon. Ions as heavy as lead are extracted from the PS and injected into the SPS for acceleration to a maximum energy at 160 GeV/nucleon. Atomic physics studies including measurements of Coulomb pair production, electron capture from the negative continuum, and K-shell ionization have been carried out with these ultrarelativistic Pb ions.

References

1. H. Geiger and E. Marsden, *Philos. Mag.* [6] **25**, 604 (1913).
2. E. Rutherford, *Philos. Mag.* [6] **21**, 669 (1911).
3. P. Richard, ed., *Methods of Experimental Physics*, Vol. 17. Academic Press, New York, 1980.
4. See, e.g., D. A. Knapp, R. E. Marrs, S. R. Elliot, E. W. Magee, and R. Zasadzinski, *Nucl. Instrum. Methods Phys. Res., Sect. A* **334**, 305 (1993).
5. See, e.g., E. D. Donets, in *The Physics and Technology of Ions Sources* (I. G. Brown, ed.), Wiley, New York, 1989; M. P. Stockli, *Z. Phys. D* **21**, s213 (1991).
6. M. Paul, *Nucl. Instrum. Methods Phys. Res., Sect. A* **328**, 330 (1993).
7. See, e.g., *NATO ASI Ser., Ser. B* **296** (1992).
8. J. A. Tanis, *AIP Conf. Proc.* **205**, 538ff. (1989).
9. J. E. Miraglia and V. D. Rodrigues, in *The Physics of Electronic and Atomic Collisions* (W. R. MacGillivray, I. E. McCarthy, and M. C. Stein, eds.), pp. 423ff. Adam Hilger Press, Bristol, England, 1991.

10. C. L. Cocke, *Phys. Rev. A* **20**, 749 (1979).
11. S. Datz, ed., *Applied Atomic Collision Physics*, Vol. 4. Academic Press, New York, 1983.
12. D. J. Pegg, in *Methods of Experimental Physics* (P. Richard, ed. Chapter 10), Vol. 17, Academic Press, New York, 1980.
13. H.-P. Hülskötter *et al.*, *Phys. Rev. A* **44**, 1712 (1991).
14. M. Breinig, S. B. Elston, S. Huldt, L. Liljeby, C. R. Vane, S. D. Berry, G. A. Glass, M. Schauer, I. A. Sellin, G. D. Alton, S. Datz, S. Overbury, R. Laubert, and M. Suter, *Phys. Rev. A* **25**, 3015 (1982).
15. J. K. Swenson, Y. Yamazaki, P. D. Miller, H. F. Krause, P. F. Dittner, P. L. Pepmiller, S. Datz, and N. Stolterfoht, *Phys. Rev. Lett.* **57**, 3042 (1986).
16. M. J. Rhoades-Brown and J. Weneser, *Phys. Rev. A* **44**, 330 (1991).
17. M. C. Güclü, J. C. Wells, A. S. Umar, M. R. Strayer, and D. J. Erust, *Phys. Rev. A* **51**, 1836 (1995).
18. K. Momberger, A. Belkacem, and A. H. Sørensen, *Phys. Rev. A* (in press).
19. *Conceptual Design of the Relativistic Heavy Ion Collider, Brookhaven Natl. Lab. [Rep.] BNL* **BNL 52195** (1989).
20. K. Kajantie, J. Kapusta, L. McLerran, and A. Mekjian, *Phys. Rev. D* **34**, 2746 (1986).
21. A. Belkacem, H. Gould, B. Feinberg, R. Bossingham, and W. E. Meyerhof, *Phys. Rev. Lett.* **73**, 2432 (1994).
22. C. R. Vane, S. Datz, P. F. Dittner, H. F. Krause, R. Schuch, H. Gao, and R. Hutton, *Phys. Rev. A* **50**, 2313 (1994).
23. S. Datz, P. F. Dittner, J. Gomez del Campo, K. Kimura, H. F. Krause, T. M. Rosseel, C. R. Vane, Y. Iwata, K. Komaki, Y. Yamazaki, F. Fujimoto, and Y. Honda, *Radiat. Eff. Defects Solids* **117**, 73 (1991).
24. S. Humphries, Jr., *Principles of Charged Particle Accelerators*. Wiley, New York, 1986.
25. S. Datz, *Nucl. Instrum. Methods Phys. Res., Sect. A* **287**, 200 (1990).
26. R. J. Van de Graaff, *Phys. Rev.* **38**, 1919A (1931).
27. C. R. Vane, S. Datz, P. F. Dittner, J. Giese, N. L. Jones, H. F. Krause, T. M. Rosseel, and R. S. Peterson, *Phys. Res. A* **49**, 1847 (1994).
28. G. D. Alton, *Nucl. Instrum. Methods Phys. Res., Sect. B* **73**, 221 (1993).
29. E. O. Lawrence and M. S. Livingston, *Phys. Rev.* **40**, 19 (1932).
30. Lawrence Radiation Laboratory, Rep. No. UCRL-16828 (1966).
31. G. Sundström, J. R. Mowat, H. Danared, S. Datz, L. Broström, A. Filevich, A. Källberg, S. Mannervik, K. G. Rensfelt, P. Sigray, M. af Ugglas, and M. Larsson, *Science* **263**, 785 (1994).
32. R. Schuch, in *Review of Fundamental Processes and Application of Atoms and Ions* (C. D. Lin, ed.) pp. 169ff. World Scientific, Singapore, 1993.

11. ION MASS ANALYZERS

Peter W. Harland

Chemistry Department, University of Canterbury, Christchurch, New Zealand

11.1 Introduction

The purpose of the mass spectrometer is to separate ions according to their mass-to-charge ratio, and the methods and technology involved have improved rapidly. This chapter will concentrate on the practical aspects of those mass analyzers in current use and, it is hoped, aid the reader in selecting the device most suited to his or her purpose. Ion sources, ion optics, and particle detectors are the subjects of other chapters in this volume and will find mention here only where they form an integral part of the mass analyzer. Specialized texts and reviews on mass spectrometry should be consulted for further background information and accounts of applications [1–18].

Without exception, mass analyzers must be immersed in a high-vacuum environment. Generally, pressures of $\leq 10^{-5}$ torr are required, with $\leq 10^{-8}$ torr being desirable. The performance of all mass analyzers degrades with increasing pressure as ion-neutral collisions scatter ions and, in some cases, change their identity through ion–molecule reaction. Resolution and ion transmission increase with decreasing collision frequency, and instruments with long ion path lengths are particularly sensitive to the vacuum environment.

An important parameter characterizing any mass analyzer is the mass resolution which defines its ability to separate ions of different mass. The mass resolution, $R_{X\%}$, is usually expressed as the ratio of the ion mass of interest, m, to the mass width of the peak, Δm, at $X\%$ of the peak height, $R_{X\%} = m/\Delta m_{X\%}$, where $X = 5$, 10, or 50. This is illustrated in Figure 1. Ideally, Δm would remain independent of m across the full mass range, equivalent to increasing resolution with increasing m. This mode of operation is called the *constant Δm mode*. In practice, this is not usually possible, and Δm increases with increasing ion mass, giving constant resolution across the full mass range; this is called the *constant $m/\Delta m$ mode*.

Typically, increased resolution can be achieved only with a sacrifice in sensitivity (ion transmission). For applications in which the mass analyzer is used simply as a mass-sensitive detector, the resolution would be

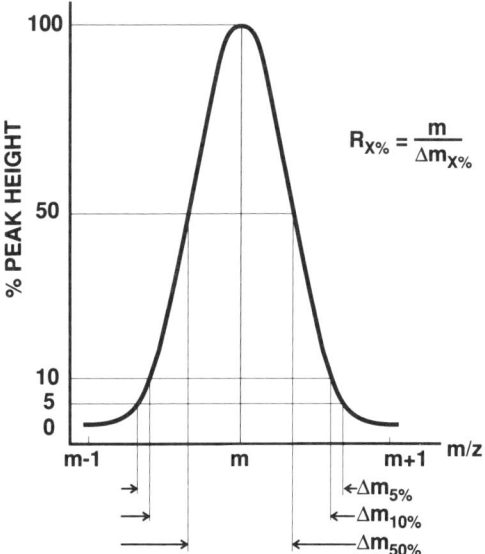

FIG. 1. Mass resolution defined in terms of ion peak width at different peak heights above the base line.

reduced to maximize transmission within the confines of the experimental objectives.

A parameter sometimes used to describe the resolution of magnetic sector instruments is the mass dispersion produced by the magnetic field which is directly proportional to the radius of the ion path through the magnet. Although the constant of proportionality is dependent on the geometry of the instrument, increased size translates into increased resolution, and most sector instruments are physically large.

11.2 Motion of Charged Particles in Electric and Magnetic Fields

Mass analyzers exploit the behavior of charged particles in electric and magnetic fields to facilitate ion separation and identification. The following sections consider the motion of positive ions in electric, magnetic, and combined electric and magnetic fields and the mass analyzers based on them. Although positive ions are used to model ion behavior, the equations apply to negative ions and electrons. Mass analyzers disperse or separate ions according to their mass-to-charge ratio, $m/(zq)$, where z is the number of unit charges q (1.602 × 10^{-19} C) on the ion. Mass scales are usually labeled as m/z in the unit daltons, where m is the ion mass in amu (g mol^{-1}).

11.3 Uniform Electric Fields

11.3.1 Axial Field

A positive ion of mass m, charge $+zq$, and initial velocity v_0 in the field direction, entering a uniform axial electric field, E_z, along the z axis will be subject to a Coulomb force, F_z, such that $F_z = zqE_z$. The ion will experience a constant acceleration and in traversing a distance, l_z, will acquire energy $F_z l_z = zqE_z l_z$ from the field. Since $E_z = V_z/l_z$, where V_z is the potential difference between the final and the initial positions, the energy gained, T_z, is given by $T_z = zqV_z = \tfrac{1}{2}m(v_z^2 - v_0^2)$, where v_z is the final ion velocity.

The kinetic energy imparted to ions accelerated through the same potential difference is independent of ion mass, although the final ion velocity is mass dependent. This forms the basis for the separation of ions in the time-of-flight mass analyzer, in which an accelerating potential typically in the range 1 to 6 kV is used to impart a constant kinetic energy to all ions for temporal separation according to their flight time over a fixed path length. High accelerating potentials are employed in order to minimize the effect of the initial velocity distribution of the ions on the mass resolution by maintaining $v_z \gg v_0$. However, some instruments, such as the radiofrequency quadrupole, employ accelerating potentials in the range of 2 to 50 V, and such devices are inherently of low to medium resolution.

Time-of-Flight (TOF) Mass Analyzers. The pulsed TOF analyzer is based on the temporal separation of a short-duration ion pulse according to m/z over a fixed flight path. Ions formed by electron impact or injected from an external source into the equipotential ion source are extracted in pulses by application of a voltage pulse on an ion extractor element. The ions are accelerated through an axial electric field and enter a field-free flight tube in which ions separate into bunches according to their m/z, lighter ions reaching the ion detector ahead of heavier ions. Either the pulse counting electronics can be electronically gated to receive the current pulses corresponding to a particular ion mass or a multichannel analyzer can be used to capture a mass spectrum in real time. Although there are other types of mass analyzer based on temporal separation, such as the Bennett radio-frequency analyzer, disadvantages in performance or construction have precluded their development as a rival to the pulsed TOF analyzer, and the reader is referred to specialized texts on mass spectrometry for further details on these variants [1–3].

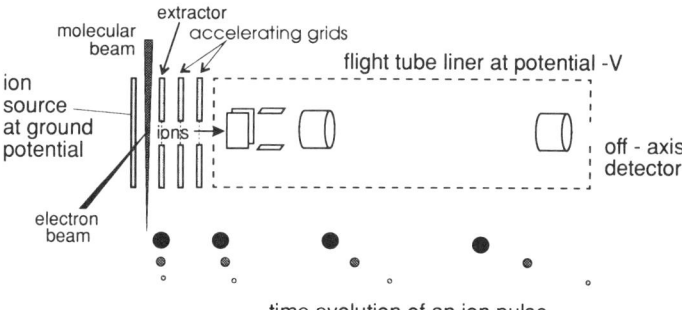

FIG. 2. A pulse linear time-of-flight mass analyzer. Positive ions formed in an equipotential ion source region at ground potential by the electron impact ionization of a molecular beam are extracted by a voltage pulse, accelerated to constant kinetic energy, and directed into the flight tube at potential $-V$.

A schematic diagram of a TOF mass analyzer is shown in Figure 2. A short burst of ions is produced using a short duration electron pulse, $\Delta t \cong 250$ nsec, or a laser pulse, $\Delta t \cong 10$–20 nsec. Assuming Δt is short compared with the flight time and ignoring the spatial and velocity distributions of ions in the source before acceleration and the time spent in axial acceleration through potential V_z, all ions, independent of m/z, enter the equipotential flight tube with equal kinetic energy. If the length of the flight tube is l_z, then the time, t, for an ion of mass m, charge zq, and velocity v_z to reach the exit plane of the flight tube is given by l_z/v_z. The ion flight time is then proportional to the square root of the mass-to-charge ratio,

$$t = \left(\frac{m}{zq}\right)^{1/2} \frac{l_z}{(2V_z)^{1/2}} \quad (11.1)$$

$$t = 72.20 \, (m/z)^{1/2} \frac{l_z}{V_z^{1/2}}. \quad (11.2)$$

In Equation (11.2), m/z is expressed in atomic mass units, l_z in meters, and V_z in volts, giving the flight time directly in microseconds. The TOF analyzer can be operated with the source at V_z and the flight tube at ground potential or with a ground-referenced source and a flight tube fabricated from mesh (to facilitate pumping), defining an equipotential flight path at potential V_z. In Figure 2 the flight tube is fitted with horizontal and vertical deflection plates for beam steering, this is particularly important if ions are formed from neutral beams traversing the source perpendicular to the flight tube axis, and cylindrical ion lenses to compensate for beam divergence. The lens close to the flight tube exit plane can be used as a retarding lens to slow down the ion pulses for separation of ions from fast neutrals produced by ion decomposition in the flight tube [19]. Instruments such as that illustrated in Figure 2 operate in constant $m/\Delta m$ mode, with $R_{10\%}$ typically around 300–500 and a mass range between 1000 and 3000 daltons.

A more rigorous analysis of the ion flight time requires that the ion motion through the ion source and acceleration regions is taken into account. An ion formed with an initial kinetic energy component T_0 along the z axis at distance l_0 from the extractor grid will experience an increase in its kinetic energy of zql_0E_0 to the grid plane on application of the pulsed extractor field, E_0. There will be a further increase in kinetic energy, T_1, as the ion traverses the accelerating region of length l_1 under the influence of the accelerating field E_1 to the flight tube entrance plane. The total kinetic energy of the ion, T_z, in the flight tube is then given by

$$T_z = T_0 + zql_0E_0 + zql_1E_1, \tag{11.3}$$

and the flight time to the exit plane of the flight tube will be given by

$$t(T_0, l_0) = t_0 + t_1 + t_z, \tag{11.4}$$

where t_z is the flight time through the flight tube of length l_z. The time an ion spends in the source depends on its initial kinetic energy, T_0, and the distance from its point of formation to the grid, l_0, according to

$$t_0 = \frac{(2m)^{1/2}}{zqE_0}[(T_0 + zql_0E_0)^{1/2} \pm T_0^{1/2}]. \tag{11.5}$$

The $\pm T_0^{1/2}$ term accounts for ions with initial axial velocity components directed toward and away from the detector. Ions initially traveling away from the detector must slow down and reverse direction on application of the extractor field. The time spent in the accelerating region, t_1, is given by

$$t_1 = \frac{(2m)^{1/2}}{zqE_1}[T_z^{1/2} - (T_0 + zql_0E_0)^{1/2}], \tag{11.6}$$

and t_z is given by

$$t_z = \frac{m^{1/2}l_z}{(2T_z)^{1/2}}. \tag{11.7}$$

Since t_0 is not uniquely defined, its inclusion in Equation (11.4) and its effect on Equations (11.6) and (11.7) are manifested as a broadening of the ion arrival time distribution at the detector and a degradation of the mass resolution. The first commercial TOF instruments had a resolution of $R_{5\%} \cong 200$. During extraction from the source, ions produced farther from the extractor grid experience a greater acceleration than those formed close and may overtake them during transit of the flight tube. An increase in resolution can be affected by adjustment of the extractor pulse amplitude to compensate for the spatial distribution of ions initially formed in the source so that ions of the same m/z arrive at the detector plane at the same time. This is called position focusing, and its introduction boosted the resolution to ~500. The additional effect of the initial ion velocity distribution is superimposed on the spatial ion distribution, and its affect on the peak shape can be used to deduce the velocity distribution of ions formed in the source for investigation of the energetics and mechanistics of ion formation [20,21]. The time spread introduced by the initial kinetic energy distribution can be minimized by energy focusing. This involves a short time delay, up to 4 μsec, between ion formation and ion extraction, allowing ions formed with their velocity components along the z axis, toward and away from the detector, to spread out and those with components off the z axis to move out of the region sampled in the extraction. Position focusing minimizes the peak width, increasing resolution to ~1000 and extending the mass range to ~4000.

The reflectron was a major innovation first proposed in 1973 [22] and subsequently refined [23–29]. The reflectron TOF mass spectrometer (MS) is a versatile high-resolution instrument competing favorably with the double-focusing mass spectrometer [17,30]. The reflectron is a retarding axial electric field assembly located at the end of the flight tube which reverses the ion motion. The reflectron can be either coaxial, with detectors arranged around the axis close to the ion source, or tilted to direct the reflected ion pulses to a detector located to one side of the source, as shown in Figure 3. The two-stage reflectron consists of two grids over which a retarding potential of $\sim\frac{2}{3}V_z$ is applied, followed by a gridless drift tube over which an adjustable retarding field is used to bring the ions to rest and then reverse or redirect their motion. Penetration of the reflectron will depend on velocity, ions with greater velocity traveling farther than those of lower velocity. The field gradient can be adjusted so that ions of all velocities arrive at the detector simultaneously. This velocity compensation coupled with increased flight paths has led to a dramatic increase in resolution and mass range [23,27,28,31], e.g., $R_{50\%} > 5000$ for $m/z = 110,000$.

The reflectron TOF mass analyzer coupled with new ion preparation techniques and multichannel array detection has found new and exciting applications in biological and medical research [16–18]. The main disadvantage of the TOF analyzer usually quoted is the loss in sensitivity inherent in its pulsed operation. However, this is compensated by wide

UNIFORM ELECTRIC FIELDS

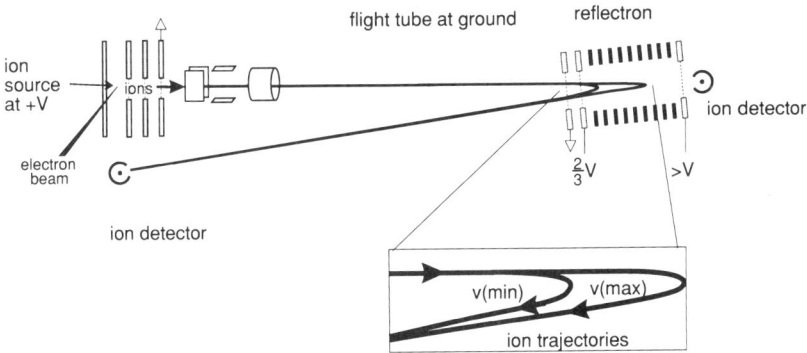

FIG. 3. A reflectron time-of-flight mass analyzer. In this diagram the ion source is held at potential $+V$, and ions are accelerated into the flight tube at ground potential. The reflectron uses a retarding axial electric field to reflect the ions back toward the detector where the arrival time is nearly independent of the initial ion position and velocity in the ion source.

apertures and an absence of slits, resulting in a sensitivity which is generally superior to double-focusing magnetic sector instruments. Its advantages include delivery of a complete single-shot mass spectrum in less than 100 μsec at a 10-kHz repetition rate, the ease of switching between positive and negative ion modes, the high mass range, and the absence of a large, expensive electromagnet.

11.3.2 Transverse Electric Fields

Ions passing through inhomogeneous transverse DC and superimposed RF fields undergo a complex oscillatory motion which is exploited for mass selection in the quadrupole and monopole mass filters and the RF ion trap [7,12].

The Quadrupole Mass Filter. The quadrupole mass filter is a nonmagnetic device that offers ion transmission up to 100%, scan rates exceeding 2000 Da sec^{-1} with low to medium resolution, and the transmission of positive or negative ions without any electrical adjustments to the analyzer. The filters are used as residual gas analyzers, beam detectors in particle beam experiments, and ion selectors for techniques such as the selected ion flow tube (SIFT). The quadrupole comprises four hyperbolic or circular rods equally spaced about the axis. They are held firmly in position by glass ceramic yokes and enclosed in a grounded stainless steel sleeve with entrance and exit apertures. The rods carry DC and RF voltages, with diammetrically opposite rods connected together. The rods

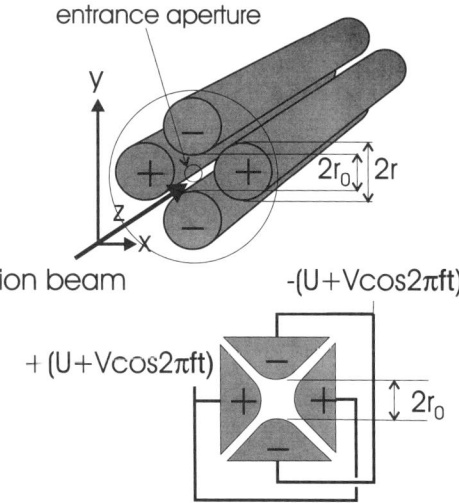

FIG. 4. Quadrupole mass filters with circular and hyperbolic rods, showing the DC and RF voltages applied to the electrodes.

in the x–z plane are held at $+(U + V\cos(2\pi ft))$, and the rods in the y–z plane are held at $-(U + V\cos(2\pi ft))$, where U is the DC voltage and V is the peak RF amplitude of frequency f, as shown in Figure 4. Hyperbolic rods offer a significant improvement in resolution, but rods of circular cross-section are usually used since the hyperbolic field is closely matched using circular rods of radius r with an inscribed radius of r_0, where $r_0 = 0.871r$. Attempts to more closely match a hyperbolic field by the inclusion of extra shim electrodes between the circular rods have been of limited success. The increasing demand imposed on the mass range and transmission of mass analyzers by analytical ionization techniques such as liquid chromatography, thermospray, and electrospray, which have been developed to ionize large biochemical molecules with minimal fragmentation, has led to an increase in the use of hyperbolic rods.

Ions are accelerated into the entrance aperture of a quadrupole analyzer with energies typically between 4 and 10 eV. They are immediately exposed to the transverse inhomogeneous DC and RF fields and undergo a complex oscillatory motion. There have been numerous theoretical investigations of the ion trajectories and the degradation in performance resulting from fringing fields at both ends of the filter [7,12,32,33]. Briefly, positive ions passing through the quadrupole, illustrated in Figure 4, will experience a repulsive DC potential in the x–z plane with a superimposed RF field that will induce an oscillation whose amplitude is greatest for light ions. This amplitude also depends on the initial transverse velocity component of the ion beam; ions injected with higher off-axis angular deviations or beams with lower initial energies will exhibit higher amplitudes and are less likely to penetrate the filter. Simultaneously, the ions sample an attractive DC potential in the y–z plane, with the superimposed RF field tending to guide lighter ions toward the axis. The voltage amplitudes are adjusted so that only ions with a single m/z execute trajectories that carry them through the filter; lighter ions are lost to the rods in the x–z plane, and heavier ions are lost to the rods in the y–z plane. Ions transmitted by the filter correspond to stable solutions of the equations of motion. Since the quadrupole acts as a filter transmitting only those ions with stable trajectories and rejecting all others, it is called a mass filter in preference to a mass spectrometer. The properties of the filter are described by a stability plot of the transformation parameters a and q' used in the Mathieu equations that describe ion motion, where a is a function of U and q' is a function of V, as shown in Figure 5. The area enclosed by the stability lines on the plot corresponds to stable solutions of the equations of motion and represents conditions under which ions are transmitted by the filter. Scan lines of slope $a/q' = 2U/V$ drawn from the origin correspond to different filter resolutions; only those ion masses corresponding to points on the scan line that lie within the stability region are transmitted. So the U/V ratio determines the resolution of the filter, and the mass spectrum is scanned by the simultaneous variation of the U and V amplitudes at a fixed RF frequency. The resolution can be changed externally by readjusting the U/V ratio. The scan line which intersects the apex of the stability region corresponding to $a = 0.23699$, $q' = 0.70600$, and $U/V = 0.16784$ represents infinite resolution (and zero transmission!) described theoretically [7] by

$$\frac{m}{\Delta m} = \frac{K}{0.16784 - (U/V)}, \tag{11.8}$$

where K is a constant.

Quadrupoles are normally operated on a scan line close to the apex. The amplitude of the x–z and y–z ion oscillations are dependent on the U/V ratio as well as the conditions at the filter entrance. The axial displacement of the ion beam, angular divergence, fringing fields, and the RF phase all influence filter resolution. As U/V is increased for higher resolution, the ion loss increases. This becomes spurious for entrance apertures exceeding $\sim 0.5 r_0$, and smaller apertures favor high resolution by minimizing angular divergence. Low-energy ions are more affected by fringing fields

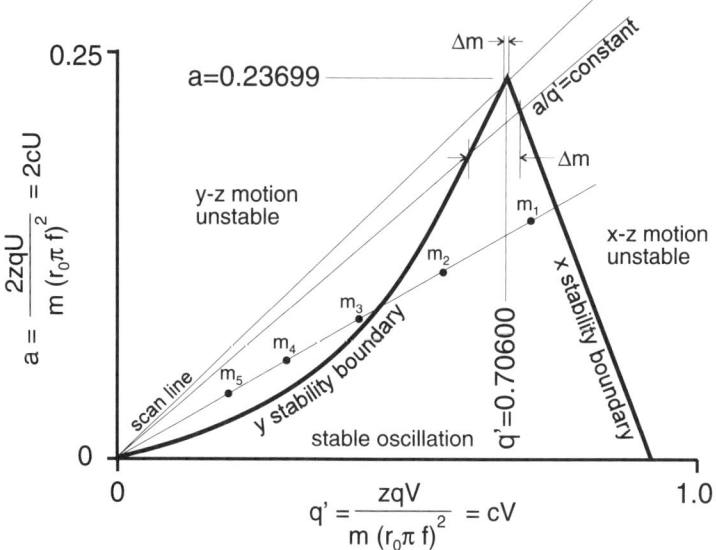

FIG. 5. Ion stability diagram for the quadrupole mass filter, showing scan lines corresponding to high and low resolution.

and are more readily defocused, giving lower transmission. High ion energies maximize transmission at the expense of resolution. Ion transmission is, however, relatively insensitive to the spacing between the entrance aperture and the rods, allowing some latitude in the design of source and optics [34–37]. It has been suggested that the addition of short coaxial quadrupoles at the entrance and exit apertures of the filter that carry only the RF component of the applied field enhances filter performance [38]. RF-only quadrupoles and hexapoles are also used as ion guides in several commercial double-focusing magnetic sector mass spectrometers to transport ion beams over extended path lengths with minimal beam degradation. Basically, the resolution depends on the number of RF cycles, N, experienced by the ion, where

$$N = fL \left(\frac{m}{2zqV_z} \right)^{1/2}. \tag{11.9}$$

The resolution increases with increasing RF frequency, f, and filter length, L, and decreases with increasing energy, V_z, according to

$$R_{10\%} = \frac{m}{\Delta m_{10\%}} = kN^n, \tag{11.10}$$

where $n \cong 2$ and $k = 0.05$. Expressing f in megahertz, m in atomic mass units, L in meters, V and V_z in volts, then $R_{10\%}$, $\Delta m_{10\%}$, and the mass range, m_M, are given by

$$\frac{m}{\Delta m_{10\%}} = 259.1(fL)^2 \left(\frac{m}{V_z}\right) \tag{11.11}$$

$$\Delta m_{10\%} = 3.86 \times 10^{-3} \frac{V_z}{(fL)^2} \tag{11.12}$$

$$m_M = \frac{7 \times 10^{-6} V}{(fr_0)^2}, \tag{11.13}$$

and the ratio of maximum mass transmitted to the peak width is independent of frequency.

The mass range is maximized for lower RF frequencies, high RF peak amplitudes, and small inscribed radii. However, high RF levels lead to problems with noise pickup, and the use of small inscribed rod radii requires tight mechanical tolerances in manufacture and mounting. An increase in the length of the filter exposes the ions to more RF cycles, but a greater fraction fails to survive to the detector. Rearranging Equation (11.11),

$$V_z = 259.1(fL)^2 \Delta m_{10\%}, \tag{11.14}$$

and, at constant ion energy qzV_z, Equation (11.14) shows that the quadrupole operates at close to constant Δm mode, the resolution increasing with increasing m. However, heavier ions will spend longer in the filter, and more are lost, giving rise to a mass discrimination effect. This can be offset by ramping V_z in step with the U and V amplitudes and pushing the filter toward constant $m/\Delta m$ mode. Such ion energy programing provides an additional external control on the filter charactersitics.

Commercial quadrupoles are available with mass ranges from <100 to ~3000 Da and resolutions up to ~1000. Modifications that involve reduction of the RF frequency have extended the mass range to 9000 Da [39]. Quadrupoles with rod lengths from 100 to 250 mm and mass ranges from 100 to over 500 Da are available premounted on $2\frac{3}{4}$-in. ConFlat flanges for use in existing vacuum systems. The simultaneous acquisition of positive and negative ion mass spectra has also been described using a modulated bipolar ion source and optics at repetition rates of 10 kHz with dual detectors, one biased for positive ion collection and the other for negative ion collection [40].

The Monopole Mass Filter. The monopole mass filter [7] is constructed from a single hyperbolic rod electrode or a circular electrode with $r_0 = 0.862r$, at potential $-(U + V\cos(2\pi ft))$, nested into a grounded

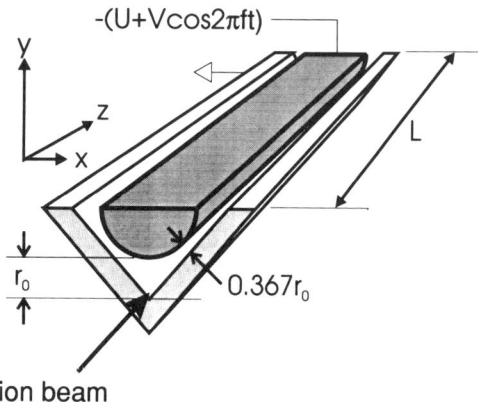

FIG. 6. Monopole mass filter.

V-electrode as shown in Figure 6. Theoretically, ions should be injected along the inside apex of the V-electrode. Since the monopole structure precludes ion oscillations in the $z-x$ plane, the ion stability region for the monopole is a narrow band of a and q' values lying along the y-stability boundary of the diagram shown in Figure 5. In consequence, resolution is almost independent of U/V, and mass scanning is achieved by sweeping the RF frequency, with U, V, and V_z held constant. Under these conditions the m/z transmitted is inversely proportional to the square of the RF frequency and dependent on the square of the RF peak amplitude, giving constant Δm resolution mode. The close proximity of the ion beam path to the V-electrode surface necessitates a small entrance orifice, a highly collimated ion beam, high accelerating potentials to minimize the effects of stray fields, and very clean operating conditions (high vacuum) to minimize contact potentials which would quickly degrade performance.

Instruments with entrance and exit apertures of an inverted triangular shape to match the apex region of the V-electrode have been described [7], although circular apertures can be used with some degradation in performance. There is a strong reciprocal relationship between the aperture sizes and the resolution, increasing aperture size tending to decreasing resolution. The resolution of the monopole is given by

$$R_{50\%} = \frac{m}{\Delta m_{50\%}} = 20 \left(\frac{U}{V}\right)^2 N = \frac{19.5 m (U/V)^2}{(r_0/L)^2 V_z}, \qquad (11.15)$$

where the frequency is expressed in megahertz and the ion mass in atomic mass units. Δm is independent of m, giving constant Δm mode, and the peak width at half height is given by

$$\Delta m_{50\%} = \frac{9.61 \times 10^{-6} V_z}{(U/V)^2 f^2 L^2}.$$ (11.16)

Lower ion energies favor high resolution at the expense of transmission, but space-charge effects at the entrance orifice diminish the gains to be made. High accelerating potentials result in rather poor resolution for low m/z but optimum transmission and mass range. Transmission becomes mass independent if ion energy programming is employed; under these conditions N remains constant and the filter operates in constant $m/\Delta m$ mode. Negative ions can be transmitted by changing the polarity of the DC bias on the rod.

The Quadrupole Ion Storage Trap (QUISTOR). The three-dimensional RF quadrupole ion trap operates on the same principles as the quadrupole mass filter. It comprises a ring electrode of hyperbolic profile with hyperbolic endcaps, as illustrated in Figure 7.

Ions are formed *in situ* by electron impact ionization, photoionization, or an external source either continuously or by a short pulse. The ionizing agent traverses the trap through holes drilled in the endcaps or through the equator of the ring electrode. Positive ions are constrained to execute a complex closed oscillatory motion within the confines of the trap by superimposed DC and RF potentials applied to the ring electrode with the endcaps held at ground potential. The stability diagram is complex and beyond the scope of this chapter. Ions can be trapped for several minutes and selectively eliminated from the trap using a notch filter mode of operation [7,12]. The DC bias is swept to scan the spectrum. Ion detection has been achieved by resonant power absorption from a reference generator operating at a fixed frequency and connected across the endcaps, with the DC bias bringing ions into resonance. Mass analysis

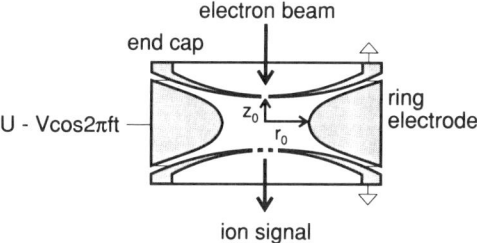

FIG. 7. Radio-frequency ion trap with electron beam ionization through one of the endcaps.

by mass-selective instability has been introduced. U, V, and f are changed singly or in combination so that the trajectories of trapped ions of consecutive m/z become successively unstable. The unstable trajectories carry the ions through a mesh window, forming part of an endcap electrode to an ion detector.

11.3.3 Radial Transverse Electric Fields

Radial or cylindrical electrostatic fields can provide velocity focusing. Mass resolution is enhanced in a double-focusing magnetic sector mass spectrometer by the incorporation of a radial transverse electric field to achieve velocity focusing prior to passage of the ion beam through a sector magnetic field. Curved field plates produce a radial field which minimizes aberrations and brings the velocity-selected beam to a well-defined focal point as shown in Figure 8.

For a radial electric field of strength E_E, the electrostatic force, F_E, imposed on an ion of mass m and charge zq is balanced by the centrifugal force, $F_c = mv_z^2/r_E$, and the particles will follow circular trajectories of radius r_E.

$$r_E = \frac{2V_z}{E_E} = \frac{m}{zq}\frac{v_z^2}{E_E}. \qquad (11.17)$$

Since the radius of the trajectory is dependent on v_z^2, velocity dispersion occurs. Ions of a specific mass and a trajectory of radius r_E, defined by the radius of curvature of the field plates and narrow slits at the object and image plane, will be monoenergetic and direction focused as shown in Figure 8. Herzog [41] showed that direction focusing for radial electrostatic velocity analyzers is restricted to sector angles given by

$$\phi = \pi/(\sqrt{2}\, n), \qquad (11.18)$$

where $n = 1, 2, 3, \ldots$. For a symmetrical analyzer, the object and image focal lengths $l_{o,E}$ and $l_{i,E}$ are equal and given by

$$l_E = (r_E/\sqrt{2})[\cot(\sqrt{2}\,\phi) + (\sin(\sqrt{2}\,\phi))^{-1}]. \qquad (11.19)$$

An ion or electron beam with energy $T_z = zqV_z$ will be focused on the image slit when

$$T_z = \frac{zq(V_{out} - V_{in})}{2\ln(r_{in}/r_{out})}, \qquad (11.20)$$

with $V_{out} = V_z(1 + 2\ln(r_{out}/r_E))$, $V_{in} = V_z(1 + 2\ln(r_{in}/r_E))$, and $r_E = (r_{out} - r_{in})/2$. V_{out} and V_{in} are the outer and inner field plate potentials, and r_{out} and r_{in} are the outer and inner field plate radii. The energy resolution

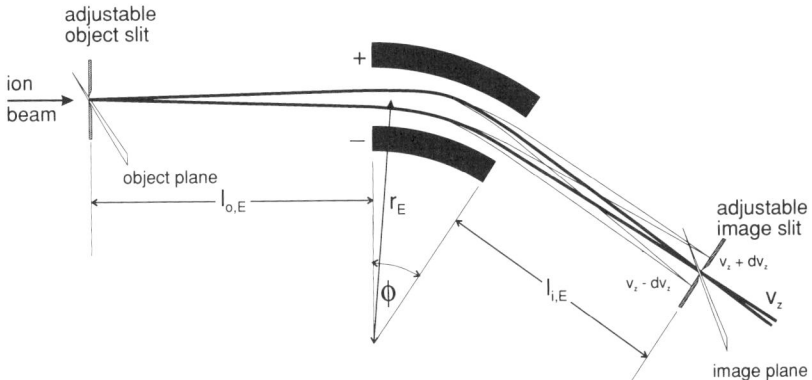

FIG. 8. The direction-focused dispersion of an ion beam according to velocity by a radial electric field established between wedge-shaped electrodes.

of the beam is then given by

$$T_z/\Delta T_z = 2r_E/s, \quad (11.21)$$

where s is the width of the object and image slits. Large radial electrostatic analyzers equipped with narrow slits can achieve energy resolutions of $\sim 5 \times 10^5$.

11.4 Uniform Magnetic Fields

11.4.1 Transverse Motion

Particles with charge zq traveling with velocity v_z through a homogeneous magnetic field of strength B experience a constant force perpendicular both to the direction of motion and to the magnetic field direction given by $\mathbf{F}_B = zq\mathbf{v}_z \times \mathbf{B}$. This Lorentz force changes the direction of motion of the particle but not the magnitude of its velocity. The particles are constrained to move in a circular orbit of radius r_B which depends on the ion mass, thereby facilitating mass analysis. Equating Lorentz and centripetal forces shows that an ion of mass m and energy zqV_z will execute an orbit of radius

$$r_B = \left(\frac{m}{zq}\right)^{1/2}\left(\frac{2V_z}{B^2}\right)^{1/2}. \quad (11.22)$$

Ions entering a magnetic field after acceleration through potential V_z will be dispersed according to their m/z, heavier ions following trajectories with a larger radius. Practical magnetic dispersion mass spectrometers utilize the direction focusing properties of uniform wedge-shaped magnetic fields in which the ion beam enters and leaves perpendicular to the field edges as shown in Figure 9. Herzog [42] showed that fields bounded by angles θ such that

$$\theta = \pi/n, \qquad (11.23)$$

where $n = 1, 2, 3, \ldots$, exhibit direction focusing. For a symmetrical analyzer, such as that illustrated in Figure 9, for which object and image focal lengths $I_{o,B}$ and $I_{i,B}$ are equal, $I_{o,B} = I_{i,B} = I_B$, and the ratio I_B/r_B is 0, 1.000, 1.732, and 2.414 for $\theta = 180°, 90°, 60°,$ and $45°$, respectively. Under these conditions the object slit, image slit, and tip of the wedge lie in a straight line.

Although the resolving power of magnetic sectors is independent of θ, it does depend on the radius of curvature of the ion beam through the field, such that $R_{\Delta m=1} = m = r_B/(s_o + s_i)$, where s_o and s_i are the slit widths or beam widths and m is the mass of resolved ions for which $\Delta m = 1$. Usually, $s_o = s_i = s$. As the slits are narrowed the upper mass limit for resolved ions increases, albeit with the usual concomitant decrease in signal intensity.

Magnetic Sector Mass Analyzers. In the absence of a velocity focusing electric sector, single-focusing magnetic sector mass spectrometers can provide only low to medium resolution, generally $R_{50\%} < 500$ with a mass range of <600 Da. Double-focusing instruments were developed to satisfy the demand for a high-resolution instrument capable of unequivocal ion identification through accurate measurement of ion mass. High resolu-

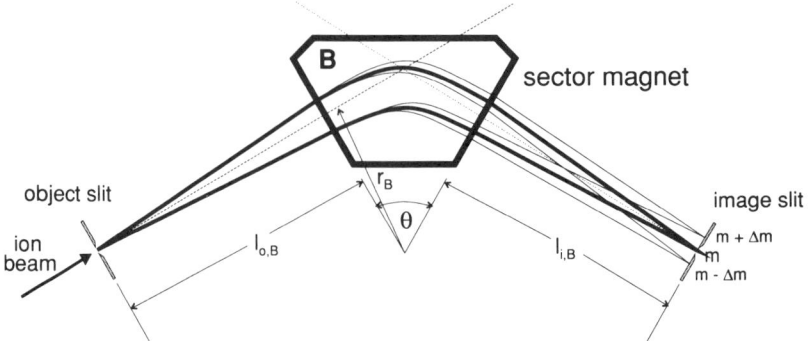

FIG. 9. The direction-focused dispersion of an ion beam according to the mass-to-charge ratio by a homogeneous magnetic sector field.

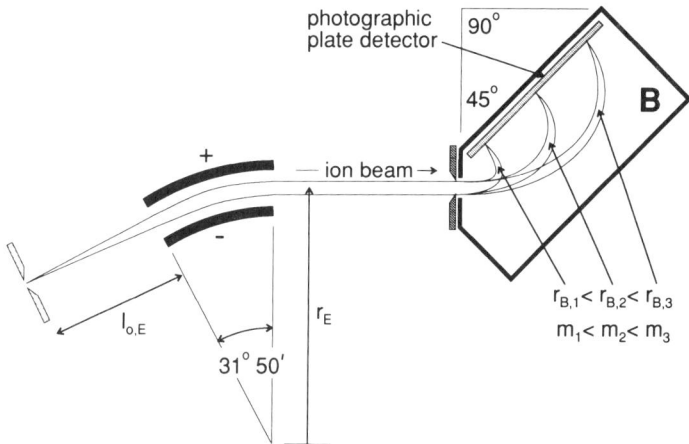

FIG. 10. The first double-focusing magnetic sector mass analyzer, designed by Mattauch and Herzog in 1934.

tion is achieved using a velocity focusing electric sector as a source of monoenergetic ions for subsequent dispersion by a magnetic sector mass analyzer.

The first double-focusing mass spectrometer was designed by Mattauch and Herzog [43] using a 31° 50′ electric sector and a 90° magnetic sector arranged so that the velocity-selected ion beam directed to the magnetic sector exhibited an infinite focal length. Ions were brought to a focal point on a photographic plate lying along the focal plane according to the radius of the ion trajectories in the field, as shown in Figure 10. This instrument delivered a mass spectrum with a resolution up to ~7500 over a mass range of 250 Da. Dempster [1–4] designed a double-focusing instrument around a 90° nonsymmetrical electric sector and a 180° magnetic sector, giving a resolution of ~7000, and Bainbridge and Jordan [1–4] introduced an instrument based on a 127° 18′ electric sector and a 60° magnetic analyzer. Many of the commercial instruments currently available are designed around a nonsymmetrical 90° electric and 90° magnetic sector design, illustrated in Figure 11, which can achieve $R_{10\%} > 80{,}000$.

A double-focusing instrument will transmit velocity-focused, direction-focused ions which satisfy

$$\frac{m}{zq} = \frac{B^2 r_B^2}{E_E r_E}. \quad (11.24)$$

Both r_B and r_E must be large in order to maximize mass and energy resolu-

338 ION MASS ANALYZERS

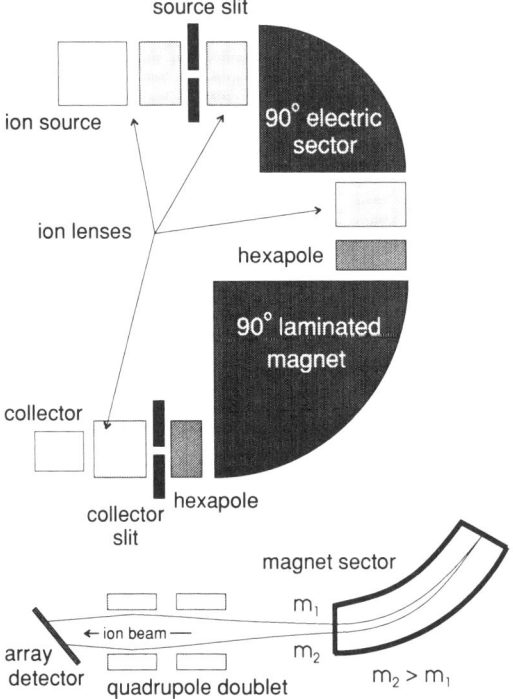

FIG. 11. A block diagram of a modern double-focusing magnetic sector mass analyzer incorporating a laminated magnet for rapid scanning and improved field homogeneity and hexapole ion lenses to minimize the effects of fringing magnetic fields on the ion beam. The application of a quadrupole doublet ion lens to direct a magnetically dispersed segment of the mass spectrum onto a diode array detector for simultaneous recording is illustrated at the bottom of the figure.

tion, respectively, with r_B usually larger than r_E. The mass range can be extended by increasing B or the ratio r_B^2/r_E.

Modern commercial double-focusing instruments may employ laminated magnets [44] to facilitate rapid scanning, down to 1 second per 100 Da at lower resolution, quadrupole or hexapole lenses to correct beam trajectories for the effects of the fringing fields at the sector entrances and exits, and multichannel array detectors for mass spectrum recording in segments for computer splicing [45] (Figure 11). Array detectors offer the advantages of shorter scan times and reduced sample requirements. The magnetic field is scanned in steps across the selected mass range; at each step the dispersed segment of the spectrum passed by the magnet is refocused onto a multichannel plate by a quadrupole double lens. Overlapping segments are normalized and spliced by computer algorithms to generate a full-scale mass spectrum. Resolution has been extended to ~100,000 and the mass range pushed to over 125,000 in the more expensive instruments. More typically, machines deliver $R_{10\%}$ from 1000 to ~50,000 and mass ranges from ~3000 at high accerating voltages to ~20,000 at low accelerating voltages. To achieve the high resolutions quoted by manufacturers, however, it becomes necessary to use scan rates as low as 30,000 seconds per 100 Da.

Double-focusing instruments in which the magnetic sector precedes the electric sector are called reverse geometry or **BE** mass spectrometers. Their principal advantage lies in the study of metastable ion decomposition. Metastable ions formed in the ion source by electron impact or photoionization may decompose en route to the detector. If this occurs between the electric and the magnet sectors in a conventional geometry **EB** mass spectrometer the daughter ions are recorded as broad peaks with an apparent mass, m^*, which is related to the parent ion mass m_P and the daughter ion mass m_D by $m^* = m_D^2/m_P$. Ions which decompose after the magnetic field are recorded as parent ions. In a **BE** instrument, ions of a selected mass and energy are transmitted by the magnetic sector and velocity focused by the electric sector. If the metastable ion dissociates in the field-free region between sectors, the daughter ion will have a reduced kinetic energy, which will preclude its transmission by the electric sector. Reduction of the transverse field to $(m_D/m_P)E_E$, where E_E is the field strength corresponding to the transmission of ions with kinetic energy zqV_z, will pass the daughter ions exclusively. A scan of E_E to lower values produces a daughter ion mass spectrum for the metastable ion transmitted by the magnetic sector. This procedure is called MIKES, or metastable ion kinetic energy spectrometry. Conventional geometry **EB** instruments can be used to provide less-specific information from metastable ion decomposition in the field-free region preceding the electric sector by scanning the accelerating potential to transmit daughter ions through the electric sector.

11.4.2 Cyclotron Motion

Low-energy particles formed or injected into high-intensity magnetic fields traveling perpendicular to the field axis execute circular orbits and remain trapped within the confines of the field. The radius of the ion orbit, r_c, the cyclotron radius, and the time per revolution, t_c, the cyclotron period, are given by

$$r_c = \frac{mv_z}{zqB}; \quad t_c = \frac{2\pi r_c}{v_z} = \frac{2\pi m}{zqB} = \frac{1}{\omega_c}, \quad (11.25)$$

where ω_c is the cyclotron frequency and is independent of the magnitude of the velocity. Ions exhibit cyclotron motion with a characteristic frequency and an orbital radius dependent on the axial velocity. The cyclotron motion of ions in strong magnetic fields forms the basis of several mass analyzers.

11.5 Superimposed Electric and Magnetic Fields

The force acting on a charged particle in field combinations is given by the sum of the Coulomb and Lorentz forces according to

$$\mathbf{F} = zq\mathbf{E} + zq\mathbf{v} \times \mathbf{B}. \tag{11.26}$$

In mass analyzers based on cyclotron motion and in the Wien filter crossed fields are employed. If an ion is injected into crossed electric and magnetic fields, perpendicular to both, it experiences opposing electric and magnetic forces. If, however, its velocity is such that $v_z = E/B$ it experiences no net force and this is used to advantage in the Wien filter which exhibits dual velocity filter–mass analyzer functions.

The Wien Filter. The basic Wien filter design illustrated in Figure 12a uses permanent magnets or electromagnets with perpendicular electric field plates [46,47]. Positive ions of velocity $v_z = E/B$ entering the filter on the z axis will pass through the filter undeflected. Since the velocity of ions entering the filter after acceleration through potential V_z depends on m/z, the electric or magnetic field strengths can be changed to sweep ions of different mass across a slit at the exit plane of the filter and obtain a mass spectrum. Alternatively, for an incident ion beam of a specific mass the fields can be swept to measure the velocity distribution or adjusted to selectively transmit ions of a narrow velocity range through the exit slit.

The mass resolution, which is typically rather low, is given by

$$m/\Delta m = \frac{El^2}{2V_z s}, \tag{11.27}$$

where l is the length of the filter and s is the width of the entrance and exit slits. The major advantages offered by the Wien filter, when low resolution is not a limiting factor, are small size, a straight-through path, and a dual role as a velocity analyzer–selector and mass analyzer. A limitation imposed on the performance of the filter, which is inherent in the design, is the focusing of the incident beam by the filter which results because the electrostatic potential and hence the ion energy are position dependent within the filter. The focusing properties can be exploited to focus ion beams with a circular cross-section to a line image on a target.

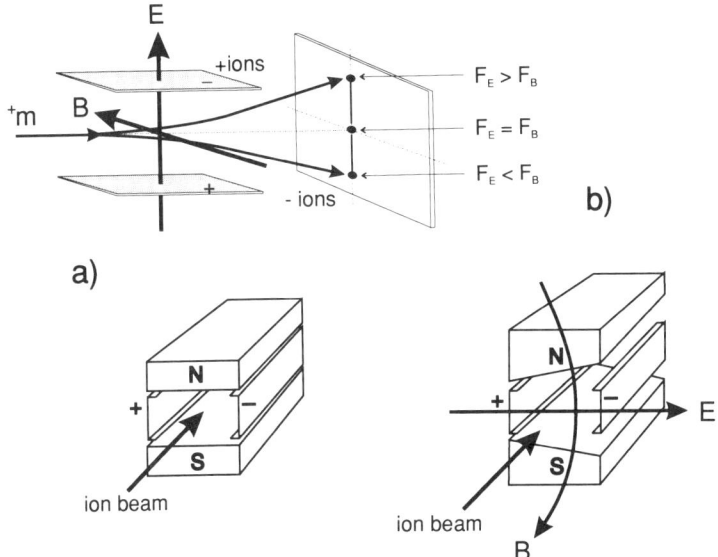

FIG. 12. (a) The Wien filter. The dispersion of ions directed transverse to mutually perpendicular electric and magnetic fields. (b) The focusing effect imposed on off-axis ion trajectories by the Wien filter, which degrades velocity resolution, and the application of tilted or tapered magnetic pole pieces for the elimination of this effect.

The velocity and mass resolutions of the filter for circular beams can be improved with the elimination of the focusing effect. This has been accomplished [48] with a 30% improvement in performance using tapered or tilted pole pieces to produce an inhomogeneous magnetic field which is stronger toward the positive field plate and weaker toward the negative field plate as shown in Figure 12b. Alternatively, slits oriented in the direction of the magnetic field can be used to minimize the beam dimensions in the electric field direction.

Omegatron, ICR MS, and FTICR MS. This group of mass analyzers or ion traps shares a common lineage. They are low-sample-pressure devices in which the ion source, analyzer, and detector share the same volume of space sandwiched between the pole faces of a high-strength magnet at which ions generated *in situ*, or injected from an external source, undergo cyclotron motion. They span a range of mass resolutions from <100, for the Omegatron, to >10^8, for the FTICR. Although the ion detection efficiency is high, space-charge effects limit the ion population to low densities, and the main application of ICR and FTICR spectrometers has been the study of ion–molecule kinetics and reaction mechanisms [8,10,11,13,49–51].

The cyclotron motion of ions in a strong magnetic field was first exploited in the *Omegatron* [52] which comprises a 20-mm equipotential stainless steel cubic cell with the top and bottom plates separated from the main body. Ions are usually formed by an electron beam passing through the center of the cube parallel to the magnetic field direction, or they are injected from an external source. Ions in the cell undergo cyclotron motion perpendicular to the magnetic field direction, the cyclotron frequency depending on the m/z according to Equation (11.25). The top and bottom plates are connected to an RF signal generator which can be swept through the cyclotron frequency range appropriate to ions in the cell. As the RF field comes into resonance with an ion cyclotron frequency, energy is absorbed from the field and the ion velocity, and the cyclotron radius increases. The ions spiral outward toward the top and bottom plates, where they are measured on an ion collector plate.

The resolution of the Omegatron is given by

$$m/\Delta m = \frac{zqd\,B^2}{2E_{RF}m}, \qquad (11.28)$$

where d is the separation between the electron beam track and the ion collector and E_{RF} is the peak RF field strength. The resolution is high for low mass and decreases below usable values with increasing ion mass quite rapidly [53].

The ion cyclotron resonance mass spectrometer (ICR MS) was developed from the Omegatron [54,55], and a schematic of a typical four-section cell is shown in Figure 13. It is particularly well suited for the study of ion–molecule chemistry [11] using techniques such as pulsed double resonance [8,11,56]. The ICR MS can operate in two modes, the trapping mode and the drift mode. In the trapping mode only the source section of the cell is used, the potentials on the top and bottom field plates or drift plates are set to ground, $V_D = 0$ V, and a small positive voltage is applied to the source trapping plates to prevent ion discharge at the trapping plate surfaces as illustrated in Figure 14a. Ions are formed in the cell at a pressure of 10^{-6}–10^{-8} Torr by an electron pulse typically 5 msec in duration and trapped in the source section of the cell for up to 5 sec while ion chemistry takes place. The magnetic field is then swept to bring ions into resonance with a marginal oscillator or capacitance bridge connected to the top and bottom field plates. Ions are detected through the power loss in the detector circuit when the applied frequency is in resonance with a cyclotron frequency corresponding to ions in the cell. In some applications the magnetic field can be fixed and the detector frequency swept.

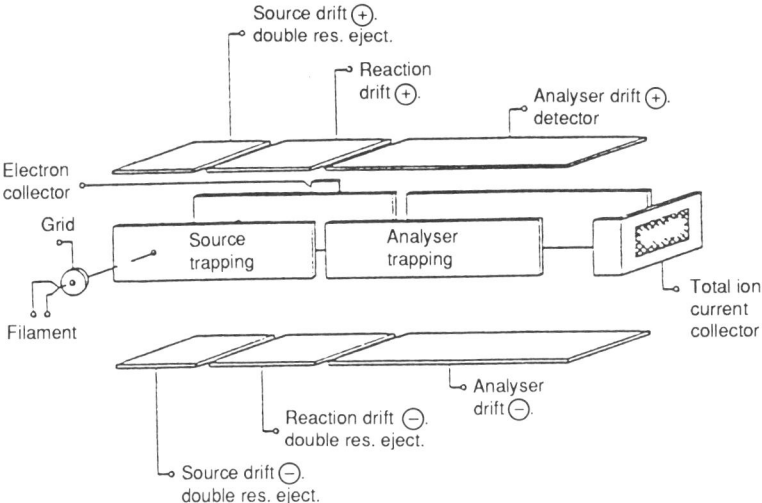

FIG. 13. A four-section ICR cell. Reprinted with permission from P. R. Kemper and M. T. Bowers, in *Techniques of Chemistry*, Vol. XX, *Techniques for the Study of Ion-Molecule Reactions*, Chap. 1. Wiley–Interscience, New York, 1988. Copyright © 1988 John Wiley & Sons, Inc.

In drift mode the inclusion of a weak electric field perpendicular to the magnetic field results in ion motion, called ion drift, mutually perpendicular to both **E** and **B** (Figure 14b), with drift velocity given by $v_{\text{drift}} = E/B$. Ions are produced *in situ* or injected into the source section of the cell continuously, with the cell pressure maintained within the range 10^{-6}–10^{-3} Torr. In practice, the electron gun cathode bias is square wave modulated so that a lock-in amplifier can be used in the detector circuitry to achieve an acceptable signal-to-noise ratio. The ions drift into the reaction section, where unwanted ions can be ejected or translationally excited by resonance absorption from an RF field between the top and the bottom field plates. In the analyzer section, reactant and product ions are detected by the power absorption in the detector circuit using a lock-in amplifier. Ions that drift out of the analyzer section are detected at a total current collector.

The analyzers described above provide mass spectra by measurement of individual ion masses sequentially on a single channel detector. Resolution is dependent on the rate at which the spectrum is acquired, with longer acquisition times leading to higher resolution and lower sensitivity. In Fourier transform ion cyclotron resonance mass spectrometry (FTICR MS), the full mass spectrum is sampled simultaneously on a single channel detector as a time-domain signal composed of superimposed sine waves,

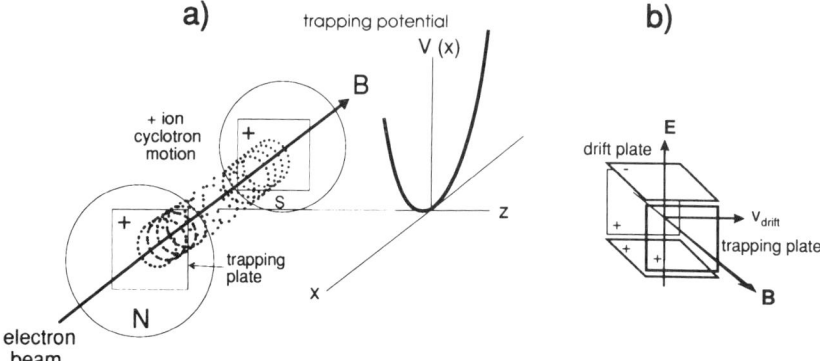

FIG. 14. (a) Magnetically constrained positive ions undergoing trapped oscillation between the trapping field plates in the source section of an ICR spectrometer. (b) Diagram of the ion source section of a drift-mode ICR cell. Ions formed along the magnetic field axis are confined in the cell by the trapping field. The perpendicular drift field results in a low-velocity, polarity-independent, ion drift of the cyclotron motion perpendicular to both the magnetic and the drift fields that carries ions into the analyzer section of the cell.

one for each ion mass. The mass spectrum is then calculated using Fourier transform techniques [51].

The cell is based on the trap mode ICR and is usually cubic as shown in Figure 15. Ions are constrained to oscillate between trapping plates perpendicular to the magnetic field as they execute cyclotron motion. Typical magnetic field strengths employed are ~3 T with fields up to 7 T

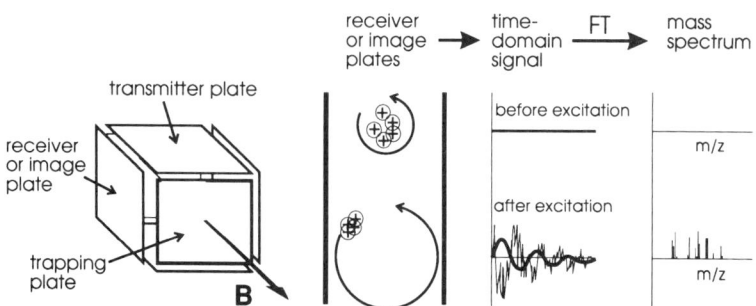

FIG. 15. Diagram of an FTICR ion trap and of the generation and Fourier transform analysis of the time–domain decay signal from the coherent ion cyclotron motion of ions in the trap upon cessation of the excitation pulse.

available with superconducting magnets. The mass-dependent concurrent cyclotron motion of ions in the strong uniform magnetic field lends ICR to the Fourier method [50,57]. Ion manipulation and detection involve the remaining four ground-referenced side plates of the cell. One diammetrically opposite pair is used as transmitter plates for an ion irradiation pulse at cyclotron frequencies. Following cessation of the irradiation pulse the remaining pair is used as receiver or image current plates, where an induced RF signal from the ions is detected as illustrated in Figure 15.

Ion detection and the measurement of a mass spectrum rely on the excitation of ions into coherent cyclotron motion with a cyclotron radius independent of m/z. Both objectives can be met by broadband irradiation over the appropriate frequency range. This can be achieved using either impulse excitation [58], in which 50- to 100-nsec, ± 100-V pulses are applied to the transmitter plates to excite all ions into larger orbits simultaneously, or chirp excitation [51], using a broadband RF sweep across the frequency range which takes ~ 100 μsec. Cyclotron orbital radii before and after excitation might be ~ 0.25 and 10 mm, respectively, with ion kinetic energies up to 10^4 eV following excitation. As the ions move into the larger orbit on excitation, random motion gives way to coherent motion as shown in Figure 15. Irrespective of m/z, ions irradiated by either impulse or chirp excitation for the same time attain the same cyclotron radius. The signal levels detected at the receiver plates will then depend only on the number of ions at each m/z, giving accurate relative ion intensities with no mass discrimination. Detection is through the composite induced image current at the receiver plates, made up of superimposed sinusoidal waveforms corresponding to all of the trapped coherent ion bunches executing cyclotron motion in the cell. The image current is amplified before mathematical transformation into a mass spectrum using the fast Fourier algorithm [59]. Fourier transform techniques provide major improvements in speed, the signal-to-noise ratio, mass resolution, the mass range, and data manipulation in comparison with other instruments.

The mass resolution, expressed as $R_{50\%}$, has been shown to depend on $\omega_c/\Delta\omega_c$, the resolution with which the cyclotron image current frequency can be measured. This depends on the time, t, over which the image current remains detectable according to

$$R_{50\%} = (m/\Delta m)_{50\%} = (\omega_c/\Delta\omega_c)_{50\%} \leq 1.7 \times 10^7 \frac{zBt}{m}, \qquad (11.29)$$

where m is expressed in atomic mass units. So, in contrast to other analyzers, for which an increase in the sampling time required for improved resolution translates into lower sensitivity, an increased image current sampling time improves both resolution and sensitivity. Since ion scattering is responsible for image current decay, gains in resolution and sensitivity are ultimately limited by vacuum quality. Ion–molecule studies require high cell pressures, $\geq 10^{-5}$ Torr, at which image signal decays are on the order of milliseconds. Measurements of accurate ion mass at ultrahigh resolution are carried out with cell pressures of $\leq 10^{-8}$ Torr, at which image current decays are tens of seconds. Resolutions of $>10^8$ for H_2O^+ at $m/z = 18$ and of 555,000 at $m/z = 866$ have been reported, and the mass difference between $^1H_2O^+$ and $^2HO^+$ has been measured to be 0.001548296 amu to an accuracy of 0.6 ppb [60]. This result has provided the most accurate absolute value for the mass of the deuteron at 2.014101764 amu.

The mass range of the FTICR in atomic mass units is given by

$$m_M \cong \frac{qB^2a^2}{8\alpha V_T} = 8.7 \times 10^6 \frac{B^2a^2}{V_T}, \qquad (11.30)$$

where a is the separation of the trapping plates in meters and α is a function of the cell geometry, usually set at 1.386 for cubic traps [51]. Using screened trapping plates to reduce V_T, mass ranges up to 63,145 amu at a field strength of 3 T and to 340,000 amu at 7 T have been achieved. Despite significant advances in trap design, electronics, software control algorithms, and mathematical data manipulation, there are still performance gains to be made by further reducing trapping field inhomogeneities [61,62], improving RF field homogeneity [63], and developing a better appreciation of nonlinear effects such as harmonics and sidebands [51].

Chirp excitation using a rapid constant amplitude frequency sweep has been the preferred method for ion cyclotron excitation. However, the amplitude profile of the chirp pulse in the frequency domain is not an ideal square-wave chirp. A perfect square-wave frequency domain chirp can be mathematically designed and synthesized using a signal generator. This technique is known by the acronym SWIFT, for stored waveform inverse Fourier transform. SWIFT waveforms can also incorporate selective ion ejection and excitation of specific ions to predetermined kinetic energies [64]. FTICR will undoubtedly continue to be improved and diversify into areas which are inaccessible to other techniques.

References

1. R. W. Kiser, *Introduction to Mass Spectrometry and its Applications*. Prentice-Hall, Englewood Cliffs, NJ, 1965.
2. S. R. Shrader, *Introductory Mass Spectrometry*. Allyn & Bacon, Boston, 1974.
3. R. Jayaram, *Mass Spectrometry, Theory and Applications*. Plenum, New York, 1966.
4. P. F. Knewstubb, *Mass Spectrometry and Ion-Molecule Reactions*. Cambridge Univ. Press, Cambridge, UK, 1969.

5. C. E. Melton, *Principles of Mass Spectrometry and Negative Ions*. Dekker, New York, 1970.
6. A. Maccoll, ed., *Mass Spectrometry*, Int. Rev. Sci., Phys. Chem., Ser. 2, Vol. 5. Butterworth, London and Boston, 1975.
7. P. H. Dawson, ed., *Quadrupole Mass Spectrometry, and its Applications*. Elsevier, Amsterdam, 1976.
8. T. A. Lehman and M. M. Bursey, *Ion Cyclotron Resonance Spectrometry*. Wiley, New York, 1976.
9. I. Howe, D. H. Williams, and R. D. Bowen, *Mass Spectrometry, Principles and Applications*, 2nd ed. McGraw-Hill, New York, 1981.
10. M. V. Buchanan, ed., *Fourier Transform Mass Spectrometry, Evolution, Innovation, and Applications*, ACS Symp. Ser. 359. Am. Chem. Soc., Washington, DC, 1987.
11. J. M. Farrar and W. H. Saunders, Jr., eds., *Techniques for the Study of Ion-Molecule Reactions*, Techn. Chem. Ser., Vol. 20. Wiley, New York, 1988.
12. R. E. March and R. J. Hughes, *Quadrupole Storage Mass Spectrometry*. Wiley, New York, 1989.
13. A. G. Marshall and F. R. Verdun, *Fourier Transforms in NMR, Optical, and Mass Spectrometry*. Elsevier, Amsterdam, 1990.
14. *Dynamic Mass Spectrometry*, Vol. 1. Heyden, London, 1970, Vol. 2 (1972).
15. R. D. Smith, J. A. Loo, C. G. Edmonds, C. J. Barinaga, and H. R. Udseth, *Anal. Chem.* **62**, 882 (1990).
16. A. L. Burlingame, D. S. Millington, D. L. Norwood, and D. H. Russell, *Anal. Chem.* **62**, 268R (1990).
17. R. J. Cotter, *Anal. Chem.* **64**(21), 1027A (1992).
18. A. L. Burlingame, T. A. Baillie, and D. H. Russell, *Anal. Chem.* **62**, 268R (1990).
19. R. N. Compton, L. G. Christophorou, G. S. Hurst, and P. W. Reinhardt, *J. Chem. Phys.* **45**, 4634 (1966); J. C. J. Thynne, P. W. Harland, and R. MacDonald, in *Dynamic Mass Spectrometry* (D. Price, ed.), Vol. 2, Chapter 6. Heyden, London, 1971; P. W. Harland and J. C. J. Thynne, *J. Phys. Chem.* **75**, 3517 (1971); *J. Chem. Soc., Chem. Commun.* **8**, 476 (1972).
20. J. L. Franklin, P. M. Hierl, and D. A. Whan, *J. Chem. Phys.* **47**, 3148 (1967).
21. P. W. Harland, J. L. Franklin, and D. E. Carter, *J. Chem. Phys.* **58**, 1430 (1973); P. W. Harland and J. L. Franklin, *ibid.* **61**, 1621 (1974).
22. B. A. Mamyrin, V. I. Karataev, D. V. Shmikk, and V. A. Zagulin, *Sov. Phys.—JETP (Engl. Transl.)* **37**, 45 (1973).
23. U. Boesl, H. J. Neusser, R. Weinkauf, and E. W. Schlag, *J. Phys. Chem.* **86**, 4857 (1982).
24. H. Kühlewind, H. J. Neusser, and E. W. Schlag, *Int. J. Mass Spectrom. Ion Phys.* **51**, 255 (1983).
25. E. W. Schlag and H. J. Neusser, *Acc. Chem. Res.* **16**, 355 (1983).
26. T. Bergmann, T. P. Martin, and H. Schaber, *Rev. Sci. Instrum.* **60**(3), 347 (1989).
27. T. Bergmann, T. P. Martin, and H. Schaber, *Rev. Sci. Instrum.* **60**(4), 792 (1989).
28. T. Bergmann, H. Goehlich, T. P. Martin, H. Schaber, and G. Malegiannakis, *Rev. Sci. Instrum.* **61**(10), 2585 (1990).
29. T. Bergmann, T. P. Martin, and H. Schaber, *Rev. Sci. Instrum.* **61**(10), 2592 (1990).
30. R. J. Cotter, *Biomed. Environ. Mass Spectrom.* **18**, 513 (1989).

31. M. Karas, U. Bahr, and F. Hillenkamp, *Int. J. Mass. Spectrom. Ion Processes* **92**, 231 (1989).
32. J. E. Campana, *Int. J. Mass Spectrom. Ion Phys.* **33**, 101 (1980).
33. J. E. Campana and P. C. Jurs, *Int. J. Mass Spectrom. Ion Phys.* **33**, 119 (1980).
34. P. H. Dawson, *Int. J. Mass Spectrom. Ion Phys.* **17**, 423 (1975).
35. J.-F. Hennequin and R.-L. Inglebert, *Int. J. Mass Spectrom. Ion Phys.* **26**, 131 (1978).
36. K. L. Hunter and B. J. McIntosh, *Int. J. Mass Spectrom. Ion Processes* **87**, 157 (1989).
37. B. J. McIntosh and K. L. Hunter, *Int. J. Mass Spectrom. Ion Processes* **87**, 165 (1989).
38. P. H. Dawson, *Int. J. Mass Spectrom. Ion Phys.* **6**, 33 (1971).
39. P. Labastie and M. Doy, *Int. J. Mass Spectrom. Ion Processes* **91**, 105 (1989).
40. D. F. Hunt, G. C. Stafford, Jr., F. W. Crow, and J. W. Russell, *Anal. Chem.* **48**, 2098 (1976).
41. R. Herzog, *Z. Phys.* **89**, 447 (1934).
42. R. Herzog, *Z. Phys.* **89**, 786 (1934).
43. J. Mattauch and R. Herzog, *Z. Phys.* **89**, 786 (1934).
44. J. C. Bill, B. N. Green, and I. A. S. Lewis, *Int. J. Mass Spectrom. Ion Phys.* **46**, 147 (1983).
45. J. A. Hill, J. E. Biller, S. A. Martin, K. Biemann, Z. Yoshidome, and K. Sato, *Int. J. Mass Spectrom. Ion Processes* **92**, 215 (1989).
46. R. L. Seliger, *J. Appl. Phys.* **43**, 2352 (1972).
47. L. Holmlid, *Int. J. Mass Spectrom. Ion Phys.* **17**, 403 (1975).
48. E. Leal-Quiros and M. A. Prelas, *Rev. Sci. Instrum.* **60**(3), 350 (1989).
49. C. L. Wilkins, *Anal. Chem.* **50**(4), 493A (1978).
50. A. G. Marshall, *Acc. Chem. Res.* **18**, 316 (1985).
51. A. G. Marshall and P. B. Grosshans, *Anal. Chem.* **63**(4), 215A (1991).
52. H. Sommer, H. A. Thomas, and J. A. Hipple, *Phys. Rev.* **82**, 697 (1951).
53. E. Y. Wang, L. Schmitz, Y. Ra, B. LaBombard, and R. W. Conn, *Rev. Sci. Instrum.* **61**(8), 2155 (1990).
54. J. L. Beauchamp, L. R. Anders, and J. D. Baldeschwieler, *J. Am. Chem. Soc.* **89**, 4569 (1967).
55. J. D. Baldschwieler, *Science* **159**, 263 (1968).
56. R. T. McIver, Jr. and R. C. Dunbar, *Int. J. Mass Spectrom. Ion Phys.* **7**, 471 (1971).
57. M. B. Comisarow and A. G. Marshall, *Chem. Phys. Lett.* **25**, 282 (1974).
58. R. T. McIver, Jr., R. L. Hunter, and G. Baykut, *Anal. Chem.* **61**, 491 (1989); *Rev. Sci. Instrum.* **60**(3), 400 (1989); *Int. J. Mass Spectrom. Ion Phys.* **89**, 343 (1989).
59. J. W. Cooley and J. W. Tukey, *Math. Comput.* **19**, 9 (1975).
60. M. V. Gorshkov, G. M. Alber, I. Schweikhard, and A. G. Marshall, *Phys. Rev. A* **47**, 3433 (1993).
61. M. Wang and A. G. Marshall, *Anal. Chem.* **61**, 1288 (1989).
62. C. D. Hanson, M. E. Castro, E. L. Kerley, and D. H. Russell, *Anal. Chem.* **62**, 520 (1990).
63. M. Wang and A. G. Marshall, *Anal. Chem.* **62**, 515 (1990).
64. A. G. Marshall, T.-C. L. Wang, and T. L. Ricca, *J. Am. Chem. Soc.* **107**, 7893 (1985).

12. ION TRAPS

Hugh A. Klein

National Physical Laboratory, Teddington, Middlesex TW11 0LW, United Kingdom

12.1 Introduction

Ion traps are useful for research in several areas. These include spectroscopy, frequency standards, tests of quantum mechanics, g-factor and mass measurements, collision studies, and lifetime measurements. It is an active field which was recognized in 1989, when Dehmelt and Paul shared the Nobel Prize for Physics with Ramsey. For more information the reader should refer to existing reviews [1-8] and compilations of papers [9-11] or consult experienced specialists directly. This chapter describes traps for ionized atoms (not electrons or other particles) and gives an introduction to some experimental techniques.

Traps can provide an almost perturbation-free and nearly field-free environment which is ideal for high-resolution spectroscopy in the microwave and optical regions. Several spectroscopic techniques are covered in a later volume in this series. Laser cooling may be used to substantially reduced Doppler effects. Long interaction times are possible so transit-time effects can be eliminated and high accuracy achieved. Transitions in various trapped ion species are being investigated as references for potential new frequency standards.

It is possible to trap single or a few ions for extended periods to study quantum jumps, photon antibunching, interference, and other quantum-optics effects. Metastable state lifetimes may be measured. Laser-cooled clouds may be condensed into ordered structures (see Section 12.2.1). Under certain conditions the static thermodynamic properties of an ion cloud are like those of a "one-component" plasma (see Section 12.2.4). With buffer gases present, various collision studies have been performed. Low-energy electron transfer to multicharged trapped ions is discussed by Church [12]. At low temperatures ion-molecule reactions have been studied [13].

Electron beam ion traps (EBITs) are the subject of another chapter of this volume. Ion traps have been used extensively as mass spectrometers [14]; ultraprecise mass spectroscopys described in another chapter. Mass and g-factor measurements will not be considered further here. This chapter is divided into sections on types of trap, trap construction, and trapped-ion detection, diagnostics, and cooling.

12.2 Types of Ion Trap

12.2.1 Paul or RF Trap

The Paul trap, in which an RF field is applied between a ring and two endcap electrodes, traps ions in a pseudopotential well and does not use a magnetic field [1]. The RF drive for a Paul trap is typically a few hundred volts amplitude at a few megahertz. It may be provided by a commercial synthesizer or function generator followed by an amplifier and a specially constructed resonant step-up transformer. Ions oscillate in the pseudopotential well with "secular" motion (or "macromotion") at what is known as the secular frequency. Ions can be very well cooled only if they are at the RF center of a Paul trap; ions trapped away from this field-free point experience nonzero RF fields, and the resulting movement is called "micromotion." When more than one ion is trapped, ion–ion collisions may couple energy from the micromotion into the secular motion in a process called "RF heating."

Paul traps are particularly attractive for single-ion optical frequency standards work because of their advantage of a field-free central point; also they can be miniaturized. They have been used with single ions for nonclassical radiation field studies [15]. Many-ion applications include microwave transition frequency measurements and standards development. Dynamical studies of clouds of laser-cooled ions in Paul traps have yielded interesting phase transition effects in which trapped ions crystallize [16]. Lifetime measurements, particularly of long-lived states with lifetimes of many seconds, have sometimes been made by observing decay following laser excitation [7] or by observing quantum jumps [18]. Collisional relaxation studies have been performed by deliberately introducing background gases [19,20]. Many examples of Paul trap applications, in particular spectroscopy and frequency standards work, are referred to in the reviews cited in Section 12.1.

12.2.2 The Paul–Straubel Trap

A very small RF trap based on a single 100-μm ring surrounded by an uncritical grounded electrode structure has been demonstrated [21]. The Paul–Straubel trap requires higher RF voltages than a conventional Paul trap for the same well depth, but offers the attraction of openness and large solid angles for single-ion fluorescence detection. Also, the ring can be heated to high temperatures to drive off unwanted surface contaminants after a single ion has been loaded, so that micromotion can be greatly reduced (see Section 12.4.3). The Paul–Straubel trap may be of interest for single-ion frequency-standard applications.

12.2.3 Linear and Ring Quadrupole Traps

Linear and ring RF traps, in which there is a two-dimensional field-free region along the axis, have now been built by several groups [22–24]. For clouds of ions the average distance from an ion to a node of the trapping fields is much less in a linear trap than in a conventional Paul trap. Thus the average magnitude of the RF trapping field is smaller. This leads to less micromotion and hence a reduced second-order Doppler shift which is an advantage for microwave standards in particular. This is the motivation behind the JPL Hg^+ microwave-standard linear trap work which is buffer gas cooled [22]. Many individual laser-cooled ions may be trapped like a string of beads along the axis of a linear trap [23]. Laser-cooled Yb^+ is also being studied in a linear trap at CSIRO [24]. Linear traps can be very sensitive to stray DC electric fields [25]. Ordered structures and phase transitions have been studied in a quadrupole storage ring [26].

12.2.4 The Penning Trap

This trap combines a magnetic field with a static electric field between ring and endcap electrodes. Ultra-high-vacuum conditions ($<10^{-7}$ Pa) are required to prevent collisional loss. An advantage is the lack of "RF heating" which is observed in Paul traps with clouds of ions. Interesting "one-component" plasma effects have been studied [27]. There are many mass measurement applications using Penning traps. A suitable magnet needs to be selected, and the trap vacuum system must be chosen to fit between the pole pieces; this may limit the trap size in high-field applications. For high-precision or high-field work, superconducting magnets are often necessary.

12.2.5 Other Types of Trap

The "combined" Paul–Penning trap [28] is of interest because species of opposite charge may be trapped simultaneously in it, perhaps, for example, to generate antihydrogen. The Kingdon trap (which uses the field around a thin wire) has not been widely used; trapping times are short, typically less than a second [8]. Arrays of "microtraps" have been suggested for optical clocks [29].

12.3 Trap Construction

12.3.1 Trap Electrodes

Traditional Penning and Paul traps both use similar ring and endcap electrodes which are hyperboloids of revolution about the z axis, given by

$$\left(\frac{z}{z_0}\right)^2 - \left(\frac{r}{r_0}\right)^2 = \pm 1.$$

The endcap separation is $2z_0$ and r_0 is the minimum ring radius. These electrodes set up a purely quadratic electrostatic potential inside the trap when a voltage is applied between the ring and the endcaps. The choice of $r_0^2 = 2z_0^2$ is discussed in Wineland et al. [3]. Figure 1 shows a view and cross-section of trap electrodes. Figure 2 shows two views of the National Physical Laboratory (NPL) Sr$^+$ trap.

Traps with a wide range of r_0 from a fraction of a millimeter to several centimeters have been constructed. Larger traps are generally used for clouds of ions. Single ions have been trapped with r_0 in traps as large as 5 mm [15,30], but usually small or miniature traps are used, particularly when very tight confinement is required to reach the "Lamb–Dicke" regime [31]. In this regime ions are confined in a region smaller than the wavelength of the radiation which is interrogating it; thus, Doppler broadening is effectively eliminated (see Section 12.4.5).

A number of groups use electrodes for RF and Penning traps which are easier to fabricate and have a more open structure for illuminating and viewing ions than the traditional hyperbolic shapes. The resulting ion

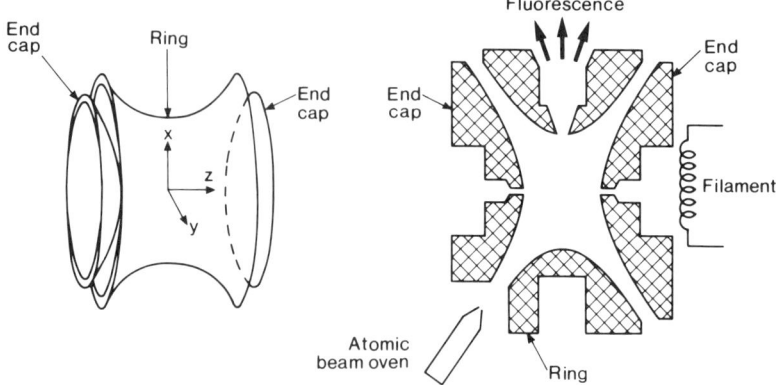

FIG. 1. View and cross-section of ion trap electrodes.

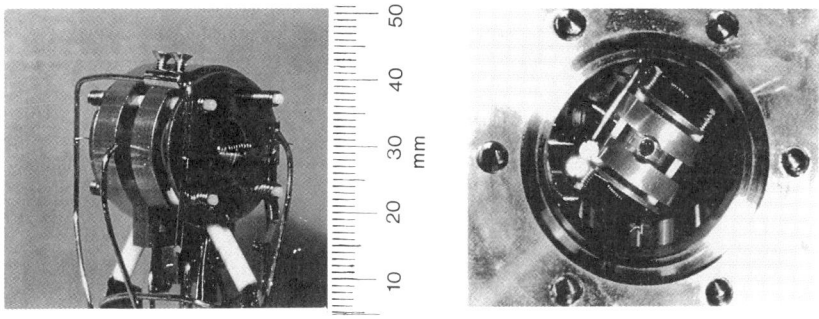

FIG. 2. Photographs of the NPL Sr$^+$ trap.

motion for an RF trap is no longer truly harmonic, and the pseudopotential contains higher order terms in r^2 and z^2. Reference [32] shows that these effects are not very significant for hemispherical electrodes, particularly close to the trap center; they are more serious for a cylindrical trap. Beaty [33] describes other simple electrodes that are cones of revolution in which high-order terms are minimized; one such design was used successfully at NIST [34]. Serious anharmonicities may be introduced by misalignment of electrodes. NRC use spherical electrodes in which the endcap radii are exactly the same as the ring electrode for ease of alignment [19].

Linear traps [22–24] consist of four parallel rods arranged with their centers on the corner of a square, with diameters and spacings of typically a few millimeters. A driving RF is applied to the electrodes so that nearest neighbors have opposite polarity. Large numbers of ions may be confined axially by central DC-biased electrodes [22,24] separated by 60 or 80 mm (e.g., see Figure 3). Alternatively further sets of capacitively coupled outer rods which are DC biased with respect to a 2.5-mm-long central set have been used for axial confinement; a simpler design using segmented rods has also been developed [23]. Both these last two arrangements allow a laser beam to pass down the axis of the trap. In a ring trap, electrodes with a cross-section similar to those described above are bent to form a closed loop. A published example had a ring diameter of 115 mm and a 2-mm gap between electrodes [26].

Trap electrodes have been made out of molybdenum, OFHC copper, beryllium–copper (BeCu), tellurium–copper (a hard alloy more machineable than BeCu), stainless steel, tungsten, and tantalum (easier to work than tungsten). Materials are chosen for their machinability, vacuum properties, magnetic properties, and susceptibility to surface contact potentials. Electrodes may be gold plated or polished to improved surface

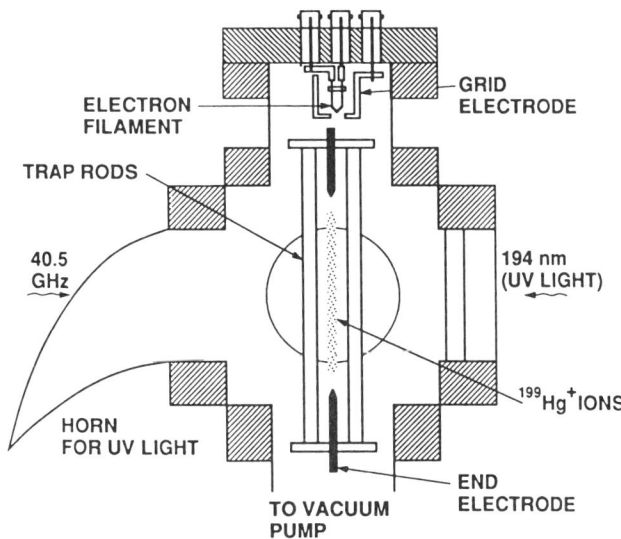

FIG. 3. A JPL linear ion trap assembly residing in its high-vacuum enclosure. State selection light from an ^{202}Hg discharge lamp enters from the right, is focused on to the central $\frac{1}{3}$ of the trap, and is collected in the horn. Fluorescence from the trapped ions is collected in a direction normal to the page (taken from Prestage et al. [22]).

properties. Macor or alumina is most commonly used as an insulating spacer material; Macor is machineable and both have good vacuum properties. Many traps are constructed by spot welding, although barrel-type screw connectors can be used.

12.3.2 Oven and Filament Design

To "load" or trap ions, atoms are usually ionized within a trap by an electron beam. The desired atoms may be simply present in the background gas; if not, an oven is heated to release atoms. Several oven types have been developed: at NPL, for example, a rolled tube of tantalum foil 25 or 50 μm thick is pinched flat at one end and filled with material. The tube is then flattened at the other end and folded, and an exit hole is pierced with a pin. The ends of the tube are spot welded to supporting wires using folded pieces of 100-μm-thick constantan foil to prevent overheating of the leads. Heating is achieved by passing up to a 10-A current directly through the oven tube, which is enclosed in an alumina cylindrical sleeve. At NPL this method has been used to produce Mg$^+$ [35], Yb$^+$ [36], and Sr$^+$ [30]. Variations are in use elsewhere [25,37,38].

Some groups use a Pt–Ir filament arrangement: a pellet of the element required is spot welded to the filament which is heated electrically to load ions, e.g., for Yb^+ [20]. At NIST 50-μm-diameter beryllium wire is heated by a 75-μm-diameter tungsten wire to load Be^+ [39]. Another approach is indirect heating of a crucible or oven containing the desired materials; molybdenum or beryllium–copper crucibles were used as NIST for Mg production [39]. Some species such as Ba and Sr are made by heating an oven containing a mixture of an aluminum compound and metal power such as nickel [37,38].

Ionizing electrons may be generated using a biased heated filament. At NPL a home-made filament wound from 100-μm-diameter tungsten wire is mounted behind a small hole in an endcap (see Figure 2). In some cases ionizing electrons are produced by the oven [37], or ions may be created directly by surface ionization [40].

A procedure for loading ions into a trap needs to be developed for each trap and ion species. Typical parameters used at NPL are mentioned below. The oven current is usually adjusted so that the time for it to heat up sufficiently is about a minute. The filament is then fired for a few seconds with a few microamperes of electron current. To load a very few or single ions the oven and filament currents are reduced and the filament bias may even be pulsed at intervals so that a check for an ion signal can be made after each attempt.

12.3.3 Vacuum System Requirements

Generally a base vacuum of few 10^{-8} Pa is achieved using an ion pump. Good vacuum is particularly required for trapping a few ions, and usually stainless steel or glass vacuum systems are used. A careful bakeout is generally necessary. To bake, the entire high vacuum apparatus should be heated typically to 200°C for several hours or even days. The bakeout temperature is determined by the properties of the various feedthroughs, windows, and oven material. During the bakeout filaments and ovens need to be "degassed" by turning them on briefly. The exact procedure may need to be developed by trial and error. Vapors emitted during bakeout are pumped away into a "dirty" pump. Although another ion pump is sometimes used, access to a turbopump is preferable for this process. A turbopump is also convenient if buffer gas is to be admitted for thermal cooling or collision studies.

A good electrical feedthrough is needed on which to mount the trap electrodes, ovens, and filaments. Various commercial products are available, but some groups make their own. One concern is the use of magnetic materials in feedthroughs which use Kovar seals. The choice of feed-

356   ION TRAPS

through conductor material may be important: copper, tungsten, stainless steel, tantalum, and nickel have been used. Vacuum windows and view ports may be needed to admit laser beams and view fluorescence; a variety of products are available for metal systems. Window material wavelength-transmission characteristics need to be considered, together with the material's flatness. Brewster or antireflection coated windows may be essential to reduced scattered laser light at laser windows. A leak valve should be fitted if work with buffer gases is planned.

The trap pressure is most easily measured by an ionization gauge, but a number of groups rely on measuring ion pump currents. One technique used to monitor very low ion pump currents is to insert a capacitor in parallel with a neon indicator bulb in the supply line; the rate of flashing is proportional to the current. The background pressure in a trap may be reduced by cryogenic cooling; systems for frequency standards work are currently being constructed as NIST [39].

12.4 Trapped Ion Detection, Diagnostics, and Cooling

12.4.1 Electronic Detection of Trapped Ions

Oscillating trapped ions may be detected electrically by measuring image currents induced in the trap endcaps. In the passive "bolometric" scheme [41] the current induced in a resistor between the trap endcaps is measured. Active detection techniques have been used but few detailed circuits have been published: an additional fixed RF frequency, ω, close to one of the ion's secular oscillation frequencies, ω_z or ω_r (see [3] for definitions) is applied to an endcap while a DC sweep voltage is applied to the ring to change ω_z or ω_r. As the ions' oscillation frequency passes through ω, the ions absorb energy; signals can be detected using a weakly coupled resonant circuit followed by amplification and demodulation. An example is shown in Figure 4 [42,43]. This technique can be used to identify different ion species in a trap. A "sum frequency" $\Omega + \omega$ detection scheme is described by Dehmelt [2]. Ions may be detected by ejection from a trap into an electron multiplier; this technique is often used in mass spectrometry [14]. Resonant ejection may be used to measure reaction rate constants [44].

12.4.2 Fluorescence Detection

Ions are often detected by observing fluorescence from a resonance transition which is driven by a lamp or a laser. Using a suitable optical system, the center of a trap may be viewed between the electrodes through

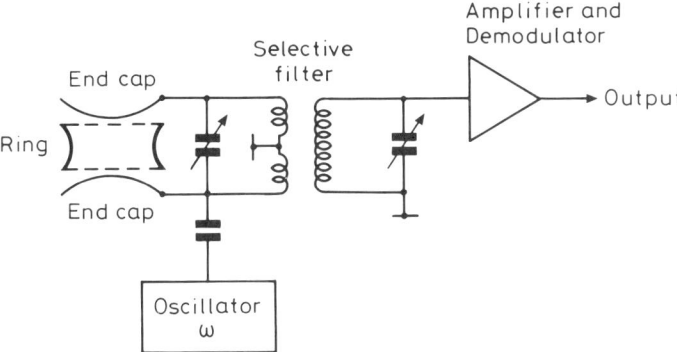

FIG. 4. Electronic detection schematic (taken from Ifflander [42]).

a special hole cut in the ring (see Figures 1 and 2) or through a mesh which is part of an endcap electrode [15].

Sometimes lenses are mounted inside the vacuum system [15,20], but they cannot easily be adjusted. Usually, after a suitable vacuum window, a spherical mirror or lens system is used to image light onto a photon detector. Near-diffraction-limited optics may be used for high resolution, and customized sytems may be needed; camera lenses may be employed for visible work. Magnification factors from 1× to 10× have been used for photomultiplier work. Pinholes or apertures are often used to shield scattered light in the image plane, whereas electrodes and other surfaces may be blackened with graphite or Aquadag [45]. Higher magnification factors up to 60× have been used for position-sensitive photon counting detectors which themselves have limited spatial resolution, typically between 50 and 100 μm. Apertures are also often used to reduce scattered light from incoming laser beams (e.g., NRC). Bandpass filters may be used to exclude background light. Two linear trap lens systems are described in References [23] and [24]; the JPL linear trap uses a 6-in. spherical mirror to image fluorescence [22].

12.4.3 Correlation Techniques

When laser-induced fluorescence is being observed, information about trapped ion motion may be gained by studying photon–photon correlation and, in a Paul trap, RF–photon correlation. The equipment needed to gather these data is a time-to-amplitude converter (TAC), a multichannel analyzer, and a photomultiplier tube. Photon–photon correlation may be used to determine ion oscillation frequencies by taking the Fourier transform of the correlation signal. This technique has been applied successfully in a combined trap [46]. If the DC center of a Paul trap is not the same as the RF center, a single laser-cooled trapped ion would be displaced from the RF center and would experience micomotion (see Section 12.2.1). An RF–photon correlation signal may be used to measure the extent of this micromotion. Often this happens when contact potentials build up due, for example, to a buildup of contaminants (deposits from the atomic beam) on trap electrodes. Voltages may be applied to the ring, endcap electrodes, or specially positioned "compensation" electrodes to move the DC center of the trap and minimize residual micromotion [47,48]. It is a particularly useful diagnostic for laser-cooled single-ion high-resolution spectroscopy experiments.

12.4.4 Cooling Trapped Ions

Two methods of cooling trapped ions are in common use: laser and buffer gas cooling. Laser cooling is the subject of a chapter in a later volume of this series. It is most commonly employed in experiments on a few or single ions for which subkelvin temperatures are required, and it has been used in both Penning and Paul traps [49]. Ions are loaded in a trap with the cooling laser tuned typically a few tens of megahertz below the line center of the resonance transition. In other applications, such as microwave transition studies, many ions need to be trapped to get a sufficiently strong signal; often buffer gas is used to cool and counter the effect of RF heating in a Paul trap. A light gas such as helium is bled into the trap region using a leak valve to give pressures of up to 10^{-2} Pa. Buffer gas may also be used to quench metastable states. Often it is useful to cool the trap with a buffer gas when first loading while parameters are being optimized.

12.4.5 Temperature Determination of Trapped Ions

Ion cloud temperatures may be obtained by measuring the width of Doppler-broadened lines. This is a technique commonly used for laser-cooled ions. Care is needed in the lineshape interpretation; often, because laser heating occurs above the line center, only half of the traditional lineshape is observed. Natural broadening is often significant, and a Voigt profile needs to be considered. Narrow transitions in very cold single ions which are in the Lamb–Dicke regime exhibit sidebands about a central carrier; the relative heights of these sidebands to the carrier may be measured to determine the effective temperature of the ion [50]. Alternatively the temperature of a cloud of ions may be determined using the "bolometric" technique [41], from the size of the noise voltage in a resistor across the trap endcaps. Another approach to temperature determination is to observed the spatial extent of the ion cloud in the trap pseudopotential well [36]. One may also extract ions from a trap and transport them through a well-defined system and observe the time structure of the pulse the ion cloud produces at a channelplate detector, from which the temperature may be computed [51].

12.4.6 Radiation Sources for Exciting and Probing Trapped Ions

Examples of lasers which have been used to excite trapped ion transitions are given in the references and reviews previously cited. In some species (e.g., Hg^+ [22]) lamp sources have been used to excite the resonance transition. The availability of particular laser wavelengths has often been a strong reason for studying a particular species. A chapter in a later volume should be consulted for details of specific lasers and various nonlinear mixing techniques. In general there is a move away from dye lasers and schemes involving high-power ion lasers to solid-state devices [52]. For frequency standards there is a need for very narrow stable lasers (see a chapter in a later volume). Zerodur [34] or ultra-low-expansion (ULE) cavities may be used to stabilize such lasers.

Acknowledgments

Many ion trappers kindly helped by answering questions; I also thank my colleagues at NPL and RC Thompson for advice and comments on the manuscript.

References

1. H. G. Dehmelt, *Adv. At. Mol. Phys.* **3**, 53 (1967).
2. H. G. Dehmelt, *Adv. At. Mol. Phys.* **5**, 109 (1969).
3. D. J. Wineland, W. M. Itano, and R. S. Van Dyck, Jr., *Adv. At. Mol. Phys.* **19**, 135 (1984).
4. P. E. Toschek, in *New Trends in Atomic Physics* (G. Grynberg and R. Stora, eds.), pp. 381–450. North-Holland Publ., Amsterdam, 1984.
5. G. Werth, *Metrologia* **22**, 190 (1986).
6. R. C. Thompson, *Meas. Sci. Technol.* **1**, 93 (1990).
7. R. Blatt, P. Gill, and R. C. Thompson, *J. Mod. Opt.* **39**, 193 (1992).
8. R. C. Thompson, *Adv. At. Mol. Phys.* **31**, 63 (1993).
9. J. C. Bergquist, J. J. Bollinger, W. M. Itano, and D. J. Wineland (eds.), *Trapped Ions and Laser Cooling I, NBS Tech. Note* (U.S.) **1086** (1985); *Trapped Ions and Laser Cooling, II NIST Tech. Note* **1324** (1988); *Trapped Ions and Laser Cooling III, NIST Tech. Note* **1353** (1992).
10. Proceedings of Workshop and Symposium on the Physics of Low-Energy Stored and Trapped Particles, *Phys. Scr.* **T22** (1988).

11. Special issue on *Physics of Trapped Ions, J. Mod. Opt.* **39**, 193–433 (1992).
12. D. A. Church, *Phys. Scr.* **T22**, 164 (1988).
13. S. E. Barlow, J. A. Luine, and G. H. Dunn, *Int. J. Mass Spectrom. Ion Processes* **74**, 97 (1986).
14. J. L. F. Todd, *Mass Spectrom. Rev.* **10**, 3 (1991).
15. F. Diedrich and H. Walther, *Phys. Rev. Lett.* **58**, 203 (1987).
16. F. Diedrich, E. Peik, J. M. Chen, W. Quint, and H. Walther, *Phys. Rev. Lett.* **59**, 2931 (1987).
17. C. Knab-Bernarbini, H. Knab, F. Vedel, and G. Werth, *Z. Phys. D* **24**, 339 (1992).
18. W. M. Itano, J. C. Bergquist, R. G. Hulet, and D. J. Wineland, *Phys. Rev. Lett.* **59**, 2732 (1987).
19. A. A. Madej and J. D. Sankey, *Phys. Rev. A* **41**, 2621 (1990).
20. A. Bauch, D. Schnicr, and C. Tamm, *J. Mod. Opt.* **39**, 389 (1992).
21. N. Yu, W. Nagourney, and H. G. Dehmelt, *J. Appl. Phys.* **69**, 3779 (1991).
22. J. D. Prestage, R. L. Tjoelker, R. T. Wang, G. J. Dick, and L. Maleki, *IEEE Trans. Instrum. Meas.* **42**, 200 (1993); J. D. Prestage, R. L. Tjoelker, G. J. Dick, and L. Maleki, *J. Mod. Opt.* **39**, 221 (1992).
23. M. G. Raizen, J. M. Gilligan, J. C. Bergquist, W. M. Itano, and D. J. Wineland, *J. Mod. Opt.* **39**, 233 (1992); *Phys. Rev. A* **45**, 6493 (1992).
24. P. T. H. Fisk, M. A. Lawn, and C. Colins, *Appl. Phys. B* **57**, 287 (1993).
25. P. T. H. Fisk, private communication (1993).
26. I. Waki, S. Kassner, G. Birkl, and H. Walther, *Phys. Rev. Lett.* **68**, 2007 (1992).
27. S. L. Gilbert, J. L. Bollinger, and D. J. Wineland, *Phys. Rev. Lett.* **60**, 2022 (1988).
28. D. J. Bate, K. Dholakia, R. C. Thompson, and D. C. Wilson, *J. Mod. Opt.* **39**, 305 (1992).
29. R. G. Brewer, R. G. DeVoe, and R. Kallenbach, *Phys. Rev. A* **46**, R6781 (1992).
30. G. P. Barwood, C. S. Edwards, P. Gill, H. A. Klein, and W. R. C. Rowley, *Opt. Lett.* **18**, 732 (1993).
31. R. H. Dicke, *Phys. Rev.* **89**, 472–473 (1953).
32. Chunn-Sing O and H. A. Schuessler. *Int. J. Mass Spectrom. Ion Phys.* **35**, 305 (1980).
33. E. C. Beaty, *Appl. Phys.* **61**, 2118 (1987).
34. J. C. Bergquist, W. N. Itano, F. Elsner, M. G. Raizen, and D. J. Wineland, in *Kinetic Effects on Atoms, Ions and Molecules* (L. Moi, S. Gozzini, C. Gabbanini, E. Arimondo, and F. Strumia, eds.), p. 291. ETS Editrice, Pisa, 1991.
35. R. C. Thompson, G. P. Barwood, and P. Gill, *Appl. Phys. [Part] B* **B46**, 87 (1988).
36. H. A. Klein, A. S. Bell, G. P. Barwood, and P. Gill, *Appl. Phys. [Part] B* **B50**, 13 (1990); A. S. Bell, Ph.D. Thesis, University of London (1992).
37. J. D. Sankey and A. A. Madej, *Appl. Phys. [Part] B* **B49**, 69 (1989).
38. G. R. Janik, Ph.D. Thesis, University of Washington, Seattle (1984).
39. D. J. Wineland, private communications (1992, 1993).
40. Huang Guilong, private communication (1992).
41. H. G. Dehmelt and F. L. Walls, *Phys. Rev. Lett.* **21**, 127 (1968).
42. R. Ifflander, Diplomarbeit, Mainz (1976).
43. J. Yoda and K. Sugiyama, *Jpn. J. Appl. Phys.* **31**, 3744 (1992).
44. F. Vedel and M. Vedel, *J. Mod. Opt.* **39**, 431 (1992).

45. J. D. Prestage, private communication (1993).
46. K. Dholakia, G. Zs. K. Horvath, D. M. Segal, R. C. Thompson, D. M. Warrington, and D C. Wilson, *Phys. Rev. A* **47,** 441 (1993); D. C. Wilson, Ph.D. Thesis, University of London (1992).
47. I. Siemers, M. Schubert, R. Blatt, W. Neuhauser, and P. E. Toschek, *Europhys. Lett.* **18,** 139 (1992).
48. A. A. Madej, K. J. Siemsen, J. D. Sankey, R. F. Clark, and J. Vanier, *IEEE Trans. Instrum. Meas.* **IM-42,** 234 (1993).
49. R. Blatt, in *Fundamental Systems in Quantum Optics, Summer Sch. Les Houches* (1990).
50. J. C. Bergquist, W. M. Itano, and D. J. Wineland, *Phys. Rev. A* **36,** 428 (1987).
51. M. D. N. Lunney, F. Buchinger, and R. G. Moore, *J. Mod. Opt.* **39,** 349 (1992).
52. G. P. Barwood, C. S. Edwards, P. Gill, H. A. Klein, and W. R. C. Rowley, *SPIE Proc.* **1837** (Freq. Stabil. Lasers Their Appl.), 271 (1992).

13. ULTRA-HIGH-RESOLUTION MASS SPECTROSCOPY IN PENNING TRAPS

Robert S. Van Dyck, Jr.

Department of Physics, University of Washington, Seattle, Washington

13.1 Introduction

The basic principle of this mass spectrometer is the same as that of all other high-precision mass spectrometers. A region of highly uniform magnetic field provides an ideal volume in which a charged particle may revolve in an orbit perpendicular to the field, B_0, at a measurable cyclotron frequency $\nu_c = \omega_c/2\pi$. The mass-to-charge ratio, m/q, for charge $q = ne$ is extracted from this frequency by applying

$$\omega_c(X^{n+}) = \frac{neB_0}{m}, \tag{13.1}$$

where X^{n+} represents element X with n electrons removed. However, for ultra-high-precision measurements, B_0 cannot be directly measured and thus must be calibrated relative to some other charged particle, preferably a carbon ion since the resulting ratio would directly yield atomic masses when suitably corrected for lost electrons and binding energies. The assumption which is implicit in any mass ratio comparison is that n is an integer and the quantum of charge is identical for all electrons, free or bound within atoms.

The spectrometer described in this chapter is characterized by its ideal environment which is made free of unwanted collisions with background gas by immersing the entire apparatus into a liquid helium bath. It is also free of Coulomb interactions from like charges since the sample is generally trapped as a single ion. In addition, the effects of the electric trapping fields are well understood and can be removed from the measurement of the cyclotron frequency. In fact, it is even possible to reshape both the electric and the magnetic fields in order to further reduce systematic errors associated with the finite energies that exist in each normal mode. All of these things help the Penning trap mass spectrometer (PTMS) have extremely high accuracy [1–7]; eventually, this accuracy will exceed 1 part in 10^{10}.

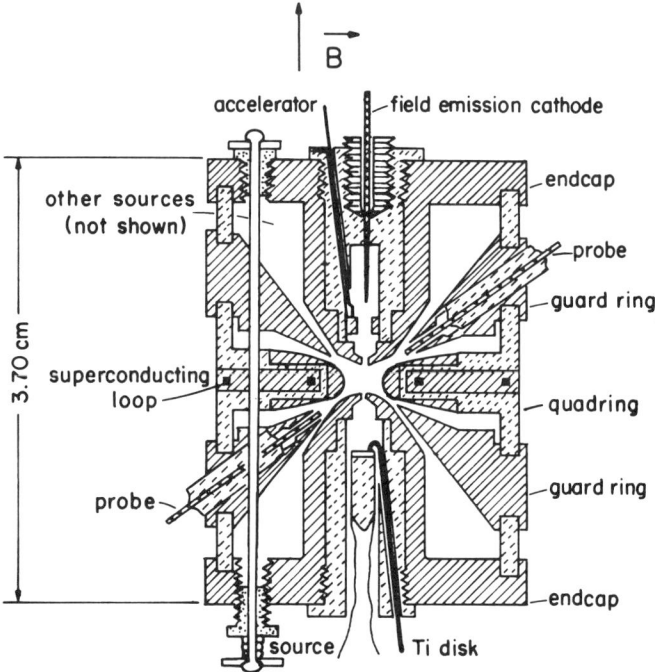

FIG. 1. Detailed scale drawing of the UW-quadring Penning trap. The main electrodes (i.e., endcaps, guard rings, and the quadring) are made of OFHC copper. Macor (a machineable glass) is used extensively in the trap, for instance, as the insulating spacer rings and as the substrate for the quadring electrode. In addition, the trap includes probes for sideband cooling (tungsten in glass), a superconducting loop (NbTi) for magnetic field diagnostics, a field emission cathode (sharp tungsten rod in Macor holder) to generate an axial electron beam, and several sources of hydrogen isotopes (titanium disks on Macor) (from Moore et al. [31]).

Figure 1 is an example of the compensated [8] Penning trap used in this type of spectrometer. A description of this device can be found in the following section. When properly biased, a charged particle will be electrostatically bound along the axis of symmetry with a harmonic oscillation frequency, $\nu_z = \omega_z/2\pi$. Radial confinement is due to the large axial magnetic field which is required to compensate for the nonbinding radial electric field. The price one pays for this confinement is the magnetron drift which corresponds to the $\vec{E} \times \vec{B}$ motion or guiding-center rotation at frequency $\nu_m = \omega_m/2\pi$. Likewise, the moving charge now sees the magnetic field from a rotating frame of reference, thus producing an observable (trap-dependent) cyclotron frequency given by $\omega_c' = \omega_c - \omega_m$. In addition, from the equations of motion for the bound charge [9,10], one can also determine the ideal magnetron frequency to be $\omega_m = \omega_z^2/2\omega_c'$, where generally $\omega_m \ll \omega_c'$. Therefore in principle, one need determine only ω_z and ω_c' in order to recover the free space cyclotron frequency, ω_c. However, there will always be a small misalignment between the electric and the magnetic axes as well as possible asymmetries in the trapping electrodes (due to construction). Thus, if one can measure all three normal mode frequencies for a trapped ion, then the following quadrature equation [11],

$$\omega_c^2 = (\omega_c')^2 + (\omega_z)^2 + (\omega_m)^2, \tag{13.2}$$

is found to be invariant to the perturbations just described.

In order to review the PTMS, the UW instrument will be extensively described in Sections 13.2–13.8. However, before summarizing the perturbative limitations of the Penning trap mass spectrometers in Section 13.11, it will be appropriate to first describe alternative detection methods because of the added versatility inherent in such diversity and also because each will have a different set of limitations.

13.2 Trap Construction and Confinement

The primary objective in machining a Penning trap for high-precision mass spectrometry is to produce an axial confinement potential which is harmonic over as large a volume as possible. This generally requires the electrodes to be members of the following family of ideal equipotentials (i.e., hyperboloids of revolution in cylindrical coordinates):

$$V = \frac{V_0}{2}\left(C_0 + \frac{z^2 - \rho^2/2}{d^2}\right). \tag{13.3}$$

Here, V_0 is the applied potential difference between the ring and the endcap electrodes and d is the characteristic dimension of the trap given by

$$2d^2 = Z_0^2 + R_0^2/2, \tag{13.4}$$

where $2R_0$ is the minimum ring diameter and $2Z_0$ is the minimum endcap separation. The constant C_0 is an unobservable parameter which depends on exactly where the potential is applied. Since the trap is usually operated with endcaps grounded, $C_0 = -Z_0^2/d^2$; also, $-V_0$ is applied to the ring for positive ions (i.e., V_0 remains a positive quantity). As a result, the axial normal mode has an angular frequency given by

$$\omega_z^2 = \frac{qV_0}{md^2}. \qquad (13.5)$$

Characteristic values for R_0 and Z_0 for the three high-precision spectrometers being reviewed are given in Table I along with typical values for V_0 and the corresponding trap constants, defined as $\nu_z/\sqrt{V_0}$ for protons.

To be useful, these electrodes require truncation of the hyperbolas, holes to be placed in the endcaps for the ionizing electron beam, and possible slots/cuts to be made in some of the electrodes for various reasons. In general, such modifications will inject only even-order perturbations to the potential distribution if made with cylindrical and reflection symmetry. To minimize their effects, a set of "guard" rings are placed in the truncation region (see Figure 1) and biased with a voltage that makes the axial frequency [given by Equation (13.5)] as independent of the axial energy as possible. A single set of guard or compensation electrodes can typically improve the axial resolution by a factor of 100 or more. This was particularly important in the small "quadring" trap [12] shown in Figure 1, in which the main ring electrode was split into four equal quadrants in order to directly excite/detect the ion's cyclotron motion. This design was later replaced by a 2× larger trap without a quadring (but containing split guards).

In the past, a hyperbolic surface was carefully fabricated from molybdenum, copper, or phosphor bronze using a special "forming" tool containing the desired contour which is formed by a lathe in a single cut. Getting both the tool and the final electrode "right" (i.e., to within 2.5 µm) was an extremely long and time-intensive process. To get around this major effort, simpler electrode geometries were introduced such as the cylindrical Penning trap [13] and the conical Penning trap [14]. In principle, any of these other geometries yield the same potential distribution near the trap center. However, in order to significantly increase the size of the working volume, more than one set of compensation electrodes is now required. All such geometries are trying to approximate the necessary hyperbolic equipotentials and thus could not *ever* be superior to starting with properly machined hyperbolic electrodes. Now with the introduction of CNC (computer numerically controlled) lathes, these hyperbolic electrodes can be fabricated using point-contact tools to greater precision than possible by older methods in about an order of magnitude less time.

The choice of material was driven by our desire to keep the associated magnetic susceptibility as low as possible, since a nonzero value would yield a residual quadratic dependence (so-called B_2 term) to the magnetic field [15]. Making large traps greatly reduces this problem, but it also

TABLE I. Experimental Parameters for High-Precision Penning Trap Mass Spectrometers

Symbol	Description	Units	UW spectrometer[a]	Mainz spectrometer	MIT spectrometer
T_A	Ambient temperature	K	4.2	300	4.2
$\omega_z/2\pi$	Axial frequency	MHz	4–5	1.14 (H^+)	0.16
B_0	Magnetic field	kG	58	58.3	85
B_1	Linear B-field	G/cm	<0.05	0.05	<0.1
B_2	Quadratic B-field	G/cm^2	1–2	~0.09[b]	0.1
V_0	Ring-endcap potential	V	20–80	10	5–10
$2Z_0$	Minimum endcap separation	mm	4	8.4	12
$2R_0$	Minimum ring diameter	mm	5	12	14
TC	Trap constant for p^+	MHz/$\sqrt{V}$	0.74	0.37	0.28
Q	Tuned circuit quality factor		~1000	—	~32,000
C	Tuned circuit capacitance	pf	~18	—	100
R	Tuned circuit resistance	MΩ	~2	—	~150
C_4	Anharmonic coefficient		~0.00003	~0.0001	≤0.00005
			Without ν_c' excitation		
kT_z	Axial thermal energy	eV	~0.001	~0.3	~0.001
E_c	Equilibrium cyclotron energy	eV	<0.01	<0.05	≤0.05
$\|E_m\|$	Equilibrium magnetron energy	eV	≤0.001	~0.001	~0.00001
			With ν_c' excitation (relative to a given ion)		
			C^{6+}	H^+	N_2^+
Z_a	Axial amplitude	μm	~5	~1000[b]	~50
R_m	Magnetron radius	μm	<10	150–400	~10
R_c	Cyclotron radius	μm	~30	~25[b]	~200

[a] For a trap 2× larger than the "quadring" trap shown in Figure 1.
[b] Estimated values.

vastly reduces the coupling to the ion's motion as described in Section 13.3. Because of the need to compromise on the size, copper is an excellent choice because of its very low susceptibility; however, it is also relatively soft and easily distorted at the 2.5-μm tolerance level. Thus, phosphor bronze has been used in the 2× larger UW trap as a compromise between hardness and susceptibility.

The vacuum envelopes also require careful design. The present choice consists of an all-metal envelope made of BeCu, with copper pin bases containing ~20 ceramic-to-metal feedthroughs [16,17]. The vacuum seal to the pin base is made via an indium corner seal with a threaded BeCu ring that applies the required spring force to the indium. The important procedure in this process is a prewetting of the indium (with solder rosin) to all mating surfaces (as well as a thorough solvent cleaning) prior to the final press. The entire envelope (plus trap) is then baked on a vacuum system at 130°C for about two days (or until the internal pressure has reached ~1 × 10^{-8} Torr).

13.3 Basic Axial Resonance

For the UW PTMS, the axial resonance provides all the information concerning the radial energy states and thus requires the highest possible resolution. Currently, this spectrometer can achieve an axial frequency resolution, $R_z = \Delta\omega_z/\omega_z \sim 10^{-8}$, which then requires the voltage source to be stable to 20 ppb. To achieve this, it was necessary to minimize the effects of thermal–electric potentials generated between the liquid-helium-cooled trap and the room temperature cable header located on top of the cryostat. Typically, one can observe thermal EMFs of ~1.5 mV on all cryogenic lines that connect to the trap; to minimize variations in this voltage, nearly identical coaxial cables connect to the ring electrode and the voltage source's local ground, located on the liquid-helium-cooled pin base. To further stabilize the thermal EMFs, the liquid helium level above the trap envelope is stabilized in position to within ~0.003 cm. This is accomplished by using an annular bore-tube dewar system which has the outer annulus overpressurized by ~10^3 Pa relative to the central bore tube that connects into a large liquid helium reservoir. A capacitive sensor and voltage-controlled valve are operated in a feedback loop to control this level.

As for the actual voltage source, no commercial instrument yet exists that can be varied from 10 to 80 V in microvolt steps and likewise maintain a stability of 2 × 10^{-8} over the entire range. The present supply for the UW PTMS consists of ~80 Weston-style unsaturated standard cells (EMF,

~1.019 V/cell) that are located inside an oven enclosure, held near 36°C (corresponding to a local minimum in voltage versus temperature). The internal resistance per cell is ~ 500 Ω, and the corresponding rms thermal noise is ≲30 nV/$\sqrt{\text{Hz}}$ for all 80 cells. In a 1-Hz detection band, this would represent a thermal noise limit of 0.4 ppb and thus is not the source of the voltage noise. The actual limitation is believed to be due to fluctuations in density within the electrolyte, possibly caused by bubbles on the electrode surfaces that break free.

The theoretical model [18] which best describes the axial motion treats the trapped charge as a series lc circuit, in which

$$l = \frac{4mZ_0^2}{(C_1 q)^2} \qquad (13.6)$$

is the inductance of the ion of mass m and charge q. The linear coefficient C_1 is defined by the relationship that relates the driving RF electric field to the difference in RF potential across the endcaps, $E_{RF} = C_1 V_{RF}/2Z_0$, and its magnitude is generally less than unity, corresponding to the attraction of some lines of force onto the ring electrode. To be observable, this lc series combination, which appears across the trap's capacitance, C, must be attached in parallel to an external inductor, L, such that the LC combination is resonant at ω_z. This is shown in Figure 2, along with the effective resistance R of the parallel circuit that provides the damping of the axial motion. As listed in Table I, R is typically 2 MΩ for the UW PTMS and is related to the axial linewidth for a single ion according to $\gamma_{z,1} = R/l_1$ (where $l_1 \sim 10^6$ H for protons in the UW quadring trap). The resistance necessarily should be large (i.e., the tuned circuit should have a high Q)

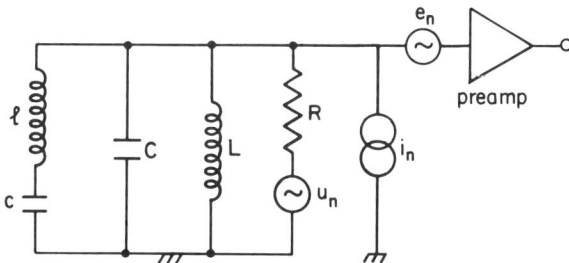

FIG. 2. Input detection circuit, with equivalent lc representing the trapped ion. The trap's capacity, C, is turned out with an external inductor, L, yielding a large parallel resistance, R, with an equivalent noise generator, u_n. The preamp is shown with an equivalent series noise generator, e_n, and a parallel current source, i_n, at the input to the preamp (adapted from Wineland and Dehmelt [18]).

in order that driven current induced in the endcaps can produce a signal voltage which is easily detected (with the minimum possible amplitude) by the preamp shown in Figure 2.

In general, for an applied RF drive of amplitude V_{RF} and angular frequency ω, the induced current for a single ion is given by

$$i = \frac{\omega V_{RF}}{l_1} \frac{\cos(\omega t + \phi_0)}{\sqrt{(\omega_z^2 - \omega^2)^2 + \gamma_{z,1}^2 \omega^2}}, \qquad (13.7)$$

where ϕ_0 is an arbitrary phase. From the available signal-to-noise ratio (S/N) in the UW PTMS, one can show that $i \sim 10^{-14}$ amperes (corresponding to a motional amplitude of $\sim 0.3\%$ of Z_0), and this produces a signal of ~ 20 nV. If $\Delta \nu_{det}$ is the detection bandwidth, then this signal can be compared with the rms noise voltage associated with the resistance R, given by

$$u_n^2 = 4kT_z R \Delta \nu_{det}, \qquad (13.8)$$

which is ~ 22 nV for a 1-Hz detection band and assuming a 4-K ambient temperature. Now, referring back to Figure 2, one can see that if the trap is ideal and the parallel LC circuit is tuned to resonate with the series lc circuit, then the latter totally shorts out the resistor R as well as its equivalent noise generator, u_n. However, the anharmonic terms in the trapping potential keeps the ion from shorting out the trap due to the random frequency fluctuations associated with the thermal noise drive. Thus, u_n is retained and any shorting from the lc circuit would equivalently replace the ambient temperature T_z with a lower effective value.

Also as shown in Figure 2, the input noise of the preamplifier, which is a GaAs field-effect transistor (FET), is modeled with a series noise generator, e_n, and a parallel current souce, i_n. Not shown, the FET has a low input resistance, R_i, and an input capacitance which is used in a impedance-matching capacitive transformer that also steps the signal down to the preamp by n (where $n < 1$). Thus, the input resistance of the FET is reflected back to the parallel tuned circuit as R_i/n^2, and if R_u is the unloaded resistance of the tuned circuit (i.e., without the preamp), then $R^{-1} = R_u^{-1} + n^2/R_i$. The resulting S/N at the FET input can then be shown to become

$$(S/N)^2 = \frac{(i_s R n)^2}{(4kTRn^2 + i_n^2(Rn^2)^2 + e_n^2)\Delta \nu_{det}}, \qquad (13.9)$$

where i_s is the amplitude of the signal current given by Equation (13.7). This optimizes when

$$n_{opt}^4 = \frac{e_n^2 (R_i/R_u)^2}{4kTR_i + (i_n R_i)^2 + e_n^2}. \qquad (13.10)$$

For the FETs used in the UW PTMS, $(e_n/i_n R_i)^2 \ll 1$ (see Richards et al. [19] for typical i_n and e_n), and some noise shorting will cause $4kTR_i$ to become negligible relative to $(i_n R_i)^2$. Thus, $n_{opt} \simeq \sqrt{e_n/i_n R_u}$ and the optimum S/N is found from

$$(S/N)^2_{opt} \simeq \frac{i_s^2 R_u}{(4kT + 2e_n i_n)\Delta\nu_{det}}. \tag{13.11}$$

Finally for typical parameters, an observed $(S/N)_{opt} \sim 1$ for a 1-Hz detection band corresponds to $i_s \sim 10^{-14}$ A as indicated.

13.4 Preparing the Ion Sample

The Penning trap mass spectrometer is not limited to singly charged ions; in fact, it operates better with higher charge states. Also, such ions are relatively easy to obtain since the ionizing electron beam can be reflected back on itself many times upon properly biasing the source in the opposite endcap. To illustrate the advantage of multiply charged ions [20], it can be shown from Equation (13.7) that the on-resonance signal voltage is just $i_s R = V_{RF}$, and for a given axial amplitude, Z_a, this signal becomes

$$V_{RF} = \frac{C_1 q_1 Q}{2C}\left(\frac{Z_a}{Z_0}\right), \tag{13.12}$$

where Q is the quality factor of the tuned circuit. Since the leading anharmonic shift in the axial frequency is proportional to $(Z_a/Z_0)^2$ (see Section 13.8) and it can be shown that driven axial line broadening is proportional to (Z_a/Z_0), it is clear that some characteristic $(Z_a/Z_0)_c$ exists which corresponds to the maximum allowed perturbation that does not significantly reduce the axial frequency resolution. Under these conditions, the maximum allowed signal is directly proportional to q_1, the ion's charge state. Also, the maximum signal can be increased by decreasing trap capacitance C and increasing tuned circuit Q (as illustrated in two notable examples [21,22]). In addition, from Equation (13.6), it follows that the ion's axial linewidth, which represents the strength of the ion's coupling to the external world, is proportional to q_1^2; therefore, the response time and cooling rates are greatly enhanced.

Due to the reflection of the electrons from the field emission cathode, the electron beam expands until it reaches electrode material inside each endcap. As a result, the trapped ion cloud can often contain a random sample of everything cryogenically adsorbed onto the electrode surfaces as well as many allowed charge states (unless the maximum ionizing energy is limited by the use of the accelerator shown in Figure 1). Before any precision measurements can be made, great care must be exercised to throw out all such unwanted ions.

Two methods have been used to eliminate these contaminants. The most practical is the application of a very strong (~0.1 V rms) white-noise broadened RF axial drive [23] which is repeatedly swept from 1 to 10 MHz (exclusive of bands containing ω_z, $2\omega_z$, and $\omega_z - \omega_m$). The ions of interest are tuned onto the resonant LC circuit in order to provide some damping during the collisional or off-resonant heating of their motion. The white-noise broadening ensures that the axial drive will be kept "on resonance" long enough (in spite of strong anharmonic frequency shifts, described in Section 13.8) to throw out the unwanted ions. A second method involves selectively applying intense (~1 V rms) cyclotron drives to the radial excitation electrodes and sweeping repeatedly in a 2-kHz band around each possible contaminant's cyclotron frequency. Because this latter scheme is far too time intensive, it is used only for ions that are protected axially by one of the exclusion bands in the white-noise axial heating scheme. Evidence for any remaining background ions comes primarily from the loss of axial S/N, the inability to observe the axial–magnetron cooling resonance described in Section 13.5, and the broadening of the observed cyclotron resonance.

Finally, in order to verify that a *single* ion is in the trap, return to the series lc model for which N identical charges correspond to the parallel combination of N series lc circuits. From this model, one can easily show that the resonant axial frequency is independent of N, whereas the width of the resonance will be $N\gamma_{z,1}$ since $\gamma_{z,N} \propto 1/l_N$. Likewise from Equation (13.7), the signal voltage on resonance is independent of the number of identical trapped ions, whereas the off-resonance signal (for $\delta\omega \gg \gamma_{z,N}$) is directly proportional to N. This suggests two ways of determining the number of trapped ions. The off-resonance driven signal can be calibrated in amplitude by loading different quantities of ions and observing the smallest unit of signal (corresponding to one ion) of which all other signals are integer multiples (as routinely done with electrons [24]). Or, the axial linewidth can be measured during the same variation in the number loaded, in order to find the smallest linewidth, for which all other linewidths are again integer multiples. The best way to actually reduce any cloud in number is to drop the potential well to less than 1% of its normal value, thus allowing the hotter ions to evaporate from the trap.

13.5 Sideband Cooling Resonances

There are two sideband resonances which are extremely important for this type of spectrometer: $\omega_z + \omega_m$ and $\omega_c' - \omega_z$. (There is in fact a third combination, $\omega'_c \pm \omega_m$, which is used by a group at CERN [25], whose spectrometer is still at a much lower precision [26].) The first sideband is used to center [24] the ion into an easily reproducible position near the trap center, where perturbations due to spatial field inhomogeneities are minimized. The second is crucial for extracting energy from the cyclotron motion, once it has been resonantly absorbed. Both have the axial frequency in common since the axial motion is strongly damped by the detection-tuned circuit. To excite such resonances, one must generate a spatially inhomogeneous RF field driven at the given sideband frequency. The probes shown in Figure 1 were installed for this purpose; however, one quadrant of the ring electrode or one half of a split guard can and has been used to generate these sideband fields, relying on the particle's motion at either ω_m or ω'_c to couple to these sideband drives. The primary effect of mixing the radial motion to the sideband field is to generate a synchronous electric field in the axial direction, oscillating at ω_z.

To further illuminate the role that such drives play in this spectrometer, consider conservation of energy for the axial–magnetron sideband excitation. When a photon of energy, $\hbar(\omega_z + \omega_m)$, is absorbed from the sideband field, $\hbar\omega_m$ is absorbed by the magnetron motion while the remaining part, $\hbar\omega_z$, is harmlessly added to the damped axial motion. Since the magnetron energy is strongly dominated by the negative radial potential hill given by $-m\omega_z^2 R_m^2/4$ (which is much greater than the kinetic energy, $m\omega_m^2 R_m^2/2$), the addition of a positive quantum of energy to the negative hill therefore has the effect of reducing the magnetron radius, R_m.

In the case of cyclotron–axial sideband, it is easiest to picture the cyclotron and axial modes as two simple harmonic motions which become coupled by the sideband field at $\omega'_c - \omega_z$. Thus, the undamped cyclotron motion now picks up the axial damping as well. As shown in Figure 3, this coupling can be observed by monitoring the amplitude of the axial resonance (as described in Section 13.7). Upon approaching $\omega'_c - \omega_z$, the amplitude is first suppressed (since the two drives interfere), followed by a ringing between the two modes as the cyclotron energy exponentially damps out. The equilibrium temperature for this type of cooling is

$$T_c = \left(\frac{\omega'_c}{\omega_z}\right) T_z \sim 20 \text{ K}, \qquad (13.13)$$

where one assumes that $T_z \sim 4$ K, although it may be greater than this due to the preamplifier's possibly higher input temperaure. In practice, the equilibrium between axial and cyclotron motion is achieved in less than 1 min, and the higher-than-ambient equilibrium temperature is not critical as long as it remains constant from measurement to measurement and represents approximately the same initial state for the corresponding

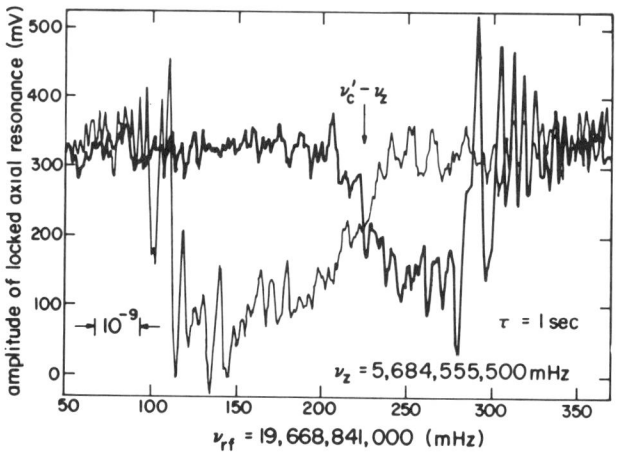

FIG. 3. Cyclotron-axial sideband resonance for a single C^{4+} ion. In addition to the suppression of the axial amplitude, a damped ringing can be seen as the energy couples between axial and cyclotron motions (from Van Dyck et al. [5]).

calibration ion (i.e., one adjusts q to keep $\omega_c' \propto q/m \sim$ constant). In addition, the initial suppression of the axial amplitude in the vicinity of $\omega_c' - \omega_z$ allows one to determine the cyclotron frequency with relatively high precision. Typical full-widths of these suppression-type resonances for $^3He^+$, C^{4+}, and O^{6+} have been observed to be ~ 1 ppb wide. The limitation to their accuracy is the basic uncertainty in the axial frequency, which in fact can be calibrated by comparison with the direct cyclotron resonance at ω_c'.

13.6 Magnetic Field Stability

The PTMS obviously requires a strong magnetic field with great temporal stability in order to fully utilize the advantages of the long-term confinement which is possible in a cryogenic environment. The use of a persistent, monofilament NbTi superconducting solenoid has long been the field of choice, although such magnets can achieve strengths of only about 6 T. Higher fields are available if the superconductor is multifilament Nb_3Sn, although, in the past, stability was not as good because of the difficulty of joining all the filaments. As with all superconductors, the solenoid offers some degree of self-shielding from environmental magnetic noise, such as that produced by moving steel tools, chairs, or equipment in the vicinity of the solenoid. However, this shielding is typically no better than a factor of 5–10, unless specially designed compensation coils are installed [27] which can increase the shielding by more than an order of magnitude.

In spite of such shielding and the great care required to keep the immediate environment free of magnetic disturbances, such solenoids often can be observed to wander by more than 10 ppb/hr. Figure 4 illustrates a very unusual week in 1990 for the UW PTMS, marked by a severe change in weather. The field had a net wander of ~5 ppm, and there was evidence for a diurnal cycle as well. The stimulus for this wander was traced to the large changes in atmospheric pressure, which subsequently affected the rate of evaporation of liquid helium in the experiment cryostat. It is surmised that the temperature dependence [28] of the magnetic susceptibility of all materials that pass through the strong-field region of the magnet is responsible for this effect. When the evaporation rate varies with pressure, the temperature distribution along these materials must change, thus causing the magnetic field to wander. To eliminate this problem for the UW PTMS, the *absolute* pressure within the cryostats is controlled to better than 10^{-3} Torr and the liquid helium boil-off rate is increased in order to maintain a continuous flow over the offending surfaces. Typical long-term field stability is now ~0.2 ppb/hr, with 2–3 ppb wander superimposed with a 3- to 4-hr average duration.

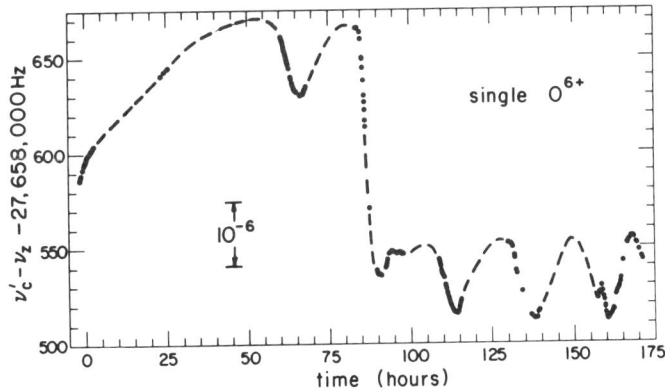

FIG. 4. Uncontrolled magnetic field wander. For approximately a week, the field was occasionally monitored using the cyclotron-axial sideband resonance of a single O^{6+} ion. The oscillation shown on the right-hand side is diurnal and apparently related to daily pressure changes. The dashed line is added as a visual aid (adapted from Van Dyck et al. [5]).

13.7 Frequency-Shift Detector

Figure 5 schematically summarizes the excitation/detection electronics used in the UW PTMS. A particularly useful experimental complication involves the use of a low-frequency modulation signal, applied to the ring electrode, such that

$$V(\text{ring}) = V_0 + V' \cos(\omega' t), \qquad (13.14)$$

where $\omega' = 2\pi(100 \text{ kHz})$ and V' is typically ≤ 10 mV. As a result, the axial frequency is now time dependent and the axial motion will be described by

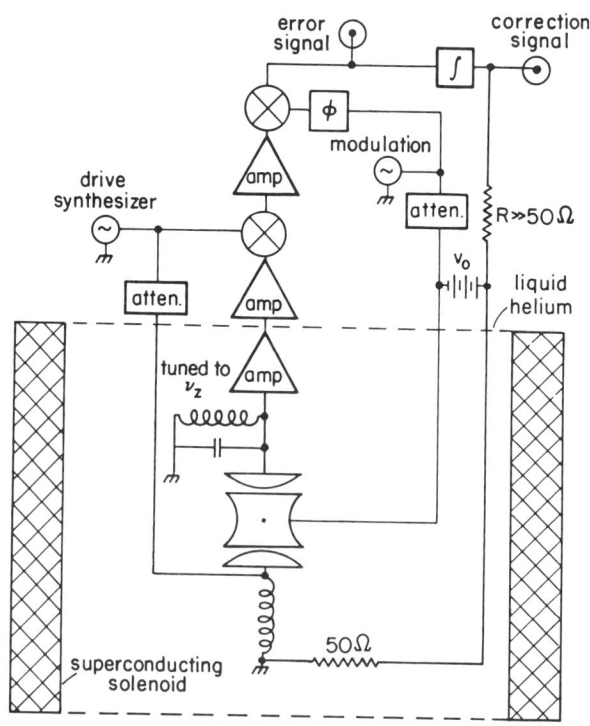

FIG. 5. Overall schematic of the axial detection electronics. The axial motion is driven on a frequency-modulated sideband via an RF drive applied to one endcap, and a liquid-helium-cooled GaAs FET preamp observes the induced currents that flow through the other endcap into a parallel LC circuit, tuned to ν_z. The amplified signal is then mixed to DC to display the phase relation between the drive and the driven axial motion. This phase relation is used as an error signal which is integrated and fed back to the ring voltage to frequency lock the axial motion to the drive source (from Moore et al. [31]).

the instantaneous phase $\phi(t)$, governed by

$$\frac{d\phi}{dt} = \omega_z + m_f \omega' \cos(\omega' t), \qquad (13.15)$$

where ω_z is now the unmodulated carrier frequency and the quantity m_f is referred to as the modulation index, equal to $(\omega_z/\omega')(V'/2V_0)$. In effect, for $m_f \ll 1$, the frequency spectrum consists of an unshifted resonance at ω_z plus the two strongest sideband resonances at $\omega_z \pm \omega'$. By driving on one of these sidebands, the strong drive power is effectively tuned away from the detection channel at ω_z. The response, however, is primarily at the fundamental frequency and is detected on the resonant LC-tuned circuit as usual. When using the upper sideband, the only change for the signal current given in Equation (13.7) is the multiplication by the factor $m_f/2$. The lower sideband has an additional phase shift of 180°.

Recall that the typical signal is 20 nV (see Section 13.3) which corresponds to the effective RF drive voltage applied to the trap. The required 150 dB of attenuation down from the 1-V rms output of a frequency synthesizer (shown in Figure 5) can now be distributed equally between the drive synthesizer and the modulation drive source. It is often wise in any RF work to *not* have too much amplitude or attenuation in any single-frequency band.

Another reason for this modulation scheme is that it fits nicely into the superheterodyne system shown schematically in Figure 5. After the amplifiers detect the driven signal at ω_z, the response is mixed with the drive synthesizer at $\omega_z + \omega'$, thus translating the response down to the 100-kHz frequency range, where it can again be amplified in a different frequency band. The amplifier plus mixer that follows is a phase-sensitive detector (PSD) with its 100-kHz reference derived from the phase-shifted modulation source. The usual advantage of coherent detection is that it allows the driven motion to be detected above the particle's thermal motion (assuming similar amplitudes) simply by setting the detection bandwidth, $\Delta\nu_{det}$, to be narrower than the trapped particle's axial linewidth. An example of the synchronous output is shown in Figure 6. This particular choice of phase is the appropriate one for the integrator whose output is fed back to the local ground of the floating battery. As long as the frequency of the drive synthesizer, plus 100 kHz, equals the ion's resonant frequency, the error (output of the PSD) is zero and the integrator output does not change. However, if something perturbs ω_z, the integrator output changes, producing a current that floats the batteries to a new potential that again zeros the PSD output. In this way, the ion's axial resonance is kept "frequency locked" to the drive synthesizer, and the resulting correction voltage, now referred to as the frequency-shift signal, gives real-time

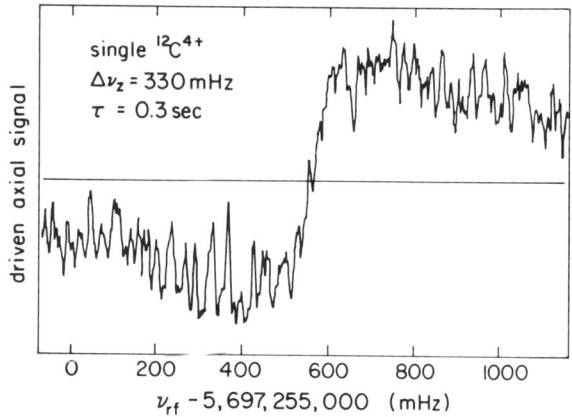

FIG. 6. The axial resonance of a single C^{4+} ion. A phase shift has been injected into the detection system to produce this dispersive shape which is appropriate as an error signal for the frequency lock (from Van Dyck et al. [1]).

information about perturbative changes in the frequency of the axial motion.

13.8 Anharmonic Detection

For the UW PTMS, the frequency shift detector is used exclusively to observe both the cyclotron and the magnetron resonances through the corresponding changes in the radial position of the excited ion. The nature of the coupling mechanism may be magnetic, electrostatic, relativistic, or a combination thereof. For the pioneering g-factor experiments on single electrons [9,24] and positrons [29], the magnetic field was deliberately modified to have a large B_2 quadratic term, resulting in a $\delta\omega_z/\omega_z \sim 2 \times 10^{-8}$ per electron cyclotron quantum level. The second-generation g-2 experiment [30] uses a combination of B_2 and relativistic coupling (about equal contributions in opposite directions) to yield a $\delta\omega_z/\omega_z \lesssim 1 \times 10^{-9}$ per electron cyclotron quantum state. However, for the current spectrometer, the coupling is dominated by the residual anharmonic term [31] given by

$$\Delta V = \frac{V_0 C_4}{2} \left[\frac{8z^4 - 24z^2\rho^2 + 3\rho^4}{8d^4} \right], \quad (13.16)$$

where C_4 is the coefficient which describes the strength of this perturbation [32]. The z^4 term induces a shift in the axial frequency proportional to

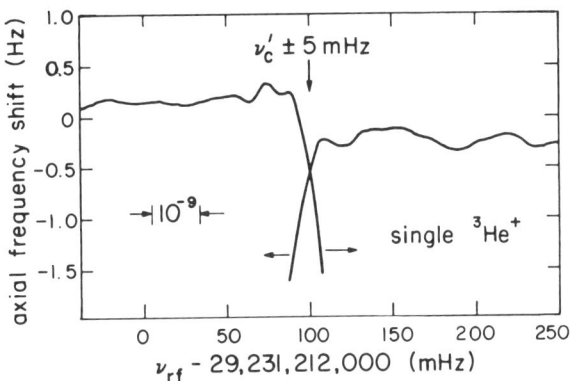

FIG. 7. Single ^{3}He$^+$ cyclotron resonance observed via anharmonicity induced axial frequency shifts. The ion is cooled between each trace using the cyclotron-axial sideband technique described in Section 13.5. The resolution is ~0.2 ppb (from Van Dyck et al. [5]).

the energy in the axial motion and is used primarily to tune the guard potential such that, for moderate increases in the axial drive, no shift occurs in the axial frequency. With some effort, one can tune C_4 nearly to zero, but, in practice, C_4 is set to ~3 × 10^{-5} in order to maintain an adequate signal strength for observing the $z^2\rho^2$ term which is proportional to radial energy.

As an example of anharmonic detection, consider Figure 7 which is a cyclotron resonance for a single ^{3}He$^+$ ion. A sweep is made in both directions, each being terminated after absorbing a detectable amount of cyclotron energy; the bracketing procedure yields a typical resolution of 0.2 ppb. Similar traces are obtained for the magnetron resonance, with a resolution about 100 times less due to the lower magnetron frequency as well as the contribution due to battery noise. However, no more than 10^{-4} of this magnetron uncertainty shows up in the corrected free-space cyclotron frequency as evident from the use of the quadrature invariance theorem shown in Equation (13.2).

From Equation (13.16), the first-order electrostatic shifts in the axial frequency can be determined [10] and expressed explicitly in position coordinates according to

$$\frac{\delta\omega_z}{\omega_z} = \frac{3C_4}{2d^2}\left(-R_c^2 + \frac{Z_a^2}{2} - R_m^2\right), \qquad (13.17)$$

where R_c, Z_a, and R_m are the amplitudes of the respective cyclotron, axial, and magnetron motions. Equation (13.17) can now be used to determine the amount of amplitude required to detect the radial modes by anharmonic coupling.

13.9 The Ejection-Detection Method

One very popular method of determining the ion's cyclotron frequency is a time-of-flight scheme that is sensitive enough to observe single ions and can easily be adapted for any q/m. This scheme was first introduced by Gräff to measure the proton–electron mass ratio [33] and has been refined to the current resolution of ~ 1 ppb by Werth [34] at the University of Mainz, Germany. Table I lists the typical parameters of the Mainz PTMS.

The ejection–detection method begins by exciting the ion to a large cyclotron radius and then follows with a ring-electrode voltage ramp, with superimposed pulses up to a potential near that of the endcaps. This ramp thus provides a means for ejecting the highest energy ions remaining in the trap at each pulse through a 1.6-mm-diameter hole in one endcap. The ions then drift along the field lines until they reach a microchannel plate detector beyond one end of the solenoid.

The actual method of detection was originally proposed by Bloch [35]. The axial velocity and therefore the time of flight will be a function of the initial axial kinetic energy at ejection and the initial orientational potential energy $-\vec{\mu} \cdot \vec{B}_0$, where $\vec{\mu} = -|\mu_z|\hat{z}$ is the orbital magnetic moment of the charge and $\vec{B}_0 = B_0\hat{z}$. If the RF field is resonant at ω'_c, the transverse orbit increases, and thus $|\mu_z|$ increases. Therefore, in the inhomogeneous field along the z axis, the charge will be further accelerated by an axial force toward the collector, given by $-|\mu_z|\partial B/\partial z$ (since $\partial B/\partial z$ is negative along $\hat{z}$). By the time the charge arrives at the detector, all the initial radial energy will be converted to axial kinetic energy, thus *reducing* the time of flight according to the size of the magnetic moment.

Figure 8 shows examples of the reduction in average time of flight for $^4\text{He}^+$ and H_2^+ as a function of the RF-drive frequency. The corresponding linewidths are 2 ppb for H_2^+ and 4 ppb for $^4\text{He}^+$. Such narrow lines are a vast improvement over earlier proton resonances [33] obtained in measurements of m_p/m_e. In addition to the usual techniques of trimming the inhomogeneities of the trapping fields, a significant reduction in linewidth arises from an adiabatic increase in trapping potential. The reason that this benefits this detection scheme is due to the necessity of averaging the time-of-flight data over an ensemble of ions that make up the points shown in Figure 8. In addition to the initial distribution of axial energies

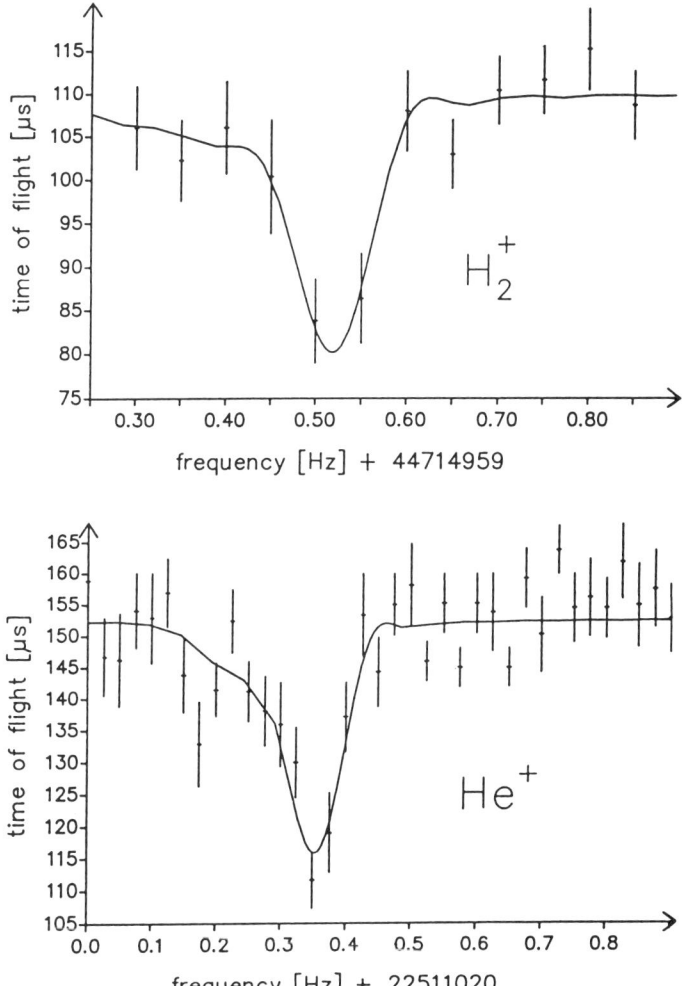

FIG. 8. Cyclotron resonances for ions H_2^+ and $^4He^+$, showing the achieved minimum linewidth in the Mainz spectrometer. The standard ejection–detection technique, in which a reduction in time of flight indicates an increase in radial energy, has been used. Experimental points are fitted to a theoretical distribution of cyclotron frequencies which incorporates the major field perturbations that exist when axial amplitude is large (>1% of Z_0) compared with radial amplitude (see solid line) (from Gerz et al. [34]).

from load to load, there are often more than one ion isolated in the trapping volume which also produces ion-ion interactions to which the ensemble is sensitive.

At the present time, the Mainz PTMS is restricted to mass doublets (atomic masses which are nearly equal for two different ions). To interrogate their mass difference [34], a precise knowledge is required only of ω'_c, and a less precise determination is required for ω_m, according to

$$\frac{\Delta m}{m} \simeq \frac{\omega'_c[2] - \omega'_c[1]}{\omega'_c[2] + \omega_m} \left(1 - \frac{\omega_m^2}{\omega'_c[1]\omega'_c[2]}\right), \qquad (13.18)$$

where the numbers in brackets refer to two different ion masses, and, to first order, ω_m is independent of mass. The detection of the magnetron frequency follows from a different criterion, since the time-of-flight method is impractical due to its smaller magnetic moment. After excitation to an orbit radius of >0.8 mm, the ions can no longer be ejected through the hole in the endcap which is closest to the channel plate detector. Thus, the total number of ions arriving at the detector versus frequency decreases at resonance, allowing ω_m to be determined typically to an accuracy better than 1 part in 10^4. From Equation (13.18), this uncertainty in ω_m would typically limit the accuracy of Δm to 5–10 ppb.

13.10 The Pulse-and-Phase Method

As another alternative detector, the PTMS at MIT utilizes a superconducting quantum interference device (SQUID) to observe the driven ion's axial motion [21]. The primary advantage of such a detector lies in its commercial availability. However, there are also some inherent limitations associated with this device that must be overcome. Its primary characteristics are its low-frequency capability and its low-input impedance. The first limitation requires the use of larger traps (approximately five times the UW quadring trap) in order that trapping voltages be much larger than possible contact potentials and patch effects on the main electrodes. Initially, this limitation also required ion masses to be greater than 10 u, but it was reported [36] that this problem has been overcome. An obvious advantage of larger traps is that the perturbations described in the following section are greatly reduced. However, coupling to the ion drops accordingly (see Section 13.3). The second limitation is overcome by using an impedance-matching transformer which also overcomes the intrinsic input current noise of the SQUID. This transformer is incorporated into a high-Q (~32,000) superconducting tuned circuit which also enhances the axial signal. As a further complication, the SQUID has an equivalent negative parallel resistance which tends to make the device somewhat unstable. This has been overcome by applying a feedback signal effectively to its input. As a final restriction, the SQUID needs to operate outside the strong magnetic field and is therefore coupled to the trap via a meter-long twisted pair of copper wires. When optimally coupled and tuned, the device can detect a single N_2^+ ion driven to $\sim 10\%$ of Z_0.

For the MIT PTMS, the guard rings are split into halves by a plane containing the axis of symmetry (as are the guards in the $2\times$ larger UW Penning trap). As described in Section 13.5, an inhomogeneous RF field at $\omega_c' - \omega_z$ can be applied to these split guards in order to couple the cyclotron motion to the axial oscillation. Since the system now acts like two strongly coupled harmonic oscillators, it is possible to completely convert all the energy in the excited cyclotron motion into the axial mode using an RF sideband pulse with the appropriate amplitude-duration product (so-called π-pulse).

In order to accurately determine the cyclotron frequency, a variant [37] of the Ramsey double-loop excitation scheme [38] is used. Initially, the cold cyclotron drive is excited at $t = 0$ by a strong pulse of RF electric field at ω_c' with some initial phase ϕ_0. This excitation generates an orbit size of 0.2 mm, or about 3% of R_0. Then, the excited motion is allowed to evolve undisturbed for a precise time, T, at which point the π-pulse at $\omega_c' - \omega_z$ is applied. This converts all the cyclotron energy to axial energy, preserving the instantaneous phase $\phi(T)$ of the cyclotron motion at the instant of conversion. The damped axial excitation (or ring-down signal) is then detected, using a Fourier transform analysis to recover both $\phi(T)$ and the frequency of the free axial motion. By plotting the phase difference $\phi(T) - \phi_0$ vs T, the frequency difference between the actual cyclotron motion (which evolves undisturbed) and the generator which is used to excite this motion is obtained from the slope of the resulting graph (see Figure 9). The undisturbed evolution of the cyclotron motion for time T is equivalent to the drift time between separated loops in the uniform field region of the Ramsey-style atomic beam machine.

The primary limitation to the precision of this method is the stability of the magnetic field during the evolution time [2]. To overcome this problem, in a field that typically drifts several ppb/hr, the spectrometer has been computer automated to alternatively load the ion of interest or the calibration ion every few minutes, thus allowing the comparison of cyclotron resonances to be made in a relatively short time. In this way, this spectrometer has succeeded in measuring [7] $M(CO^+)/M(N_2^+)$ to about 1 part in 10^{10}. In addition, it was reported [36] that the spectrometer can now measure atomic masses to this same accuracy. The UW PTMS is also near this same limit.

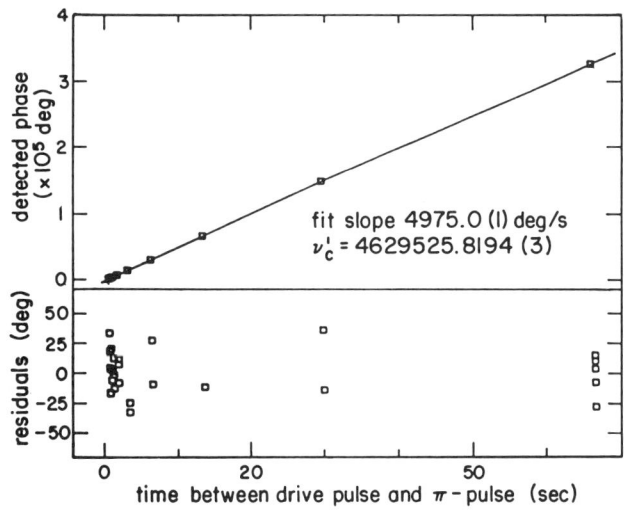

FIG. 9. Phase vs. dwell time between initial excitation of a cold ion and the π-pulse of $\omega'_c-\omega_z$, which converts its heated cyclotron motion to axial motion. As the ion's axial motion rings down, its phase is detected for each of the fixed dwell times. The appropriate multiple of 360° is added, and a line is fitted to the resulting plotted points. The slope of this line represents the offset between the generator's frequency and the ion's cyclotron frequency (from Cornell et al. [2]).

13.11 Perturbations to the Cyclotron Frequency

In any real trap and nonideal environment, some perturbations will always be present. The least serious perturbation is the one discussed in Section 13.1 which is associated with possible misalignments between electric and magnetic axes and possible asymmetries in the trapping electrodes. However, there are several other perturbations for which summing the three normal modes in quadrature [as suggested by Equation (13.2)] does not help.

One type of perturbation which has often plagued the PTMS is associated with the number of ions trapped. Depending on the severity of this dependency, the causes may be quite different from trap to trap. For large clouds, the effect is probably rooted in the space-change repulsion of like ions in the nonharmonic part of the potential well. This is an argument for using only single ions. However, even after reducing a cloud to one ion and making the trap as harmonic as possible, there will still be a shift in the cyclotron frequency dependent on the total charge present inside the trap. This effect has been extensively investigated [39] and is associated with the image charges induced in the real metal walls that surround the trapped ion. The effect yields an axial frequency shift which scales like q_1^2/mR_0^3 and therefore shows up in both ω_m and ω_c'. In the small UW quadring trap, the relative shift was 0.25(4) ppb for a single proton and 3.7(5) for a single O^{6+} ion. Thus, the R_0^{-3} factor prevents one from reducing trap size further, and, in fact, for the 2× larger UW trap, this effect is eight times less. For the Mainz and MIT spectrometers, this shift should be nearly two orders of magnitude smaller than that found in the small quadring trap. However, the Mainz spectrometer does see a serious number dependency for the upper magnetron sideband of ω_c', amounting to ~10 ppb/ion [34], and, for this reason, choose to measure ω_c' for their determination of the ion's cyclotron frequency.

The next most serious problem is associated with the leading higher-order terms in the trapping fields (i.e., the C_4 and B_2 terms). Linear magnetic gradients can also cause shifts if an ion of interest and its calibration ion have different centers of motion in the trap, due to position shifts from possible nonzero (but constant) electric fields. This problem is minimized by comparing ions in nearly the same potential well and by minimizing local surface potential variations. Likewise, the linear magnetic gradient can be shimmed to the point at which the effect is negligible. However, the intrinsic diamagnetism of the trapping electrodes will often form a B_2 term in the strong magnetic field (due to the cylindrical symmetry) which dominates the residual B_2 from the solenoid. Additional shims can remove even this problem *in situ* if ω_c' is measured as the ion is translated axially in the trap by the application of an antisymmetric potential on the endcaps [12,40]. As the dominant magnetic perturbation to the quadratic invariance equation, the first-order shifts to the free-space cyclotron frequency can be expressed in position coordinates according to

$$\frac{\delta\omega_c}{\omega_c} = \frac{B_2}{2B_0}\left[\left(\frac{\omega_m}{\omega_c'}\right)^2 R_c^2 + Z_a^2 - R_m^2\right]. \quad (13.19)$$

For the MIT PTMS and the UW PTMS, this easily corresponds to $<10^{-10}$ in relative shift. In the case of the Mainz spectrometer for doublets, Equation (13.18) requires only ω_c', whose relative shift is essentially the same as that in Equation (13.19), except that the coefficient of the R_c^2 term is replaced by -1. This larger cyclotron term in $\delta\omega_c'/\omega_c'$ is canceled by the corresponding contribution from $\delta\omega_z/\omega_z$ if both frequencies are measured with the same cyclotron energy. In the case of the UW PTMS, a residual $(\omega_m/\omega_c')R_c^2$ term is also canceled by a similar term in $\delta\omega_m/\omega_m$, if again these measurements are made with the same initial conditions.

The systematic shifts due to electrostatic perturbations are also of concern. The magnitude of C_4 can of course be varied with guard potential and one can check for the lack of sensitivity to this perturbation (within the level of resolution of ω_c') by measuring cyclotron frequency versus guard potential in the immediate vicinity of the C_4–null point. The corresponding first-order shift in ω_c, as a perturbation to the quadrature invariance theorem, can again be expressed in position coordinates according to

$$\frac{\delta\omega_c}{\omega_c} = \frac{3C_4}{2d^2}\left(\frac{\omega_m}{\omega_c'}\right)\left[-R_c^2 - Z_a^2 + \left(\frac{\omega_m}{\omega_c'}\right)R_m^2\right]. \tag{13.20}$$

In the case of the UW and MIT spectrometers, the values in Table I show that all the relative shifts are again $<10^{-10}$. The assumption made in all of these shift formulas is that the three normal-mode frequencies are measured for the same energies in all the modes and then recombined with the quadrature invariance equation to determine ω_c. This is not currently done in the Mainz PTMS, and one is referred to Gerz [34] for a discussion of the corresponding shift in this case.

Finally, the last systematic to be discussed is associated with relavitistic mass shifts when the various mode energies change. In all cases, before excitation, the relativistic shift is much less than $<10^{-10}$. However, detection of this mode typically requires between 0.2 and 1.0 ppb relative shift in ω_c to occur according to

$$\frac{\delta\omega_c}{\omega_c} = -2\left(\frac{\omega_c'R_c}{2c}\right)^2 - \left(\frac{\omega_z Z_a}{2c}\right)^2 - \left(\frac{\omega_m R_m}{c}\right)^2, \tag{13.21}$$

where, evidently, the last two terms will always be negligible compared to the cyclotron term. For the Mainz PTMS, this should be no problem because of the mass-doublet formula [Equation (13.18)]. However, to keep this term from affecting the atomic mass measurements in the MIT PTMS, one tries to keep $\omega_c'R_c$ constant for the comparison. In the case of the UW PTMS, the anharmonic detection scheme is basically a trigger method, in which only the cooled equilibrium state is important. The minimum detectable cyclotron energy does *not* correspond to a systematic shift in the nonequilibrium absorption of energy when one sweeps onto the very narrow cyclotron resonance. Equilibrium linewidths are typically less than 1×10^{-11} due to the dominant axial noise coupled through the B_2 and C_4 terms shown in Equations (13.19) and (13.20). It is also interesting to note that the minimum detectable cyclotron orbit can be obtained for the UW PTMS from the minimum cyclotron energy that produces a detectable axial frequency shift at the limit of resolution of the axial resonance according to

$$E_c = mc^2 \cdot \frac{R_z}{3C_4}\left(\frac{\omega_c' d}{c}\right)^2, \tag{13.22}$$

from which it follows that $R_c \sim d/67 \sim 30$ μm, independent of which ion is trapped.

To conclude this discussion of systematics, it should be pointed out that there are several ways of testing the accuracy of these spectrometers. In the case of the UW PTMS, it is convenient to compare ions which have different charge states for the same element. As an example, the ratio of free-space cyclotron frequencies for C^{5+} relative to C^{4+} has yielded a three-run average value of 1.250,057,113,296(509) which can be compared with the expected value of 1.250,057,112,946(1), whose uncertainty is limited only by the additional electron's atomic mass. For other spectrometers which do not utilize multiply charged ions, a comparison of atom versus molecule is possible. For instance, N_2^+ has been compared to N^+ in the MIT PTMS, and H_2^+ has been compared with H^+ in the Mainz PTMS. Such comparisons should always precede any new high-precision measurements of atomic masses in order to distinguish between available resolution and real accuracy.

Acknowledgments

This work is supported by the National Science Foundation under the "Mono-Ion Research" grant.

References

1. R. S. Van Dyck, Jr., F. L. Moore, D. L. Farnham, and P. B. Schwinberg, in *Frequency Standards and Metrology* (A. De Marchi, ed.), p. 349. Springer-Verlag, Berlin, 1989.
2. E. A. Cornell, R. M. Weisskoff, K. R. Boyce, R. W. Flanagan, Jr., G. P. Lafyatis, and D. E. Pritchard, *Phys. Rev. Lett.* **63**, 1674 (1989).
3. C. Gerz, D. Wilsdorf, and G. Werth, *Z. Phys. D* **17**, 119 (1990).
4. D. Hagena and G. Werth, *Europhys. Lett.* **15**, 491 (1991).
5. R. S. Van Dyck, Jr., D. L. Farnham, and P. B. Schwinberg, *J. Mod. Opt.* **39**, 243 (1992).
6. R. S. Van Dyck, Jr., D. L. Farnham, and P. B. Schwinberg, in *Nuclei Far from Stability/Atomic Masses and Fundamental Constants, 1992* (R. Neugart and A. Wöhr, eds.), p. 3. Inst. Phys., Bristol, 1993.
7. V. Natarajan, K. R. Boyce, F. DiFilippo, and D. E. Pritchard, in *Nuclei Far from Stability/Atomic Masses and Fundamental Constants, 1992* (R. Neugart and A. Wöhr, eds.), p. 13. Inst. Phys., Bristol, 1993.

8. R. S. Van Dyck, Jr., D. J. Wineland, P. A. Ekström, and H. G. Dehmelt, *App. Phys. Lett.* **28**, 446 (1976).
9. R. S. Van Dyck, Jr., P. B. Schwinberg, and H. G. Dehmelt, *Phys. Rev. D* **34**, 722 (1986).
10. L. S. Brown and G. Gabrielse, *Rev. Mod. Phys.* **58**, 233 (1986).
11. L. S. Brown and G. Gabrielse, *Phys. Rev. A* **25**, 2423 (1982).
12. R. S. Van Dyck, Jr., P. B. Schwinberg, and S. H. Bailey, in *Atomic Masses and Fundamental Constants 6* (J. A. Nolen, Jr. and W. Benenson, eds.), p. 173. Plenum, New York, 1980.
13. G. Gabrielse and F. C. Mackintosh, *Int. J. Mass Spectrom. Ion Processes* **57**, 1 (1984); see also J. Tan and G. Gabrielse, *Appl. Phys. Lett.* **55**, 2144 (1989).
14. E. C. Beaty, *J. Appl. Phys.* **61**(6), 2118 (1987).
15. G. Gabrielse and H. Dehmelt, *Bull. Am. Phys. Soc.* [2] **26**, 598 (1981).
16. G. Gabrielse and H. Dehmelt, *NBS Spec. Publ. (U.S.)* **617**, 219 (1984).
17. R. S. Van Dyck, Jr., P. B. Schwinberg, and H. G. Dehmelt, in *Atomic Physics* (R. S. Van Dyck, Jr. and E. N. Fortson, eds.), Vol. 9, p. 53. World Scientific, Singapore, 1984.
18. D. J. Wineland and H. Dehmelt, *J. Appl. Phys.* **46**, 919 (1975).
19. M. G. Richards, A. R. Andrews, C. P. Lusher, and J. Schratter, *Rev. Sci. Instrum.* **57**, 404 (1986).
20. R. S. Van Dyck, Jr., H. A. Schuessler, R. D. Knight, D. Dubin, W. D. Phillips, and G. Lafyatis, *Phys. Scr.* **T22**, 228 (1988).
21. R. M. Weisskoff, G. P. Lafyatis, K. R. Boyce, E. A. Cornell, R. W. Flanagan, Jr., and D. E. Pritchard, *J. Appl. Phys.* **63**, 4599 (1988).
22. S. R. Jefferts, T. Heavner, P. Hayes, and G. H. Dunn, *Rev. Sci. Inrum.* **64**, 737 (1993).
23. F. L. Moore, Thesis Dissertation, University of Washington, Seattle (1989).
24. R. S. Van Dyck, Jr., P. B. Schwinberg, and H. G. Dehmelt, in *New Frontiers in High Energy Physics* (B. Kursunoglu, A. Perlmutter, and L. F. Scott, eds.), p. 159. Plenum, New York, 1978.
25. L. Schweikhard, M. Blundschling, R. Jertz, and H.-J. Kluge, *Rev. Sci. Instrum.* **60**, 2631 (1989).
26. G. Bollen, H.-J. Kluge, M. König, H. Hartmann, T. Otto, G. Savard, H. Stolzenberg, G. Audi, R. B. Moore, G. Rouleau, and the ISOLDE Collaboration, in *Nuclei Far from Stability/Atomic Masses and Fundamental Constants, 1992* (R. Neugart and A. Wöhr, eds.), p. 19. Inst. Phys., Bristol, 1993.
27. G. Gabrielse and J. Tan, *J. Appl. Phys.* **63**, 5143 (1988).
28. G. L. Salinger and J. C. Wheatley, *Rev. Sci. Instrum.* **32**, 872 (1961); J. M. Lockart, R. L. Fagaly, L. W. Lombardo, and B. Muhlfelder, *Physica B (Amsterdam)* **165 & 166**, 147 (1990).
29. R. S. Van Dyck, Jr., P. B. Schwinberg, and H. G. Dehmelt, *Phys. Rev. Lett.* **47**, 1679 (1981); **59**, 26 (1987).
30. R. Mittleman, F. Palmer, G. Gabrielse, and H. Dehmelt, *Proc. Natl. Acad. Sci. U.S.A.* **88**, 9436 (1991).
31. F. L. Moore, L. S. Brown, D. L. Farnham, S. Jeon, P. B. Schwinberg, and R. S. Van Dyck, Jr., *Phys. Rev. A* **46**, 2653 (1992).
32. G. Gabrielse, *Phys. Rev. A* **27**, 2277 (1983).
33. G. Gräff, H. Kalinowski, and J. Traut, *Z. Physik A—Atoms Nuclei* **297**, 35 (1980).

34. C. Gerz, D. Wilsdorf and G. Werth, *Nucl. Instrum. Methods Phys. Res., Sect. B* **47**, 453 (1990).
35. F. Bloch, *Physica (Amsterdam)* **19**, 821 (1953).
36. D. E. Pritchard, private communication (APS Meeting, Washington DC) (1993).
37. E. A. Cornell, R. M. Weisskoff, K. R. Boyce, and D. E. Pritchard, *Phys. Rev. A* **41**, 312 (1990).
38. N. Ramsey, *Molecular Beams*. Oxford Univ. Press, London, 1956.
39. R. S. Van Dyck, Jr., F. L. Moore, D. L. Farnham, and P. B. Schwinberg, *Phys. Rev A* **40**, 6308 (1989).
40. G. Gabrielse, *Phys. Rev. A* **29**, 462 (1984).

14. ELECTRON BEAM ION TRAPS

Roscoe E. Marrs

Lawrence Livermore National Laboratory, Livermore, California

14.1 Introduction

The electron beam ion trap (EBIT) is a device for producing, trapping, and studying highly charged ions within the small volume of a highly compressed electron beam. Highly charged ions are of interest for several reasons. First, their atomic structure and interactions with other particles challenge our understanding of basic atomic physics, particularly because of the strong enhancement of the contributions of relativity, QED, and forbidden transitions at high atomic number. Second, highly charged ions have important applications in controlled fusion, astrophysics, X-ray lasers, and other hot plasmas; in all cases highly charged ion data are making a significant impact on the understanding of these plasmas. Finally, trapped highly charged ions can be used for experiments in other fields of physics such as nuclear physics, fundamental interactions, and strongly coupled plasmas. The EBIT is being used for new types of measurements in all of these areas.

The most important feature of the EBIT, which sets it apart from any other method of studying highly charged ions, is the ability to obtain high resolution X-ray spectra from stationary ions excited by monoenergetic electrons. In the following sections we will see how this powerful feature plays out in several different types of experiments. Another remarkable feature of the EBIT is the range of charge states available for study. EBIT experiments have been done with ions ranging from Ne^{8+} to bare U^{92+}.

The EBIT is not the only device for studying highly charged ions. Highly charged ions can be produced in several types of ion sources and in hot plasmas. At higher energy, very highly charged ion beams are available at a few large accelerator facilities. The techniques for studying highly charged ions that are closest to the EBIT in terms of the atomic physics that can be studied are: (1) the electron beam ion source, which shares the same operating principle and can be used for collision experiments, but is not designed for X-ray spectroscopy; (2) storage rings, which can store (at high velocity) many of the same ion species as the EBIT and perform selected X-ray and electron–ion cross-section measurements by collisions with gas jets or merged electron beams; (3) beam-foil spectroscopy, which can be used to measure transition energies and lifetimes; and (4) X-ray spectroscopy of hot plasmas, which can be used to measure transition energies.

In the remainder of this chapter the EBIT and its operating principles are first described in general terms. Then specific EBIT techniques for several different kinds of atomic physics measurements are discussed.

14.1.1 General Features of the EBIT

The basic idea of the EBIT is shown in Figure 1. A narrow, high-energy ($\sim$10–200 keV) electron beam passes through three drift tubes. Near thermal ions are confined within the electron beam in the center drift tube. In the radial direction, the ions are confined by the $\sim$10-V space charge potential of the electron beam itself; in the axial direction they are confined by a $\sim$100-V barrier applied to the two end drift tubes. The two superconducting coils have only a minor effect on the ion motion. Their role is to

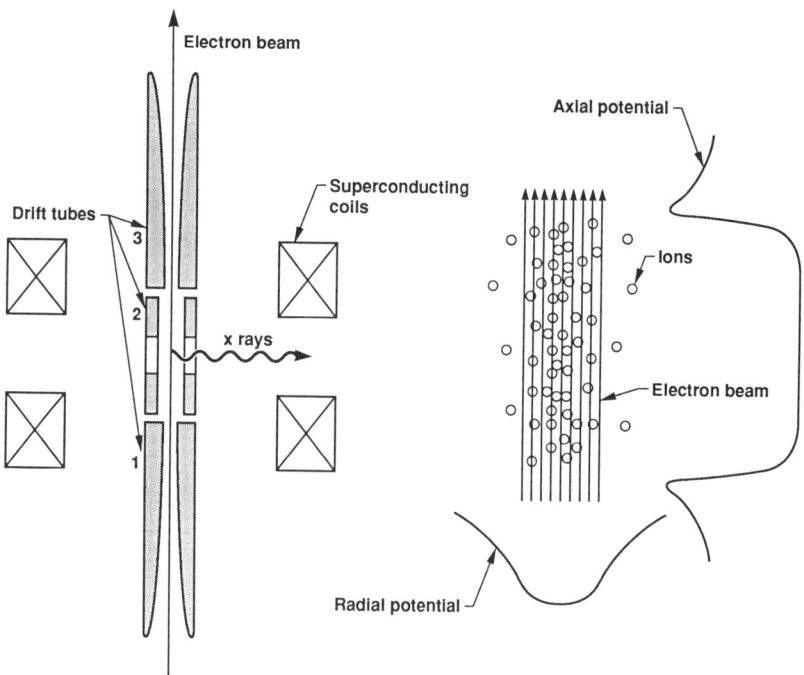

FIG. 1. Key features of the EBIT. (left) Scale drawing of the trap. (right) Schematic enlargement of the trapping potentials. Ions are confined in drift tube No. 2.

compress the electron beam to the highest possible density, up to 5000 A/cm^2 in a 70-μm-diameter beam. The split coil geometry with a vertical electron beam was chosen to facilitate X-ray measurements through radial ports in the horizontal plane. The short (2-cm) length of the ion trapping region is sufficient for X-ray spectroscopy and avoids the expense and difficulty of the meter-long traps used in the electron beam ion sources.

14.1.2 Development of the EBIT

The first EBIT was built for the specific purpose of X-ray measurements of trapped ions [1,2]. The development of the EBIT was inspired by the electron beam ion sources that had been built in several laboratories as sources of highly charged ions for injection into accelerators. The most notable of these was the KRION II device, which demonstrated the production of ion charge states up to Xe^{54+} [3]. This proved the concept of very highly charged ion production in an electron beam. The EBIT was foreshadowed by two preceding observations of X rays from an electron beam ion source. In one case, dielectronic recombination X rays from argon ions were observed in an Si(Li) detector looking along the electron beam axis through a hole in the electron gun [4]. In the second case a grazing-incidence mirror was used to image the soft X-ray emission in a radial direction and obtain a measure of the electron-ion overlap [5].

At present there are two cryogenic (4-K) EBITs in operation at the Lawrence Livermore National Laboratory (LLNL). One of them, an identical copy of the original EBIT, operates at electron energies up to approximately 30 keV. The other EBIT operates at electron energies up to approximately 200 keV and is often referred to as "Super EBIT." It is an upgrade of the first (30-keV) EBIT, as discussed below. The operating parameters for these EBITs are summarized in Table I. Several other laboratories are also building EBITs, in some cases with new designs, and initial results can be expected in the near future.

14.2 Principles of EBIT Operation

14.2.1 Electron Beam Propagation and Collection

The electron beam is at the heart of the EBIT: its space charge is what traps ions in the first place; successive ionization by the electron beam produces high charge states; and electon beam excitation drives the X-ray emission used in most EBIT experiments. In all three of these roles it is desirable to have the highest possible electron beam current density in the ion trapping region.

TABLE I. Operating Parameters for the 30- and 200-keV EBITs at LLNL

Parameter	Value
Electron energy	0.5–200 keV
Electron energy spread	15–100 eV (FWHM)
Electron current	2–200 mA
Electron beam radius	35 μm
Electron current density	50–5000 A/cm^2
Electron density	10^{11}–10^{13} cm^{-3}
Trap length	2 cm
Number of ions	10^4–10^6
Ion density	10^8–10^{11} cm^{-3}
Ion temperature	~10 eV per charge
Magnetic field	3 tesla
Base Pressure	~10^{-12} torr

The electron current density is determined by the local magnetic field and by parameters associated with the launching of the beam. The maximum electron beam compression that can be obtained in a magnetic field (neglecting space-charge neutralization) is known as Brillouin flow. It occurs when the space charge and centrifugal forces acting on each electron are exactly balanced by the Lorentz force. Perfect Brillouin flow requires beam launching conditions of zero cathode magnetic field, laminar electron flow (i.e., no trajectory crossing and zero electron temperature), and electron optics that avoid scalloping or other distortion of the beam envelope. The Brillouin current density has a square profile and is given in convenient units by

$$J_B = 1.5 \times 10^3 \, B^2 \sqrt{E_e} \quad (A/cm^2), \tag{14.1}$$

where B is the magnetic field in teslas and E_e is the electron energy in kilo-electron volts. Note that J_B is independent of both the total current and the current density at the cathode. The Brillouin radius is given by

$$r_B = (I_e/\pi J_B)^{1/2}, \tag{14.2}$$

where I_e is the total beam current. For a 3-T EBIT field and a 10-keV electron energy, the Brillouin density is $J_B = 4.3 \times 10^4$ A/cm^2, and the radius of a 200-mA beam is $r_B = 12$ μm. Unfortunately, Brillouin flow at such high current densities has not been achieved. In the LLNL EBITs the actual current density is about 10 times less than the Brillouin limit due to the effect of finite electron temperature.

A theory of magnetic beam compression that accounts for both finite electron temperature and cathode magnetic field has been developed by

Herrmann [6,7]. In this theory the characteristic beam radius, r_0, is given in terms of the Brillouin radius by

$$r_0 = r_B \left\{ \frac{1}{2} + \frac{1}{2} \left[1 + 4 \left(\frac{8mkT_c r_c^2}{e^2 B^2 r_B^4} + \frac{B_c^2 r_c^4}{B^2 r_B^4} \right)^{1/2} \right] \right\}^{1/2}, \quad (14.3)$$

where r_c, kT_c, and B_c are the cathode radius, temperature, and magnetic field, respectfully; m is the electron mass; e is the electron charge; and B is the local magnetic field. When the cathode temperature term dominates, the beam profile is approximately Gaussian and r_0 corresponds to the radius enclosing about 80% of the total current [7]. These are the conditions that prevail in an EBIT. The observed electron current density in the LLNL EBITs agrees with the Herrmann formula prediction with $B_c = 0$ and $kT_c = 0.1$ eV, which is the thermal temperature of the electron emitting surface [8,9]. Current densities up to 5000 A/cm² have been achieved.

The beam radius in the existing EBIT traps is determined by the cathode radius and temperature and is almost independent of the total current. In fact, the total current and current density can be varied together over a wide range with a constant beam radius by adjusting the anode voltage on the electron gun. This feature has been used in electron–ion cross-section measurements to change the ionization or recombination rate for ions already trapped and stripped to a high charge state. At LLNL, currents ranging from approximately 2 to 200 mA have been used in atomic physics experiments.

The magnetic field and electric potential profiles along the axis of the 30-keV EBIT at LLNL are shown in Figure 2. The cathode of the electron gun is operated at ground potential, and all other electrodes are operated at a positive potential. The electron–ion interaction energy is determined by the voltage applied to the drift tube (i.e., trap) assembly. A bucking coil surrounding the electron collector is used to maintain $|B| < 50$ G in the collector so that the beam is not magnetically confined and spreads over the interior surface of the collector. The electron beam is collected at 1.5-keV energy with an efficiency of 99.99% (typically 20 μA lost out of 200 mA).

14.2.2 Ion Injection and Trapping

Three different methods have been used to inject ions into an EBIT. In one method, neutral gasses are continuously injected through a radial port in the vacuum vessel and directed at the electron beam through collimating apertures. The injection rate is controlled by differential pumping of the space between the apertures and by regulation of the gas feed

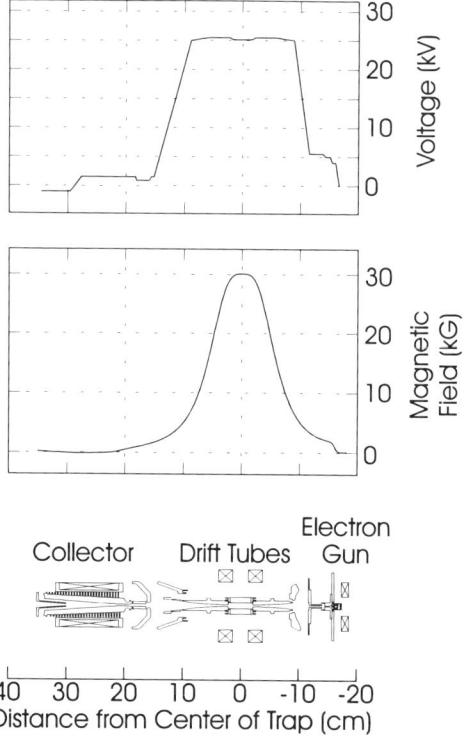

FIG. 2. Arrangement of the electrodes and magnets in the 30-keV EBIT at LLNL. The electric potential and magnetic field profiles along the axis are also shown. The (relatively small) space charge potential of the electron beam is not included.

rate. Some of the neutral atoms that intercept the electron beam are ionized and captured by the space charge of the beam. The trap is easily filled with neutral densities in the range of 10^5–10^6 cm^{-3} at the electron beam. This same method has been used with high-vapor-pressure solids such as sulfur. In this case the injection rate can be controlled by heating the solid.

A second filling method makes use of background elements that happen to be present in the vacuum chamber. The most notable of these is barium evaporated from the dispenser cathodes of the electron guns used with the LLNL EBITs. Tungsten, mercury, osmium, and lead have also been trapped and ionized in this way. These elements are accreted into the trap very slowly with typical filling times on the order of many seconds. These background elements can be avoided in other measurements by frequent dumping and refilling of the trap with other ions.

The most commonly used method of filling an EBIT, and also the most complicated, is to inject ions in low charge states from a metal vapor vacuum arc (MEVVA) source [10] located externally to the EBIT apparatus. The MEVVA source works by triggering a spark between anode and cathode electrodes. Ions in charge states ranging from 1+ to 4+ are produced from small amounts of material eroded from the cathode, so the element to be injected into EBIT is determined by selecting the MEVVA cathode material. The cathode must be a conductor; however, powdered nonconductive oxides of some elements have been successfully used as MEVVA cathodes by mixing and press forming them with copper powder. The (unmeasured) extracted ion current from the "micro MEVVA" used at LLNL is believed to be in the range of 1 to 5 mA [10] in a pulse lasting for 20 to 40 μsec, so approximately 10^{11} ions are emitted in each MEVVA pulse.

MEVVA ions are injected along the EBIT axis antiparallel to the electron beam and are focused into the electron beam with an einzel lens. A block diagram of an electronic arrangement used for capturing MEVVA ions is shown in Figure 3. The key feature is the raising of the axial

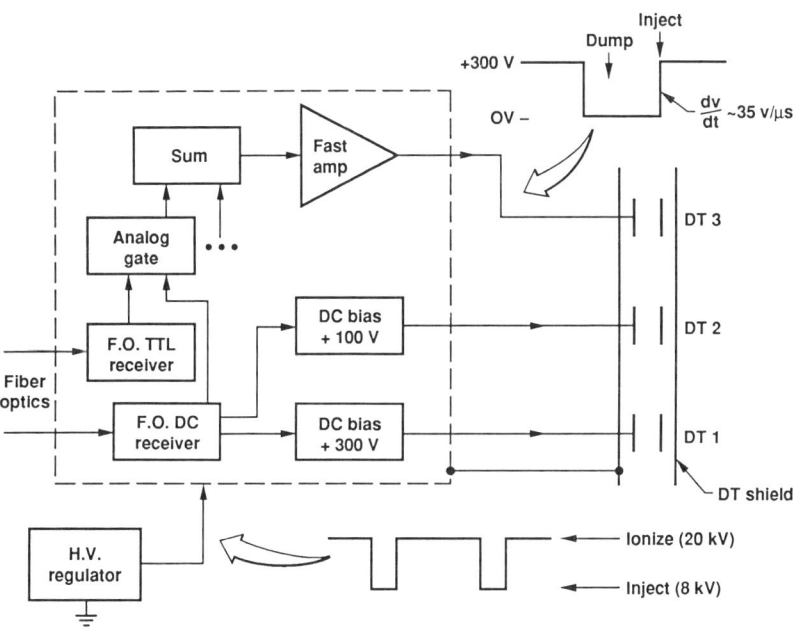

FIG. 3. Block diagram of the control electronics used at LLNL for injecting ions into an EBIT trap from an MEVVA source (not shown).

trapping potential on DT3 at a rate of ~35 V/μsec. Ions with energies slightly above the instantaneous potential of DT3 are reflected from DT1 and trapped by the now slightly greater potential of DT3. The emittance and energy spread of the MEVVA ions are much larger than the acceptance of the EBIT trap. However, roughly 10^5 ions are sufficient to fill the trap, and many elements throughout the periodic table have been successfully injected from a MEVVA.

At LLNL the MEVVA source is triggered ~30 μsec before the rising edge of the dump/inject pulse applied to DT3. Drift tube timing pulses and DC levels are fed to a floating high voltage rack through fiber optics. They are combined through two or more analog gate channels and a summing amplifier to generate the waveform applied to DT3. A high-voltage regulator sets the potential of the whole drift tube assembly to match that of the MEVVA (usually +8 kV) during injection and changes it to a different (usually higher) value for ionization.

14.2.3 Ion Heating and Cooling

For very highly charged ions or for long trapping times of any ion, heating by small-angle Coulomb collisions with the electron beam is a very important consideration. The heating rate can be obtained from energy transfer formulas in plasma physics [11]. In convenient units, the heating rate per ion is

$$dE/dt = \frac{0.442 \ln \Lambda \, q^2 j_e}{E_e A} \quad \text{(eV/sec)}, \tag{14.4}$$

where $\ln \Lambda$ is the Coulomb logarithm (normally $\ln \Lambda \approx 15$), q is the ion charge, j_e is the effective electron beam current density in A/cm^2, E_e is the electron energy in kilo-electron volts, and A is the ion atomic mass number. The heating rate of U^{82+} ions in a 25-keV electron beam of 4000 A/cm^2 (typical of the EBIT beam that produces these ions) is 3×10^4 eV/sec. Thus, in the absence of a cooling mechanism, these ions would acquire a temperature greater than q times the space-charge potential of the electron beam and move outside the beam in ~30 msec.

The process of evaporative cooling (by escape over the axial trap barriers) of highly charged ions by low-Z ions is used in the EBIT to balance the electron beam heating [2,12,13]. Evaporative cooling is essential for the retention of highly charged ions in an EBIT. Background gasses present in the trap and low-density jets of nitrogen or neon gas introduced radially are used to provide a source of low-Z ions for evaporative cooling. The density of neutrals in the trap and the height of the axial barrier determine the cooling power, which in turn determines the equilibrium electron–ion interaction rate for the highly charged ions. Usually the trap is initially overfilled by MEVVA injection and high-Z ions are lost until an equilibrium between heating and cooling is reached. This results in a reproducible highly charged ion inventory that is determined by the low-Z neutral density. In other words, the high-Z X-ray emission rate from an EBIT is determined by the neutral cooling gas density and not by the number of high-Z ions injected into the trap. A detailed time-dependent computer model of the ion energy balance, including energy exchange between different ion species, is now available to provide guidance in the choice of EBIT parameters [13].

14.2.4 Charge State Selection

Almost any charge state of any element can be selected for study in an EBIT. In some cases, particularly for the lighter elements, the fraction of the desired charge state can be greater than 90%. The evolution rate of the ionization balance is of critical importance for the ion output of the electron beam ion sources. However, in the case of the X-ray measurements for which EBIT was developed, it is more useful to consider the equilibrium ionization balance. The equilibrium ionization balance is determined by the competing relative rates of ionization, radiative recombination, dielectronic recombination, and charge exchange recombination with neutral atoms. For example, the abundance ratio of hydrogenlike and bare ions is

$$N_H/N_{bare} = \frac{\sigma^{RR}_{bare \to H}}{\sigma^{ion}_{H \to bare}} + \frac{(e/j_e) n_0 v \sigma^{CX}_{bare \to H}}{\sigma^{ion}_{H \to bare}}, \quad (14.5)$$

where the two terms on the right-hand side account for radiative recombination (RR) and charge exchange (CX) recombination, respectively. Here n_0 is the neutral gas density and v is the average ion-neutral collision velocity (approximately the ion thermal velocity). The dielectronic recombination term has been omitted since it contributes only at certain resonant electron energies that are easily avoided. Usually the charge exchange recombination rate can be made smaller than that of radiative recombination by running at a low neutral density (i.e., minimal injection of evaporative cooling gas). However, as explained above, a low neutral density implies a low highly charged ion inventory. Three-body recombination is negligible at EBIT electron energies and densities.

The observed equilibrium ionization balance for two very different cases is listed in Table II. In one case a large fraction of heliumlike Ti^{20+} was obtained by selecting an electron energy just below the Ti^{20+} ionization potential. In the other case a very high electron energy was used in an

TABLE II. The Observed Equilibrium Ionization Balance in an EBIT for Two Very Different Atomic Numbers and Electron Energies

Ionization stage	Titanium ($Z = 22$) ($E_e = 6.0$ keV) (%)	Uranium ($Z = 92$) ($E_e = 200$ keV) (%)
Bare	0	0.02
H-like	0	1.0
He-like	97	17
Li-like	3	34
Be-like	<1	31
B-like	—	15
C-like	—	3

Note. The ionization potential of the He-like charge state is 6.2 keV for titanium and 130 keV for uranium.

attempt to produce the highest possible charge states of uranium, for which the ionization cross-sections are very much smaller and the radiative recombination cross-sections are larger than those for titanium.

14.3 The 200-keV EBIT

The production of a large fraction of a particular ionization stage in an EBIT requires an electron beam energy well above the ionization threshold for producing that ion; usually electron energies roughly twice the threshold energy are most appropriate since ionization cross-sections rise slowly from zero at threshold. This requirement means that many interesting ions are well beyond the reach of the 30-keV EBIT. For example, the threshold for production of hydrogenlike U^{91+} is 130 keV. In order to reach these ions a 200-keV "Super EBIT" was constructed at LLNL.

The layout of the LLNL 200-keV EBIT is shown in Figure 4. The 200-keV EBIT is actually a conversion of the original 30-keV EBIT. Except for some minor changes that allow the drift tube assembly to operate at voltages up to 40 keV, the trap, liquid helium dewar, and vacuum chamber are those of the original EBIT and are of the same design as those of the operating 30-keV EBIT at LLNL. The new features are an electron gun and a collector assembly that float on top of a potential as large as -160 kV, as well as a series of transport magnets that transport the beam to the new collector. Except for the electron energy, the operating parameters are essentially the same as those of the 30-keV EBIT (see Table I). A more detailed description of the 200-keV EBIT design can be found elsewhere [9].

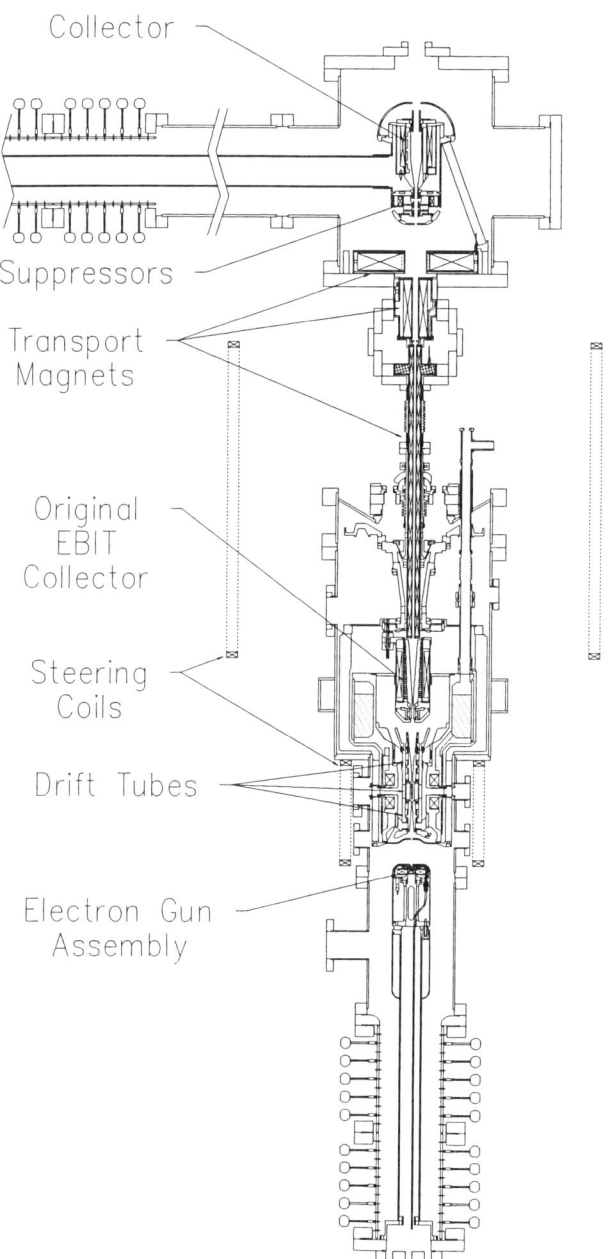

FIG. 4. Layout of the 200-keV EBIT at LLNL. From Knapp *et al.* [9].

An X-ray spectrum from uranium ions in a 198-keV electron beam is shown in Figure 5. At these high electron beam energies and high ion charge states bremsstrahlung from electron–ion collisions dominates the spectrum at X-ray energies below the electron beam energy. X rays at energies above the electron beam energy are due to radiative recombination into the indicated open shells of highly charged uranium ions. The K-shell recombination lines at 330 keV demonstrate the production of hydrogenlike and bare uranium ions in the trap. The radiative recombination spectrum is a good indicator of the ionization balance in the trap. For comparison, a spectrum from titanium at 6-keV electron energy is also shown in Figure 5. In this case the spectrum is dominated by the bound–bound transitions. These are the two cases for which the ionization balance is listed in Table II.

14.4 X-ray Spectroscopy

X-ray spectroscopic measurements place the fewest demands on EBIT operation: usually the electron beam energy can be left at one constant value, and the ionization balance is not too important as long as there is enough X-ray intensity from the transitions of interest. One of the most interesting aspects of X-ray spectroscopy with the EBIT is the way in which different kinds of spectrometers have been adapted to take advantage of the unique features of the EBIT. In the following sections several different classes of EBIT spectroscopy experiments will be discussed with an emphasis on how the EBIT machine and the X-ray spectrometer function together.

14.4.1 Transition Energy Measurements above 2 keV

Unlike measurements at lower energies, X rays above roughly 2 keV in energy can be observed through vacuum windows. Above roughly 3 keV 125-μm-thick beryllium vacuum windows can be used, and the X-ray spectrometers can often operate in air. At LLNL X-ray spectrometers of the von Hamos type [14] have been used in a series of EBIT wavelength measurements in the X-ray energy range of 2–15 keV [15–21]. These spectrometers have a cylindrically bent crystal that focuses X rays in the vertical (nondispersive) plane as shown in Figure 6. Because the diffraction crystals are flat in the dispersion plane, the small diameter of the EBIT electron beam (actually less than the resolution of most position sensitive detectors) is an important factor in obtaining good wavelength resolution.

The calibration of EBIT X-ray spectrometers with the Lyman-series transitions of appropriate hydrogenlike ions has become a very powerful

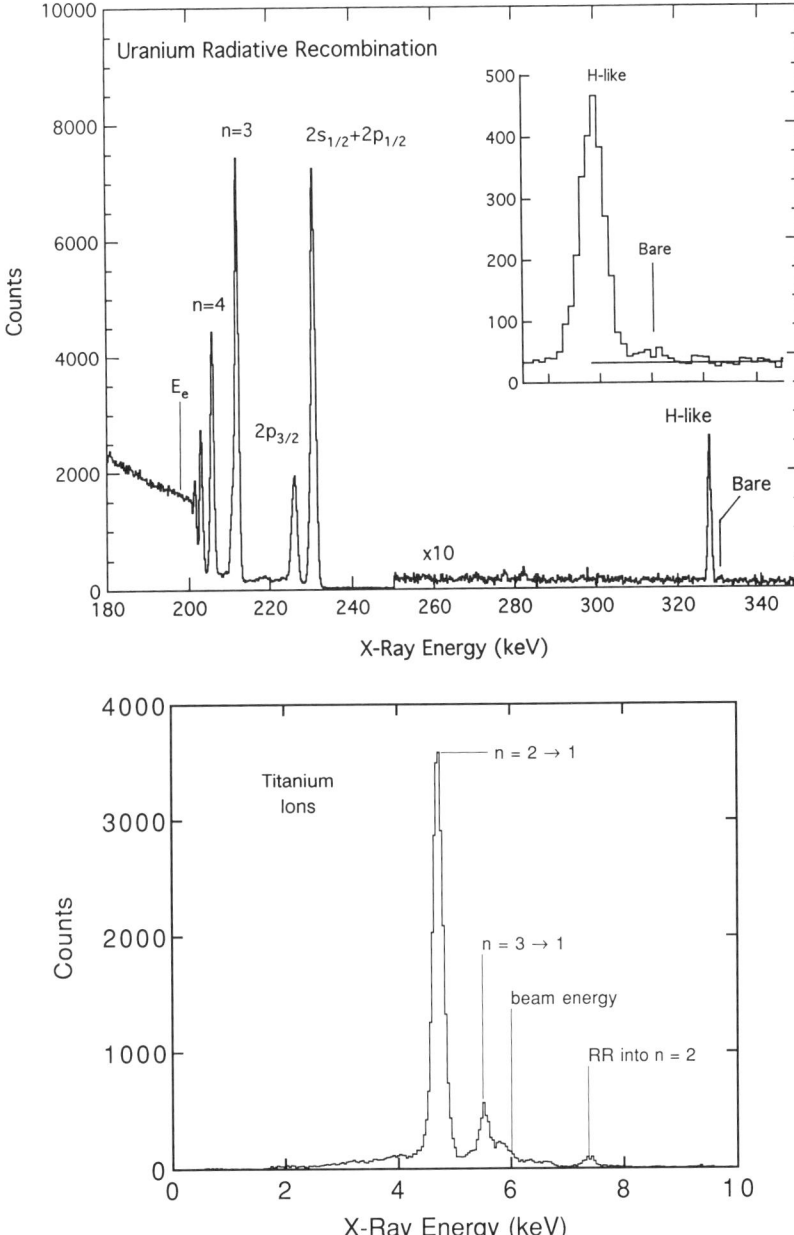

FIG. 5. X-ray spectra illustrating extremes of EBIT operating parameters. (top) X-ray spectrum from uranium ions at a 198-keV electron energy obtained in a 2-cm-deep by 3-cm-diameter germanium detector. The inset enlargement is from a second similar detector. From Marrs et al. [34]. (bottom) X-ray spectrum from titanium ions at a 6.0-keV electron energy obtained in a 1-cm-deep detector.

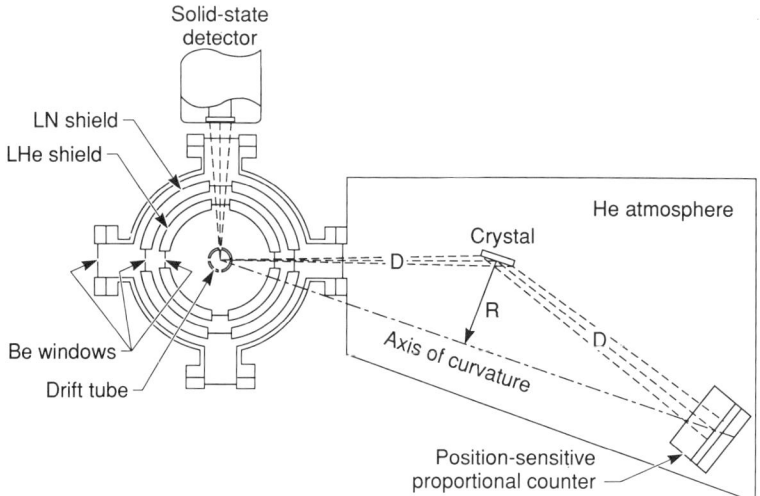

FIG. 6. Diffraction-plane layout of a high-resolution von Hamos spectrometer used with an EBIT. The electron beam direction is out of the page. The axis of curvature of the cylindrically bend crystal is the line joining the X-ray source and X-ray detector; the source-to-crystal and crystal-to-detector distances are equal. A solid-state detector used for monitoring the EBIT X-ray emission is also shown. From Beiersdorfer et al. [14].

and commonly used technique. These transition energies are well known, and the hydrogenic ions are easily produced in an EBIT; hence, all X rays come from the same location (the EBIT electron beam). This is essential with spectrometers having the von Hamos geometry because the location of the source directly affects the position of the observed X-ray lines. The $2s_{1/2}-2p_{3/2}$ transitions at approximately 4.5 keV in lithiumlike U^{89+} through neonlike U^{82+} were measured with respect to the Lyman series of hydrogenlike K^{18+} with a von Hamos spectrometer [19]. In this case transition energies were determined with an accuracy as high as 37 ppm. Since the contributions of both relativity and QED to atomic energy levels scale like Z^4, measurements such as these in highly charged uranium ($Z = 92$) ions are particularly important for understanding QED in strong fields.

Similar EBIT experiments have studied $n = 2$ to 3 transitions in the neonlike isoelectronic sequence for atomic numbers ranging from 56 to 92 [20]. In combination with lower-Z spectroscopic measurements from tokamak plasmas, these results reveal systematic inaccuracies in theoretical transition energies on the order of 1 or 2 eV that extend over a wide range of Z.

Another application of transition energy measurements that is almost ideally suited for the EBIT is the identification of line-pair candidates for photopumped X-ray lasers [21]. In this case what is needed is the precise energy separation of two nearly coincident transitions in different elements; it is not necessary to measure the absolute energies.

14.4.2 Transition Energy Measurements below 2 keV

For X-ray spectroscopy below an energy of roughly 2 keV it is necessary for the X-ray spectrometer to be coupled to the same vacuum as the EBIT since leakfree X-ray windows capable of reliably supporting an atmosphere of pressure and transmitting radiation below 2 keV are not available. An ultra-high-vacuum spectrometer of the Johann type, shown in Figure 7, has been used to obtain a precision measurement of the 653-eV $3s_{1/2}$–$3p_{3/2}$ transition energy in Na-like Pt^{67+} [22]. In this experiment the spectrometer was positioned so that the column of trapped platinum ions was located on the Rowland circle and oriented perpendicular to the (horizontal) diffraction plane. Since the EBIT X-ray source is then a small (70-μm) spot on the Rowland circle, only one X-ray wavelength satisfies the Bragg diffraction condition; those X-rays are refocused onto a position-sensitive microchannel plate detector.

To obtain an X-ray spectrum over a broad range, the entire spectrometer was rotated back and forth about the center of the Rowland circle, in effect moving the EBIT along the Rowland circle. Multiparameter data

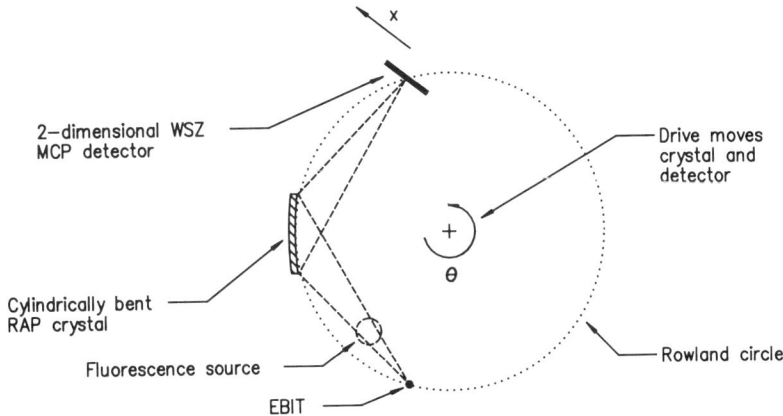

FIG. 7. Schematic diagram of the soft X-ray Johann spectrometer used to measured the $3s_{1/2}$–$3p_{3/2}$ transition energy in Na-like Pt^{67+}. The RAP crystal, MCP detector, and fluorescence source all rotate as a unit about the center of the Rowland circle. From Cowan et al. [22].

consisting of the x and y detector position and the spectrometer scan angle were analyzed with software cuts to reduce background, resulting in spectra of the type shown in Figure 8. The 1-eV-FWHM resolution obtained in these data is dominated by the resolving power of the rubidium-acid-phthalate (RAP) crystal.

The wavelength scale of the spectrometer was determined in two steps. First, the spectrometer dispersion was determined with manganese K X rays from a fluorescence source as shown in Figure 8c. Second, the absolute wavelength scale was calibrated with the Ly_α transition in hydrogenlike oxygen (see Figure 8b), which is within 0.2 eV of the $3s_{1/2}-3p_{3/2}$ Pt^{67+} transition. Hydrogenlike oxygen was produced from oxygen gas injected into the trap, and several alternate runs with platinum (injected from an MEVVA) and oxygen were taken to ensure an accurate calibration. A final value of 653.44 ± 0.02(stat) ± 0.05(syst) eV was obtained for the energy of the $3s_{1/2}-3p_{3/2}$ Na-like Pt^{67+} transition, corresponding to a 1% measurement of the QED contribution to this transition energy [22]. This result is the highest Z in the sodium isoelectronic sequence for which this transition has been measured. It is in significant disagreement with theories available at the time; however, a more recent theoretical calculation has succeeded in reproducing the experimental value [23,24].

Other kinds of spectrometers have also been used for EBIT measurements in the soft X-ray band. For example, a vacuum flat-crystal spectrometer with a thin-window position sensitive proportional counter has been used in the 0.7- to 3-keV energy range [25]. One of the most exciting future developments will be the extension of EBIT wavelength measurements into the VUV and visible bands.

14.4.3 X-ray Polarization

Since the X-ray emission from an EBIT is excited by a directed electron beam, the X rays can be linearly polarized and their emission can be anisotropic. This has two important consequences: First, cross-section measurements performed by counting X rays emitted at 90° to the beam direction must account for the X-ray angular distribution and, if a Bragg-crystal spectrometer is used, the polarization dependence of Bragg diffraction. Second, EBIT measurements of X-ray polarization can be used to obtain additional atomic-physics information beyond that available from an unpolarized source.

FIG. 8. Spectra obtained with the setup of Figure 7 for (a) Pt $3s_{1/2}-3p_{3/2}$ transitions excited at a 18.5-keV electron energy; (b) O Ly_α excited at a 3.6-keV electron energy; and (c) 9th-order Mn K_α and 10th-order Mn K_β produced by X-ray fluorescence of a solid target. From Cowan et al. [22].

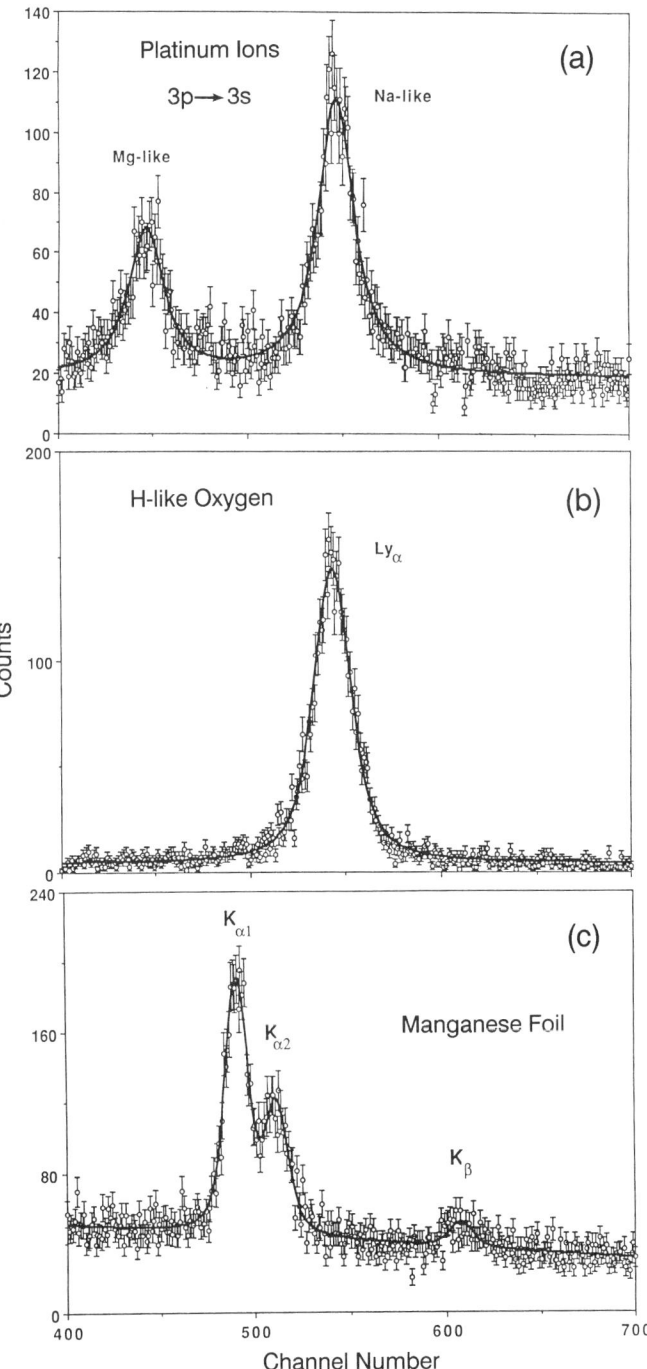

One example of an EBIT X-ray polarization measurement is the study of the effect of the hyperfine interaction on atomic X-ray transitions. In one such experiment the polarization of the $n = 2 - 1$ lines in heliumlike $^{45}\mathrm{Sc}^{19+}$ (nuclear spin, $I = \frac{7}{2}$) was measured with a rotatable Bragg-crystal spectrometer [26]. The reflectivity ratio for the two polarization components was ≤ 0.003, so the spectrometer was essentially blind to one polarization component. The X-ray polarization was determined by setting the diffraction plane of the spectrometer either horizontal (perpendicular to the electron beam) or vertical and comparing line intensities. An unpolarized reference line, the Ly_{α_2} ($1s^2S_{1/2}-2p^2P_{1/2}$) transition in hydrogenlike scandium, was used to determine the relative efficiency of the spectrometer in the two orientations. Spectra obtained with the two orientations of the spectrometer are shown in Figure 9 along with a level diagram that identifies the observed transitions. The dramatic differences in the relative intensities of the Sc^{19+} X-ray lines between the two orientations demonstrates that these lines have very different polarizations.

14.4.4 Lifetime Measurements

The EBIT provides an opportunity to measure lifetimes of metastable levels over a range of many orders of magnitude, including five decades that have not been accessible with other techniques. Lifetime measurements require a slightly more complicated mode of EBIT operation than the X-ray spectroscopy measurements discussed above because the feeding of the metastable levels must be switched on and off.

A very nice example of an EBIT lifetime experiment is the measurement of the ~ 90-μsec lifetime of the $1s2s\,^3S_1$ level in heliumlike Ne^{8+} at LLNL [27]. The EBITs at LLNL are equipped with precision high-voltage regulators (see Figure 3) that can quickly switch the electron beam energy between two predetermined values. For the Ne^{8+} measurement the EBIT electron energy was switched between 960 and 750 eV in approximately 20 μsec. The higher energy is just above the ~ 910-eV threshold for excitation of the $1s2l$ levels, but below threshold for excitation of all higher levels. The lower energy is just below threshold for excitation of all of the $1s2l$ levels. (A diagram of the $1s2l$ levels in the heliumlike isoelectronic sequence may be found in Figure 9.)

FIG. 9. (top) Diagram of the $n = 2 - 1$ transitions in heliumlike Sc^{19+}. Polarizations were measured for transitions w, x, y, and z. The 4293-eV decay of the $^3P_0$ level occurs only through the hyperfine interaction and is blended with line y. (bottom) Spectra of the Sc^{19+} $n = 2 - 1$ transitions excited at a 4.36-keV electron energy (just above threshold) for the two different spectrometer orientations. The horizontal spectrum is a composite of two separate spectra. From Henderson et al. [26].

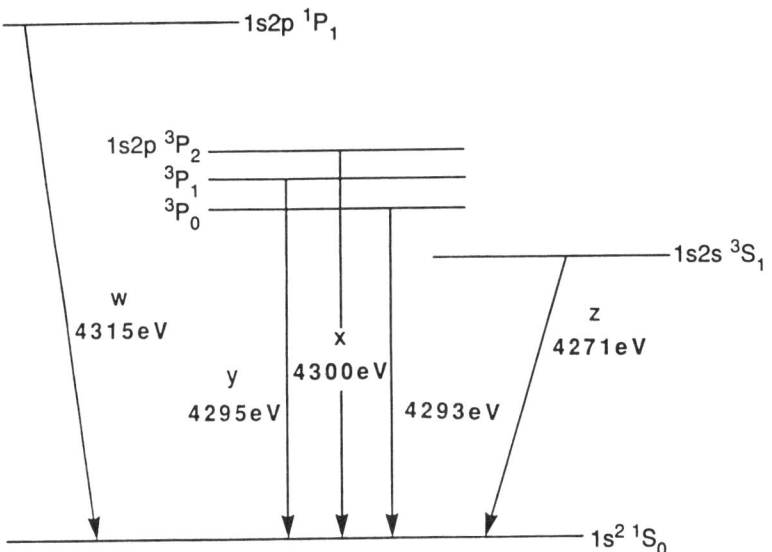
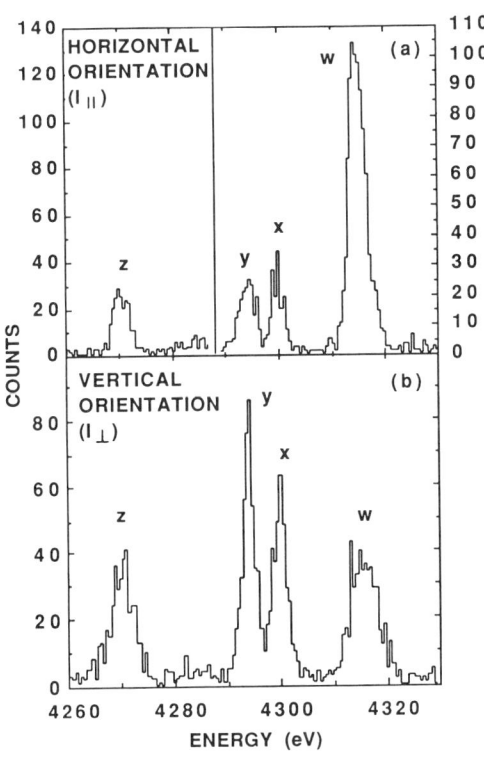

The $1s2l$–$1s^2$ transitions were observed with a vacuum crystal spectrometer and a thin-window position-sensitive proportional counter. The $^3S_1$ lifetime was determined by counting the decay X rays as a function of time after switching the electron beam energy below threshold. Sample spectra and the measured decay curves are shown in Figure 10. The exponential decay of the $^3S_1$ level (line z) can be clearly seen, and a value of 90.4 ± 1.4 μsec was determined for its lifetime [27]. Line w decays promptly, as expected for an allowed transition. Lines x and y, both of which have lifetimes less than 1 μsec, appear to have the longer decay time of line z because of collisional transfer of population from z to x and y.

The observation of collisional depopulation of the $^3S_1$ level by the EBIT electron beam offers a future opportunity to measure the collisional depopulation rate, especially since the EBIT electron density can be varied over a range of two orders of magnitude. Collisional depopulation measurements and lifetime measurements are important for understanding the density dependence of X-ray spectral lines emitted from hot plasmas; they will undoubtedly be the subject of future EBIT experiments.

14.5 Electron–Ion Collisions

In the preceding sections we have seen how an EBIT can be used for several different kinds of spectroscopic measurements of trapped highly charged ions. In most cases very little manipulation of the trapped ion population is required. By contrast, EBIT measurements of electron–ion collision cross-sections often require substantial manipulation of the trapped ion population. Usually this involves preselecting a particular ionization balance with one electron beam energy and current and then measuring collision cross-sections at other energies.

With a few exceptions involving extracted ions (see Section 14.6), EBIT electron–ion collision cross-sections have been determined by counting X rays associated with the collision process. In order to measure an absolute cross-section it is necessary to know the number of target ions in the trap, the effective electron current density (including the electron–ion overlap), and the absolute efficiency of the X-ray detectors. (Note that absolute ionization and recombination cross-sections can be determined from the abundance *ratio* of the two charge states involved rather than the absolute number of ions.)

Since the effective current density is difficult to determine, most EBIT cross-section measurements have been normalized to known cross-sections, usually to radiative recombination. Radiative recombination cross-sections provide an almost ideal normalizer: they can be accurately

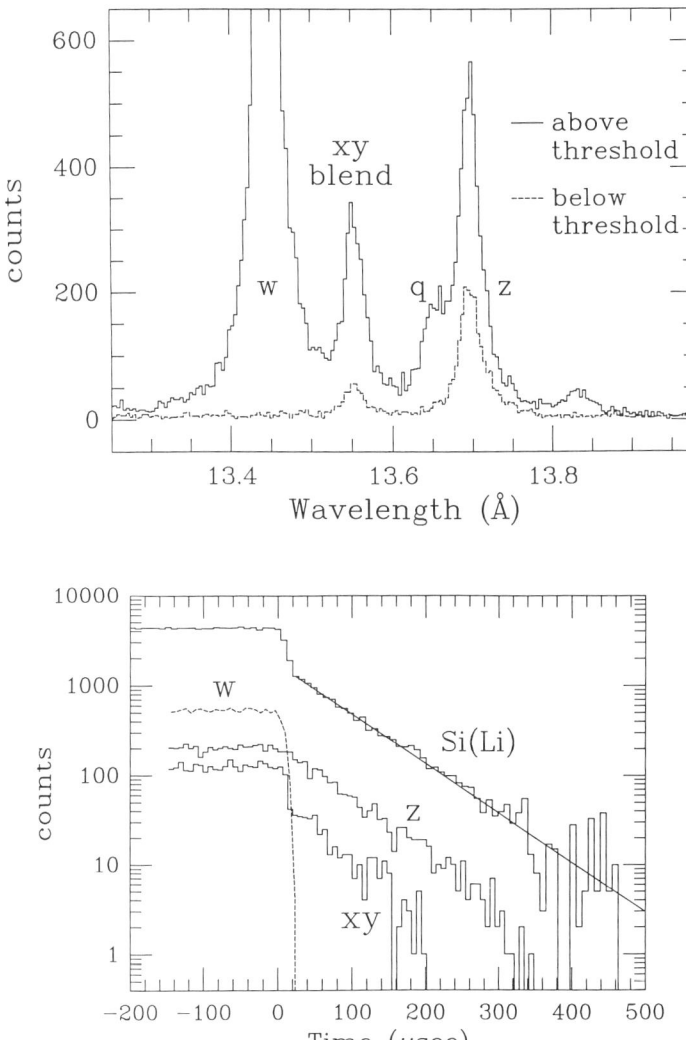

FIG. 10. Data for the measurement of the $1s2s\ ^3S_1$ lifetime in heliumlike Ne^{8+}. (top) Bragg crystal spectra obtained at electron energies above and below threshold. The feature labeled q is a satellite line from a small percentage of lithiumlike neon also present in the trap. (bottom) Line intensities as a function of time. The zero of the time axis is when the EBIT electron energy starts to drop from 960 to 750 eV. The decay curve labeled Si(Li) is for an unresolved blend of w, x, y, and z counted in an Si(Li) detector. From Wargelin et al. [27].

calculated for the range of electron and X-ray energies used in EBIT experiments, and there are extensive cross-section measurements for the inverse process of photoionization that support theoretical calculations at a level of 3% or better [28,29].

14.5.1 Excitation

The excitation of an X-ray line in a highly charged ion is a surprisingly complex process to which several different feeding mechanisms can contribute. These include direct excitation of the radiating level, cascade feeding from higher levels, and resonant excitation. Inner-shell ionization, radiative recombination, and charge exchange recombination of adjacent ionization stages can also contribute. The EBIT is able to unravel all of these different feeding mechanisms and measure their separate cross-sections through a combination of charge state preselection, monoenergetic electron excitation, and high-resolution X-ray spectroscopy. At present, the EBIT is the only technique for the measurement of electron impact excitation cross-sections for very highly charged ions.

The first experiment done with an EBIT was a measurement of electron impact excitation cross-sections in neonlike Ba^{46+} at two electron energies [1]. Subsequent measurements obtained a complete excitation function (vs electron energy) for the lowest excited state in neonlike barium, including partial cross-sections for many of the processes mentioned above [30].

A measurement of electron impact excitation cross-sections for heliumlike titanium provides a particularly nice illustration of the EBIT technique [31]. In this experiment the electron energy was first set at 6.0 keV (just below the 6.2-keV ionization potential of heliumlike Ti^{20+}) for 250 msec to drive the titanium ions into the heliumlike charge state (a 97% heliumlike fraction was achieved as listed in Table II). The electron energy was then switched to a designated value, and the $n = 2 - 1$ transitions (denoted w, x, y, and z) were observed with a Bragg-crystal spectrometer. To maintain the ionization balance the electron energy was dithered between the 6.0-keV ionization energy (for 14 msec) and the excitation energy (for 6 msec). A complete electron excitation function was constructed from a series of measurements at different electron energies.

The separation of different level feeding mechanisms with an EBIT is dramatically illustrated in Figure 11 for the $n = 2 - 1$ transitions in heliumlike iron, an important element in astrophysical X-ray sources. Such measurements have significant implications for plasma diagnostics since differences among available theories correspond in some cases to factor-

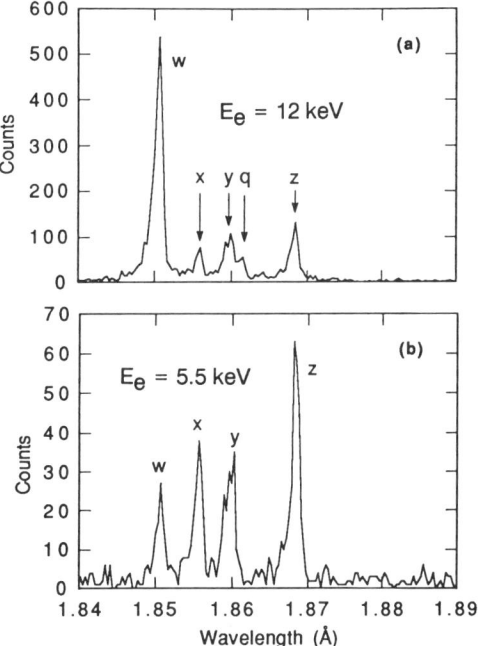

FIG. 11. Separation of different population mechanisms for heliumlike Fe^{24+} levels. (a) 12-keV incident electrons. Population is by direct excitation and cascade feeding. (b) 5.5-keV incident electrons, below threshold for collisional excitation. Population is by radiative recombination of hydrogenlike target ions.

of-2 differences in plasma temperatures inferred from X-ray line ratios in spectra from hot plasmas.

14.5.2 Ionization

Several entirely different techniques have been used to obtain ionization cross-sections from EBIT X-ray emission. In one technique, similar to the excitation cross-section measurement discussed above, the intensity of the $n = 2 - 1$ X ray in the reaction

$$Cr^{21+}(1s^22s) + e \rightarrow Cr^{22+}(1s2s^3S_1) + 2e \rightarrow Cr^{22+}(1s^2) + 2e + \gamma \quad (14.6)$$

was used to determine the partial cross-section for the inner-shell ionization of the $1s$ electron in lithiumlike Cr^{21+} [32]. Using another technique, the total ionization cross-section for lithiumlike ions of Ti, V, Cr, Mn, and Fe was determined by setting the EBIT electron energy to a fixed value corresponding to a strong dielectronic recombination resonance of the heliumlike ionization stage. The intensity of the dielectronic recombination X rays was then used as a measure of the (much slower) reionization rate of the lithiumlike charge state [33].

A third technique, based on the condition of steady-state ionization balance, was used to measure the ionization cross-section of hydrogenlike uranium, for which the threshold energy is 132 keV [34]. In this case the ionization cross-section was determined at an electron energy of 198 keV from the data shown in Figure 5a. The relative intensity of K-shell radiative recombination X rays was used to determine the U^{92+}/U^{91+} abundance ratio, and the U^{91+} ionization cross-section was then determined using Equation (14.5). A correction for the charge-exchange-recombination term in Equation (14.5) was made by taking runs at several different neutral-gas densities and extrapolating to zero.

14.5.3 Dielectronic Recombination

Dielectronic recombination is the resonant capture of an electron into a doubly excited state with the subsequent emission of a photon:

$$A^{q+} + e \rightarrow [A^{(q-1)+}]^{**} \rightarrow A^{(q-1)+} + \gamma. \tag{14.7}$$

At the resonance energies at which they occur, dielectronic recombination cross-sections are usually much larger than those of any competing process. Hence dielectronic recombination strongly affects both the ionization balance of a plasma and its X-ray emission spectrum. The development

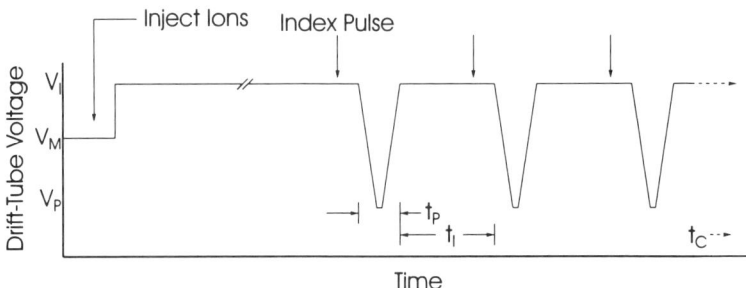

FIG. 12. Illustration of the drift-tube-voltage (i.e., electron energy) timing pattern used for dielectronic recombination measurements. Data are acquired while the electron energy is being slewed between the two extreme values. Typical values for Mo^{40+}: V_I, ionization energy (21 keV); V_P, minimum probe energy (11 keV); V_M, MEVVA injection energy (8 kV); t_I, ionization time (200 ms); t_P, probe time (10–20 ms); and t_C, cycle time (~20 s).

of the EBIT has provided the first opportunity to measure the $\Delta n = 1$ dielectronic recombination cross-sections of importance in many controlled fusion and astrophysical plasmas.

A very elegant method has been used at LLNL to obtain cross-sections for dielectronic recombination as well as several other processes in one comprehensive data set [35]. In this method, after injection and production of the desired target ionization stage, the electron beam energy begins a repetitive timing pattern in which most of the time is spent at a high (ionization) energy to maintain the ionization balance, but for part of the time cycle the electron energy is slewed down and up through an energy range encompassing the dielectronic recombination resonances. The slew rate is chosen so that the ionization balance does not change significantly during an energy sweep, in spite of the large dielectronic recombination cross-sections encountered. Each detected X ray is saved as a multiparameter event that consists of X-ray energy, instantaneous electron beam energy (strobed at the time of X-ray detection), and time in the machine cycle. A typical timing pattern is shown in Figure 12.

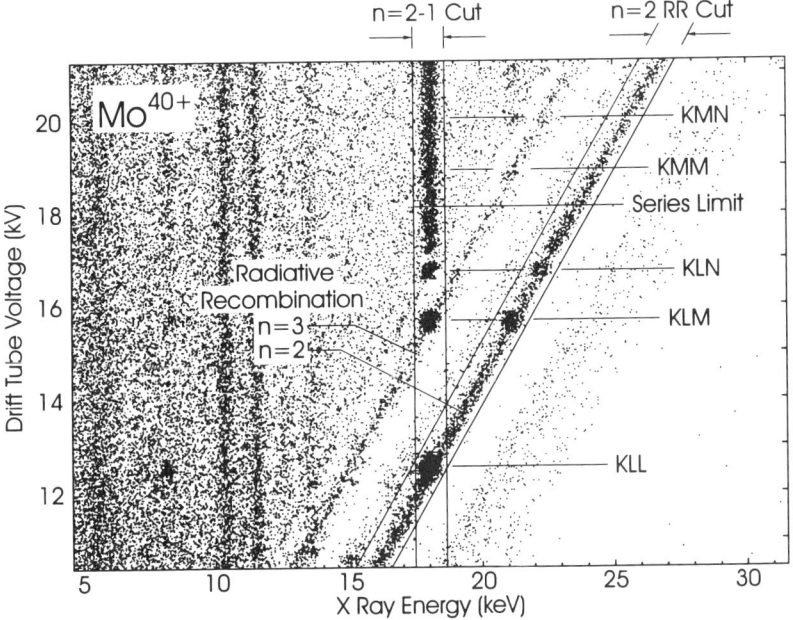

FIG. 13. Scatter plot of event-mode data for dielectronic recombination of helium-like Mo^{40+}. The bright spots are dielectronic recombination resonances, and the diagonal bands are radiative recombination photons, whose energy equals the sum of the electron beam energy and the binding energy. The vertical band above the series limit corresponds to direct excitation. From Knapp et al. [35].

Several ions in the heliumlike [35] and neonlike [36] isoelectronic sequences have been studied with this method. For a heliumlike target ion the electron configurations are

$$(1s^2) + e \rightarrow (1snln'l') \rightarrow (1s^2 2s) + \gamma. \tag{14.8}$$

In this case exactly one K X ray is emitted for each dielectronic recombination event, and the cross-sections may be determined by counting these X rays relative to the number of radiative recombination X rays appearing in the same data set. This is illustrated in Figure 13, which shows a scatter plot of data obtained for heliumlike Mo^{40+} target ions with a germanium X-ray detector and an electron energy slew rate of 1.2 eV/μsec. Some structure within the KLL resonances is resolved with the 50-eV-FWHM energy spread of the electron beam. The use of a Bragg-crystal spectrometer, with much higher X-ray energy resolution, allows complete separation

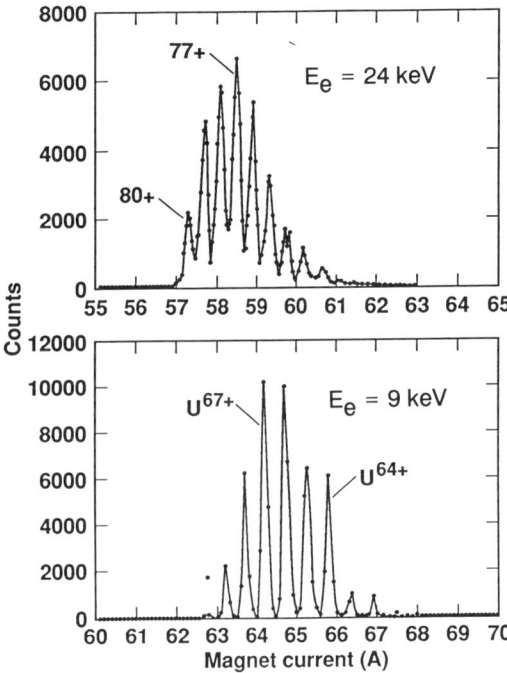

FIG. 14. Spectra of charge-state analyzed ions extracted from the 30-keV EBIT at LLNL plotted as a function of current in the magnet used to separate the charge states. (top) Thorium ions produced at an electron energy of 24 keV. (bottom) Uranium ions produced at an electron energy of 9 keV. From Schneider et al. [40].

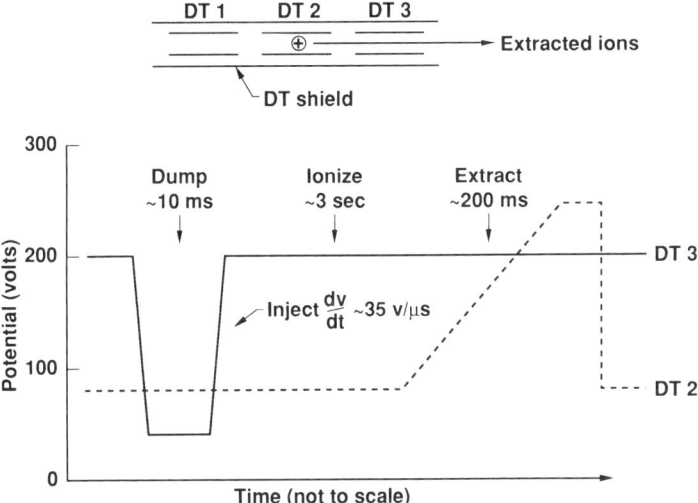

FIG. 15. Schematic diagram of the individual drift-tube voltage waveforms used to obtain the extracted ion spectra shown in Figure 14. Voltages are with respect to the drift-tube shield. The time axis is distorted to show all the features of one machine cycle. Ions are trapped when DT3 is at a higher potential then DT2. The actual potentials at the ion position (i.e., on the axis) are depressed with respect to the drift tube voltages due to the space charge of the electron beam.

of the individual resonances, and level-specific cross-sections can be obtained [37,38].

14.6 Extraction of Highly Charged Ions from an EBIT

The EBIT was developed for X-ray measurements of trapped ions. However, all of the ions produced within the EBIT trap can be extracted. At LLNL this is routinely done with the 30-keV EBIT, although not yet with the 200-keV EBIT. Charge states up to Th^{80+} have been extracted, as shown in Figure 14. In this case the total number of thorium ions extracted was approximately 10^4 every 3 sec [39].

Ions are extracted from an EBIT trap by removing the axial trapping potential. The ions then follow the space charge of the electron beam to the collector, where they are separated from the electron beam by the bias voltages on the collector and extractor electrodes. At LLNL the ions then pass through an einzel lens, a 90° electrostatic bender, a second einzel lens, and an analyzing magnet. The analyzing magnet separates the ions by charge state so that they can be either counted to obtain a measure of their abundance in the trap or used for external collision experiments [40].

In most experiments it is desirable to extract the ions over a period of 10 to 100 msec in order to minimize detector dead time and saturation. This is accomplished with the drift-tube voltage waveform shown in Figure 15. The rapidly switched voltage waveform on DT3 is used for dumping and injection of ions as discussed in Section 14.2.2. The new feature for extraction is the slow sawtooth waveform on DT2. Because the rise time of the sawtooth is much longer than the axial bounce time of the trapped ions, each ion leaves the trap at an energy corresponding to the fixed potential of DT3. The ion spill time is given by $(kT_i/q)/(dV/dt)$, where kT_i is the ion temperature, typically $\sim 10q$ eV for high-Z ions, and dV/dt is the voltage slew rate on DT2. It should be noted that the emittance of EBIT, corresponding to ions with a temperature of $kT_i \sim 10q$ eV in a 70-μm-diameter volume, is much smaller than that of most other ion sources.

Ion energies are determined by the bias voltage on the drift-tube assembly at extraction time, which can be different from the bias voltage (i.e., electron energy) that produced the ions in the first place. For example, Th^{75+} ions produced at an electron beam energy of 24 keV have been used in ion surface interaction experiments at kinetic energies less than $4q$ keV [41].

References

1. R. E. Marrs, M. A. Levine, D. A. Knapp, and J. R. Henderson, *Phys. Rev. Lett.* **60**, 1715 (1988).
2. M. A. Levine, R. E. Marrs, J. R. Henderson, D. A. Knapp, and M. B. Schneider, *Phys. Scr.* **T22**, 157 (1988).
3. E. D. Donets, in *The Physics and Technology of Ion Sources* (I. G. Brown, ed.), p. 245. Wiley, New York, 1989.
4. J. P. Briand, P. Charles, J. Arianer, H. Laurent, C. Goldstein, J. Dubau, M. Loulergue, and F. Bely-Dubau, *Phys. Rev. Lett.* **52**, 617 (1984).
5. M. A. Levine, R. E. Marrs, and R. W. Schmieder *Nucl. Instrum. Methods Phys. Res., Sect A* **237**, 429 (1985).
6. G. Herrmann, *J. Appl. Phys.* **29**, 127 (1958).
7. K. Amboss, *IEEE Trans. Electron Devices,* **ED-16**, 897 (1969).
8. M. A. Levine, R. E. Marrs, J. N. Bardsley, P. Beiersdorfer, C. L. Bennett, M. H. Chen, T. Cowan, D. Dietrich, J. R. Henderson, D. A. Knapp, A. Osterheld, B. M. Penetrante, M. B. Schneider, and J. H. Scofield, *Nucl. Instrum. Methods* **B43**, 431 (1989).
9. D. A. Knapp, R. E. Marrs, S. R. Elliott, E. W. Magee, and R. Zasadzinski, *Nucl. Instrum. Methods Phys. Res., Sect. A* **334**, 305 (1993).
10. I. G. Brown, J. E. Galvin, R. A. MacGill, and R. T. Wright, *Appl. Phys. Lett.* **49**, 1019 (1986).

11. I. P. Shkarofsky, T. W. Johnston, and M. P. Bachynski, *The Particle Kinetics of Plasmas*. Addison Wesley, Reading, MA, 1966.
12. M. B. Schneider, M. A. Levine, C. L. Bennett, J. R. Henderson, D. A. Knapp, and R. E. Marrs, *AIP Conf. Proc.* **188,** 158 (1989).
13. B. M. Penetrante, J. N. Bardsley, D. DeWitt, M. Clark, and D. Schneider, *Phys. Rev. A* **43,** 4861 (1991).
14. P. Beiersdorfer, R. E. Marrs, J. R. Henderson, D. A. Knapp, M. A. Levine, D. B. Platt, M. B. Schneider, D. A. Vogel, and K. L. Wong, *Rev. Sci. Instrum.* **61,** 2338 (1990).
15. P. Beiersdorfer, M. H. Chen, R. E. Marrs, and M. A. Levine, *Phys. Rev. A* **41,** 3453 (1990).
16. R. Hutton, P. Beiersdorfer, A. L. Osterheld, R. E. Marrs, and M. B. Schneider, *Phys. Rev. A* **44,** 1836 (1991).
17. S. MacLaren, P. Beiersdorfer, D. A. Vogel, D. Knapp, R. E. Marrs, K. Wong, and R. Zasadzinski, *Phys. Rev. A* **45,** 329 (1992).
18. P. Beiersdorfer, J. Nilsen, A. Osterheld, D. Vogel, K. Wong, R. E. Marrs, and R. Zasadzinski, *Phys. Rev. A* **46,** R25 (1992).
19. P. Beiersdorfer, D. Knapp, R. E. Marrs, S. R. Elliott, and M. H. Chen, *Phys. Rev. Lett.* **71,** 3939 (1993).
20. P. Beiersdorfer, *AIP Conf. Proc.* **274,** 365 (1993).
21. S. Elliott, P. Beiersdorfer, and J. Nilsen, *Phys. Rev. A* **47,** 1403 (1993).
22. T. E. Cowan, C. L. Bennett, D. D. Dietrich, J. V. Bixler, C. J. Hailey, J. R. Henderson, D. A. Knapp, M. A. Levine, R. E. Marrs, and M. B. Schneider, *Phys. Rev. Lett.* **66,** 1150 (1991).
23. K. T. Cheng. W. R. Johnson, and J. Sapirstein, *Phys. Rev. Lett.* **66,** 2960 (1991).
24. S. A. Blundell, *Phys. Rev. A* **46,** 3762 (1992).
25. P. Beiersdorfer and B. J. Wargelin, *Rev. Sci. Instrum.* **65,** 13 (1994).
26. J. R. Henderson, P. Beiersdorfer, C. L. Bennett, S. Chantrenne, D. A. Knapp, R. E. Marrs, M. B. Schneider, K. L. Wong, G. A. Doschek, J. F. Seely, C. M. Brown, R. E. LaVilla, J. Dubau, and M. A. Levine, *Phys. Rev. Lett.* **65,** 705 (1990).
27. B. J. Wargelin, P. Beiersdorfer, and S. M. Kahn, *Phys. Rev. Lett.* **71,** 2196 (1993).
28. E. B. Saloman, J. H. Hubbell, and J. H. Scofield, *At. Data Nucl. Data Tables* **38,** 1 (1988).
29. J. H. Scofield, *Phys. Rev. A* **40,** 3054 (1989).
30. P. Beiersdorfer, A. L. Osterheld, M. H. Chen, J. R. Henderson, D. A. Knapp, M. A. Levine, R. E. Marrs, K. J. Reed, M. B. Schneider, and D. A. Vogel, *Phys. Rev. Lett.* **65,** 1995 (1990).
31. S. Chantrenne, P. Beiersdorfer, R. Cauble, and M. B. Schneider, *Phys. Rev. Lett.* **69,** 265 (1992).
32. D. Vogel, P. Beiersdorfer, R. Marrs, K. Wong, and R. Zasadzinski, *Z. Phys. D* **21,** Suppl., S193 (1991).
33. K. L. Wong, P. Beiersdorfer, M. H. Chen, R. E. Marrs, K. J. Reed, J. H. Scofield, D. A. Vogel, and R. Zasadzinski, *Phys. Rev. A* **48,** 2850 (1993).
34. R. E. Marrs, S. R. Elliott, and D. A. Knapp, *Phys. Rev. Lett.* **72,** 4082 (1994).
35. D. A. Knapp, R. E. Marrs, M. B. Schneider, M. H. Chen, M. A. Levine, and P. Lee, *Phys. Rev. A* **47,** 2039 (1993).
36. M. B. Schneider, D. A. Knapp, M. H. Chen, J. H. Scofield, P. Beiersdorfer,

C. L. Bennett, J. R. Henderson, R. E. Marrs, and M. A. Levine, *Phys. Rev. A* **45,** R1291 (1992).
37. P. Beiersdorfer, T. W. Phillips, K. L. Wong, R. E. Marrs, and D. A. Vogel, *Phys. Rev. A* **46,** 3812 (1992).
38. M. B. Schneider, D. A. Knapp, P. Beiersdorfer, M. H. Chen, J. H. Scofield, C. L. Bennett, D. R. DeWitt, J. R. Henderson, P. Lee, M. A. Levine, R. E. Marrs, and D. Schneider, *AIP Conf. Proc.* **257,** 26 (1992).
39. D. Schneider, M. W. Clark, B. M. Penetrante, J. McDonald, D. DeWitt, and J. N. Bardsley, *Phys. Rev. A* **44,** 3119 (1991).
40. D. Schneider, D. DeWitt, M. W. Clark, R. Schuch, C. L. Cocke, R. Schmieder, K. J. Reed, M. H. Chen, R. E. Marrs, M. Levine, and R. Fortner, *Phys. Rev. A* **42,** 3889 (1990).
41. J. W. McDonald, D. Schneider, M. W. Clark, and D. DeWitt, *Phys. Rev. Lett.* **68,** 2297 (1992).

15. DC CURRENT MEASUREMENTS

Thomas J. Mego

Keithley Instruments, Inc., Solon, Ohio

15.1 Introduction

Many types of detectors produce a current or charge output in proportion to the number of charged particles and photons incident upon them, and examples of such devices are illustrated in Figure 1. Thus an understanding of techniques and considerations for making reliable, accurate low-current and charge measurements is essential in many experiments. This chapter will outline the techniques used to amplify and measure small current and charge signals.

15.1.1 The Ideal Ammeter

The function of ammeter A in the arrangements shown in Figure 1 is to give an accurate and immediate indication of the current that would flow through a conductor of zero resistance at that point in the circuit. In practice, however, several characteristics of the ammeter limit the accuracy of the measurement. Ammeters have finite resistance between their terminals, resulting in a potential difference across their terminals (referred to as voltage burden). Ammeters have finite bandwidth and dynamic range. Offset currents and noise limit their sensitivity.

15.1.2 The Current Measurement System

While we conventionally refer to the current measurement device as an ammeter, it is in fact a current measurement system consisting of (1) amplification, (2) quantification and recording, and (3) connections, fixturing, and environmental control. The output from the experiment (typically a detector) is a low-current signal that is highly susceptible to measurement error and interference. This signal is amplified to convert it to a high-level signal representative of the input, usually a voltage on the order of volts and able to drive several milliamps. This high-level signal is in turn converted for display or storage. In the following sections we will discuss considerations for each aspect of this measurement system in terms of how they affect the overall system performance.

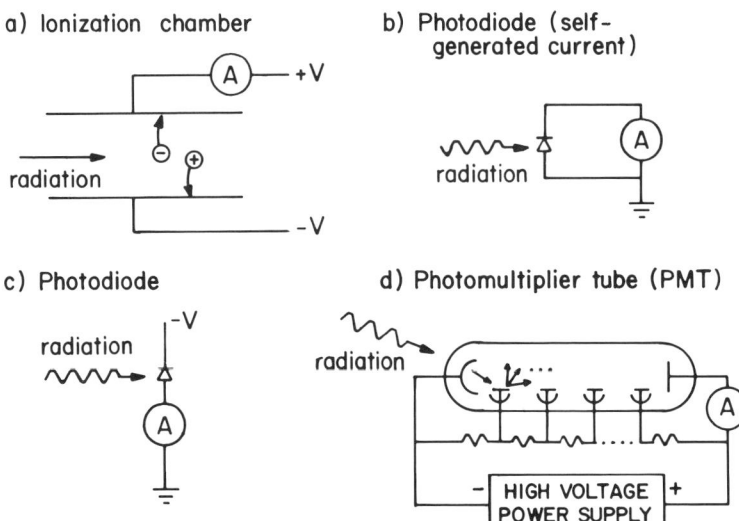

Fig. 1. Low-current measurement applications.

15.2 Description and Comparison of Current Amplifiers

15.2.1 Overview

In part, the difference among current amplification techniques lies in the choice of the current sensing element. The amplification methods discussed here are shown in Figure 2.

Most ammeters use a resistor to sense current and a voltage measuring circuit to quantify the voltage drop across the sense resistor. Ammeters using a resistor to sense current have the advantages of simplicity, linear response, and accurate and stable gain. However, they have limited bandwidth at low levels, and sensitivity is limited by the Johnson noise of the resistor. The two most common versions of this type of ammeter are the resistive shunt ammeter and transresistance ammeter.

Logarithmic ammeters use the nonlinear current–voltage characteristics of a diode or transistor to measure current over wide dynamic range. The basic logarithmic ammeter measures only one polarity of current. Because it is nonlinear, it is most appropriate for strictly DC measurements. Gain accuracy and stability are affected by the strong temperature dependence of the diode or transistor $I-V$ characteristics.

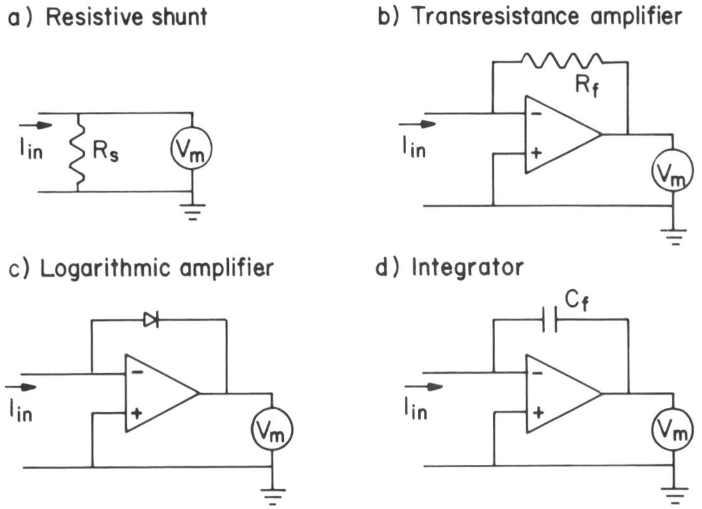

a) Resistive shunt   b) Transresistance amplifier

c) Logarithmic amplifier   d) Integrator

FIG. 2. Schematic diagram of low-current measurement methods.

A capacitor can also be used to sense current by integrating to measure charge. This can be useful when the total charge is of interest. By measuring the charge input in a given period, current waveforms can be averaged. Because the capacitor does not contribute Johnson noise, an integrator can provide the highest current sensitivity.

15.2.2 Resistive Shunt Ammeter

The resistive shunt ammeter technique is the most basic and is used in most digital multimeters (DMMs). As shown in Figure 2a, a shunt resistor of known value R_s is inserted into the circuit. A voltmeter measures the resistive voltage drop V_m, and the current is determined from $I_{in} = V_m/R_s$.

In practice, several effects limit the performance of the shunt ammeter. The first limitation of the shunt ammeter is voltage burden. The specific value of R_s must be chosen to generate a large enough voltage for the voltmeter to measure accurately. This means that the shunt ammeter has a significant voltage drop across its terminals. In measurement applications such as those employing a photodiode that use a small bias voltage in series with the ammeter and photodiode, voltage burden can cause a significant error by changing the photodiode bias. In that situation the voltage appearing across the photodiode is reduced to $V_{bias} - V_{burden}$.

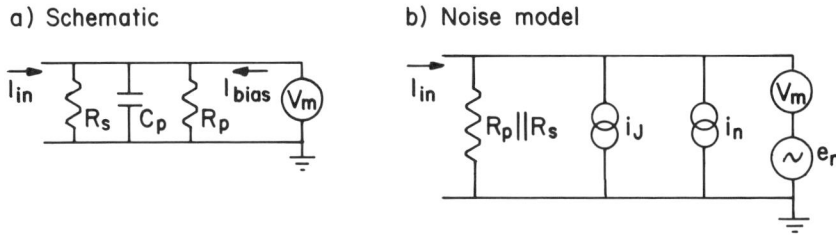

FIG. 3. Resistive shunt ammeter.

Additionally, the connection of the voltmeter to the shunt resistor places the impedance of the voltmeter input and connecting cables in parallel with the shunt resistor. This impedance is represented in Figure 3a by R_p and C_p. In order to achieve adequate sensitivity to low currents very high values of R_s may be required, perhaps as high as 100 GΩ (1 × 10^{11} Ω). Ordinary voltmeters typically have an input resistance between 1 MΩ and 10 GΩ which when placed in parallel with R_s would dramatically reduce the effective shunt resistance value. With a voltmeter input resistance, R_p, in parallel with R_s, the voltmeter will measure

$$V_m = I_{in} \frac{R_s R_p}{(R_s + R_p)}, \qquad (15.1)$$

and the apparent current will be $I_m = V_m/R_s \neq I_{in}$. Thus, unless $R_p \gg R_s$, neglect of R_p can result in a significant measurement error. For high values of R_s, an electrometer voltmeter may be used because it typically offers very high input resistance (as high as 1 × 10^{16} Ω).

A related effect that limits shunt ammeter performance is that the input current I_{in} must charge any stray capacitance in parallel with R_s in order for the resistive voltage drop to settle to its final value. Thus, the response of the shunt ammeter will be limited by the RC time constant associated with R_s in parallel with R_p and C_p so that the frequency at which the signal is attenuated by 3 dB is

$$f_{3\,dB} = \frac{(R_s + R_p)}{2\pi R_s R_p C_p}. \qquad (15.2)$$

With large values of R_s, stray capacitance and the input capacitance of the voltmeter can cause settling times of many seconds. For example, to sense 1 pA with a voltmeter measuring 0.1 V requires R_s = 0.1 V/1 pA = 100 GΩ. If 1 m of coaxial cable is used to connect the shunt and voltmeter to the experiment, a stray capacitance of about 100 pF would appear across R_s, resulting in an RC time constant of 10 sec. It would take 4.6 time constants (46 sec) for this signal to settle to within 1% of its final value.

The input bias current of the voltmeter itself will cause an offset error in the measured current. The input bias current I_{bias} is caused by bias voltages and leakage in the voltmeter input. This current directly adds or subtracts from I_{in}, the input current to the ammeter.

A noise model for the shunt current amplifier is shown in Figure 3b. The voltmeter is modeled as a noiseless voltmeter with a current noise source, i_n, and voltage noise source, e_n, as shown. The resistive current shunt and any parallel resistance also contribute Johnson thermal noise i_j given by [1]

$$i_j = \sqrt{\frac{4kTB}{R_s \| R_p}}, \qquad (15.3)$$

where k is Bolzman's constant, T is the absolute temperature, $R_s \| R_p$ is the equivalent resistance of R_p in parallel with R_s, and B is the noise bandwidth of the circuit. For a first-order system, $B = (\pi/2)f_{3\,dB}$ [1]. Since these three noise sources are typically uncorrelated and random, total noise will be given by the root of the sum of their squares, i.e.,

$$\text{total current noise} = \sqrt{(e_n/R_s \| R_p)^2 + i_n^2 + i_j^2}. \qquad (15.4)$$

15.2.3 Transresistance Ammeter

It is clear that for low current levels the shunt ammeter has significant limitations. A simple but very effective technique to avoid many of these limitations is to use a transresistance ammeter, also known as a feedback ammeter. This approach, shown in Figure 4a, uses an operational amplifier (op amp) to create a current-to-voltage converter. The op amp ideally has infinite voltage gain and zero input current. With feedback from the output to the inverting input, the op amp produces whatever output is necessary to produce a zero differential input signal.

In contrast to the shunt ammeter, the transresistance ammeter better achieves the characteristics of an ideal ammeter. The resistive current sensing element R_f is placed in the feedback path of the operational amplifier. While the output voltage is still proportional to the input current, the op amp holds the input at virtual ground. The voltage burden is given by

$$V_{burden} = I_{in}R_{in} + V_{OS} = \frac{I_{in}R_f}{A_{OL}} + V_{OS} = \frac{V_{out}}{A_{OL}} + V_{OS}, \qquad (15.5)$$

where A_{OL} is the op-amp open-loop voltage gain, typically on the order of 1 million, V_{OS} is the op-amp offset voltage, typically millivolts or less,

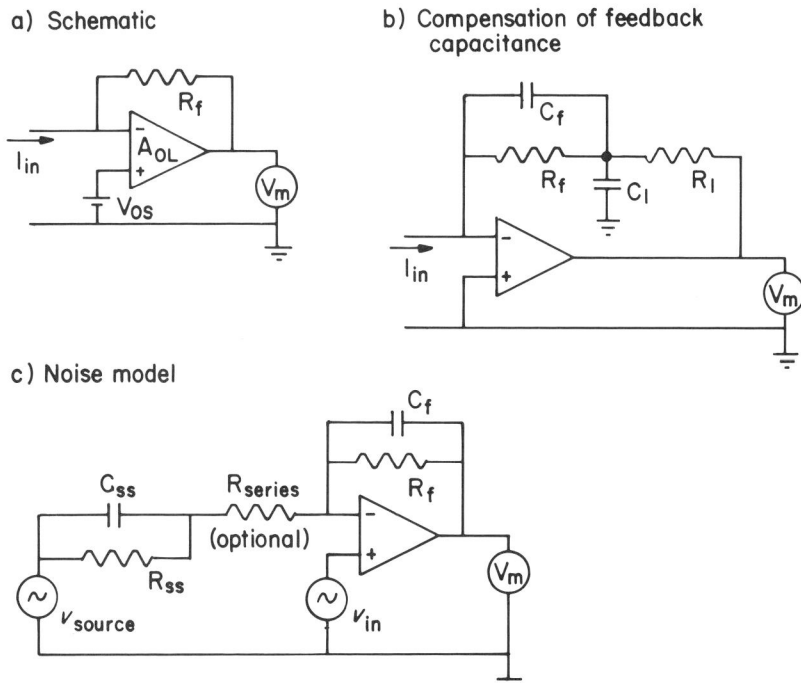

Fig. 4. Transresistance ammeter.

and $R_{in} = R_f/A_{OL}$ is the effective input resistance. Another benefit of this approach is that the shunt resistance and capacitance of connecting cables and the voltmeter do not appear across the sensing element. Thus the gain is simply $V_{OUT} = I_{in}R_f$, and the risetime of the circuit is more tightly controlled such that

$$f_{3\,dB} = \frac{1}{2\pi R_f C_f}, \qquad (15.6)$$

where C_f is the stray feedback capacitance across R_f as indicated in Figures 4b and 4c.

The risetime of the transresistance amplifier is set by $R_f C_f$. In order to achieve high sensitivity, very large values of R_f are required. Even for small values of C_f settling times can be long. For instance, a current gain of 1 V/pA requires $R_f = 1 \times 10^{12}$ Ω, and a stray capacitance of even 1 pF across R_f will result in a time constant of 1 sec. To achieve high speed the stray capacitance C_f should be minimized through careful mechanical design. For further improvement, the remaining stray capacitance can be neutralized using a special circuit configuration as shown in Figure 4b. By matching the time constant of R_1C_1 to equal R_fC_f the response becomes that of R_f in parallel with effectively "zero" capacitance.

For the case of a resistive signal source of resistance R_{ss}, both R_{ss} and R_f will contribute Johnson current noise so that, similar to the resistive shunt ammeter,

$$\text{total current noise} = \sqrt{(e_n/R_f\|R_{ss})^2 + i_n^2 + i_j^2}. \qquad (15.7)$$

The noise model shown in Figure 4c illustrates how the noise gain of the transresistance amplifier is affected by the impedance of the input signal source, which we can model as a voltage source V_s with output impedance Z_{ss}. An optional series resistance R_{series} may be inserted between the signal source and the ammeter to limit noise gain [2]. If v_{in} is the input amplifier voltage noise and v_{source} is the noise of V_s, the output voltage noise v_{out} is given by

$$v_{out} = v_{in}\left(1 + \frac{Z_f}{(Z_{ss} + R_{series})}\right) + v_{source}\left(\frac{Z_f}{Z_{ss} + R_{series}}\right), \qquad (15.8)$$

where

$$Z_f = \frac{R_f}{\sqrt{(2\pi f R_f C_f)^2 + 1}} \quad \text{and} \quad Z_{ss} = \frac{R_{ss}}{\sqrt{(2\pi f R_{ss} C_{ss})^2 + 1}}. \qquad (15.9)$$

When the current signal is produced by a detector with high capacitance, the noise on the bias supply for $f \gg [1/(2\pi f R_f C_f)]$ approaches

$$v_{out} = v_{in}\left(\frac{C_{ss}}{C_f}\right).$$

Using a larger value for C_f will lower this noise gain but will also slow response. Although inclusion of R_{series} reduces noise gain at high frequencies, it also increases voltage burden to $I_{in}R_{series}$. An alternative approach is to place a pair of low leakage diodes in parallel with opposite polarities (back to back) in place of R_{series}. In this manner, when large currents are flowing to charge the capacitance of the device on the input, the diodes will have a low impedance and will not cause voltage burden beyond one diode drop. Once the signal has settled and noise is most critical, the diode impedance is higher and the diodes reduce high-frequency noise gain.

The operational amplifier must be carefully chosen for this ammeter circuit. Its input bias and noise currents directly sum with the input signal, and its input offset and noise voltages create voltage burden and error in

V_m. When selecting or building an ammeter that relies on a resistive current sensor, additional error sources must be considered. These include excess resistor noise above the ideal Johnson thermal noise, drift in the resistor value due to the time or temperature coefficient of resistance, nonlinearity due to the voltage coefficient of resistance, and leakage resistance and stray capacitance due to the mechanical mounting or switching of the sensing resistor.

15.2.4 Logarithmic Ammeter

In situations in which a wide range of currents is anticipated, a logarithmic amplifier might be preferable. Figure 2c illustrates the basic transdiode logarithmic amplifier. This circuit generates an output voltage [3],

$$V_{out} = \frac{mkT}{e} \ln\left(\frac{I_{in}}{I_0}\right) + I_{in} R_B, \qquad (15.11)$$

where e is the electronic charge, m is the empirical diode conduction factor (between 1 and 2), I_0 is the extrapolated diode current at zero voltage, and R_B is the ohmic bulk resistance of the diode. While the transdiode circuit gives an output proportional to the natural log of the current input, its range of use is limited by I_0, R_B, and m. At low current levels, the ratio of I_{in} to I_0 becomes too small to accurately measure. At high signal levels the term $I_{in} R_B$ causes deviation from logarithmic behavior. The factor m introduces additional uncertainty in the measurement.

A transdiode circuit that eliminates the effect of m and R_B is shown in Figure 5a. The input current flows into the collector of the feedback transistor, and the output of the op amp is connected to the emitter. Since the collector current is an accurate exponential function of V_{BE}, the op

a) Basic transdiode amplifier

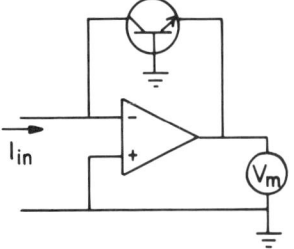

b) Compensated transdiode amplifier using a matched transistor pair

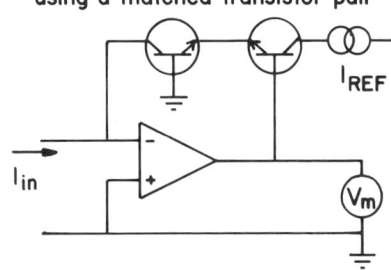

FIG. 5. Logarithmic current amplifier.

amp output will be approximately one base-emitter diode drop below ground and will change (mkT/e) ≈60 mV per decade of collector current. V_{BE}, however, is also an exponential function of temperature, so circuits of this type are often temperature compensated by adding a second matched base emitter junction under a constant bias and a polarity opposite that of the feedback loop, as shown in Figure 5b.

Since the two V_{BE} drops will match when $I_{in} = I_{REF}$, this circuit measures the ratio (I_{in}/I_{REF}) according to (for $h_{FE} \gg 1$)

$$V_{out} = \frac{kT}{e} \ln\left(\frac{I_{in}}{I_{REF}}\right). \tag{15.12}$$

Even more sophisticated compensation schemes which reduce errors due to mismatch between transistors and which allow operation over a wider dynamic range have been designed [4]. When selecting or building an ammeter that relies on a diode or transistor as a current sensor, it is important to also consider shot noise due to current flow through the semiconductor junction [1] given by $I_{rms} = \sqrt{qI_{DC}B}$, where B is the measurement bandwidth, and to match the gain of transistor pairs and mount them on a common heat sink to minimize drift with changing ambient temperature.

15.2.5 Current Integrators

The current integrator employs a current measurement method distinctly different from those described in earlier sections. A current integrator is essentially a charge measuring device. For applications in which the fundamental measurement is of charge instead of current, this is a superior approach. The circuit configurations used, shown in Figure 6, are similar to those of the resistive shunt and transresistance ammeters

a) Shunt integrator   b) Feedback integrator

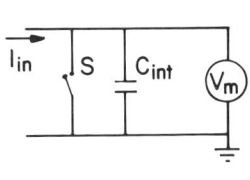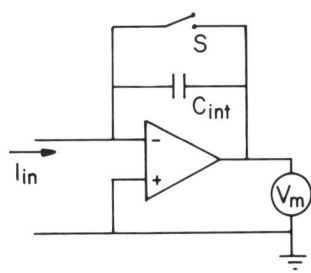

FIG. 6. Current integrators.

except that the sensing element is a capacitor. The use of a capacitor eliminates the Johnson thermal noise of resistive sensors.

In essence, each circuit in Figure 6 consists of an integrating capacitor and a voltmeter to measure the voltage V_{int} across it, given by

$$V_{int}(t = t_1) = \frac{1}{C_{int}} \int_{t=0}^{t=t_1} i_{in}(t)\, dt = \frac{Q_{in}}{C_{int}}. \tag{15.13}$$

Note that the integrating capacitor must be "reset" by discharging at $t = 0$ and before its voltage exceeds the range of the voltmeter used or permissible output swing of the op amp. This necessitates the reset switch, S, shown in Figures 6a and 6b, which must have high isolation characteristics. When it is open and the circuit is integrating, the off resistance of S discharges the integrator with time constant $R_{off}C_{int}$. Spurious leakage through S can lead to measurement errors. The behavior of S as it is in transition from being closed to being open is also important. When opening, a switch such as a reed relay will actually generate a charge pulse due to the change in the relay coil potential coupled through stray capacitance and a change in mechanical stress as physical components of the switch move relative to one another.

There are two principal methods for determining the average current, i_{avg}, between $t = 0$ and $t = t_1$. In the first, V_{int} is measured at regular intervals, Δt, when $i_{avg} = C_{int}\Delta V_{int}/\Delta t$. Alternately, the input current can be integrated from $t = 0$ to some time, t_1, at which V_{int} exceeds a predetermined $V_{threshold}$ when $i_{avg} = C_{int}V_{threshold}/t_1$.

For a shunt integrator, ideally $V_m = V_{int}$. However, the shunt integrator is limited by many of the same problems as the shunt ammeter. When the experiment and voltmeter are connected to the integrating capacitor, the stray capacitance C_{stray} directly adds to C_{int} and increases its value (reducing output signal magnitude). Further, the stray resistance of connecting cables and the voltmeter input will discharge the integrating capacitor with a time constant, $R_{stray}(C_{int} + C_{stray})$. If this method is used an electrometer voltmeter will minimize shunt errors. It provides very high input resistance (up to 1×10^{16} Ω), low input capacitance (as low as 2 pF), and low input bias current (as low as 0.01 fA). However, during a measurement the voltage across the shunt integrating capacitor builds up, resulting in voltage burden. Further, this voltage burden error will accumulate over time and may be very difficult to correct.

For these reasons a feedback integrator configuration as shown in Figure 6b is strongly recommended. In this case, the feedback loop senses the voltage across the integrating capacitor so that $V_{out} = -V_{int}$. This configuration maintains low input voltage burden (approximately constant at V_{OS}). Most importantly, it moves the current sensing element from being a shunt across the input to the feedback loop. Thus, even when cable and detector capacitance is added to the input and when voltmeter input capacitance is added to the output, the value of the integrating capacitor is precisely known and remains C_{int}.

As with a transresistance ammeter, when the current signal is produced by a detector with capacitance C_d, input voltage noise is amplified according to

$$v_{out} = v_{in}\left(\frac{C_d}{C_f}\right). \tag{15.14}$$

In this case, however, the signal gain is also inversely proportional to C_f and is independent of frequency over a wide range. This can be important for accurately digitizing charge vs time waveforms. When selecting or building an ammeter that relies on a capacitor as a current sensor, it is important to consider potential errors due to changes in the capacitance of the sensing capacitor with temperature or applied voltage and due to capacitor dielectric absorption and leakage.

15.3 Signal Quantification and Conversion

Each of the methods discussed so far converts an input current to a voltage. The output voltage is often measured using an analog-to-digital converter (ADC) or some recording device. Common types include voltage-to-frequency (V-to-f) converters, sampling ADCs, integrating ADCs, or waveform recording devices such as an oscilloscope or a chart recorder.

In experiments such as those conducted with synchrotron radiation sources, it is common to connect the current amplifier output to a V-to-f converter. The V-to-f convertor provides an output signal of digital pulses whose frequency is proportional to the input voltage. Thus, the combination of a current amplifier (current-to-voltage converter) and a V-to-f converter forms a current-to-frequency converter. This approach has the advantage that the "frequency" output signal can be transmitted over long distances, through hostile environments, or to regions at very different potential using fiber optics.

The output of the V-to-f converters is measured using a frequency counter. The instantaneous current is given by the instantaneous pulse frequency. The average value during some interval is simply the number of pulses during that period divided by its duration. The total count for that period also represents total charge input during that interval. By recording the entire digital waveform, each of these parameters can be obtained through postprocessing.

Several circuit configurations and the wide offering of commercial V-to-f converters (also called voltage controlled oscillators, or VCOs) are summarized elsewhere [1]. A number of V-to-f converters offer linearity of better than 0.01% at a frequency of 10 kHz and have the additional advantage of sampling the signal continuously during the measurement.

For measurements requiring higher speed, a sampling ADC may be a good solution. Such converters are popular in personal computer data acquisition systems and systems designed for high-speed digitization. They often require a sample and hold stage to prevent the input voltage level from changing during the conversion which can lead to error. ADCs of this type, such as flash or successive approximation converters, periodically sample the waveform and produce digital signals representative of the waveform during the brief sample and hold or conversion periods. Unlike the V-to-f converter, sampling ADCs can miss portions of the signal between conversions. This can be a problem when digitizing noisy signals or when the integrated current is of interest.

For measurements in which resolution is more critical than high speed and in which periodic noise such as power line interference is a problem, an integrating ADC offers a significant improvement over sampling ADCs. The input signal is "integrated" for a period of time to produce an output relative to a reference signal. The integration period is chosen to comprise an integral number of cycles of the periodic noise waveform. Thus, the noise signal integrates to zero and is rejected, limited mainly by the match between the noise period and the integration time.

15.4 Practical Experimental Considerations

Even after selection and implementation of the most appropriate measurement method, several practical considerations need to be addressed. DC current measurements, especially in the range of nanoamperes or less, are particularly susceptible to error. Minimizing current measurement error requires identification of sources of potential interference, careful apparatus design, and measurement of spurious signals.

The environment in which an experiment is conducted can strongly affect measurement results. The principal factors to be considered are temperature, light, ionization interference, and humidity. Temperature control is important to allow accurate compensation of leakage currents. Any semiconductor junctions connected to the ammeter input (such as protection diodes or JFETs in the electrometer–op amp circuit) have leakage currents which are exponentially dependent on temperature. In the vicinity of room temperature, a common rule of thumb is that such

PRACTICAL EXPERIMENTAL CONSIDERATIONS   433

leakage currents double with every 10°C increase in temperature. If the temperature at such junctions is held constant, leakage will contribute a constant current offset to the measurement which can easily be measured with zero signal input and subtracted from subsequent measurements. If, however, the temperature varies, the leakage currents will change, causing drift and error. Temperature changes can also cause mechanical expansion and contraction of insulator materials that can generate currents due to movement of charge in the insulator.

Another environmental factor to consider is light. Many parts of the current measurement circuit use semiconductor junctions to provide high isolation. If light is allowed to shine on these junctions, even through glass seals surrounding component leads in the device package, photocurrents can be generated of picoamperes or higher.

Ionization of the air can also generate small spurious currents. This can be caused by radioactive decay of components of lead–tin solder or ceramic packages. Ions produced in this way can be attracted to the measurement node and deposited. Such effects can be minimized by not using large quantities of solder containing lead or tin near the measurement node and keeping the local electric field to a minimum.

High humidity can also increase the severity of leakage currents. In a humid environment, some insulators will experience a drop in surface resistivity due to moisture absorption. This increases the leakage in response to a bias voltage. In addition, moisture can combine with ionic contaminants from solder flux residue or body oil contamination to create electrochemical reactions that source error currents even in the absence of external bias. The best way to prevent such errors is to minimize insulator contamination initially and to clean any unavoidable contamination with fresh methanol or deionized water. Allow surfaces to dry completely before resuming measurements. Then maintain humidity in critical areas to <50%.

Triboelectric charge generation is a limiting factor for insulators such as Teflon. Charge is generated triboelectrically when surfaces rub against one another, creating a charge imbalance. Piezoelectric effects in crystalline insulators and stored charge effects in polymers cause charge generation in response to mechanical deformation of the insulator. To minimize these difficulties, avoid vibration or relative movement of components of the experiment. It is advisable to tape cables down to prevent flexing and avoid taut cable routing. Both triboelectric and mechanical stress effects can be minimized in cables by choosing a "low-noise"-type cable. Such cables have a conductive lubricant between the shield braid and insulators that reduces friction and provides a local discharge path for triboelectrically generated charge.

Dielectric absorption can cause spurious current when insulators are stressed with high voltage for a period of time. Dielectric absorption causes the insulator to store minute amounts of charge via long time-constant processes, so that even after removal of the high voltage the insulator will only gradually discharge, sourcing an error current as it does so. This can be minimized by avoiding stressing critical insulators.

Stray capacitance coupling to the current measurement node can also result in unintended displacement current $I = CdV/dt + VdC/dt$, where V is the potential of some local electrode and C is the capacitance between that electrode and the measurement node. Such effects can be reduced by minimizing C, keeping high voltages away from sensitive measurement leads, and keeping the physical configuration (C) and electric field (V) stable.

The sensitivity of the circuit to external electric fields can be minimized through careful shielding. To accomplish this, the ammeter input and any high resistances in the test circuit should be surrounded with a solid conductive shield maintained at or near the potential of the measurement input. When the voltage burden is low, shields should be connected to ammeter common.

Care must be taken when connecting the experiment and ammeter and when configuring shields to avoid creating ground loops. A ground loop is formed whenever the experiment is connected to a current-carrying conductor (such as a power-line ground bus) at more than one location. When that occurs, the experiment is placed in parallel with the current-carrying conductor and part of that current flows through measurement leads, possibly creating offset voltages and currents. To avoid ground loops, connect the experiment to the ground bus at only one point.

Some low-current measurements involving, for example, ionization chambers or photomultiplier tubes involve high voltage. It is important to ensure that in a fault condition in the experiment the ammeter is not subjected to signal levels that would severely overload or even damage its sensitive circuits. Typically commercial ammeters and op amps will be rated for a maximum voltage and current input. To protect the ammeter it may be necessary to add external protection such as a series resistance, R_{limit}, to limit the maximum current into the ammeter to V_{max}/R_{limit}, diodes across the ammeter input to limit the maximum input voltage to $\pm V_{forward}$, or both.

References

1. P. Horowitz and W. Hill, *The Art of Electronics*, 2nd ed. pp. 293, 431–453. Cambridge Univ. Press, New York, 1989.

2. *Keithley Low Level Handbook,* 4th ed., p. 3-30. Keithley Instruments, Cleveland, Ohio, 1993.
3. *Model 617 Programmable Electrometer Instruction Manual.* Keithley Instruments, Cleveland, Ohio, 1984.
4. M. N. Ericson, K. G. Falter, and J. M. Rochelle, *IEEE Trans. Instrum. Meas.* **IM-41,** 968 (1992).

16. SIGNAL ENHANCEMENT

John R. Willison

Stanford Research Systems, Inc., Sunnyvale, California

16.1 Introduction

Many experimental techniques rely on the quantitative measurement of charged particles. Often, the signal of interest is obscured by noise. The noise may be fundamental to the process: discrete charges are governed by Poisson statistics which give rise to shot noise. On the other hand, the noise may be from more mundane sources, such as microphonics, thermal EMFs, or inductive pickup. This article describes methods for making measurements of weak signals, even in the presence of large interfering sources, emphasizing the electronic aspects of the measurement.

Figure 1 shows the elements of a typical experiment in which the photo ionization cross-section for a gas is determined by passing a laser through the gas and measuring the number of photoions which are created. The photoions are collected and detected by a particle multiplier. The problem is to distinguish the photoion signal from background sources and noise.

One important source of noise is the amplifier used to amplify the output of the detector before the signal is analyzed. Often, the amplifier noise will be the limiting factor in determining the S/N ratio in a measurement, especially in situations in which the charge detector has no gain.

There are two broad categories of signal analysis, depending on whether or not the source is modulated. Modulating the source, i.e., making an AC measurement, allows the signal to be distinguished from the background. Often, source modulation is inherent to the measurement. For example, when a pulsed laser is used to induce a current, the signal of interest is present only after the laser fires. Other times, the modulation must be "arranged," as when a CW source is chopped. Pulsed measurements improve the S/N ratio by measuring the signal over the brief time when the S/N is favorable, for example, when the pulsed laser is "on," excluding noise at other times. Chopped measurements improve the S/N ratio by measuring the signal in an arbitrarily narrow bandwidth around the chopping frequency, excluding noise at other frequencies.

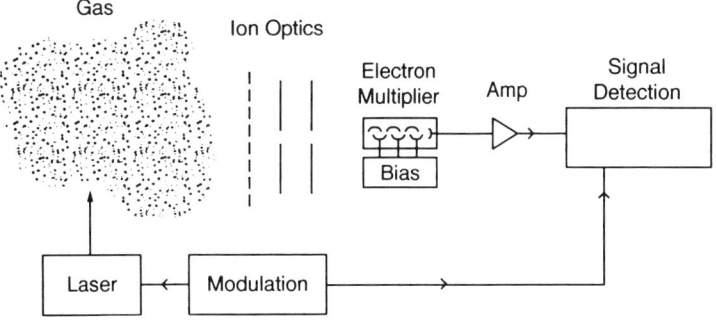

FIG. 1. Prototype experiment.

16.2 Noise Sources

An understanding of noise sources in a measurement is critical to achieving signal-to-noise performance near theoretical limits. The quality of a measurement may be substantially degraded by a trivial error. For example, a poor choice of termination resistance for an electron multiplier may increase current noise by several orders of magnitude [1].

16.2.1 Shot Noise

Light and electrical charge are quantized, and so the number of photons or electrons which pass a point during a period of time are subject to statistical fluctuations. If the signal mean is M photons, the standard deviation (noise) will be $\sqrt{M}$; hence, the S/N = $M/\sqrt{M}$ = $\sqrt{M}$. The mean, M, may be increased if the rate is higher or the integration time is longer. Short integration times or small signal levels will yield poor S/N values. Figure 2 shows the S/N ratio which may be expected as a function of current level and integration time for a shot-noise limited signal.

Since electrical current is due to the motion of quantized charges, there is a current noise associated with the statistical fluctuation of the charge motion. The rms noise current in the bandwidth Δf due to a "constant" current, I (in amperes), is given by

$$I_{\text{shot noise}} = \sqrt{2qI\Delta f},$$

where $q = 1.6 \times 10^{-19}$ C.

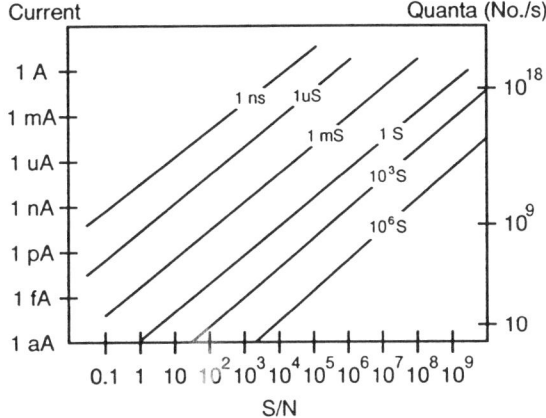

FIG. 2. Signal to noise vs. flux and measurement time.

16.2.2 Johnson Noise

The electrons which allow current conduction in a resistor are subject to random motion which increases with temperature. This fluctuation of electron density will generate a noise voltage at the terminals of the resistor. The rms value of this noise voltage for a resistor of R (in ohms), at a temperature of T (in kelvins), in a bandwidth of Δf is given by

$$V_{\text{johnson rms}} = \sqrt{4kTR\Delta f},$$

where k is Boltzmann's constant. The noise voltage in a 1-Hz bandwidth is given by

$$V_{\text{johnson, rms}} \text{ (per } \sqrt{\text{Hz}}) = 0.13 \text{ nV} \times \sqrt{R}.$$

Since the Johnson noise voltage increases with resistance, large value series resistors should be avoided in voltage amplifiers. For example, a 1-kΩ resistor has a Johnson voltage of about 4.1 nV/$\sqrt{\text{Hz}}$. If detected with a 100-MHz bandwidth, the resistor will show a noise of 41 μV rms, which has a peak-to-peak value of about 200 μV. When a resistor is used to terminate a current source or is used as a feedback element in a current-to-voltage converter, it will contribute a noise current equal to the Johnson noise voltage divided by the resistance. Here, the noise current in a 1-Hz bandwidth is given by

$$I_{\text{johnson, rms}} \text{ (per } \sqrt{\text{Hz}}) = 130 \text{ pA}/\sqrt{R}.$$

440   SIGNAL ENHANCEMENT

Since the Johnson noise current increases as R decreases, small value resistors should be avoided when terminating current sources. Unfortunately, small terminating resistors are required to maintain a wide frequency response. If a 1-kΩ resistor is used to terminate a current source, the resistor will contribute a noise current of about 4.1 pA/$\sqrt{\text{Hz}}$, which is about 1000× worse than the noise current of an ordinary FET input operational amplifier.

16.2.3 1/f Noise

The voltage across a resistor carrying a constant current will fluctuate because the resistance of the material used in the resistor varies. The magnitude of the resistance fluctuation depends on the material used: carbon composition resistors are the worst, metal film resistors are better, and wire-wound resistors provide the lowest $1/f$ noise. The rms value of this noise source for a resistance of R, at a frequency of f, in a bandwidth of Δf is given by

$$V_{1/f,\,\text{rms}} = IR \times \sqrt{A\Delta f/f},$$

where the dimensionless constant A has a value of about 10^{-11} for carbon. In a measurement in which the signal is the voltage across the resistor (IR), the S/N ratio is about $3 \times 10^5 \sqrt{f/\Delta f}$. Often, this noise source is a troublesome source of low-frequency noise in voltage amplifiers.

It is important to note the impact noise sources have on the S/N ratio may be reduced by narrowing the bandwidth of the signal detection. The noise bandwidth may be reduced at the expense of increasing the duration of the measurement.

16.2.4 Nonessential Noise Sources

There are many discrete noise sources which must be avoided in order to make reliable measurements of low level signals. Figure 3 shows a simplified noise spectrum on log–log scales. The key features in this noise spectrum are frequencies worth avoiding: diurnal drifts (often seen via input offset drifts with temperature); low-frequency ($1/f$) noise; power line frequencies and their harmonics; switching power supply and CRT display frequencies; commercial broadcast stations, including AM, FM, VHF, and UHF TV; special services such as cellular telephones and pagers; and a variety of microwave sources. Strategies to reduce these noise sources include: single-point grounding or shielding to reduce pickup, the use of differential inputs to reject noise between the source and the amplifier ground, matching the amplifier bandwidth to the spectrum of the expected signal, operating at a quiet part of the noise spectrum (in

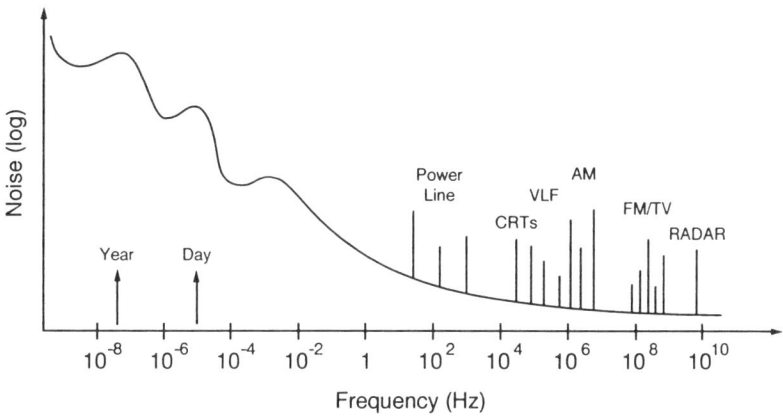

FIG. 3. Simplified noise spectrum.

chopped source experiments), and triggering synchronously with noise sources (in pulsed experiments).

Common ways for extraneous signals to interfere with a measurement are illustrated in Figures 4a–4f. Noise may be injected via a stray capacitance as in Figure 4a. The stray capacitance has an impedance of $1/j\omega C$. Substantial currents may be injected into low-impedance systems (such as transconductance inputs), or large voltages may appear at the input to high-impedance systems.

Inductive pickup is illustrated in Figure 4b. The current circulating in the loop on the left will produce a magnetic field which in turn induces an EMF in the loop on the right. Inductive noise pickup may be reduced by reducing the areas of the two loops (by using twisted pairs, for example), by increasing the distance between the two loops, or by shielding. Small skin depths at high frequencies allow nonmagnetic metals to be effective shields; however, high-permeability materials must be used as shields for DC and low-frequency magnetic fields.

Resistive coupling, or a "ground loop," is shown in Figure 4c. Here, the detector senses the output of the experiment plus the IR voltage drop from another circuit which passes current through the same ground plane. Cures for ground loop pickup include grounding everything to the same point, using a heavier ground plane, providing separate ground return paths for large interfering currents, and using a differential connection between the signal source and amplifier.

Mechanical vibrations can create electrical signals (microphonics) as shown in Figure 4d. Here, a coaxial cable is charged by a battery through a large resistance. The voltage on the cable is $V = Q/C$. Any deformation

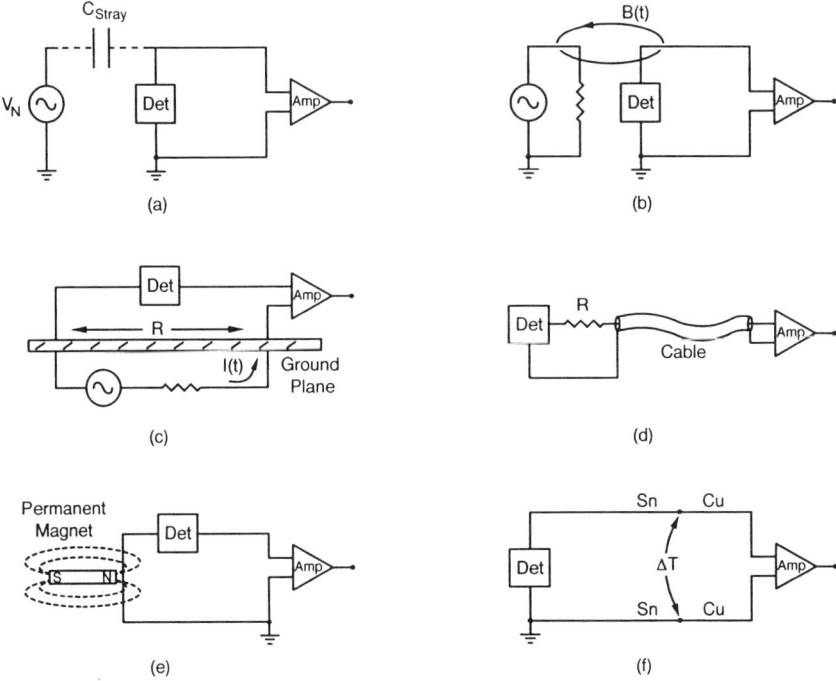

FIG. 4. Coupling of noise sources.

of the cable will modulate the cable's capacitance. If the period of the vibration which causes the deformation is short compared with the RC time constant, then the stored charge on the cable, Q, will remain constant. In this case, a 1 ppm modulation of the cable capacitance will generate an AC signal with an amplitude of 1 ppm of the DC bias on the cable, which may be larger than the signal of interest.

The case of magnetic microphonics is illustrated in Figure 4e. Here, a DC magnetic field (the Earth's field or the field from a permanent magnet in a latching relay, for example) induces an EMF in the signal path when the magnetic flux through the detection loop is modulated by mechanical motion.

Unwanted thermocouple junctions are an important source of offset and drift. As shown in Figure 4f, two thermocouple junctions are formed when a signal is connected to an amplifier. For typical interconnect materials (copper and tin) one sees about 10 μV/°C of offset. The impact of such slow drifts is eliminated by AC measurements.

16.3 Amplifiers

Several considerations are involved in choosing the correct amplifier to optimize the S/N ratio for an particular application. Often, these considerations are not independent, and compromises will be necessary. The best choice for an amplifier depends on the electrical characteristics of the detector and on the desired gain, bandwidth, and noise performance of the system.

Charge counting and fast-gated integration require amplifiers with wide bandwidth. A 350-MHz bandwidth is required to preserve a 1-nsec rise time. The input impedance to these amplifiers is usually 50 Ω in order to terminate coax cables into their characteristic impedance. When PMTs or charge multipliers (which are current sources) are connected to these amplifiers, the 50-Ω input impedance serve as the current-to-voltage converter for the anode signal. Unfortunately, the small termination resistance and wide bandwidth lead to a lot of current noise [2].

It is important to choose an amplifier with a very high input impedance and low input bias current when amplifying a signal from a source with a large equivalent resistance. Commercial amplifiers designed for such applications typically have a 100-MΩ input impedance. This large input impedance will minimize attenuation of the input signal and reduce the Johnson noise current drawn through the source resistance, which can be an important noise source. Field effect transistors (FETs) are used in these amplifiers to reduce the input bias current to the amplifiers. Shot noise on the input bias current can be an important noise component, and temperature drift of the input bias current is a source of drift in DC measurements [3].

The bandwidth of a high input impedance amplifier is often determined by the RC time constant of the source, cable, and termination resistance. For example, an anode with 1 m of RG-58 coax (about 100 pF) terminating in a 1-MΩ resistor will have a bandwidth of about 1600 Hz. A smaller resistance would improve the bandwidth, but increase the Johnson noise current.

Bipolar transistor inputs offer an input noise voltage which may be several times smaller than the FET inputs of high input impedance amplifiers, as low as 1 nV/$\sqrt{\text{Hz}}$. Bipolar transistors have larger input bias currents, hence a larger shot noise current, and so should be used only with low impedance (<1 kΩ) sources.

Transformer coupling is advantageous when measuring AC signals from very low impedance sources. The transformer steps up the voltage by its turns ratio, and its secondary may be connected directly to the input of a bipolar transistor amplifier.

Conventional bipolar and FET input amplifiers exhibit input offset drifts on the order of 5 μV/°C. For a case in which the detector signal is a small DC voltage, such as that from a bolometer, this offset drift can be the dominant noise source. A different amplifier configuration, chopper-stabilized amplifiers, essentially measures its input offsets and subtracts the measured offset from the signal. A similar approach is used to "autozero" the offset on the input to sensitive voltmeters. Chopper-stabilized amplifiers exhibit very low input offsets with virtually no input offset drift.

The use of "true-differential" or "instrumentation" amplifiers is advised in cases in which there may be a difference between the signal source ground and the amplifier ground. These devices amplify the difference between two inputs, one of which may be the source ground, unlike a single-ended amplifier, which amplifies the difference between the signal input and the amplifier ground. In high-frequency applications, where good differential amplifiers are not available or are difficult to use, a balun, which is a common mode choke, may be used to isolate disparate grounds.

16.3.1 Transconductance Amplifiers

When the detector is a current source (or has a large equivalent resistance) then a transconductance amplifier should be considered. Transconductance amplifiers (current-to-voltage converters) offer the potential of lower noise and wider bandwidth than a termination resistor and a voltage amplifier; however, some care is required in their application [4].

A typical transconductance amplifier configuration is shown in Figure 5. FET input op amps are used for their low input bias current. (Op amps with input bias currents as low as 50 fA are readily available.) Assuming an ideal op amp, the transconductance gain is $A = V_{out}/I_{in} = R_f$, and the input impedance of the circuit is R_{in} to the op amp's virtual null. (R_{in} allows negative feedback, which would have been phase shifted and attenuated by the source capacitance at high frequencies, to assure stability.) Commercial transconductance amplifiers use R_fs as large as 10 MΩ, with R_{in}s which are typically $R_f/1000$. A low input impedance will ensure that current from the source will not accumulate on the input capacitance.

The configuration shown in Figure 5 has several important limitations that degrade its gain, bandwidth, and noise performance. The overall performance of the circuit depends critically on the source capacitance, including that of the cable connecting the source to the amplifier input. The "virtual null" at the inverting input to the op amp is approximately R_f/A_v, where A_v is the op amp's open loop gain at the frequency of interest. While op amps have very high gain at frequencies below 10 Hz (typically a few million), these devices have gains of only a few hundred at 1 kHz.

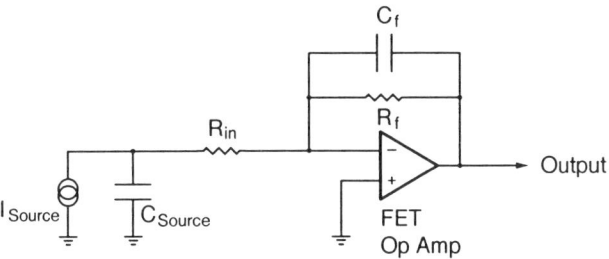

FIG. 5. Typical transconductance amplifier.

With an R_f of 1 GΩ, the virtual null has an impedance of 5 MΩ at 1 kHz, hardly a virtual null. If the impedance of the source capacitance is less than the input impedance, then most of the AC input current will go to charging this capacitance, thereby reducing the gain. The configuration provides high gain for the voltage noise at the noninverting input of the op amp. At high frequencies, at which the impedance of the source capacitance is small compared with R_{in}, the voltage gain for noise at the noninverting input is R_f/R_{in}, typically about 1000. Because FET input op amps with very low bias currents tend to have high input voltage noise, this term can dominate the noise performance of the design.

Large R_fs are desired to reduce the Johnson noise current; however, large R_fs degrade the bandwidth. If low values of R_f are used, the Johnson noise current can dominate the noise performance of the design. To maintain a flat frequency response, the size of the feedback capacitance must be adjusted to compensate for different source capacitances.

Since many undesirable characteristics of the transconductance amplifier can be traced to the source capacitance, it is beneficial to integrate the amplifier into the detector, thereby eliminating interconnect capacitance. This approach is followed in many applications, from microphones to CCD imagers.

16.4 Signal Analysis

A variety of noise sources are avoided by AC measurement of the signal. When making DC measurements, the signal must compete with large low-frequency noise sources. However, when the source is modulated, the signal may be measured at the modulation frequency, away from these large noise sources.

Techniques for pulsed AC measurements include gated integration, boxcar averaging, transient digitizers, gated counters, and multichannel scalers. Chopped AC measurement techniques include lock-in amplifiers and spectrum analyzers.

16.4.1 Gated Integrators

Gated integrators measure the integral of a signal during a short period of time while the S/N ratio is high. Commercial devices allow gates from about 100 psec to several milliseconds. A gated integrator is typically used in pulsed laser experiments, with the integration of the signal taking place during the short interval that the laser is on. The device can provide shot-by-shot data which are often recorded by a computer via an A/D converter. The gated integrator is recommended in situations in which the signal has a very low duty cycle, low pulse repetition rate, and high instantaneous count rates [5].

The noise bandwidth of the gated integrator depends on the gate width: short gates will have wide bandwidths and so will be noisy. This would suggest that longer gates would be preferred; however, the signal of interest may be very short lived, and using a gate which is much wider than the signal will not improve the S/N.

The gated integrator also behaves as a filter: the output of the gated integrator is proportional to the average of the input signal during the gate, so frequency components of the input signal which have an integral number of cycles during the gate will average to zero. This characteristic may be used to "notch out" specific interfering signals. It is often desirable to make gated integration measurements synchronously with an interfering source. (This is the case with time-domain signal detection techniques and not the case with frequency-domain techniques such as lock-in detection.) For example, by locking the pulse repetition rate to the power-line frequency (or to any submultiple of this frequency) the integral of the line interference during the short gate will be the same from shot to shot, which will appear as a fixed offset at the output of the gated integrator.

16.4.2 Boxcar Averaging

Shot-by-shot data from a gated integrator may be averaged to improve the S/N ratio. Commercial boxcar averagers provide linear or exponential averaging. The averaged output from the boxcar may be recorded by a computer or used to drive a strip chart recorder. Figure 6 shows a gated integrator with an exponential averaging circuit. Boxcar averages may be used to recover repetitive waveforms by moving the gate on successive triggers. The method suffers from its low duty cycle, and so transient digitizers and signal averagers are preferred in this application.

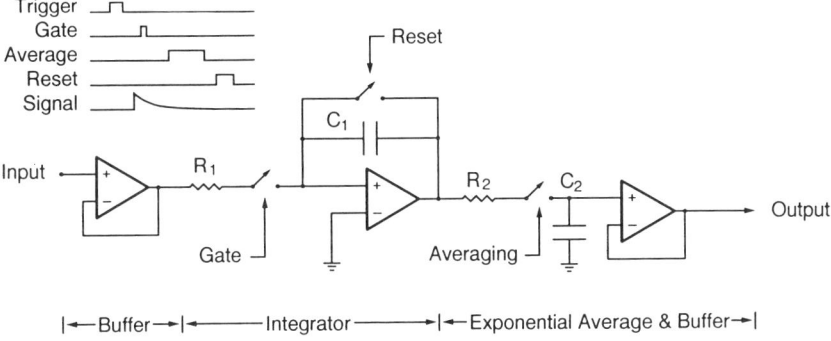

FIG. 6. Gated integrator and exponential averager.

16.4.3 Transient Digitizers and Signal Averagers

These are used to record and average waveforms. A block diagram of a typical transient digitizers is shown in Figure 7. It consists of an input amplifier and a high-speed A/D converter which writes the converted data to a buffer memory. A digital signal processor (DSP) can read the data from a single scan and sum it with data from other scans. Single records or their averages may be displayed or transferred to a host computer for further analysis.

Important features of transient digitizers include resolution, bandwidth, sample rate, record length, processing speed, and transfer rate. There is a direct trade-off between speed and resolution: commercial instruments provide up to 16 bits at 250 kHz, but only 8 bits at 1 GHz. Quantization errors, due to the finite resolution of the A/D converter, are an important source of noise in these devices.

Signals below the level of a least significant bit (LSB) of the A/D converter may be seen in the average of several scans. For this to work,

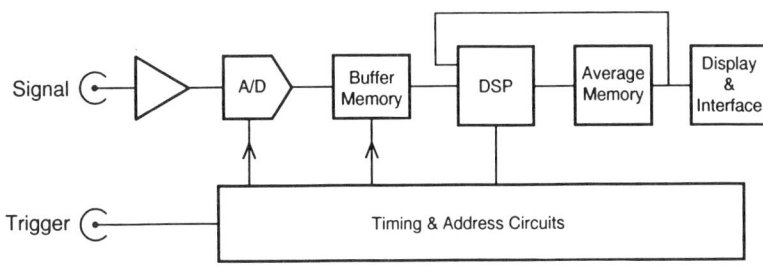

FIG. 7. Transient digitizer.

the signal needs to be able to toggle an LSB occasionally. Broadband noise of several LSBs is sometimes added to the input signal to allow this to occur. A drawback is that more averaging will be required to achieve the same S/N ratio.

Transient digitizers are preferred to multichannel scalers in the high-count-rate case (instantaneous rates exceeding several megahertz), in which there is a significant probability that particles will not be resolved by a counter. However, in the low-count-rate case, multichannel scalers are preferred for the noise reduction offered by the discriminator which precedes the counter.

16.4.4 Lock-In Amplifiers (LIAs)

Lock-in amplifiers use phase-sensitive detection to recover small signals in the presence of noise which may be much greater than that of the signal of interest. In a typical application, the signal of interest will be modulated (or chopped) by some means. It is important that only the signal of interest, and not the noise or background, be modulated. The lock-in amplifier is used to measure the amplitude and phase of the signal of interest relative to the modulation source [6].

Figure 8 shows a simplified block diagram for a lock-in amplifier. The input signal is AC coupled to an amplifier whose output is mixed with (multiplied by) the output of a phase-locked loop which is locked to the reference input. The operation of the mixer may be understood through the trigonometric identity:

$A \cos(\omega_1 t + \phi) * B \cos(\omega_2 t)$
$= AB/2[\cos((\omega_1 + \omega_2)t + \phi) + \cos((\omega_1 - \omega_2)t + \phi)].$

When $\omega_1 = \omega_2$ there is a DC component of the mixer output, $AB \cos(\phi)/2$. The output of the mixer is passed through a low-pass filter to remove the

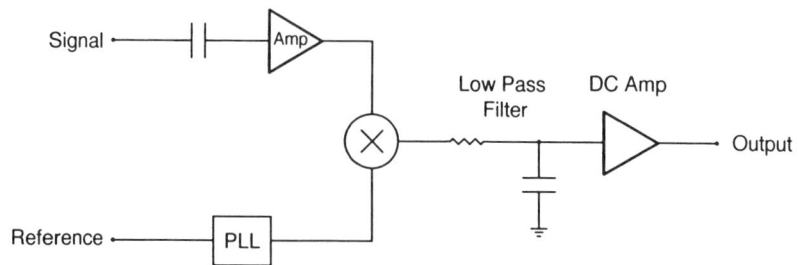

FIG. 8. Lock-in amplifier block diagram.

sum frequency component. The time constant of the filter is selected to reduce the equivalent noise bandwidth: selecting longer time constants will improve the S/N at the expense of longer response times.

The simplified block diagram shown in Figure 8 is for a "single-phase" lock-in amplifier, which measures the component of the signal at one set phase with respect to the reference. A single-phase lock-in amplifier can measure the amplitude and phase of a signal by adjusting the phase of the lock-in amplifier's reference oscillator so that the detected signal is zero ($\cos \phi = 0$) and then by jumping 90° to maximize the signal ($\cos \phi = 1$). A dual-phase lock-in amplifier has another channel which measures the component of the signal at 90° relative to the first channel, which allows simultaneous measurement of the amplitude and phase of the signal.

16.4.5 Digital Signal Processing Techniques

Digital signal processing techniques are rapidly replacing the older analog techniques for synchronous detection in lock-in amplifiers. In these instruments the input signal is digitized by a fast, high-resolution A/D converter, and the signal amplitude and phase are determined by high-speed computations in a DSP. To meet the Nyquist criterion and maintain the 100-kHz bandwidth common to analog designs, digital designs must convert the analog signal to a digital form at a rate of about 250 kHz. To perform the synchronous detection and to meet the filtering requirements of the application, the DSP needs to perform about 20 million multiply-and-accumulate operations each second. Many artifacts of the analog designs are eliminated by the DSP approach; for example, the output drift and dynamic range of the instruments are dramatically improved [7].

Counting techniques offer several advantages in the measurement of low-level signals: very high sensitivity (count rates as low as one per minute can be a usable signal level), a large dynamic range (signal levels as high as 100 MHz can be counted, allowing a 195-dB dynamic range), discrimination against low-level noise (analog noise below the discriminator thresholds will not be counted), and an ability to operate over widely varying duty cycles. Key elements of a particle counting system include a high-gain charge multiplier operated at a sufficiently high voltage so that a single charged particle will generate an anode pulse of several millivolts into a 50-Ω load, a fast discriminator to generate logic pulses from anode signals which exceed a set threshold, and fast gated counters to integrate the counts [8].

In situations in which the time evolution of a signal must be measured (LIDAR, lifetime measurements, chemical kinetics, etc.) multichannel scalers allow the entire signal to be recorded for each event. In these instruments, the discriminated pulses are summed into different bins depending on their timing with respect to a trigger pulse. Commercial instruments offer 5-nsec resolution with zero dead time between bins. The time records from many events may be summed together in order to improve the S/N ratio [9].

The time behavior of the signal is used to choose between the various signal analysis techniques outlined above. If the signal is fixed in frequency and has a 50% duty cycle, lock-in detection is best suited. This type of experiment commonly uses an electronic or mechanical chopper to modulate the signal at some low frequency. Signal events occur at random times during the "open" phase of the chopper. The lock-in amplifier detects the average difference between the signal during the open phase and the background during the "closed" phase.

If the signal is confined to a very short amount of time, then gated integration is usually the best choice for signal recovery. A typical experiment might be a pulsed laser excitation in which the signal lasts for only a short time (100 psec to 1 μsec) at a repetition rate up to 10 kHz. The duty cycle of the signal is much less than 50%. By using a narrow gate to detect the signal only when it is present, noise which occurs at all other times is rejected.

Particle counting may be used to analyze the signal of virtually any duty cycle. The use of a particle counter is recommended at very low signal intensities or if the noise rejection of a pulse height discriminator will improve the S/N ratio.

In the case in which waveform information is required, a multichannel scaler or transient digitizer can greatly reduce the time required to make a measurement. The multichannel scaler should be used in low-count-rate situations, providing that there is sufficient particle multiplier gain to allow the discrimination of individual events. A transient digitizer should be used if the peak particle flux rate exceeds several megahertz or if there is insufficient multiplier gain to discriminate among individual events.

References

1. P. Horowitz and W. Hill, *The Art of Electronics,* pp. 428–447. Cambridge Univ. Press, Cambridge, UK, 1989.
2. *Model SR445 Fast Preamplifier,* Operation and Service Manual. Stanford Research Systems, Sunnyvale, CA, 1990.
3. *Model SR560 Low Noise Preamplifier,* Operation and Service Manual. Stanford Research Systems, Sunnyvale, CA, 1990.
4. *Model SR570 Low Noise Current Amplifier,* Operation and Service Manual. Stanford Research Systems, Sunnyvale, CA, 1992.

5. *Fast Gated Integrators and Boxcar Averagers*, Operation and Service Manual. Stanford Research Systems, Sunnyvale, CA, 1990.
6. *Model SR510 Lock-in Amplifier*, Operation and Service Manual. Stanford Research Systems, Sunnyvale, CA, 1987.
7. *Model SR850 DSP Lock-in Amplifier*, Operation and Service Manual. Stanford Research Systems, Sunnyvale, CA, 1992.
8. *Model SR400 Gated Photon Counter*, Operation and Service Manual. Stanford Research Systems, Sunnyvale, CA, 1988.
9. *Model SR430 Multichannel Scaler/Averager*, Operation and Service Manual. Stanford Research Systems, Sunnyvale, CA, 1989.

Index

A

ABAP, see Accelerator-based atomic physics
AC, see Alternating current
Accelerator-based atomic physics
  accelerators, see also Electrodynamic accelerator; Electrostatic accelerator
   hybrid facilities, 318–319
   ion sources, 301
   tuning, 300–301
   types, 306–307
  applications, 299–306
  collision energy for shell electron ionization, 301–302
  electron capture, 305–306
  facilities, 299–300
  ion penetration in solids, 302–303
Accelerator mass spectroscopy, sensitivity for isotopic selection, 301
Acceptance, ion beam, 73–74
ADC, see Analog-to-digital converter
Airy disk, diameter, 53
Alternating current, measurement
  boxcar averaging, 446
  digital signal processing, 449–450
  gated integration, 446
  lock-in amplifier, 448–449
  transient digitizer, 447–448
Ammeter
  analog-to-digital converters
    integrating converter, 431–432
    sampling converter, 431–432
    voltage-to-frequency converter, 431–432
  components, 421
  current integrator
    feedback integrator, 430–431
    noise model, 431
    principle, 423, 429
    reset switch, 430
    schematic, 423, 429
    shunt integrator, 430

ideal properties, 421
logarithmic ammeter
  output voltage, 428–429
  principle, 422
  schematic, 423, 428
  transdiode amplifier, 428
  maximum input voltage, 434
  performance-affecting factors
    air ionization, 433
    dielectric absorption, 434
    ground loops, 434
    humidity, 433
    light, 433
    room temperature, 432–433
    triboelectric charge generation, 433
  resistive shunt ammeter
    noise model, 425
    performance-limiting effects, 423–424
    principle, 422
    schematic, 423–424
  transresistance ammeter
    noise model, 427
    operational amplifier, 425, 427–428
    principle, 422
    risetime, 426
    schematic, 423, 426
    voltage burden, 425
Amplifier
  boxcar averaging, 446
  digital signal processing, 449–450
  gated integration, 446
  lock-in amplifier, 448–449
  operational, for transresistance ammeter, 425, 427–428
  selection for noise minimization, 443–444
  transconductance amplifier
    noise minimization, 444–445
    schematic, 445
  transdiode, for logarithmic ammeter, 428
  transient digitizer, 447–448
AMS, see Accelerator mass spectroscopy
Analog-to-digital converter
  integrating converter, 431–432

453

sampling converter, 431–432
voltage-to-frequency converter, 431–432
Angular divergence
  circular aperture, 83
  dependence on perveance, 85–86
  slit aperture, 83
Arc plasma discharge source, see also Duopigatron; Duoplasmatron; Penning discharge source
  Calutron sources
    Calutron/Bernas/Nier configuration, 113–114
    design, 112–113
    lifetime, 113–114
  diffusion coefficient across the field, 78
  external field penetration, 92–93
  filament materials, 100–101
  Freeman source, 111–112
  geometry, 75–76
  hollow-cathode ion sources, 105–106
  hot cathode lifetime
    estimation, 98–101
    mechanisms of wear, 98
  ion current density
    atomic mass relationship, 96–97
    extraction voltage relationship, 95
    loss at column ends, 78–79, 100
    sheath edge, 77, 99–100
  ion extraction
    design of system, 93–94
    intensity of ion beam, 95–98
  plasma
    analytical approximation, 81–84
    computational simulation, 85–91
    solid emitters, analytical approximation, 80–81
  ionization, 76–77
  ionization efficiency, 96–98
  ion optical effects in the plasma boundary, 93–95
  material utilization efficiency, 98
  minimum pressure requirements, 79
  multicusp, magnetic-field confinement design, 114–115
  source types, 115
  perveance
    concentric cylinder geometry, 81
    cylindrical emission boundary, 81
    dependence of angular divergence, 85–86
    optimum for minimum angular divergence, 84–85
    parallel plate geometry, 80, 82
    spherical emission boundary, 81
  plasma potential, 92
  plasma sheath formation, 91–92
  principle, 74–75
  stable discharge criteria, 77
  thermionic emission, 75–76
  three-electrode ion extraction system, 84
  voltage breakdown in a vacuum, 98

B

Beam angle
  aperture radius relationship, 191–192
  consideration in combining lenses, 202–203
Beam-foil spectroscopy, applications, 303
Boxcar averaging, signal-to-noise improvement, 446
Brightness, ion beam, 73

C

Calutron source, see Arc plasma discharge source
CCD, see Charge-coupled device
Cesium
  activation of negative-electron-affinity semiconductor photocathode, 15–17
  restoration of photocathode efficiency, 21–22
  sources, 15
  thermal emission of ions, 158–159
Charge-coupled device, camera imaging of microchannel plate output, 260–261
Charged particle optics program, three-dimensional
  electrostatic lens system design, 206–207
  features, 206
CHORDIS, see Cold and hot reflex discharge ion source
Cockroft–Walton accelerator, see Electrostatic accelerator
Cold and hot reflex discharge ion source, properties, 115

CPO-3D, see Charged particle optics program, three-dimensional
CRESU, see Reaction kinetics in uniform supersonic flow technique
Current integrator, see Ammeter
Current measurement, see Ammeter
Cyclotron accelerator, see Electrodynamic accelerator

D

Direct current measurement, see Ammeter
Duopigatron, see also Arc plasma discharge source
  filament lifetime, 105
  geometry, 104–105
  ion-beam densities, 105
  single-aperture source, 105
Duoplasmatron, see also Arc plasma discharge source
  filaments, 102–103
  geometry, 102–104
  ion extraction region, 101–102
  potential distribution within the plasma discharge, 102–103
Dynamitron, see Electrostatic accelerator

E

EBIS, see Electron-beam ion source
EBIT, see Electron-beam ion trap
ECRIS, see Electron–cyclotron resonance ion source
Electrodynamic accelerator, see also Storage ring
  cyclotron accelerator
    development, 312–313
    isochronous cyclotron, 313–314
    maximum energy, 313
    principle, 312
    superconducting cyclotron, 314
  principle, 306–307
  radiofrequency linear accelerator
    applications, 314–315
    principle, 314
  synchrotron accelerator, 315–316

Electron
  impact ionization of ions
    confinement time, 172
    theory, 170–172
  spin-polarized sources, 1, 30–32
Electron-beam ion source
  design, 172–173
  excitation, 172
  fractional abundance of ion charge states, 175
  operation, 174
  performance, 174–177
Electron-beam ion trap
  apparatus, 392–393
  beam
    collection, 395
    energy, 177, 393–395
    radius, 394–395
  charge state
    range, 391
    selection, 399–400
  compactness, 177
  development, 393
  electron–ion collision cross-section measurement, 410, 412
    dielectric recombination, 414–417
    excitation of an X-ray line, 412
    ionization measurement techniques, 413–414
  extraction of highly charged ions, 417–418
  ion heating and cooling, 398–399
  ion injection methods, 395–398
  ion trapping, 397–398
  metal vapor vacuum arc source, 397–398
  operating parameters, 393–394
  performance, 176, 391–392
  super 200-keV instrument
    apparatus, 400–401
    X-ray sectroscopy, 402
  X-ray spectroscopy
    lifetime measurements, 408, 410
    transition energy measurements
      above 2 keV, 402, 404–405
      below 2 keV, 405–406
    X-ray polarization, 406, 408
Electron–cyclotron resonance ion source
  comparison with other sources, 126–127, 184
  design, 178

development, 123, 177–178
electron velocity distribution, 179–180
gas pressure, 181
ion beam intensity, 182–183
ionization mechanism, 124–125
lifetime, 127
low-charge-state sources, 125–127
low-emittance, high-intensity source, 129
magnetic field and microwave frequency relationship, 179
maximum plasma electron density, 183
microwave
  frequency, 183
  power, 182
multicusp-field source, 127
operation, 178–182
performance, 182–185
properties, 123–124
Electron energy analyzer
construction, 227–228
efficiency and electron energy, 303–304
electrostatic deflection devices
  applications, 215
  configurations, 213–214
  energy resolution, 217–218, 228
  figure of merit and luminosity, 218–220
  focusing conditions, 216–217
  fringing field minimization, 220–221, 227
  input lens
    configurations, 223
    design, 224–225
    multielement lenses, 225
    optimization of voltages, 226
  operating modes, 221–223
  operating voltage, 222
  preretardation, 215
  principle, 214–215
electrostatic shielding, 220–221, 227–228
imaging, 226–227
magnetic field, maximum tolerable in construction, 227–228
multichannel energy detection, 226–227
reflection/transmission device, 213
retarding grid devices, 212
time-of-flight devices
  components, 210
  design, 211
  detection system response time, 211
  sources, 209–210
  space-charge effects, 210
Electron polarimeter, *see also* Mott polarimeter
accuracy, 248–249
calibration
  calculation of Mott asymmetry, 232, 235, 243–244
  double-scattering measurements, 244
  measurement of analyzing power, 232, 235, 244
  use of electrons with known polarization, 244–245
classes, 231–232, 248
degree of polarization, 231–232
Mott scattering error sources
  background, 247
  extrapolation error, 245
  instrumental asymmetry, 245–247
optical electron polarimetry
  error sources, 247
  instrumentation, 242
  theory, 234, 242
polarized low-energy electron diffraction polarimeter, 240
secondary electron polarimeter, 242
selection for experiments, 247–250
Electrostatic accelerator
Cockroft–Walton accelerator, 307
Dynamitron, 312
principle, 306
Van de Graaff accelerator
  principle, 307
  resonant coherent excitation experiments, 309, 311–312
  sectional view, 309–310
  tandem accelerator, 307, 309
Electrostatic lens
aberrations
  significance, 190
  spherical aberration effects, 199–200
  types, 199
alignment, 204
asymptotic trajectory of particle, determination, 194
circular aperture lens, 195–196
combining lenses, 201–204

computer simulation in design
  advantages, 204–205
  CPO-3D, 206–207
  SIMION, 205–206
cylindrical lens, 195–196
data sources for various electrode configurations, 195
design rules, 201
disc of least confusion, 200–201
electron energy analyzer input lens
  configurations, 223
  design, 224–225
  multielement lenses, 225
  optimization of voltages, 226
four-cylinder lens, 199
materials, 204
optical lens analogy, 189–191
thick lens representation, 193–195
three-cylinder lens
  focal properties, 197–198
  voltage, 197–199
two-cylinder lens, 196
voltage
  measurement, 192
  ratio, 201
Emittance, ion beam, 71–73
β-Eucryptite
  ion beam emission, 158
  synthesis, 158

F

FA, see Flowing afterglow
FALP, see Flowing afterglow/Langmuir probe
FDT, see Flow drift tube
Field evaporation, see also Liquid-metal ion source
  brightness of sources, 142–143
  ion current density, 141
  ion formation mechanism, 140–141
  microfocused beam applications, 140, 143–144
Field ionization
  ion current low-field ionization, 136–138
  lifetime of transitions, 136
  positive ion formation, 134

potential energy of atoms, 135
sources
  applications, 138
  types, 138–139
tunneling mechanics, 134–135
Figure of merit, electron energy analyzer
  configuration effects, 218–219
  luminosity effects, 218–220
Flow drift tube
  apparatus, 291–292
  ion-neutral reaction studies, 291–292
Flowing afterglow
  apparatus, 280–282
  applications, 282–283
  ion detection, 282
Flowing afterglow/Langmuir probe
  apparatus, 283–285
  applications, 283–284
  carrier gad, 284
  electron attachment rate determination, 284–285, 291
  ionic recombination coefficient determination, 285
  operation, 283
  spectroscopic experiments, 284
Fourier transform ion cyclotron resonance mass spectrometer, see Mass spectrometer
Fowler–Nordheim relation, current density calculation, 131
Freeman source, see Arc plasma discharge source

G

Gallium arsenide, negative-electron-affinity semiconductor photocathode
  applications, 32
  comparison of polarized electron sources, 30–32
  development, 1–2
  diffusion length, 4, 6
  electron escape, 4–5
  electron gun
    brightness values, 25–26
    configuration, 27
    electron-optical considerations, 24–25

low-energy electron diffraction gun, 27–28
spin rotation, 28, 30
stability, 26
incident radiation, 22–23
optical spin orientation, 3
photocathode
  anodization, 12–13, 34
  heat cleaning, 13–14
  lifetime, 21–22
  limitations of response, 20–21
  material, 11–12
  photocurrent monitoring, 15
  surface activation, 15–18
  surface cleaning, 12–15, 33–34
polarization
  circular polarization of incident radiation, 22
  film growth, 7, 9, 12
  formula, 2–3, 6
  optimization, 6–7, 9–10
quantum efficiency
  carbon contamination and quantum efficiency, 15
  factors affecting, 11
  formula, 18
  yield curve, 19
steps in photoemission, 3
thickness and electron polarization, 6–7, 9–10
Gated integrator
  noise minimization, 446
  schematic, 447

acceptance, 73–74
brightness, 73
emittance, 71–73
selection, 70
thermal emitters, 157–159
pupil, 192
window, 191
Ion cyclotron resonance mass spectrometer, *see* Mass spectrometer
Ion mass analyzer, *see* Mass spectrometer
Ion storage ring, *see* Storage ring
Ion trap, *see also* Electron-beam ion trap; Penning trap mass spectrometer
  applications, 349
  construction
    electrodes, 352–354
    filament, 355
    oven, 354
    vacuum system requirements, 355–356
  cooling trapped ions, 358
  correlation techniques, 357–358
  detection of trapped ions
    electronic detection, 356
    fluorescence detection, 356–357
  Paul–Penning trap, 351
  Paul–Straubel trap, 350–351
  Paul trap, 350
  Penning trap, 351, 364
  quadrupole traps, 351
  radiation sources for exciting trapped ions, 359
  temperature determination of trapped ions, 358–359

H

Helmholtz–Lagrange law
  lens configuration in electron energy analyzer, 224–225
  pencil angle determination, 192, 202

I

Ion beam
  applications, 69
  positive ion sources, *see also specific source*

J

Johnson noise, *see* Noise

L

Langmuir probe, *see* Flowing afterglow/Langmuir probe
Langmuir space-charge-limited flow formula, planar diode, 91–92

Laser-produced-plasma ion source
  highly charged ion generation, 185
  laser types, 185–186
Lens, see Electrostatic lens
LIA, see Lock-in amplifier
LINACS, see Radiofrequency linear accelerator
Liouville's theorem, 70–71
  brightness relationship, 73
  emittance relationship, 71–73
Liquid-metal ion source
  brightness, 142–143
  lifetime, 139
  metals and alloys, 146–147
  microfocused beam applications, 140, 143–144
  needle-type source, 144–145
  space-charge effects, 144
Lithium aluminum silicate, see β-Eucryptite
LMIS, see Liquid-metal ion source
Lock-in amplifier
  noise reduction, 449
  principle, 448
  schematic, 448
Logarithmic ammeter, see Ammeter
Luminosity, electron energy analyzer
  effect on figure of merit, 218–220
  optimization, 219–220

M

Magnetic sector mass analyzer, see Mass spectrometer
Mass spectrometer, see also Penning trap mass spectrometer
  cyclotron motion, 339–340
  Fourier transform ion cyclotron resonance mass spectrometer
    apparatus, 341, 344
    ion cyclotron excitation, 346
    mass range, 346
    mass resolution, 345–346
    principle, 343–345
  ion cyclotron resonance mass spectrometer
    apparatus, 341–343
    drift velocity of ions, 343
    operation, 342

magnetic sector mass analyzer
  double-focusing instrument, 337–339
  geometry, 339
  resolution, 336–337, 339
  mass resolution, 321–322, 325–326
  mass-to-charge ratio, 322–323, 363
monopole mass filter
  apparatus, 331–332
  resolution, 332–333
Omegatron
  apparatus, 341–342
  resolution, 342
quadrupole ion storage trap
  apparatus, 333
  operation, 333–334
quadrupole mass filter
  apparatus, 327–328
  mass range, 331
  operation, 328–329
  resolution, 329–331
radial electrical field
  particle trajectory, 334
  velocity focusing, 334–335
time-of-flight instrument
  apparatus, 324–325
  axial field, 323
  ion flight time, 324–325
  kinetic energy of ions, 325
  principle, 323
  reflectron, 326
  resolution, 325–326
  sensitivity, 326–327
transverse particle motion in uniform magnetic field, 335–336
vacuum requirements, 321
Wien filter
  apparatus, 340–341
  mass resolution, 340–341
  focusing properties, 340–341
MCP, see Microchannel plate
Metal-organic chemical vapor deposition, film growth for negative electron affinity, 7
Microchannel plate
  composition, 253
  configuration and output, 255–256
  detection efficiency
    degradation, 258
    determination, 255
  uniformity, 258

electron detection, 210
gating, 259
output pulse amplitude and count rate, 257–258
position-sensitive particle detection
  interpolative anode arrays, 261, 263
  optical systems, 260–261
  resistive anode
    accuracy, 265
    noise sources, 266–268
    resistance parameters, 263, 265
    resolution, 266
  resolution, 253, 260
  wedge-and-strip anode
    configuration, 268
    error sources, 269–270
    resolution, 268–269
principle, 253–254
response time, 211–212, 258–259
stacking, 254–255
MOCVD, see Metal-organic chemical vapor deposition
Monopole mass filter, see Mass spectrometer
Mott polarimeter
  atomic-target polarimeter, 239–240
  calibration
    calculation of Mott asymmetry, 232, 235, 243–244
    double-scattering measurements, 244
    measurement of analyzing power, 232, 235, 244
    use of electrons with known polarization, 244–245
  concentric-electrode configuration
    advantages of design, 237
    configuration, 236
    geometry optimization, 237–239
    retarding-field analyzer, 239
  diffuse scattering polarimeter, 240
  Mott scattering error sources
    background, 247
    extrapolation error, 245
    instrumental asymmetry, 245–247
  performance, 248–250
  standard configuration polarimeter
    configuration, 234
    energy range, 236
    operation, 234–235

N

NEA, see Negative-electron-affinity semiconductor photocathode
Negative-electron-affinity semiconductor photocathode
  comparison of polarized electron sources, 30–32
  electron gun
    brightness values, 25–26
    configuration, 27
    electron-optical considerations, 24–25
    low-energy electron diffraction gun, 27–28
    spin rotation, 28, 30
    stability, 26
  gallium arsenide
    applications, 32
    development, 1–2
    diffusion length, 4, 6
    electron escape, 4–5
    optical spin orientation, 3
    polarization
      film growth, 7, 9, 12
      formula, 2–3, 6
      optimization, 6–7, 9–10
      steps in photoemission, 3
  incident radiation, 22–23
  photocathode
    anodization, 12–13, 34
    heat cleaning, 13–14
    lifetime, 21–22
    limitations of response, 20–21
    material, 11–12
    photocurrent monitoring, 15
    surface activation, 15–18
    surface cleaning, 12–15, 33–34
  principle of negative electron affinity, 4–5
  quantum efficiency
    carbon contamination and quantum efficiency, 15
    factors affecting, 11
    formula, 18
    yield curve, 19
Noise
  ground loop pickup, 441
  inductive noise pickup, 441
  Johnson noise, 439–440

mechanical vibration sources, 441–442
minimization, 438–440
1/f noise, 440
shot noise, 438
thermocouple junction noise, 442

O

Omegatron, *see* Mass spectrometer
Oxygen, degradation of hot-cathode filaments, 126–127

P

Paul trap, *see* Ion trap
Pencil angle
  aperture radius relationship, 191–192
  consideration in combining lenses, 202–203
  determination, 192, 202
Penning discharge source
  cold-cathode sources
    cathode potential, 108
    discharge types, 107
    geometry, 106–107
  hot-cathode sources
    applications, 109, 111
    high-intensity sources, 111
Penning trap mass spectrometer
  accuracy, 363, 387
  anharmonic detection, 378–380
  axial frequency resolution, 368
  cyclotron frequency, perturbation sources
    electrostatic perturbation, 386
    ion position shift, 385
    misalignment, 384
    mode energy change, 386–387
    number of ions, 384–385
  ejection-detection method, 380, 382
  experimental parameters for various instruments, 366–367
  frequency-shift detector, 376–378
  input detection circuit, 369–370
  ion sample preparation, 371–372

magnetic field stability, 374–375
pulse-and-phase method, 382–383
sideband cooling resonances, 372–374
signal-to-noise ratio, 370–371
trap
  construction, 366, 368
  electrode voltage, 365
  performance, 364–365
  voltage sources, 368–369
Photocathode, *see* Negative-electron-affinity semiconductor photocathode
Photoionization source, highly charged ion generation, 187
PIG, *see* Penning discharge source
Plasma, temperature versus particle density, 169–170
Pockels cell, negative-electron-affinity semiconductor photocathode, 22–23
Poisson equation, computational simulation of ion extraction from plasma, 85, 87–89
Poisson–Vlasov equation, computational simulation of ion extraction from plasma, 89–91
Polarization
  degree of polarization, equation, 231
  negative-electron-affinity semiconductor photocathode
    circular polarization of incident radiation, 22
    film growth, 7, 9
    optimization, 6–7, 9–10
    polarization formula, 2–3, 6
Position-sensitive particle detection, *see* Microchannel plate
Positron
  β decay sources
    debunching, 55
    emitter generation, 40–41, 43–44
    isotopes, 40, 45
    maximum energy, 40, 42
    mirror nuclei, 40–41
    principle, 39–40
  lifetime measurement, 44–45
  moderators
    efficiency of moderation, 46–47
    energy distribution from single-crystal negative-affinity moderator, 47–48
    geometry, 49–50
    negative affinity solids, 47–49

physical properties, 46
preparation, 48–49
wide band gap solids, 49
pair-production sources, 45–46
physical properties, 39
remoderation
  large sources, 52
  microbeams, 51–52
  principle, 51–52
tagging, 55
transport
  electrostatic transport, 51
  imaging optics, 50
  magnetic transport, 51
trapping
  applications, 53–54
  gaseous moderators, 54
  harmonic potential bunching, 55
  radiofrequency bunching, 54–55
  resistive traps, 54
Positronium
  acceleration, 61–62
  cooling methods, 61
  formation
    activation energy, 59, 61
    gas sources, 56
    powder sources, 57
    probability as a function of temperature, 60
    temperature and mechanism, 57–59
    vacuum source, 57–61
  lifetime and collision, 56–57
  physical properties, 55–56
  work function, 58
Ps, see Positronium
Pupil
  consideration in combining lenses, 202–203
  definition of ion beam, 192

Q

Quadrupole mass filter, see Mass spectrometer

R

Radiofrequency discharge ion source
  discharge types, 116–119

lifetime, 117
mechanism of discharge, 117–118
types
  capacitatively coupled source, 119–120
  high-frequency source, 119–121
  multicusp-field source, 121–122
  proton beam source, 120–121
  waveguide sources, 123
Radiofrequency linear accelerator
  applications, 314–315
  principle, 314
RCE, see Resonance coherent excitation
Reaction kinetics in uniform supersonic flow technique
  gas cooling, 293–295
  ion sampling, 294
  operation, 294
Resistance per unit length, resistive anode, 263, 265
Resistive shunt ammeter, see Ammeter
Resonance coherent excitation, applications, 303
Resonant transfer and excitation, accelerator applications, 304–305
RTEA, see Resonant transfer and excitation

S

SA, see Stationary afterglow
SDT, see Static drift tube
Selected ion flow drift tube
  apparatus, 293
  applications, 293
  principle, 292–293
Selected ion flow tube
  apparatus, 286–287
  applications, 289, 295
  operation, 287, 289
  principle, 287
Shot noise, see Noise
SIFDT, see Selected ion flow tube
SIFT, see Selected ion flow tube
Signal-to-noise ratio
  amplifier selection and maximization, 443–444
  improvement by pulsed measurement, 437
  Penning discharge source, 370–371

SIMION
  electrostatic lens system design, 205–206
  operation, 205
S/N, see Signal-to-noise ratio
Sputtering ratio, approximation, 99
Static drift tube, electron attachment rate determination, 290–291
Stationary afterglow
  apparatus, 276
  applications, 276–278
  Cavalleri technique, 280
  ion monitoring, 279
  negative ion reactions, 278–279
  plasma formation, 276
  pulsed radiolysis technique, 280
Storage ring
  electron cooling, 318
  facilities, 317–318
  schematic, 315–316
  strength, 317
Surface ionization source
  capillary-type cesium ion source, 156–157
  curvilinear-cesium-surface ion source, 153, 156
  ion current density, 148
  ionized materials, 147
  porous tungsten source, 148–149, 151
  probability of positive ion formation, 147–148
  self-extraction spherical-geometry source, 151–153
  theory of ionization, 147–148
Swarm techniques, see also specific technique
  applications, 273, 275–276, 295
  comparison to beam experiments, 273–274
  drift tube operation, 290
  ideal experiments, 274–275
  nonideal experiments, 275
  rate coefficient determination, 274–275
Synchrotron accelerator, see Electrodynamic accelerator

T

Thermal ionization source
  ionization
    efficiency, 159–160
    equilibrium, 159
    theory, 159–160
    types, 160–161
Time-of-flight mass spectrometer, see Mass spectrometer
Transient digitizer
  noise reduction, 448
  speed, 447
Transresistance ammeter, see Ammeter

V

Vacuum-arc ion source
  electrode lifetime, 131–132
  low-voltage vacuum-arc discharge, 129–130
  principle of operation, 129
  triggered, high-voltage vacuum-arc discharge, 130–133
Van de Graaff accelerator, see Electrostatic accelerator

W

Wedge-and-strip anode
  configuration, 268
  error sources, 269–270
  resolution, 268–269
Wien filter, see Mass spectrometer
Window
  consideration in combining lenses, 202–203
  definition of ion beam, 191

X

X-ray spectroscopy, electron-beam ion trap
  lifetime measurements, 408, 410
  transition energy measurements
    above 2 keV, 402, 404–405
    below 2 keV, 405–406
  X-ray polarization, 406, 408

ISBN 0-12-475974-2